Interior Rift Basins

Edited by
Susan M. Landon
Thomasson Partner Associates
Denver, Colorado, U.S.A.

AAPG Memoir 59

Chief Editor of World Petroleum Basin Memoirs:
Anny B. Coury

Published by
The American Association of Petroleum Geologists
Tulsa, Oklahoma, U.S.A.
Printed in the U.S.A.

Dust jacket photos: Background—Landsat image of the east-central margin of the Gulf of Suez rift, showing structural elements associated with the synthetic transfer zone between the towns of Abu Zenima and Abu Durba. Inset—photo portraying a surface view of one of the north-oblique transfer faults shown in the southernmost portion of the accompanying Landsat image. The photo looks south along the fault bounding the northern termination of Gebel Abu Durba, and is taken from within the fault zone at Gebel Ekma (photo by T. L. Patton).

Published 1994

ISBN: 0-89181-339-X

Association Editor: Kevin T. Biddle
Science Director: Gary D. Howell
Publications Manager: Cathleen P. Williams
Special Projects Editor: Anne H. Thomas
Production: Custom Editorial Productions, Inc., Cincinnati, Ohio

Basin Memoirs and their editors:

Interior Cratonic Basins
Morris W. Leighton
Dennis R. Kolata
Donald F. Oltz
J. James Eidel

Divergent/Passive Margin Basins
J.D. (Jack) Edwards
P.A. Santogrossi

Active Margin Basins
Kevin T. Biddle

Foreland Basins and Foldbelts
Roger W. Macqueen
Dale A. Leckie

Interior Rift Basins
Susan M. Landon

These publications, and all other AAPG titles, are available from:
The AAPG Bookstore, P.O. Box 979, Tulsa, OK 74101-0979
Telephone (918) 584-2555; (800) 364-AAPG (USA—book orders only)
FAX: (918) 584-0469; (800) 898-2274 (USA—book orders only)

Dedication

James W. Vanderbeek
(1926–1993)

This memoir is dedicated to James W. Vanderbeek. Jim was instrumental in the initial discovery of significant production in the Gulf of Suez and maintained an interest in the geology and petroleum potential of rift basins throughout his career. He was my mentor and, as the Vice President of Amoco's Denver office, encouraged me to accept the job of editor for this book.

Susan M. Landon

AAPG
Wishes to thank the following
for their generous contributions
to

Interior Rift Basins

◆◆◆◆

ARCO Foundation

ARCO Exploration and Production Technology

Lee R. Russell

◆◆◆◆

Contributions are applied against the production costs of the publication, thus directly reducing the book's purchase price and making it available to a greater audience.

About the Editor

Susan Landon is an independent petroleum geologist with Thomasson Partner Associates in Denver, Colorado. Ms. Landon joined Amoco Production Company in 1974 after acquiring a Bachelor of Arts degree in geology from Knox College, Galesburg, Illinois, and a Master of Arts in geology from State University of New York at Binghamton. Her career with Amoco included exploration in the Rocky Mountains, Basin and Range, Midcontinent, and Alaska. In 1987 she was transferred to Houston as Exploration Training Manager. In 1989, she left Amoco and returned to Denver to establish an independent and consulting business.

Ms. Landon has been actively pursuing exploration in the United States and is evaluating the potential of the Precambrian Midcontinent Rift System. Her interest in such a poorly known rift began in the mid-1980s and is the basis for interest in productive rift analogs worldwide.

Foreword

The World Petroleum Basins project was conceived in order to aid explorationists in deciding what basins, or parts of basins, show the greatest promise for the most profitable return on investment. One way of achieving this goal is to supply these explorationists with a broad comparable base of data and concepts to improve their forecasts through analog techniques. Several volumes have been compiled to emphasize the geological factors that combine to make a basin productive and therefore assist in evaluating the geological risks of a new venture. Ultimate rankings, risk assessments, and business decisions are usually tempered, and often overridden, by nongeological factors, such as economic climate, geography, and political stability, which lie beyond the scope of this project.

Based on such consideration, the Publications Committee of the American Association of Petroleum Geologists resolved in 1983 to publish a series of volumes on world petroleum basins, with each volume being devoted to one major class of basins. An ad-hoc committee,[1] headed by Anny B. Coury, U.S.G.S., Denver, was charged with the overall planning of the project, with deciding the optimum number of volumes, and with the general organization of each volume. It was further decided that in each volume (or major basin class) a significant portion should be devoted to the detailed description of one maturely explored "type" or "model" basin, while additional summary papers would discuss other basins or provinces of similar type.

Studies comparing frontier provinces, basins, or undrilled prospects with productive counterparts of similar tectonostratigraphic setting have always played an important role in the evaluation of the hydrocarbon potential of new ventures. Such "analogs," when properly documented and of the same general class as the new target, can assist in establishing plausible limits to as-yet-unknown geological factors, such as reservoir thickness and distribution, trap size and kind, and maturation history, within the framework of the target that is to be assessed. Statistically, such exercises in comparative geology raise the confidence level of the assessment. They also serve to provide concepts to be evaluated and tested in similar areas.

To use analogs in basin assessment without bias or danger of miscomparison or misapplication, a geologist needs to have access to a broad-based worldwide data base that must be integrated with innovative geological concepts. The assemblage of such concepts and a data base is, however, a time-consuming and costly task that most smaller organizations cannot afford. The entry of many smaller oil companies and independents into the international exploration theater during recent decades has therefore increased the demand for a readily available catalog of analogs.

[1]Kaspar Arbenz, Anny B. Coury, Michael A. Fisher, James A. Helwig, David R . Kingston, H. Douglas Klemme

The committee in charge of the petroleum basin series wrested long and hard over the issues of classification, choice of analogs, and number of volumes needed to do justice to the overall objective. Several recently published classification schemes of petroleum basins and provinces (e.g., Bally, 1975; Bally and Snelson, 1980; Kingston et al., 1983a, 1983b; Klemme, 1975, 1980, 1986) were discussed to arrive at a compromise for volume topics and principal analogs. The ad-hoc committee proposed and the full Publications Committee of the Association voted in 1984 to proceed with the preparation and publication of five volumes, one for each of the following classes:

Divergent/Passive Margin Basins

Interior Cratonic Basins

Active Margin Basins

Foreland Basins and Foldbelts

Interior Rift Basins

For each volume one or more editors were selected, who then proceeded to solicit contributions for that volume. Each editor was given a generalized table of contents and a suggested list of illustrations so that the series would have internal coherence and so that the basin descriptions would be more easily comparable. We hope that these volumes will become a welcome and valuable collection of analogs assisting explorationists for many years to come.

REFERENCES CITED

Bally, A.W., 1975, A geodynamic scenario for hydrocarbon occurrences: Proceedings of the 9th World Petroleum Congress, Tokyo, v. 2: Essex, England, Applied Science Publication, p. 33–34.

Bally, A.W., and S. Snelson, 1980, Realms of subsidence, *in* D. A. Miall, ed., Facts and principles of world petroleum occurrences: Canadian Society of Petroleum Geologists Memoir 6, p. 9–94.

Kingston, D.R., C.P. Dishroon, and P.A. Williams, 1983a, Global basin classification system: AAPG Bulletin, v. 67, p. 2175–2193.

Kingston, D.R., C.P. Dishroon, and P.A. Williams, 1983b, Hydrocarbon plays and global basin classification: AAPG Bulletin, v. 67, 2194–2198.

Klemme, H.D., 1975, Giant oil fields related to their geologic setting—a possible guide to exploration: Bulletin of Canadian Petroleum Geology, v. 23, p. 30–66.

Klemme, H.D., 1980, Petroleum basins—classifications and characteristics: Journal of Petroleum Geology, v. 3, p. 187–207.

Klemme, H.D., 1986, Field size distribution related to basin characteristics, *in* D.D. Rice, ed., Oil and gas assessment—methods and applications: AAPG Studies in Geology 21, p. 85–89.

Table of Contents

About the Editor vii

Foreword ix

Introduction
Susan M. Landon 1

PART I TYPE BASIN: GULF OF SUEZ 5

Introduction to Exploration in the Gulf of Suez 7
James W. Vanderbeek

Chapter 1
Tectonic Evolution and Structural Setting of the Suez Rift 9
T.L. Patton, A.R. Moustafa, R.A. Nelson, and S.A. Abdine

Chapter 2
Structure and Stratigraphy of the Gulf of Suez, Egypt 57
Karsten I. Schütz

Chapter 3
Ramadan Field—Egypt: Gulf of Suez Basin 97
A. Shawky Abdine, Wafik Meshref, A. Nabil Shahin, Paul Garissino, and Salah Shazly

PART II EXAMPLES OF OTHER RIFT BASINS 123

Chapter 4
Structure, Stratigraphy, and Petroleum Geology of the Pripyat and Dnieper-Donets Basins, Byelarus and Ukraine 125
G.F. Ulmishek, V.A. Bogino, M.B. Keller, and Z.L. Poznyakevich

Chapter 5
Recôncavo Basin, Brazil: A Prolific Intracontinental Rift Basin 157
A.M.F. de Figueiredo, Editor, J.A.E. Braga, J.C. Zabalaga, J.J. Oliveira, G.A. Aguiar, O.B. Silva, L.F. Mato, L.M.F. Daniel, L.P. Magnavita, and C.H.L. Bruhn

Chapter 6
Structural Style and Tectonic Evolution of the Albuquerque Basin Segment of the Rio Grande Rift, New Mexico, U.S.A. 205
L.R. Russell and S. Snelson

Summary 259
Susan M. Landon

Index 271

Introduction

Susan M. Landon
Thomasson Partner Associates
Denver, Colorado, U.S.A.

The *World Petroleum Basins* series provides one more manifestation of the great unifying theory that plate tectonics has provided the science of geology. The concept of grouping descriptions of geology and petroleum systems of similar basins to provide a collection of exploration analogs has been made possible by the development of basin classification schemes built on a plate tectonic framework (Bally and Snelson, 1980). Conversely, particular sedimentary sequences in frontier areas may allow the geologist to infer the basin setting in which deposition took place, allowing for modeling of past tectonic and heat-flow history. Therefore, as Arbenz et al. note in the foreword to each of the volumes in this series, these analogs are designed to provide the explorationist with predictive insights into areas with similar geology. This volume presents examples of the fundamental plate tectonic unit: interior rift basins. Basins discussed in this volume are shown in Figure 1.

Although they are the subject of the last volume in the *World Petroleum Basins* project, interior rift basins are certainly not least in importance. Rifts represent a portion of an evolutionary sequence of crustal and upper mantle stretching that is responsible for normal faulting resulting from extension, the most commonly occurring structural style (Lowell, 1985). In reaching this conclusion, Lowell included the wide spectrum of extension (and the resulting normal faults) which occurs in oceanic spreading, in passive continental margins (nearly 62,000 mi or 99,750 km), in intraplate rifting (the subject of this volume), and in behind-arc spreading. All of these elements are a part of a sequence that culminates in the formation of ocean basins (Figure 2). Inasmuch as mature extended areas, i.e., passive margins, began with an initial rift phase, one must understand rift mechanisms to unravel a very large portion of the earth's geology. Rift basins occur throughout the geologic record and examples discussed in this volume range from the Paleozoic, through the Mesozoic, and into the Cenozoic (Tertiary).

Klemme (1980) pointed out that rifts, as a distinct geologic category, account for only about 5% of the area of the basins of the world. However, this geographically minor category has been intensely studied because of the broader implications to global tectonics. Rosendahl (1987) summarized the volume of literature in existence (and rapidly expanding, as this book is evidence) and indicated that 10,000 pages in 700 papers have been published. Not only are rift basins the foundation for much of the geologic history of the earth, but they also are very attractive areas for hydrocarbon accumulations. Klemme stated that this geographic area has provided significant hydrocarbon reserves: "By area, these basins represent slightly over 5% of the world's basins (50% productive). However, high recovery has resulted, as they contain 10% of the world's present reserves (12% of the oil reserves and 4% of the gas reserves)" (Klemme, 1980, p. 194). The rift basins discussed in this volume are only a few of the productive and, more importantly, potentially productive rift basins in the world.

The term "rift" was coined by Gregory (1896) for the graben that now bears his name in the Kenyan portion of the East African rift system. The study of geology of rift basins began in the Rhine graben in Europe prior to Gregory's work in Africa. Near-surface fault geometries and facies distribution in rift basins were relatively well known before the beginning of this century, although there was a continuing discussion regarding the development of rifts prior to plate tectonic theory. The past century has been spent refining those descriptions and attempting to identify viable mechanisms for the initiation of rifting. The discovery of hydrocarbons in rift basins about the turn of the century provided new motivation for understanding these basins. This volume, hopefully, will further the understanding of these basins for petroleum geologists.

Part I of this volume discusses the tectonic setting, structural style, stratigraphy, and characteristic hydrocarbon production of a type rift basin. A number of basins characterized by prolific production could have been chosen for the type basin of this volume or the subject of one of the other basins presented, including the Sirte, Viking Central graben, Yemen, etc. These have not been included because of the timing, data availability, or extensive documentation elsewhere. The Gulf of Suez, where hydrocarbons have been produced for more than 80 years, was chosen as the type rift basin. Extensive exploration by the Egyptian General Petroleum Corporation (EGPC) and many other companies has provided a very large data base from surface and subsurface work.

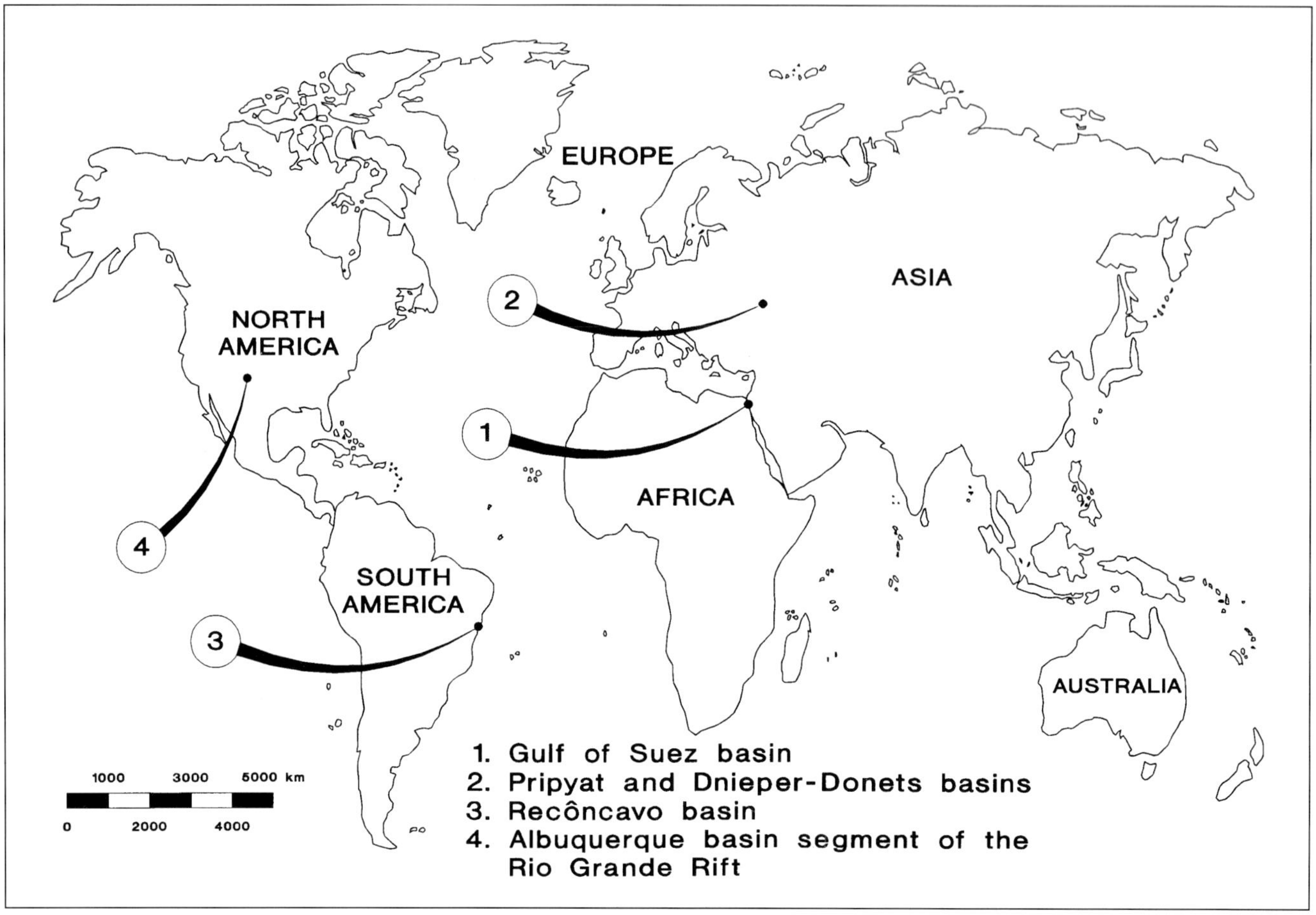

Figure 1. Location of basins discussed in this volume.

Aeromagnetics, gravity, seismic, and well control have allowed the development of a well-understood stratigraphy, evolving tectonic framework, and an understanding of the petroleum system. Much of the data have been published in Egyptian publications not readily accessible.

Part I is introduced by Jim Vanderbeek, who, as an Amoco vice president, was instrumental in initiating his company's successful exploration program in the Gulf of Suez in partnership with EGPC. He provided the historical perspective for exploration and production in the gulf. Following the introduction, Patton et al. (this volume) describe the tectonic evolution and structural configuration of the gulf. The authors develop an integrated model using surface exposures and subsurface information. This chapter illustrates the efforts and interpretations of one major company actively exploring in the gulf in partnership with EGPC. From the viewpoint of another active major company, a more-descriptive approach to the stratigraphy and structure of the gulf is presented by Schütz (this volume). This chapter presents an interpretation of 100 exploration wells and over 2000 km of seismic data. A discussion of the hydrocarbon system and primary exploration approach in the gulf are presented in a field study on Ramadan field (Abdine et al., this volume), reprinted from *Structural Traps VI* in the *AAPG Treatise on Petroleum Geology Atlas of Oil and Gas Fields* compiled by Foster and Beaumont (1992). For reference, a list of those fields from rift basins included in the *Atlas of Oil and Gas Fields* is given in the summary of this volume.

Three other rift basins are described in Part II of this volume. These chapters provide broader models of, and additional analogs for, rift basins. The first chapter in Part II describes the oldest rift basin discussed in this volume (and probably the oldest productive rift basin), the Devonian Dnieper-Donets and Pripyat basins in Byelarus and Ukraine (both of them segments of the same trend) (Ulmishek et al., this volume). Although oil was not discovered until 1964, significant gas was discovered in 1950. Gas accounts for about 80% of the reserves in this basin because of the advanced maturity of the source rocks. The largest field had original recoverable reserves in excess of 18 tcf. This basin is considered to be in a mature stage of exploration, but hydrocarbon potential exists in deeper areas and in stratigraphic traps. Associated with the breakup of Gondwana, the Recôncavo basin in Brazil is considered a classic rift basin. The Recôncavo basin (de Figueiredo et al., this volume) is also considered to be in a mature phase of exploration and production. The first production was established in 1937 and, as of 1990, 4.79 billion bbl (761 million m^3) of oil in place and 2.9 tcf (82 billion m^3) of gas have been discovered. As in the Dnieper-Donets and Pripyat

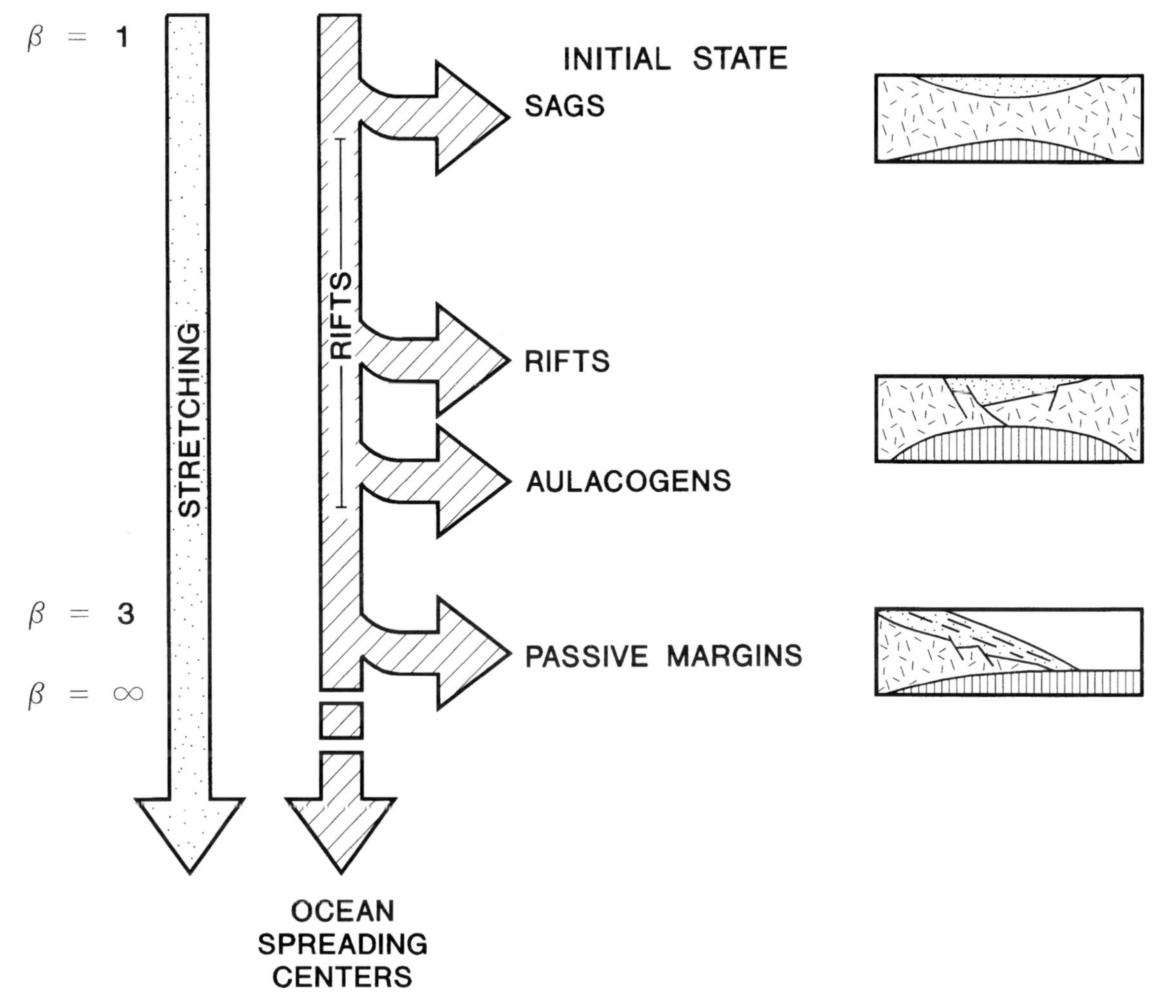

Figure 2. Relationships among sags, rifts, aulacogens, passive margins, and ocean basins, where β is the amount of stretching for the crust and lithosphere (McKenzie, 1978). After Allen and Allen, 1990.

basins, significant potential may still remain in stratigraphic traps.

Production has not been established in the Albuquerque segment of the Rio Grande rift, the subject of the final chapter, but Russell and Snelson (this volume) present an analysis of the Albuquerque segment specifically directed at some of the most basic questions of structural style and geometry associated with the initial stages of continental breakup and passive margin development. Edwards (1992, personal communication), coeditor of the first volume in the *World Petroleum Basins* series (*Divergent/Passive Margin Basins*, 1989), indicated that most of the feedback received regarding that volume was expressions of disappointment in the lack of discussion on the initiation of passive margin basins. Hopefully, some of those questions will be answered by the Russell and Snelson chapter. The summary (this volume) also presents a tabulation of mechanisms causing extension and characteristics of rift basins that form the foundation of every passive margin. This compilation of theories regarding mechanisms is designed to provide petroleum geologists with a basic understanding of how rifting occurs and the impact on stratigraphic architecture, structure, and heat-flow history of the various mechanisms. Rifting is, as mentioned, a portion of a spectrum of extension and each stage is characterized by a specific geologic history that will affect the hydrocarbon potential. Characteristics of each of the four basins presented in this volume are summarized. This summary is used to develop the key elements of rift basin geology and petroleum potential.

I especially wish to thank the contributing authors. They have provided a valuable perspective for exploration geologists working on interior rift basins, as well as providing an excellent foundation for understanding the extensional process and initial development of passive margin basins. I thank the authors for their patience as this volume developed a life of its own over the years of preparation. Editors have come and gone and authors have changed. Patton et al. were the first authors to submit their manuscript at the end of 1991 and I would like to thank these authors, in particular, for their patience.

I would also like to thank the following individuals who, primarily as reviewers, have assisted in preparing and publishing this volume: Michael Steckler, John Harms, Robert Raynolds, Jim Lowell, Charles Chapin, Steve Cather, Rick Lozinsky, Denise Stone, Anny Coury, and Dick Dietz.

REFERENCES CITED

Allen, P.A., and J.R. Allen, 1990, Basin analysis: principles and applications: Oxford, England, Blackwell Scientific Publications, 451 p.

Bally, A.W., and S. Snelson, 1980, Realms of subsidence, *in* A.D. Miall, ed., Facts and principles of world petroleum occurrence: Canadian Society of Petroleum Geology Memoir 6, p. 9–75.

Edwards, J.D., and P.A. Santogrossi, eds., 1989, Divergent/passive margin basins: AAPG Memoir 48, 252 p.

Foster, N.H., and E.A. Beaumont, comps., 1991–1992, Treatise of petroleum geology, Atlas of oil and gas fields: AAPG.

Gregory, J.W., 1896, The Great Rift Valley: London, John Murray, 405 p.

Harding, T.P., and J.D. Lowell, 1979, Structural styles, their plate tectonic habitats and hydrocarbon traps in petroleum provinces: AAPG Bulletin, v. 63, p. 1016–1058.

Klemme, H.D., 1980, Petroleum basins—classifications and characteristics: Journal of Petroleum Geology, v. 3, p. 187–207.

Lowell, J.D., 1985, Structural styles in petroleum exploration: Tulsa, Oklahoma, Oil and Gas Consultants International, Inc., 477 p.

McKenzie, D.P., 1978, Some remarks on the development of sedimentary basins: Earth and Planetary Science Letters, v. 40, p. 25–32.

Rosendahl, B.R., 1987, Architecture of continental rifts with special reference to east Africa: Annual Review Earth Planetary Science, v. 15, p. 445–503.

Type Basin: Gulf of Suez

Introduction to Exploration in the Gulf of Suez

James W. Vanderbeek
Friday Harbor, Washington, U.S.A.

This review of the geology and hydrocarbon potential of the Gulf of Suez makes sense on two counts. First, a knowledge of rifts is essential to an understanding of the tectonics of the earth and second, hydrocarbon exploration and production in the gulf have been highly successful producing a vast amount of subsurface geological data. This has greatly improved the geologic understanding of the Gulf of Suez rift basin.

Historically, from a petroleum standpoint, the earliest Egyptians no doubt used tar from seeps to caulk boats, grease wagon wheels, etc. The presence of fault controlled seeps, onshore, led to the current offshore petroleum exploration efforts.

Much has been learned about rift behavior during exploration of the Gulf of Suez and many pet theories have been discarded and others developed during the process. Strike slip movement, originally thought to be 50–80 mi, is now considered to be insignificant.

Until quite recently, source rocks were considered to be the Miocene Globigerina marls whose presence is restricted to the rift. Modern studies have shown that instead the dominant oil source is the Late Cretaceous (Campanian) "Brown Limestone" Member of the Sudr Formation. This formation is regionally extensive and rifting had no depositional effect other than to cause it to be buried deeply enough to reach thermal maturity.

Several major cycles of hydrocarbon exploration have occurred as follows:

Just south of Gebel Zeit (which means "oil mountain") on the western shore near the mouth of the gulf, oil was found in 1886 on the Gemsa Peninsula by the Egyptian Oil Trust Company after 39 years of surface mapping (Abdine and Kanes, 1983). The field was put into production in 1911. The total production from Gemsa field, which was abandoned more than 60 years ago, was about 1.3 million bbl from Miocene dolomitic reefal limestones. In 1913, Hurghada field was the first major discovery and has produced about 48 million bbl. The reservoirs in this fault block trap have been Miocene lagoonal dolomites and Carboniferous sandstones usually referred to as "Nubian."

The next major oil discovery was by Anglo-Egyptian, a Shell affiliate, in 1938 at the Ras Ghareb on the western shore about 100 mi south of Suez. Ras Ghareb field, small in areal extent but with 700 ft of Nubian sandstone reservoir, has turned out to be one of the richest fields in the world on a per acre basis with 250 MMBO within a 250–300 acre area.

In 1945, oil was found in the northeastern part of the rift by Mobil at the Sudr and Asl fields. Then, in 1950–1952, a group of California independents commenced exploration in the Belayim area 100 mi south of Suez on the eastern shore of the gulf. Their efforts led to the discovery of a number of significant fields. The first offshore seismic line was shot in 1955 by Cope (now Petrobel) and, by the mid 1950s, they and their partners Agip and Petrofina had seismically mapped several large structural prospects immediately offshore from the Belayim field production. This led to the discovery of the Belayim Marine field in 1961.

The next major producing area to be found was at El Morgan in the south central part of the gulf in 1965 as a result of aeromagnetic and reflection seismic surveys. Although many students of Red Sea rifting predicted that there would be no oil in the offshore Gulf of Suez because structures such as Morgan represented volcanic activity similar to that along the axis of the Red Sea where oceanic crust is being formed, this successful exploration effort was the result of a joint project of EGPC and Amoco. In 1970, El Morgan field was producing at a peak rate of 300 MBOPD. Ultimate reserves are estimated to be 700 million bbl (Abdine and Kanes, 1983) making it the largest field in the rift. Together, El Morgan, July, and Ramadan fields account for 39% of the oil reserves in Egypt (Abdine et al., 1992.) These discoveries established the Gulf of Suez rift as a prolific oil province which led to multiple subsequent discoveries establishing Egypt as a net oil exporting nation.

REFERENCES CITED

Abdine, S., and W. Kanes, 1983, Egyptian exploration, background, and future potential: Oil and Gas Journal, Aug. 29, p. 71–72.

Abdine, A.S., et al., 1992, Ramadan Field—Egypt, Gulf of Suez Basin, in N. Foster and E.A. Beaumont, eds., AAPG Treatise of Petroleum Geology Atlas of Oil and Gas Fields—Structural Traps VI, p. 113–139.

Tiratsoo, E.N., 1984, Oilfields of the world: Beaconsfield, England, Scientific Press, 392 pp.

Chapter 1

Tectonic Evolution and Structural Setting of the Suez Rift

T.L. Patton[1]
Gulf of Suez Petroleum Company
Cairo, Egypt

A.R. Moustafa[2]
Department of Geology, Faculty of Science
Ain Shams University, Abbassia
Cairo, Egypt

R.A. Nelson
Amoco Production Company
Houston, Texas, U.S.A.

S.A. Abdine[3]
Gulf of Suez Petroleum Company
Cairo, Egypt

ABSTRACT

Six distinctive tectonic episodes punctuate the stratigraphic record of the Gulf of Suez area. These episodes include the Pan-African event (late Proterozoic), which resulted in the development of the continental lithosphere of the area, a Cambrian extensional event, the Hercynian event (late Paleozoic), the Neo-Tethyan rift event (Jurassic), the Syrian Arc event (Late Cretaceous–early Tertiary) and the Gulf of Suez rift event (Oligocene[?]–Miocene). The structural fabrics imparted to the continental crust during the late Proterozoic and Cambrian appear to have played an important role in controlling the subsequent structural development of the Gulf of Suez area.

The first evidence of Tertiary rifting that led to the present-day expression of the Gulf of Suez is manifest by Oligocene–Miocene basaltic volcanism and in poorly age-constrained, continental to shallow-marine clastics of the Abu Zenima and Nukhul formations. The Nukhul Formation was deposited over much of the present-day extent of the rift basin, suggesting that subsequent extension experienced by the crust was constrained to the initial area of deformation. Subsidence during the initial phase of extension was slow.

[1]Present address: Amoco Production Company, Research Center, Tulsa, Oklahoma.
[2]Present Address: Department of Geology, University of Kuwait, Kuwait.
[3]Present Address: Kufpec Egypt Ltd., Cairo, Egypt.

Accelerated subsidence and extension is recorded in the deep-marine lithologies of the Rudeis Formation. This subsidence was disrupted midway through the deposition of the Rudeis Formation by the "mid-Rudeis" event, which resulted in the structural reorganization of the rift. This event correlates roughly in time with the onset of significant motion on the Dead Sea wrench system and marks the progressive abandonment of the Gulf of Suez as a site of active extension. After the mid-Rudeis event, variable and reduced tectonic subsidence rates prevailed in the gulf. The onset of significant evaporite deposition is recorded in the Belayim Formation, and evaporites subsequently become the major lithology of the South Gharib and Zeit formations as tectonic subsidence diminishes and basin restriction increases. Open-marine conditions and a change from dominantly Mediterranean to Indo-Pacific fauna took place in the Pliocene as a link was established to the Indian Ocean through continued extension in the Red Sea.

Four distinct fault populations are documented in the rift and show varying amounts of both strike-slip and dip-slip motion, depending on their orientation relative to the principal direction of extension. The major faults establish domains in their dip direction, subdividing the Gulf of Suez rift into three major structural subbasins with alternating structural asymmetry along the axis of the rift. Extension increases along the axis of the rift from northwest to southeast. Associated with this increase are increases in fault-block dip, the number of faults with large throw, and geothermal gradient. Fault-block size decreases with increasing extension to the south along the rift axis.

INTRODUCTION

After 100 years of exploration activity in the Gulf of Suez rift basin (Figure 1), there are still many aspects of the tectonic evolution and structural configuration of the gulf that are not fully understood.

One of the primary obstacles to a more in-depth knowledge of these topics in the gulf is the presence of thick evaporite sequences that lie high in the synrift stratigraphic package. On the one hand the evaporites are an essential component of oil accumulations in that these lithologies provide the ultimate seal for hydrocarbon entrapment. On the other hand, they obscure the structural geometries of the underlying prerift and synrift units due to the seismic energy attenuation and multiples they create. Although the exploration geophysical techniques of seismic, gravity, and magnetics play an essential role in exploration in the gulf, the subsurface, presalt section remains very much the domain of the geologist.

The Gulf of Suez, however, provides the exploration geologist with a unique opportunity to address rift-related structural and stratigraphic problems below the salt because immediately adjacent to the producing areas of the gulf are superbly exposed outcrop examples of the structures and stratigraphy encountered in the subsurface of the rift basin. The responsibility falls to the explorationist in the gulf to formulate from these outcrop data structural and stratigraphic models that may be applied to the subsurface, and to identify the differences between the geology in the central, producing portion of the rift and the geology accessible through exposures along the margins of the rift.

In this chapter, we review efforts by both academic and industry investigators to meet this end: to provide an integrated surface and subsurface model for the tectonic evolution and structural development of the Gulf of Suez Basin. Of course, one cannot discuss the tectonic evolution of the area without knowledge of the prerift and synrift stratigraphy, a topic addressed in more detail elsewhere in this volume. We have tried to touch only on those aspects of the stratigraphy pertaining directly to understanding the tectonic and structural development of the gulf.

Our discussion begins by briefly summarizing the evolution of the basement complex, the brittle substrate of the basin. Characterization of compositional variations and fabric of the basement is essential for

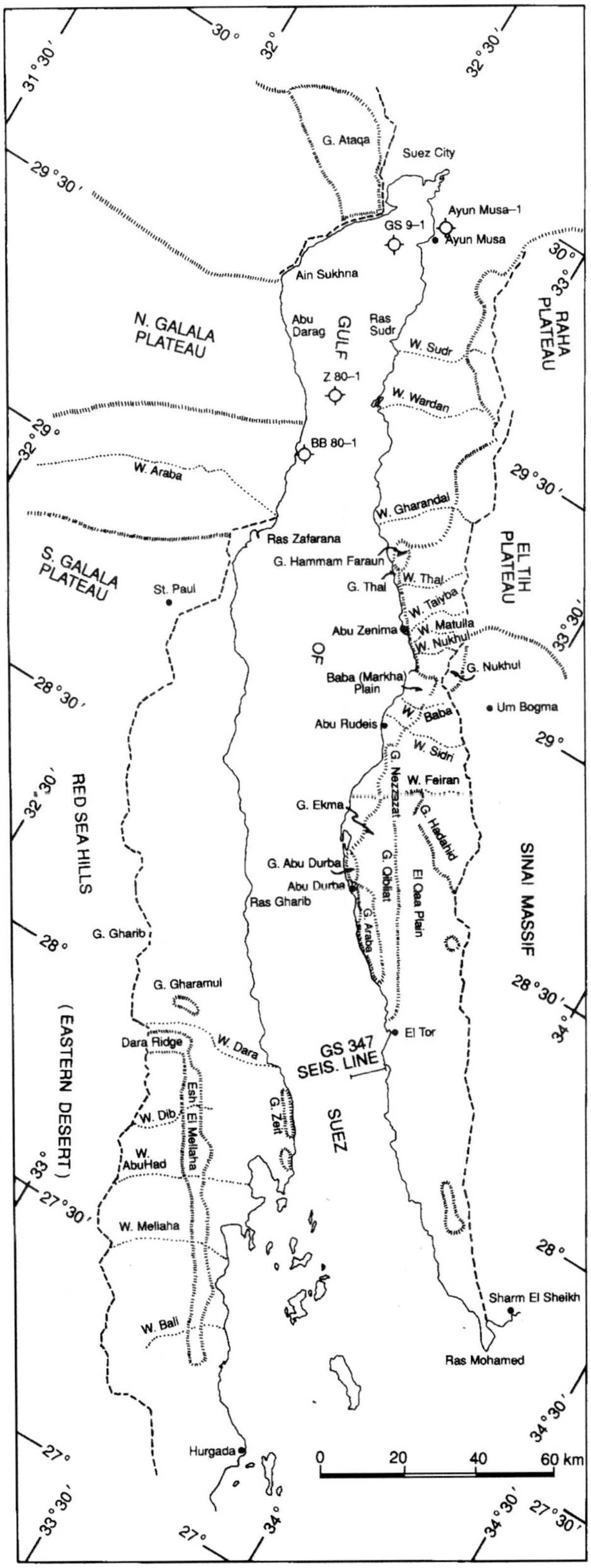

Figure 1. Gulf of Suez location map. On location names, W = wadi (a dry stream bed that captures runoff occasionally during the rainy season), indicated by dotted lines; G = gebel (Arabic for mountain or high place). Areas of significant topographic relief are indicated by hachure; the heavy dashed line indicates the location of the rift-bounding faults.

interpreting magnetic data in the gulf, a geophysical tool that (again due to the presence of thick evaporite sequences) is more effective in subsurface interpretations in the gulf than gravity data. More importantly, however, the structural fabrics imparted to the basement during the Precambrian and Cambrian appear to have played an active role in subsequent tectonic events in the area, especially during the later history of the gulf from the Jurassic to the present. Understanding the distribution, character, and extent of reactivation of these fabric elements is essential in correctly characterizing controls on deformation in the subsurface of the gulf.

Next, we review the tectonic evolution recorded in the pre-Miocene (prerift) sedimentary rocks in the gulf. A clear understanding of the distribution and thickness of these units and the pre-Miocene deformation they experienced is critical not only because they contain significant oil reserves, but also because these units define the basal rift geometry. This portion of the stratigraphic record demonstrates that the pre-Miocene of the gulf can be subdivided into two distinct tectonic phases: the Paleozoic interval, which is dominated by relatively gentle epirogenic movements, and the Mesozoic and early Tertiary interval, which is characterized by a southern stable shelf area and a northern unstable shelf.

The third section of the chapter deals with Miocene rift tectonics. Although the Gulf of Suez continues to experience localized structural movements and gentle subsidence, it is best classified as an abandoned rift basin. Two distinct periods of subsidence exist prior to the onset of the abandonment phase of rift history: an initial phase of gentle subsidence followed by a second phase of rapid subsidence, during which the maximum structural development of the rift took place. During this second tectonic phase, the majority of Miocene clastic reservoirs and productive structures developed. During and immediately after the cessation of rapid subsidence of the rift, thick sequences of Miocene evaporites and post-Miocene clastics filled the central trough and obscured the structural features developed in the immediately preceding phase of tectonism. Although gentle subsidence continued throughout the remaining history of the basin, with possible subtle reactivation in the latest Miocene or earliest Pliocene, the tectonic subsidence data argue strongly that the presalt and salt stratigraphies have each experienced markedly different structural histories. Therefore, one needs to realize that the structural configuration seismically imaged at the base of the evaporite section (the last reliable industry seismic reflector) represents only a small portion of the true structural complexities of the basin geology below the salt.

The interpretation of the presalt structural configuration is presented in the final section, in which we attempt to integrate surface and subsurface structural data. The discussion focuses on the character of faulting, describing the different fault populations defined in outcrop and examining their interactions and their roles in the rift process. Characterization of the major faults in the basin and their associated rotated blocks establishes the gross structural framework and identifies a number of fault domains in the basin. From these data, we make several generalizations about systematic changes in the structure of the rift along its axis.

Finally, many of the geologic events in the immediate gulf area can be related to the regional tectonic evolution of the Arabian Peninsula, Turkey, and north Africa. Throughout the discussion of different periods of evolution of the gulf, we attempt to put specific structural events of the gulf into the context of regional tectonic interpretations, providing a framework for local events.

PRERIFT TECTONIC EVOLUTION

In our treatment of the prerift tectonic evolution of the gulf, we break out three major periods of development. The first period extends from the late Proterozoic through the early Paleozoic and documents the development of the crystalline basement and records the first major marine transgression to inundate the area. The second period discussed, the late Paleozoic, follows a hiatus in the rock record of approximately 150 m.y. and documents structural disturbances likely associated with the Hercynian orogeny. The final section presents developments during the Mesozoic through the lower Tertiary (Eocene) and records the effects of the opening and closing of the Neo-Tethys Ocean and the events leading up to the opening of the Gulf of Suez.

Upper Proterozoic–Lower Paleozoic

The Gulf of Suez rift lies within the Arabian-Nubian shield, a segment of upper Proterozoic to lower Paleozoic continental crust formed during the widespread Pan-African tectono-thermal event; this crust forms the basement of much of northeast Africa and western Saudi Arabia. This segment of crust developed through the progressive cratonization and accretion of numerous intraoceanic island arcs and Andean-type magmatic arcs during the interval of 900–550 Ma (Engel et al., 1980; Gass, 1981; Vail, 1985; Bentor, 1985; Kroner et al., 1987; Dixon and Golombek, 1988). The development of the Arabian-Nubian shield can be subdivided into an orogenic collisional phase characterized by calc-alkaline plutonic and volcanic suites with associated eugeosynclinal sediments, and a postorogenic phase characterized by anorogenic, alkaline and peralkaline plutonic and volcanic suites with associated terrigenous sediments (Gass, 1981; Vail, 1983; Bentor, 1985).

Studies of the upper Proterozoic to early Paleozoic crystalline and sedimentary units in the Gulf of Suez support the general orogenic-anorogenic subdivisions proposed for the Arabian-Nubian shield, and we use this two-part subdivision as a framework from which to discuss the earliest events recorded in the gulf (Figure 2). Following the lead of other investigators, we interpret the orogenic phase to represent the cratonization of the Arabian-Nubian shield and its joining the African craton, and the anorogenic phase of development to represent an extensional event, possibly a local manifestation of fragmentation of the Precambrian supercontinent.

Orogenic Phase

Outcrop studies of basement exposures around the gulf document significant differences in the units recording the orogenic phase of development both between the two sides of the gulf, as well as with outcrops immediately south in the Eastern Desert. Stern and Manton (1987) proposed that the area of the intersection of the Gulf of Aqaba, Red Sea, and the southern Gulf of Suez corresponds to the juncture of three different basement terranes. The central Eastern Desert is separated from the northern Eastern Desert (the area along the western margin of the gulf) by a major northeast-trending, structural discontinuity extending from the great bend in the Nile, across the Eastern Desert, and projecting into the Gulf of Aqaba. North of this line, along the western margin of the Gulf of Suez, the basement is composed of a greater abundance of plutonic rocks relative to the abundance of volcanics, and sediments (Stern et al., 1984; Stern and Hedge, 1985) than the Pan-African terranes to the south. In particular, a noted absence of the ophiolites and melanges is found in the northern Eastern Desert (Stern et al., 1984). The northern Eastern Desert also demonstrates substantially reduced deformation relative to that recorded in the southern and central Eastern Desert, as well as slightly younger igneous activity (Stern et al., 1984; Stern and Hedge, 1985).

The basement rocks along the eastern margin of the gulf also differ from those on the west in several respects. Stern and Manton (1987) argued that the older (782 ± 7 Ma) orogenic intrusives along the eastern basement outcrops of the gulf (Wadi Feiran area) are the westernmost portion of a 780 ± 50 Ma terrane that extends across the Sinai into Jordan. Further data in support of older basement along the east is found in the comparison of initial strontium isotope ratios from synchronous dikes along opposite sides of the gulf. The higher ratios from the eastern side of the gulf suggest participation of this older basement in the evolution of the melts that generated the dikes (Stern and Manton, 1987).

Another distinctive difference between the basement of the two sides of the gulf is the degree of deformation they have experienced. As mentioned, the northern Eastern Desert outcrops along the western side of the gulf show little evidence of pervasive deformation. This contrasts markedly with the Sinai basement outcrops, which record significant deformation as manifest by the abundance of gneissic outcrops (see Stern and Manton, 1987). Outcrops of gneiss along the eastern margin of the gulf at Wadi Feiran (Figure 1) have been radiometrically dated by Bielski (1982, in Stern and Manton, 1987) and Stern and Manton (1987) at 643 ± 41 Ma (Rb-Sr whole rock) and 632 ± 3 Ma (U-Pb zircon), respectively. Stern and Manton (1987) argued that this date and the associated deformation

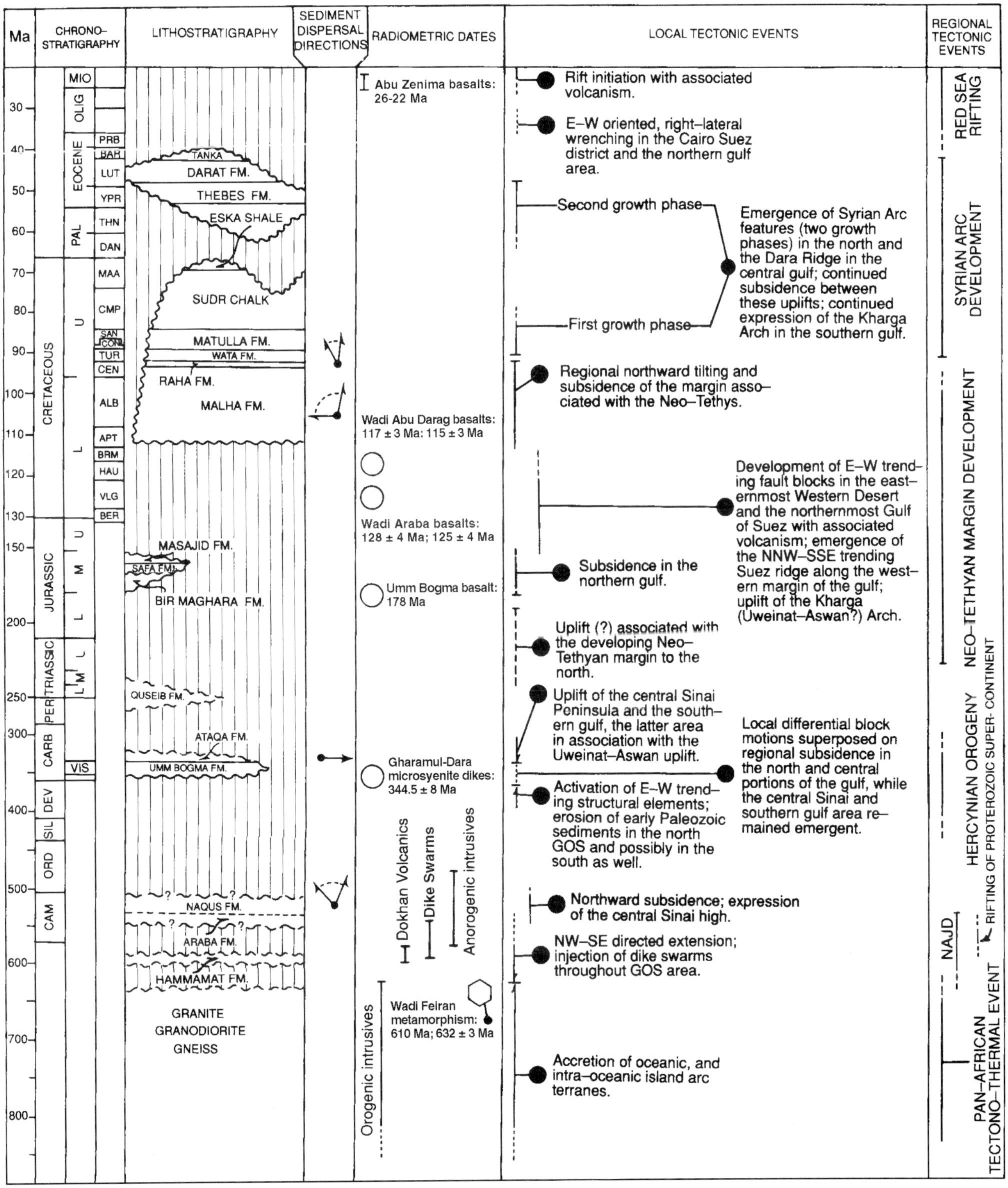

Figure 2. Pre-Miocene, tectonostratigraphic events of the Gulf of Suez. Note scale changes in "Ma" column at the Paleozoic–Mesozoic boundary and the Jurassic–Cretaceous boundary. In "Local Tectonic Events" column, GOS = Gulf of Suez.

records the juxtaposition of the Feiran gneisses against the 780-Ma-old basement to the east. The northwest-trending foliation of the Feiran gneiss has likely been exploited by the Tertiary rifting of the gulf (El Tarabili and Adawy, 1972). Moreover, the zone of juxtaposition proposed by Stern and Manton (1987) may have provided the fundamental zone of weakness along which the Gulf of Suez rift ultimately formed (R. Stern, personal communication, 1990). The presence of such a suture zone along the axis of the gulf, with probable gulf-parallel structure fabrics and units of varying composition within and adjacent to the zone, has significant ramifications for interpreting the present-day structural configuration of the gulf from data of potential fields.

Bentor (1985) argued that during the final phase of orogenic igneous activity (calc-alkaline batholithic, phase III of Bentor, 1985) cratonization of the Arabian-Nubian shield occurred. He further stated that the end of this period is marked by strong uplift and erosion, which he attributed to the joining of the Arabian-Nubian shield to the African craton.

Anorogenic Phase

The transition to the anorogenic phase of development of the Arabian-Nubian shield (Figure 2) was abrupt and follows the orogenic phase closely (Stern et al., 1984; Stern, 1985). The anorogenic phase began with a period of regional, northwest-southeast–directed extension as recorded by the extrusion of the Dokhan Volcanics, deposition of the Hammamat Formation into east-west– to northeast-southwest–elongated troughs in the north Eastern Desert, and the intrusion of the Pink Granites. Ages for the rhyolites and andesites of the Dokhan Volcanics constrain these units to lie within the relatively narrow time span of 600–580 Ma (Stern, 1985; Stern and Hedge, 1985). Deposition of the texturally and mineralogically immature sediments of the Hammamat Formation took place synchronously with the extrusion of the Dokhan Volcanics. Willis et al. (1988) stated that the Hammamat Formation was deposited around 590 Ma (Figure 2) based on Rb-Sr whole-rock analyses and K-Ar ages of the coarse clay fractions.

Further evidence of extension is manifest by the injection of numerous compositionally bimodal dike swarms (Stern et al., 1984, 1988) with generally northeast-southwest orientations and less common north-south, east-west, and north-northwest–south-southeast orientations (e.g., maps in Schurmann, 1966). Stern and Hedge (1985) stated that dike injection took place in the Eastern Desert during an interval between 595–540 Ma, a time span supported for the Suez area by a Rb-Sr date of 591 ± 9 Ma on a dike from the Wadi Feiran area (Stern and Manton, 1987). This extensional period imparts a strong fracture-strength anisotropy to the basement; this anisotropy becomes important in deformation during the Mesozoic and Tertiary.

For the intrusive rock assemblages, Stern et al. (1984) assigned all of the Younger (Gattarian, Pink, or Pan-African) Granites and associated dikes to this extensional, anorogenic phase of development following arguments presented by Greenberg (1981). The peralkaline batholithic rocks of the Sinai massif (580–550 Ma), including the Gebel Katharina assemblages, also have been assigned to the anorogenic phase (Bentor, 1985). Abdel-Rahman and Doig (1987) and Abdel-Rahman and Martin (1987) argued that a significant volume of the Younger Granites found at gebels Gharib and Dara are I-type granites, developed in an active continental margin environment, a contention disputed by Stern and Gottfried (1989); however, the Rb-Sr whole-rock dates (550–475 Ma) of Abdel-Rahman and Doig (1987) extend the period of igneous activity well into the Ordovician.

An important regional tectonic element associated with the tectonic development of the Arabian-Nubian shield is the northwest-southeast–trending Najd fault system (Brown and Jackson, 1960; Brown and Coleman, 1972; both cited in Stern, 1985). This feature, located southeast of the Gulf of Suez and east of the Red Sea, is a 2000-km-long (1240 mi), 400-km-wide (250 mi) system of predominantly left-lateral strike-slip faults extending from the Eastern Desert of Egypt (Abdel-Gawad, 1969; Sultan et al., 1988), through Saudi Arabia possibly as far south as South Yemen (Stern 1985; Sultan et al., 1988). The northwest-southeast–trending shears characteristic of the Najd fault system (Stern, 1985; Stern et al., 1984; Sultan et al., 1988) found in the central Eastern Desert are absent in the immediate area of the gulf along the northern Eastern Desert and the southwestern Sinai. However, the slightly more northerly-oriented (north-northwest–south-southeast) clysmic trend (present-day Gulf of Suez trend) was expressed in the western Gulf of Suez very early in the history of the area as manifest in the dike swarm orientations between gebels Rabul and Mellaha (Schurmann, 1966).

The timing of motion on the Najd fault system (630 and 530 Ma, Husseini, 1988; Sultan et al., 1988) is synchronous with the period of extension experienced in the gulf area. However, interpretations relating the motion on the Najd system to this extension are controversial.

Collisional models for the origin of the Najd system have been proposed by Schmidt et al. (1979), Davies (1984), and Burke and Sengor (1986). The Burke and Sengor (1986) "continental escape" model is attractive in that it accommodates contraction in the south and extension of the Sinai Peninsula in the north as this area is "extruded" northward out of the convergent zone in the south. Extensional models for the region (Stern, 1985; Husseini, 1988) use the Najd system as an intracontinental wrench system linking centers of extension along the northern African margin with extensional systems along the southeast and eastern margin of the Arabian Peninsula.

Precambrian plate reconstructions are controversial, and we present no data here that address this subject; however, the reconstructions and tectonic events proposed by Piper (1983, 1987) provide a framework that would appear consistent with the extensional models presented by Stern (1985) and Husseini (1988). Piper (1983, 1987) proposed that by the end of the Proterozoic, Kazakhstan was juxtaposed against north-central Africa, and Siberia and North America were joined to the north and northeast margin of the Arabian Peninsula (Figure 3). Piper (1983) argues that this Proterozoic supercontinent was fragmented by a 560 Ma rift event, with associated alkaline magmatism, which followed the present-day northern African margin and crossed the Arabian Peninsula (Figure 3). Such a model is consistent with the events recorded in the gulf area and corresponds well with the extension models proposed by Stern (1985) and Husseini (1988). We favor the extensional model for this period of development of the Gulf of Suez.

Further support for an early Cambrian extensional event for the north margin of Africa is found in the Cambrian units immediately overlying the crystalline basement. We propose that the extensional event likely resulted in regional subsidence in this area, making way for a southward-directed marine transgression and the deposition of the Araba Formation (Figure 2) (Issawi and Jux, 1982; Bhattacharyya and Dunn, 1986). Klitzsch and Wycisk (1987) indicated that this marine transgression extended as far south as the juncture of the southern gulf and the Red Sea. A thick sequence of fluvial sandstones of the northerly prograding Naqus Formation followed, disrupting marine deposition (Bhattacharyya and Dunn, 1986; Issawi and Jux, 1982). An Early Cambrian age, based on ichnofossils, has been assigned to the Araba Formation (Seilacher, 1990; Klitzsch and Squyres, 1990). The Early Cambrian age assignment is further supported by stromatolites and archaeocyathids from Abu Durba (Omara, 1972) and lithostratigraphic correlations to Cambrian units far to the east in Israel (Weissbrod, 1969a; Beyth, 1981). The Naqus was assigned a probable Cambrian age by Hermina et al. (1989).

Because of the fragmentary nature of the Cambrian sedimentary record, it is difficult to obtain an accurate picture of the structural elements expressed in the gulf during this time. Sand dispersal directions in the Naqus Formation are in a northwest to north-northeast direction (Figure 2) (Soliman and El-Fetouh, 1969, in Gvirtzman and Weissbrod, 1984; Weissbrod and Nachmias, 1986; Bhattacharyya and Dunn, 1986) and have prompted Bhattacharyya and Dunn (1986) to propose that east-west–trending highs were active during Naqus deposition. Beyth (1981) proposed a structural high east of the gulf in the central Sinai, based on facies and isopach data. We have assigned a speculative northeast trend to this high (Figure 4a) to correspond with the structural trend proposed by Bhattacharyya and Dunn (1986).

The model proposed here for the transition from what have been conventionally considered "basement" rocks to the depositional setting of the Araba and Naqus formations suggests a more immediate relationship between the anorogenic events and these two formations than has been proposed by other investigators. Klitzsch (1986), basing his conclusions on the nonconformable relationship between the Araba Formation and the underlying igneous and metamorphic units, and Weissbrod and Nachmias (1986), basing their conclusions on textural and mineralogical data from the lower Paleozoic, argue for peneplanation of the Pan-African terrane in the gulf area. Klitzsch (1986) used these relationships and the Cambrian age of the Araba and Naqus formations as evidence that Pan-African tectonism and associated

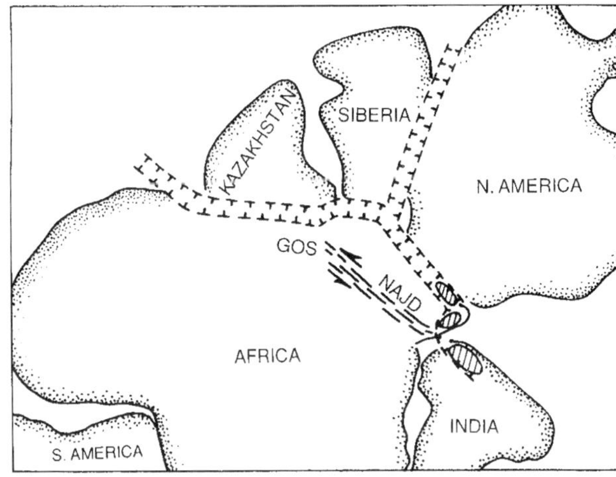

Figure 3. The Najd fault system and the Arabian Gulf, Oman, and Pakistan salt basins superimposed on the late Precambrian supercontinent. The early Cambrian reconstruction is after Piper (1983; 1987; 1985, in Donovan, 1987). The rift trends (depicted by dashed lines with tick marks) are from Piper (1983), who dated this event at approximately 560 Ma.

igneous activity ceased long before the Cambrian, and questioned the validity of Cambrian radiometric age dates cited. However, Stern (personal communication, 1990), in defense of the Cambrian radiometric age for the Gharib Granite, cited the presence of a contact-metamorphosed outcrop of "Nubia Sandstone" in contact with the eastern flank of the granite. Near this locality, Stern and Hedge (1985) obtained samples yielding a Rb-Sr model age of 544 Ma for the Gharib Granite. As argued by Stern (personal communication, 1990), acceptance of this radiometric date demands that this portion of the "Nubia Sandstone" must be at least Lower Cambrian in age. If the age of 476 Ma (Abdel-Rahman and Doig, 1987) is accepted for the Gharib Granite, the minimum age of the "Nubia Sandstone" could be as young as Ordovician. Therefore, although much of the igneous activity in the gulf had ceased by the Cambrian, local igneous activity likely continued into the early Paleozoic.

Upper Paleozoic

Although Ordovician, Silurian, and Devonian units have been documented in the subsurface of the Western Desert of Egypt (Keely, 1989) and to the northeast in the Levant (Gvirtzman and Weissbrod, 1984), no record of these units has been found in the immediate area of the Gulf of Suez. The absence of a pronounced angular unconformity between the Cambrian sequence and the overlying Carboniferous units argues against anything other than gentle epirogenic uplifts being active during the intervening period of time. The pre-Carboniferous structural movements that can be inferred from the limited stratigraphic record are impossible to accurately date. We have looked to regional considerations to assign

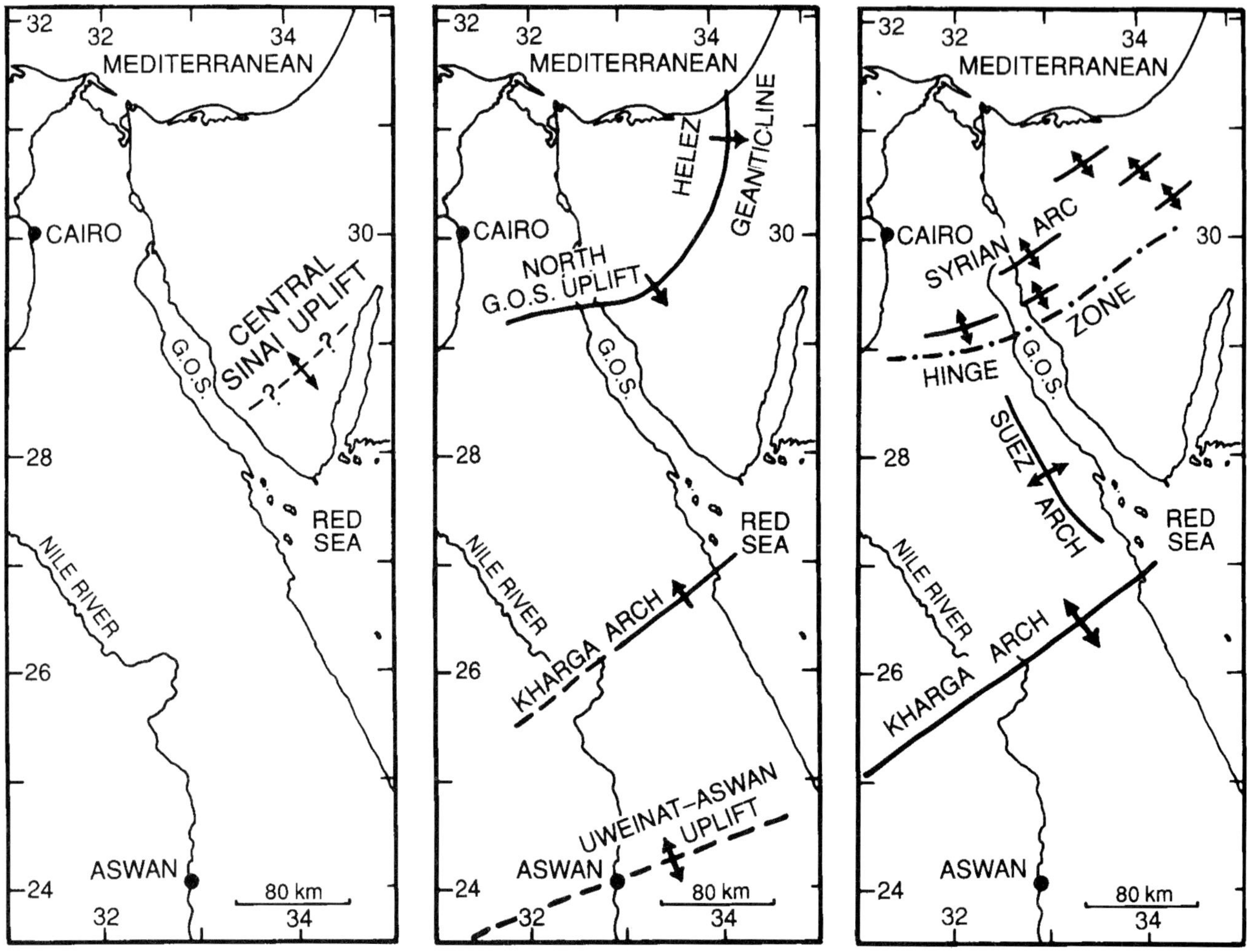

a. Cambrian b. Late Paleozoic c. Mesozoic

Figure 4. Prerift, tectonic elements in the Gulf of Suez area. G.O.S. = Gulf of Suez.

ages to these pre-Carboniferous movements in the gulf. We interpret these movements to be late Paleozoic in age and to reflect distal lithospheric adjustments to the first of two phases of the Hercynian orogeny to the west. A second phase of Hercynian activity is recorded following a marine transgression in the Visean.

First Phase of Hercynian Growth

Prior to the Early Carboniferous, the northernmost and southernmost portions of the Gulf of Suez appear to have undergone basement-block uplifts, across which the Naqus and Araba formations either were significantly reduced in thickness or were eroded away entirely (Figure 5). Beyth (1981) documented the Wadi Khaboba–Gebel Nukhul High in the Umm Bogma area, in which the Araba and Naqus formations have been eroded away and Lower Carboniferous units (Umm Bogma Formation) rest nonconformably on basement. Beyth (1981) proposed a north-south trend for the structure and suggests relief of approximately 130 m (425 ft). Bhattacharyya and Dunn (1986) indicated that the Araba-Naqus interval is not present in the Ayun Musa and Ataqa wells. However, Beleity et al. (1986) argued for the presence of Araba and thinned Naqus in these areas, based on gulf-wide electric-log correlations and outcrop data (Figure 5). Beleity et al. (1986) also used isopach data (Figure 6a) of the Naqus Formation to argue for a Paleozoic structurally high block in the north gulf with 300 m (1000 ft) of relief. The southern boundary of their block trends east-west, and lies just south of the Gebel Nukhul section investigated by Beyth (1981).

North and east of the gulf in Saudi Arabia, Jordan, Israel, and Syria, Gvirtzman and Weissbrod (1984) recorded the regional uplift of the Helez geanticline during the late Paleozoic (Figure 4b). Their interpretation of isolated data points showed an angular unconformity between conformable Cambrian through Devonian units and the overlying Carboniferous, constraining the time of the uplift to the later part of the Devonian. Such timing is compatible with the record of events in the Gulf of Suez and we have chosen to constrain the timing of the north gulf block uplifts to

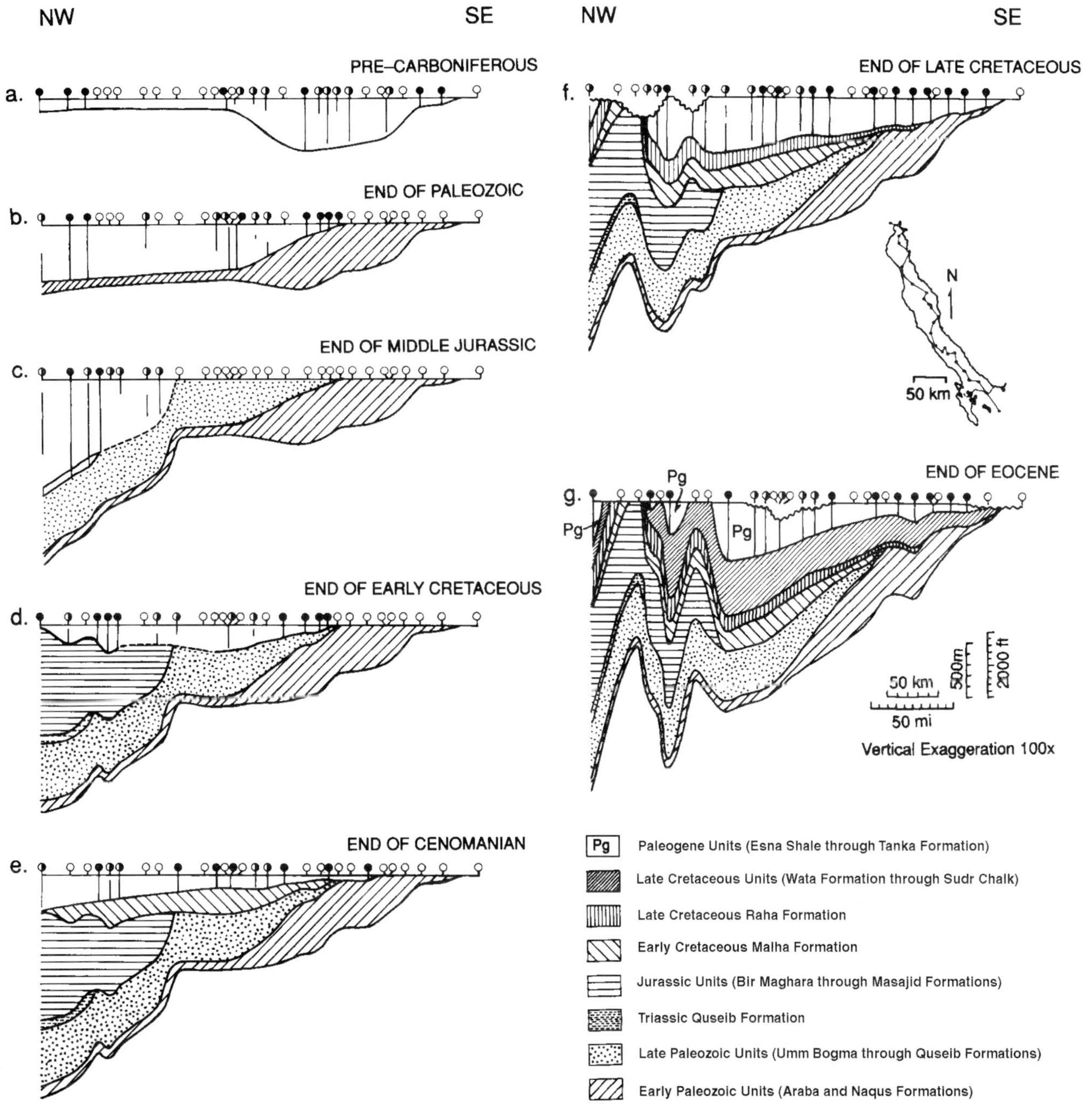

Figure 5. Pre-Miocene evolution along an axial line of section down the Gulf of Suez. Empty circles indicate that the section of interest was not present; half-filled circles mark wells in which an incomplete section of the interval of interest was present; and the solid circles designate wells interpreted to contain complete sections. The symbols representing the rocks deposited immediately prior to the time portrayed in the cross section are applied to the sections following the documentation of these units so that the stick segments, representing how much of the section was present, may be displayed. The unconformity lines in sections f and g indicate areas where erosion after the initiation of rifting has obscured the record. The stratigraphy for the northwesternmost well in the section is derived from both surface and subsurface data.

this time interval as well. Gvirtzman and Weissbrod (1984) associated these events with a first phase of Hercynian tectonics (Figure 2). Ziegler (1989) cited a late Devonian age for the initial contact of the Aquitaine-Cantabrian terrane with the north Sahara platform in Algeria.

The stratigraphic record pertaining to this event in the southernmost gulf is poorly preserved, but we have assumed uplift in the south of the gulf at this time as well. Klitzsch (1986) stated that the east-west–trending Uweinat-Aswan uplift (Figure 4b) was activated as a manifestation of Hercynian tectonism.

The uplift in the southern gulf may have been associated with the Uweinat-Aswan uplift; however, note that Klitzsch (1986) did not record Hercynian-associated growth on this feature until after the Visean. In support of earlier uplift in the southern gulf, Bhattacharyya and Dunn (1986) recorded the presence of Carboniferous sediments overlying basement in the Hurghada area, suggesting either nondeposition during Araba-Naqus deposition or erosion prior to the Carboniferous as documented in the northern gulf. A further possible manifestation of the Hercynian event in the gulf is the presence of the Gebel Gharib microsyenite dikes, which were intruded along the clysmic orientation. These dikes have a radiometric age (Burollet et al., 1982) of 344.5 ± 8 Ma (Figure 2).

After the first phase of Hercynian movements in the gulf, an extensive marine transgression took place across the northern margin of the craton during the Early Carboniferous (Hassan, 1967; Weissbrod, 1969b; Issawi and Jux, 1982; Brenckle and Marchant, 1984; Eames, 1986; Bhattacharyya and Dunn, 1986) depositing the Umm Bogma and Abu Durba formations throughout much of the Gulf of Suez area (Figures 2, 5). Facies and isopach data of the Umm Bogma and Abu Durba formations (Figure 6b) (Beleity et al., 1986) suggest that the structural high in the northern Gulf of Suez was still expressed during the deposition of these two formations. A relatively thin sequence of sandy dolomites, limestones, and marls of the Umm Bogma Formation gives way to the south to the thicker calcareous shales of the Abu Durba Formation.

Second Phase of Hercynian Growth

Marine deposition was interrupted by northward-prograding fluvial systems of the lower part of the Ataqa Formation (Bhattacharyya and Dunn, 1986), a possible manifestation of accelerated growth on the Uweinat-Aswan uplift, which Klitzsch (1986) stated was post-Visean in age. The Ataqa Formation shows a stronger influence of marine conditions in the north. The prevalence of marine environments in the north, in conjunction with the general northward-thickening wedge of sediments represented by the combined Umm Bogma–Ataqa thicknesses (Figure 5), argues for the collapse of the pre-Carboniferous high in the northern gulf. This collapse, coupled with the uplift in the south of the Uweinat-Aswan High, resulted in the northward tilt of the Gulf of Suez area as part of the margin of the evolving Pangean gap or Paleo-Tethys Ocean Basin (Sengor, 1984; Sengor et al., 1984b).

Another significant hiatus in the rock record occurs during the Late Carboniferous to Permian (Figure 2), as documented through palynological evidence in subsurface wells in the northern gulf (Eames, 1986) and unconformity surfaces in outcrop surrounding the gulf (Issawi and Jux, 1982; Bhattacharyya and Dunn, 1986; Barakat et al., 1986). Gvirtzman and Weissbrod (1984) documented a second and final phase of Hercynian uplift in the Helez geanticline to the east during this time, suggesting that similar movements may be responsible for the hiatus in the gulf area.

Throughout the northern portion of the gulf, dominantly fluvial sandstones of the Quseib Formation (Abdallah and El Adindani, 1963; Bhattacharyya and Dunn, 1986; Barakat et al., 1986) were deposited unconformably over Carboniferous units. A Permian age is frequently assigned to this unit (e.g., Issawi and Jux, 1982; Bhattacharyya and Dunn, 1986) based primarily on stratigraphic arguments due to lack of age-diagnostic fauna in the sandstones. Barakat et al. (1986) argued that the Quseib Formation extends into the Triassic based on fauna extracted from a limestone capping the red beds of this formation.

Mesozoic Through Early Tertiary (Eocene)

Mesozoic through early Tertiary events recorded in the Gulf of Suez area are dominated by the formation and ultimate closure of the Neo-Tethys Ocean. During this time, the future site of the Gulf of Suez was partitioned into two domains: a tectonically unstable area to the north and a tectonically stable area to the south.

Neo-Tethyan Opening

The Triassic commonly is cited as a period of active rifting in the development of the southern margin of the Neo-Tethys (e.g., Bernoulli and Jenkins, 1974; Biju-Duval et al., 1979; Argyriadis et al., 1980; Sengor et al., 1984b). Indeed, Triassic rocks in the Levant northeast of the Gulf of Suez clearly document Middle Triassic to Early Jurassic rifting in the Helez area (Garfunkel and Derin, 1984; Druckman, 1984) with an associated thermal pulse (Kohn and Eyal, 1981, in Garfunkel and Derin, 1984) and the igneous intrusion of the Gevim quartz porphyry (Druckman, 1984). Thin occurrences of Triassic units (Quseib Formation) are reported (Figure 5) in the northernmost gulf by Said (1962), Abdallah et al. (1963), Beleity et al. (1986), and Barakat et al. (1986). However, the absence of significant thicknesses of Triassic rocks in the Gulf of Suez area and northwest of the gulf (Sestini, 1984), despite a long-term rise in sea level from the Late Permian to later Triassic (Haq et al., 1987), argues that the effects of the initial Triassic rifting of the Neo-Tethys took place north of the limit of subsurface penetrations in the Western Desert and the Gulf of Suez. An unconformity surface terminates the Quseib Formation (Figure 2).

Jurassic rocks are recorded around the flanks of the northern one-third of the gulf (Barakat et al., 1986) and in the subsurface in wells as far south as the Baba plain (Renolds, 1979; Wasfi and Azazi, 1979). Wasfi and Azazi (1979) constrained the subsurface units to the Lower and Middle Jurassic (possibly as old as Early Triassic). Upper Jurassic (Oxfordian) limestones are recorded in the Ayun Musa area by G.A. Moustafa (1961, in Renolds, 1979). The alluvial, fluvial, deltaic, and shallow-marine facies (Renolds, 1979; Wasfi and Azazi, 1979, Barakat et al., 1986) preserved in the Lower and Middle Jurassic record a phase of accelerated tectonic subsidence in the northern Gulf of Suez, and are likely an expression of the development of the southern, rifted margin of the

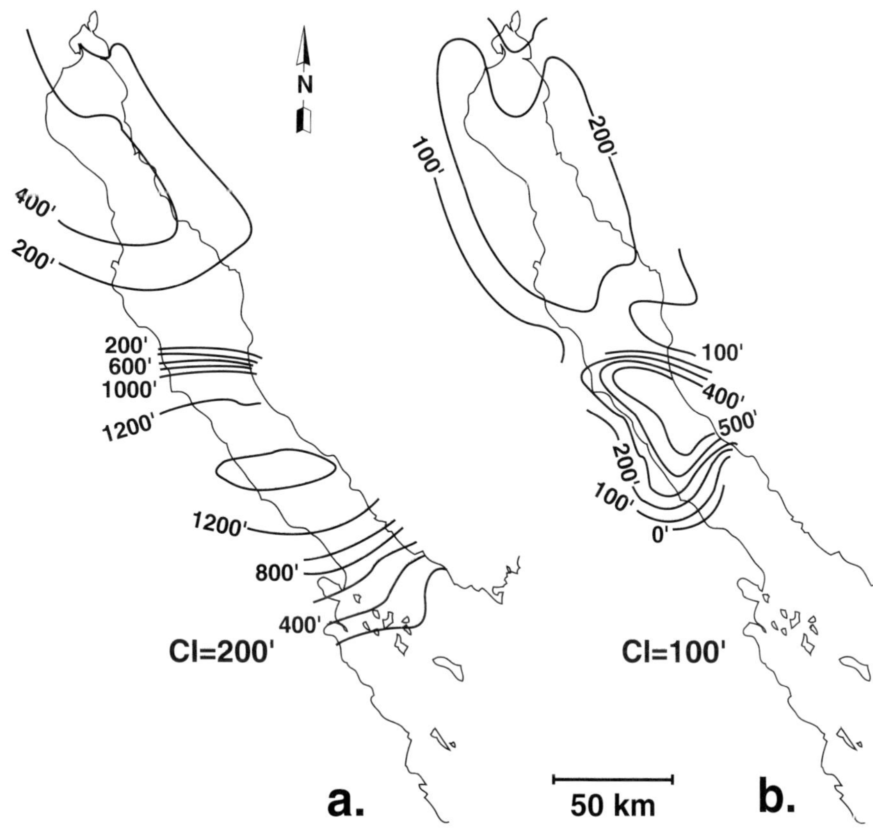

Figure 6. Isopach maps of (a) the Naqus Formation and (b) the Umm Bogma–Abu Durba formations. After Beleity et al. (1986).

Neo-Tethyan Ocean. Middle Jurassic basalts (178 Ma; Weissbrod, 1969b) documented in the Umm Bogma area (Figure 7) are probably associated with this extensional event.

Paleoreconstructions proposed for the northern African margin during the Neo-Tethyan rifting (Figure 8) (Robertson and Dixon, 1984; Şengor et al., 1984b) argue for fragmentation of the margin into a series of rhombic-shaped, continental slivers that separated to the north. This layered separation might have been aided by possible east-west, left-lateral, Triassic–Jurassic wrench faults impinging from the west (Dewey et al., 1973; Dercourt et al., 1986; Ziegler, 1987), preconditioning the margin by cutting it into elongated, east-west strips. A small faulted, but unseparated, rhombic-shaped block associated with this event may underlie what is now the Nile Delta.

The extension experienced by the northern African margin was likely felt well to the east in the Arabian block as expressed by the Triassic through Middle Jurassic subsidence documented in the Palmyra and Sinjar troughs in Syria (Lovelock, 1984). These troughs are described as broad downwarps with no obvious faults controlling their flanks (Lovelock, 1984). The Levant transform (Robertson and Dixon, 1984; Sengor et al., 1984b) likely precluded the Palmyra and Sinjar troughs from experiencing complete rifting. The transform separated the more highly extended lithosphere to the west from the partially extended lithosphere on the Arabian Platform. The Levant transform likely played an important role in localizing compressional deformation later in the history of the northern African margin, a topic we discuss in a following section.

The southern limit of the Jurassic wedge of sediments in the gulf area (Figure 5) marks the southern limit of extended lithosphere associated with the Neo-Tethyan rift event and defines the southern boundary of a zone (hinge zone in Figure 4c) that experienced recurring tectonism throughout the rest of the Mesozoic and into the early Eocene.

Renewed uplift was first expressed north of the hinge zone during the Late Jurassic and Early Cretaceous. Outcrops of Middle Jurassic rocks along the northern Galala Plateau are cut by pre-Aptian, east-west–trending faults (Bhattacharyya and Dunn, 1986). Van Houten (1980) stated that similarly oriented (elongate in the east-west direction) basins and uplifts initiated in the Western Desert during the Late Jurassic. Associated Early Cretaceous basaltic igneous activity occurs in the northern portion of the gulf with dates of 125 ± 4 Ma (El Shazly, 1977) along Wadi Araba and 115 ± 3 Ma (El Shazly, 1977) at Abu Darag (Figure 7). The hiatus in the rock record of the Gulf of Suez that resulted from this period of tectonism extends from the latest Jurassic into the Aptian.

As sedimentation returned to the Gulf of Suez area during the Aptian, fluvial deposits (Van Houten et al.,

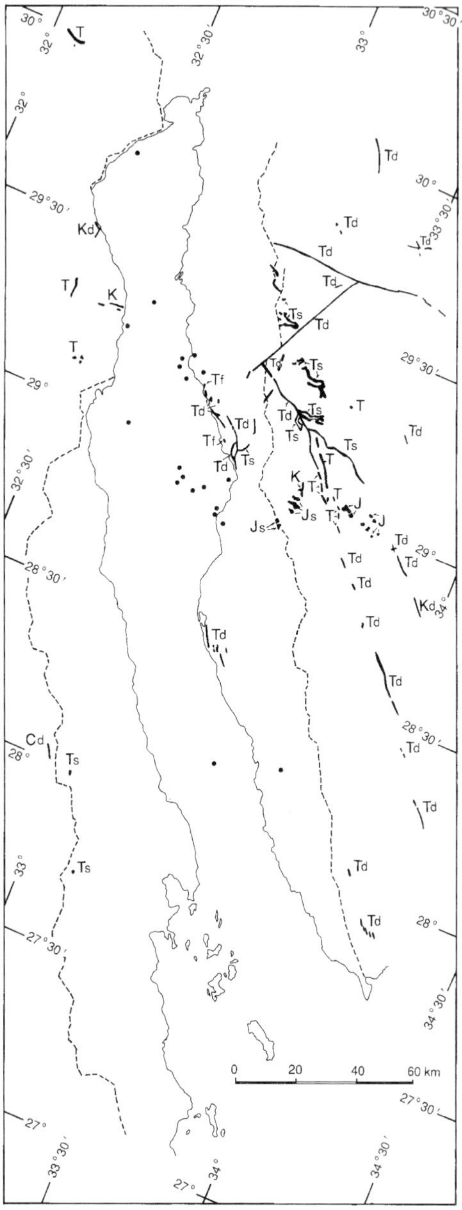

Figure 7. Post–early Paleozoic igneous rocks in the Gulf of Suez. Surface occurrences of igneous rocks are represented by a two-part symbol. The first letter denotes the age of the igneous rock: C = Carboniferous (Gebel Gharamul), J = Jurassic (Umm Bogma area), K = Cretaceous (Wadi Araba and Abu Darag), and T = Tertiary (mostly Oligocene–Miocene) igneous rocks, which have the widest distribution. The second letter indicates the nature of the occurrence: d = indicates dikes, s = sill, and f = flow. The locations of the rift-bounding faults are marked by dashed lines. Circles represent volcanics or volcaniclastics encountered in industry wells, with those in the central gulf commonly Tertiary in age. (Sources: Weissbrod, 1969b; Meneisy and Kreuzer, 1974, in El Shazly, 1977; Garfunkel and Bartov, 1977; Bartov et al., 1980b; Burollet et al., 1982; Steen, 1984; Henry et al., 1986; Ott d'Estevou et al., 1986a; Prat et al., 1986; Chowdhary and Taha, 1987; Moustafa and Abdeen, in press.)

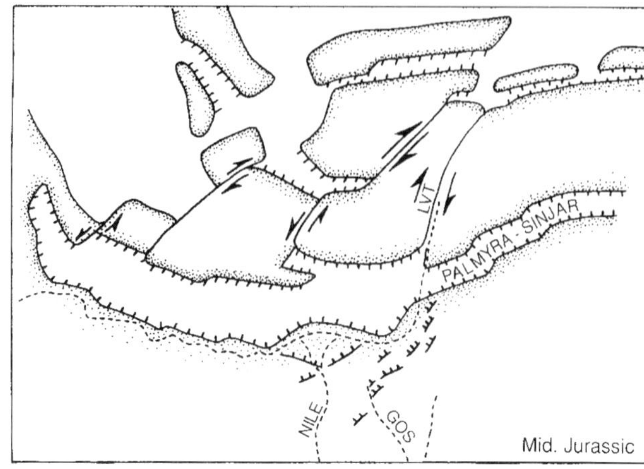

Figure 8. Middle Jurassic plate reconstruction of the eastern Mediterranean (modified after Robertson and Dixon, 1984). The blocks rifting away from the north African coast eventually will reassemble, forming present-day Turkey. Although the Palmyra-Sinjar trough is experiencing extension, rifting has not gone to the stage of developing oceanic crust as it has along north Africa because the right-lateral Levant transform is transferring extension to the north of the Arabian Peninsula. For additional details on the plate configurations in the Neo-Tethys, see Robertson and Dixon (1984). GOS = Gulf of Suez, LVT = Levant transform.

1984; Barakat et al., 1986) of the Malha Formation began filling in the topographic lows associated with the preceding phase of tectonism (Figure 5). The Malha Formation is deposited over rocks ranging in age from Jurassic to Precambrian (Soliman and Amer, 1972; Garfunkel and Bartov, 1977; Van Houten et al., 1984) attesting to the relief developed during the preceding tectonic disturbance. Sediment dispersal directions in the Malha Formation (Figure 2) of northeastern Egypt can be subdivided into two domains; in the Gulf of Suez area, northerly and westerly dispersal prevailed, whereas farther south in the Natash basin and Quseir area, southerly and westerly dispersal dominated (Van Houten et al., 1984). Van Houten et al. (1984) attributed these patterns, along with isopach data of the Malha Formation, to the presence of two regional uplifts in northeastern Egypt (Figure 4c). The first, the Kharga arch (the Kharga–Aswan–Red Sea High of Klitzsch, 1986), is a northeast-trending ridge that projects into the southern end of the Gulf of Suez area and may represent continued expression of the late Paleozoic Uweinat-Aswan High. The second uplift proposed by Van Houten et al. (1984) is the Suez arch, which is a northwest-trending ridge that paralleled the western margin of the present-day rift.

Van Houten et al. (1984) proposed that the Kharga and Suez arches intersected in the area of the northern Red Sea to form a basement uplift from which the sediments of the Malha Formation were derived. This uplift appears to have persisted throughout the Mesozoic and early Tertiary in this southern area of the gulf. Another high is suggested at the northern end of the gulf by the presence of thin Malha Formation at Ayun Sukhna (70 m or 230 ft) and Gebel Ataqa (22 m or 72 ft) (El Akkad and Abdallah, 1971), and is probably one of a series of such highs that follow an easterly trend from the northern Western Desert (Van Houten, 1980; Van Houten et al., 1984). The area to the north is also characterized by an increased marine affinity due to the influence of the Neo-Tethys Ocean.

After this Late Jurassic–Early Cretaceous disturbance, subsidence of the northern African margin returned in the late Cenomanian, possibly reflecting the return to subsidence associated with the earlier Neo-Tethyan rift event. A marine transgression occurred across the northern margin of the African continent in the Gulf of Suez area initiating a period of dominantly marine deposition in the gulf; this period persisted until the late Eocene. From the Cenomanian to Santonian, widespread marginal marine clastics and carbonates of the Raha, Wata, and Matulla formations (Tewfik and Ebeid, 1975) were deposited (Figures 2, 5). Throughout the central portion of the gulf the Raha Formation conformably overlies the Malha Formation; however, along the flanks of the rift these rocks may overlie older units (e.g., Soliman and Amer, 1972; Van Houten et al., 1984).

Neo-Tethyan Closure

The Neo-Tethyan Ocean reached its maximum extent in the Early Cretaceous (Sengor and Yilmaz, 1981), and the Late Cretaceous marks the onset of its closure. This regional readjustment of the tectonic setting was expressed on the northern margin of Africa and the Levant through the disruption of sedimentation during the Late Cretaceous caused by the development of Syrian Arc folds (Figures 2, 5). In the Gulf of Suez area, Syrian Arc folds follow a east-northeast–west-southwest trend and extend from the Abu Roash area of Egypt, across the northern Gulf of Suez, and into the northern Sinai and Israel. The folds have their most pronounced geologic expression in the northern portions of the gulf. However, Montenat et al. (1986) and Ott d'Estevou et al. (1986a) proposed that the Dara ridge (Figure 1) might also be the site of a more subtly expressed Syrian Arc feature. The folds have a geophysical expression displayed in magnetic data in northern Egypt, but their magnetic signature decreases in strength to the south (Meshref, 1984).

Moustafa and Khalil (1987a) argued for two major phases of deformation associated with the Syrian Arc features (Figures 2, 5). The earliest documented growth for the first phase of deformation occurs in two widely separated structures: one in the Abu Roash area (Moustafa, 1988) southwest of Cairo and the other in the northern Negev Desert (Braun et al., 1987). A Turonian age for initiation of structural growth in both of these areas has been assigned by these authors. Bartov et al. (1980a) documented minor differential uplift during Cenomanian–Turonian at the Areif El Naqa structure in the eastern Sinai, but significant anticlinal development was not expressed at this location until the Coniacian. According to Moustafa and Khalil (1987a), deformation associated with this first growth phase had ceased prior to the deposition of the Sudr Chalk (late Senonian).

The second period of deformation associated with the growth of the Syrian Arc structures initiated in the upper Paleocene (Velascoensis event of Strougo, 1986) and continued through the early Eocene. Moustafa (1988) documented a final phase of growth on the Abu Roash structure during the early Tertiary. In the northern Gulf of Suez, middle Eocene rocks unconformably overlie Cretaceous rocks (e.g., Said 1962; Soliman et al., 1965; El-Akkad and Abdallah, 1971; Robson, 1971) and it is not until south of the Wadi Araba structure that the more complete Cretaceous through Eocene stratigraphic sequence, typical of the central and southern gulf, is again established (Soliman et al., 1965). Also in the northern gulf, in the Hammam Faraun (Wadi Thal) area, Abul-Nasr (1986) and Abul-Nasr and Thunell (1987) attributed the presence of lower Eocene breccias and slump deposits (Figure 9) to an expression of the latest stages of Syrian Arc activity immediately to the north (Figure 5). In the St. Paul Monastery area on the west side of the gulf, similar facies are documented by Ismail and Abdallah (1966) and Travis (1984) (both in Abul-Nasr, 1986) and assigned an Eocene age by these authors. However, recent studies of planktonic foraminifera yield a late Paleocene age for these units (Strougo, 1986). Mass-transport features of latest early Eocene age are also found along the northward extension of the Syrian Arc trend in Israel and Lebanon as documented in Buchbinder et al. (1988). These authors did not associate the mass transport deposits with specific Syrian Arc anticlines, but related the deposits to flexure of the margin.

The Paleocene–lower Eocene Esna Shale was not deposited (or possibly eroded) in Gebel Gharamul and the northernmost part of Esh El Mellaha. (North and south of this area the early Tertiary is found.) Again, as in the case for the Cretaceous, the absence of the early Tertiary sequence in this area may be due to renewed growth of or residual Cretaceous relief on an isolated Syrian Arc feature in the Dara ridge area. In the Negev Desert, a middle Eocene age is recorded as the time of final growth of the Syrian Arc features (Bartov et al., 1980a; Zilberman, 1981). Marine shales and limestones (Tanka Formation, Figure 2) continued to be deposited locally in the gulf area through the late Eocene (Boukhary and Abdelmalik, 1983). West of the gulf in the West Qarun block of the eastern Abu Gharadig basin (Western Desert of Egypt) a minor compressional wrench event is recorded in the lower Tertiary (M. Crews, personal communication, 1987).

In the literature, the Syrian Arc structures frequently are interpreted to have formed as forced folds through reactivation of earlier formed rift-margin faults (Freund et al., 1975; Reches et al., 1981; Eyal and Reches, 1983; Moustafa and Khalil, 1987a). Stratigraphic arguments for the reactivation of Tethyan rift faults in Israel derive from the structural inversion of Triassic–Jurassic depositional thicks (developed

Figure 9. Lower Eocene slope breccias along Wadi Thal in Gebel Hammam Faraun.

along the downthrown sides of Mesozoic normal faults) during Cretaceous compression (Freund et al., 1975; Reches et al., 1981). Structural arguments for the reverse-fault bounded, uplifted blocks derive from observations of surface fold geometries and faults (Freund et al., 1975; Reches et al., 1981) and subsurface penetrations of reverse faults (Reches et al., 1981) documenting fault dips between 60 and 80°. Further arguments for reactivation focus on the fact that the main structure-bounding faults are not correctly oriented to most efficiently accommodate crustal shortening associated with the principal stress direction inferred from outcrop observations of mesoscopic deformation features. Nonparallel alignment is found between the principal direction of shortening derived from mesoscopic features (e.g., stylolites, fractures, slickensides) and that inferred from the axial traces of folds and fault attitudes (Eyal and Reches, 1983).

Note that the basement blocks proposed to be responsible for the structures are nowhere exposed in any of the north Sinai structures. In the northern Gulf of Suez several wells have penetrated basement either on a subsurface-defined Syrian Arc feature (i.e., Ayun Musa-1, GS 9-1; Figure 1) or crossed faults into basement along the subsurface extension of a surface Syrian Arc feature (i.e., BB 80-1, Z 80-1; Figure 1) supporting the idea that the Syrian Arc folds in the northern Gulf of Suez are basement involved. The Abu Roash folds also are cored by basement at shallow depths (1900 m (6238 ft)).

As mentioned previously, the onset of growth of the Syrian Arc features corresponds closely with the onset of closure of the Neo-Tethys. The primary site of Tethyan oceanic lithosphere consumption was located along the northern margin of the Tethys throughout most of the Mesozoic (Figure 10), as evidenced in the tectonic and stratigraphic record of the continental fragments that make up Turkey (e.g., Sengor and Yilmaz, 1981; Robertson and Dixon, 1984; Gorur et al., 1984) and the evidence of obducted ophiolites all along the northern and northeastern margin of the Arabian plate from Syria to Oman (e.g., Stocklin, 1968; Glennie et al., 1973; Murris, 1980; Sengor and Yilmaz, 1981; Hempton, 1985; Dercourt et al., 1986; Patton and O'Connor, 1988). These continental fragments record complex thrust and strike-slip fault histories associated with the closure of the Neo-Tethys and the continued, present-day collision of the Arabian plate with Eurasia.

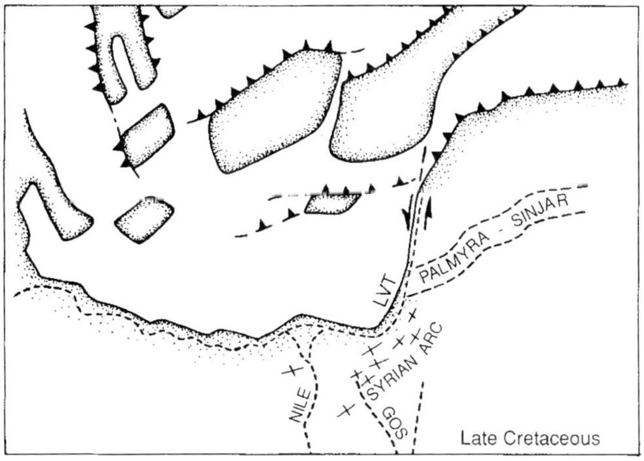

Figure 10. Late Cretaceous plate reconstruction of the eastern Mediterranean (modified after Robertson and Dixon, 1984). The Palmyra-Sinjar trough is inactive during this time. The Levant transform is now experiencing left-lateral strike-slip, transferring southward-directed motion of the oceanic crust to the inside corner of the Levant. Earlier normal fault networks in the northern Gulf of Suez and Sinai are inverted giving the Syrian Arc system of folds. For additional details of the plates in the closing Neo-Tethys, see Robertson and Dixon (1984).

One such example of these complexities is seen along the eastern margin of the Arabian Peninsula in the Oman Mountains. Mesozoic rifting in this area resulted in the development of a transform margin (Dibba line) along the southeastern margin of the Musandam Peninsula (Sengor, 1990; Stoneley, 1990). This feature had a similar, but oppositely facing, configuration as that proposed for the Levant to the west. The Semail Ophiolite was obducted onto this transform margin during the Cretaceous. During the obduction and associated thrusting, the left-lateral Dibba extensional transform margin was reactivated as a convergent, right-lateral, strike-slip fault, altering the amount and character of shortening documented to either side of its trace (Robertson et al., 1990).

As with the Dibba transform, the large Levant transform also could have been reactivated during the consumption of the Tethys (J. Smewing, personal communication, 1988). Thrust advancement west of the Arabian promontory reversed motion on the right-lateral extensional transform of the Levant, making it a left-lateral, slightly convergent wrench fault (Figure 10). Shortening and compression could have been transferred southward toward the "inside corner" of the original transform, with relatively weak, brittle, continental crust on the outside and stronger, relatively more ductile oceanic crust on the inside. Deformation on the continental side of the corner resulted in new basement faulting and substantial reactivation of old basement faults into high-angle reverse faults, structurally draping the overlying sedimentary packages across the fault-block edges. The arcuate trend of these folds roughly parallels, and is restricted to, the original Levant crustal corner. Although the relative motions of oceanic and continental components of this model are not known, the west-northwest–east-southeast orientation for the maximum compressive horizontal stress (Eyal and Reches, 1983) and the component of right-lateral slip interpreted along the flanks of the Syrian Arc features in the northern Sinai (Moustafa and Khalil, 1987a) imply a significant component of easterly travel of the oceanic plate relative to the continental corner.

Regionally, during the late Paleocene through Eocene the assembly of the Anatolian landmass and the initial collision of the Arabian plate against Eurasia occurred (Sengor and Yilmaz, 1981; Gorur et al., 1984; Sengor et al., 1984a, 1984b; Hempton, 1985, 1987). The second phase of growth on the Syrian Arc structures may reflect structural adjustments in Egypt and Sinai in response to collisions taking place along the northern margin of the Tethys. Alternatively, compression would continue until sufficient oceanic crust was subducted under the Alpine front to allow for gravitational "slab pull" to occur and reduce the strength and areal extent of the compressional front. If the latter argument is accepted, from timing of the Syrian Arc structures (Late Cretaceous to Eocene), the change to subduction pull may have occurred in the Eocene.

The final tectonic event recorded in the Gulf of Suez area, prior to the onset of active Gulf of Suez rifting, is manifest in the Cairo-Suez district and the northwestern Sinai (Figure 2). Two major fault sets are in the desert between Cairo and Suez: a pervasive east-west set and a subordinate northwest-trending set (Said, 1962). Moustafa et al. (1985) stated that the east-west faults are actually elongated belts of smaller, left-stepping, en echelon normal faults of west-northwest–east-southeast to east-west orientation. They argued that these faults formed by right-lateral, divergent wrenching on deep-seated, east-west–oriented basement faults. In most cases, the north-northwest- to northwest-oriented faults are restricted to the blocks between the en echelon fault belts and are considered by Moustafa et al. (1985) to have been formed by intrablock extension forming north-northwest- to northwest-oriented pull-apart grabens.

Moustafa and Abd-Allah (1992) indicate that other northwest-oriented normal faults with subordinate right-lateral strike-slip components form a pervasive fault fabric in the central part of the Cairo-Suez district. These faults extend southeastward across the Gulf of Suez to the Hammam Faraun-Wadi Baba area on the eastern side of the gulf. Moustafa and Abd-Allah (1992) state that the northwest faults and the east-west en echelon fault belts are interconnected forming a sigmoidal pattern. Said (1962) stated that the east-west and northwest-southeast fault systems of the Cairo-Suez district are contemporaneous. Movement on these faults is Oligocene in age and preceded the extrusion of Oligocene–Miocene (22 ± 2 Ma; Meneisy and Abdel Aal, 1984) basalts of the district (Moustafa et al., 1985). Moustafa and Abd-Allah (1991) identified a second phase of faulting that followed the deposition of middle Miocene(?) marine sediments overlying the Oligocene–Miocene basalts in the central part of the Cairo-Suez district.

MIOCENE TO HOLOCENE RIFT TECTONICS

Subsidence studies in the Gulf of Suez rift document three major phases of subsidence history: an initial period of gentle subsidence, followed by a period of rapid subsidence, and, finally, a return to gentle, irregular subsidence. This three-part subsidence history has been interpreted to correspond to three stages of rift development, which include a rift initiation stage, a stage of active rifting, and a final stage of rift abandonment. Our discussion of the tectonic evolution of the Suez rift will follow this three-part evolution scheme.

The uppermost Oligocene(?) and Miocene rocks of the Gulf of Suez record the rift event. The synrift facies in the gulf are highly variable, and thus vary greatly in terms of lithostratigraphic nomenclature (Sadek, 1959; Egyptian General Petroleum Corporation, 1964; Stratigraphic Subcommittee of the National Committee of Geological Sciences, 1976). The variety of lithostratigraphic nomenclature continues to be a topic of active debate in the recent literature of the gulf (e.g., Samuel et al., 1988; Hamza, 1988). For our purposes, we use the nomenclature established by the Egyptian General Petroleum Corporation (1964) stratigraphic committee (Figure 11). However, a number of important statements made by different investigators regarding the tectonic evolution of the gulf refer to different or more refined subdivisions of this nomenclature. When we include statements using nomenclature different than that shown in Figure 11, we refer to the alternate nomenclature given in Figure 12. Figure 12 lists only those names not found in Figure 11 and does not represent a complete treatment of the breadth of stratigraphic nomenclature for the gulf.

Rift Initiation

The first possible stratigraphic evidence of rifting is represented by the red beds in Wadi Taiyba (Figure 13). They form a characteristic unit in the Abu Zenima Formation (Figure 11) of Hanter (1965) that consists of unfossiliferous red siltstones and mudstones with several sandstone and conglomerate beds, especially at Wadi Nukhul. This unit has a maximum outcrop thickness of 110 m (360 ft) in Wadi Nukhul, but thins northward and is absent at and north of Wadi Thal and south of the Hammam Faraun block. Chenet and Letouzey (1983), Montenat et al. (1988), and Richardson and Arthur (1988) considered the Abu Zenima Formation to be early, rift-related clastics. The age of this unit is controversial and has received a variety of age assignments, including late Eocene (Hume et al., 1920), post-late middle Eocene and pre-Miocene (El Heiny and Morsi, 1986), Eocene or early Miocene (Sellwood and Netherwood, 1984), and Oligocene (Viotti and El Demerdash, 1969; Garfunkel and Bartov, 1977; Chenet and Letouzey, 1983; Abul-Nasr, 1987, 1990; Richardson and Arthur, 1988). Garfunkel and Bartov (1977) reported basalt pebbles near the middle of the Abu Zenima Formation at Wadi Nukhul, and Montenat et al. (1988) documented basalt pebbles with radiometric dates of 25–26 Ma at the base of the red beds in Wadi Nukhul. Although the presence of igneous pebbles demands a younger age for the Abu Zenima Formation than the radiometric dates, for the most part the Abu Zenima Formation is itself overlain by a basalt flow (Figure 13); therefore, the deposition of the Abu Zenima Formation is likely synchronous with the early Gulf of Suez igneous episode. These data, along with an Oligocene age assignment for the red beds at Wadi Taiyba by Abul-Nasr (1987), argue strongly for late Oligocene to early Miocene age for the Abu Zenima Formation.

These poorly age-constrained continental units contrast sharply with data derived immediately south of the gulf. Palinspastic restoration of the Midyan Peninsula to the southern mouth of the Gulf of Suez indicates that marine facies of the Musayr Formation (Purser and Hotzl, 1988) were deposited in this area during the Chattian, suggesting a significant difference between the northern Red Sea Basin and the Gulf of Suez Basin during this time.

The Oligocene(?)–Miocene magmatism along the margins of the rift commonly is cited as evidence for the initiation of rifting. Basaltic dikes, sills, and flows are exposed on both sides of the Suez rift and have been encountered in a number of exploratory wells (Figure 7). Basalt outcrops most commonly occur along the eastern side of the rift. Many of these bodies extend from the shoulders out into the rift. Dike orientations are mainly northwest to north-northwest, although other less common, synchronous trends exist (e.g., east-west, north-northeast to northeast, and north-south). The longest dike is discontinuously exposed in the west Sinai for about one-half the length of the rift (Figure 7), and is roughly parallel with the eastern rift boundary fault.

The best exposures of rift-related magmatic rocks exist in the Hammam Faraun–Abu Zenima area (Figure 13). In this area, basalt dikes, sills, and flows affect rocks older than the Nukhul Formation (see the following discussion for Nukhul Formation age). Rocks of unequivocal Miocene age in this area are never affected (intruded, baked, or colored) by these magmatic rocks. Also, eroded fragments of these basalts are deposited in the basal conglomerate beds of the Miocene synrift sediments. Exposures of basalt-bearing boulder conglomerates are found at the mouth of Wadi Taiyba. Exceptionally large (up to 1 m or 3 ft long) angular basalt boulders exist in channel-fill deposits at the base of the Nukhul Formation east of Gebel Hammam Faraun (Figure 14). Red coloration and contact metamorphic effects are seen in the pre-rift (pre-Miocene) rocks (e.g., at the mouth of Wadi Taiyba and Wadi Nukhul).

Stratigraphic relationships date these basalt occurrences as post-late Eocene and pre- or earliest

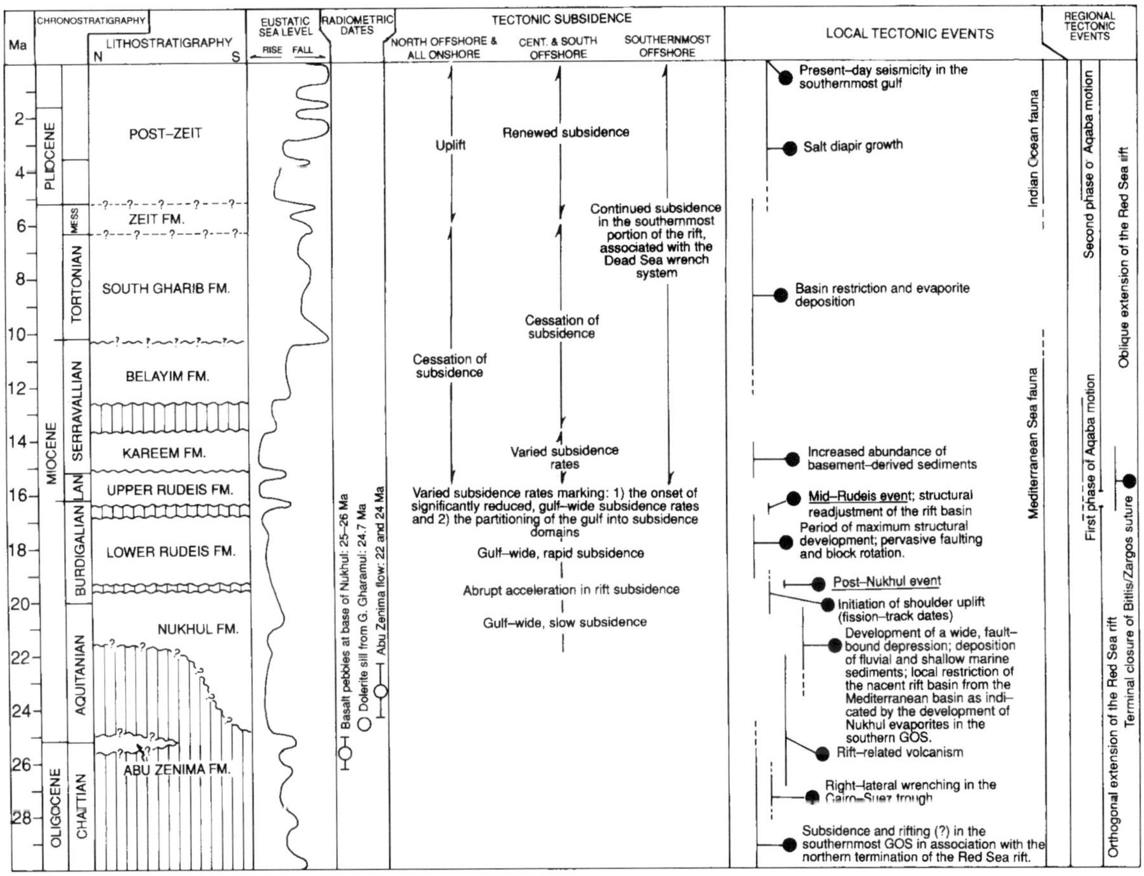

Figure 11. Post-Oligocene, tectonostratigraphic events of the Gulf of Suez. Sea level data and chronostratigraphy are from Haq et al. (1987). GOS = Gulf of Suez.

Miocene (Sadek, 1959; Moustafa and Abdeen, 1992). Absolute ages of flows range between 22 and 24 Ma (Steen, 1984; Montenat et al., 1986; Moussa, 1987). Montenat et al. (1986) recorded somewhat earlier ages (26–25 Ma) from the basalt pebbles in the base of early rift sediments at Wadi Taiyba. Moussa (1987) characterized these basalts as having subalkaline (tholeiitic) to alkaline affinities that were intruded in an intraplate, tensional environment. He also noted, however, that the chemistry of some of the samples suggests oceanic affinities.

All the basalt flows in the rift are lenticular in shape and pinch out laterally over short distances. They have maximum exposed thicknesses of up to 30 m (100 ft). We have interpreted the top of the basalt sheet at the mouth of Wadi Taiyba to have been extruded subaerially in that it is spheroidally weathered. This interpretation conflicts with that of Sellwood and Netherwood (1984) who interpreted these features to be pillow lavas.

The first undisputed stratigraphic evidence of rifting in the Gulf of Suez occurs in the poorly age-constrained Nukhul Formation. This unit demonstrates a wide variety of facies, ranging from fluvial and shallow-marine clastics at its base, to marls, anhydrites, and limestones, deposited in shallow- and open-marine environments, toward its top (Saoudi and Khalil, 1986). The ubiquitous

Epoch	Age	Formation	Member	Description
MIOCENE	SERRAVALLIAN	BELAYIM FORMATION	HAMMAM FARAUN MEMBER	The Hammam Faraun and Sidri members are composed of sands, shales and carbonates; the Feiran and Baba members consist of evaporites.
			FEIRAN MEMBER	
			SIDRI MEMBER	
			BABA MEMBER	
		KAREEM FORMATION	SHAGAR MEMBER	Upper member of Kareem Formation; consists of shales and marls with subordinate sandstones.
			RAHMI MEMBER	Lower member of Kareem Formation; consists primarily of anhydrites with interbeds of shale and sand.
	BUR.-LANG.	RUDEIS FORMATION	(LOWER AYUN MUSA FORMATION)	
			ASL FORMATION	Sandy limestones and marls.
			(HAWARA FORMATION)	
			(MHEIHERRAT FORMATION)	
	AQU.	NUKHUL FORMATION	GHARA MEMBER — Consists primarily of anhydrites and marls. / OCTOBER MEMBER — Consists primarily of sandstones and conglomerates. / GHARAMUL MEMBER — Composed dominantly of reefoidal limestones.	
OLIG. (?–?)			SHOAB ALI MEMBER	Composed primarily of sandstone, with subordinate shales; occurs only in the southern gulf.

Figure 12. Supplement for nomenclature in Figure 11. This diagram lists only Miocene nomenclature used in the text and not listed in Figure 11. This figure does not represent a complete treatment of the stratigraphic nomenclature of the gulf.

Figure 13. Stratigraphic succession between Abu Zenima and Wadi Taiyba. The view in the photograph looks north along the eastern side of the rift. The southwest-dipping block displays the Eocene Darat and Tanka formations, which are overlain by the lower rift sequence of the Abu Zenima Formation, a basalt flow, and the Nukhul Formation, respectively. The front of the outcrop is terminated by a high-angle normal fault that juxtaposes the described succession against marls of the Rudeis Formation.

distribution of the Nukhul Formation throughout the gulf (Figure 15) (Saoudi and Khalil, 1986; Richardson and Arthur, 1988) and its presence adjacent to the present-day rift shoulders argue that the outline of the rift basin seen today was established very early in the basin's structural history.

Initial rifting was not accompanied by uplift in that complete sequences of the prerift stratigraphy are preserved in the subsurface along the axis of the rift below the Nukhul Formation. Indeed, uplift of the rift shoulders has been documented to have initiated at or slightly after the onset of rifting (Omar et al., 1989) and has been attributed to the lateral transfer of heat associated with the passive rise of hot mantle beneath the rift (Steckler and Watts, 1980) and small-scale convection beneath the rift (Steckler, 1985; Buck, 1986).

The relief proposed for the early rifting in the gulf is variable, ranging from tens to a few hundred meters (Garfunkel and Bartov, 1977; Evans, 1990) to over 1000 m (Chenet et al., 1986). However, the isopach data presented in Saoudi and Khalil (1986) and Richardson and Arthur (1988) argue for maximum relief in the southern gulf of around 700 m (2300 ft) and significantly less relief in the central and northern gulf, on the order of 150–200 m (500–650 ft).

In addition to the obvious relief manifest in isopach data, we find evidence that significant block rotations had occurred during the deposition of the Nukhul. Composite standard graphing of wells in the October field area suggests that the base of the Nukhul section youngs onto the crest of the structure (F. Miller, W. Hosny, and B. Pierce, personal communication, 1988). Furthermore, significant angular relationships (5–20° discordance) found at surface outcrops are recorded between the Nukhul Formation and underlying pre-Miocene sections at Abu Zenima and gebels Zeit (Evans and Moxon, 1986), Hammam Faraun, and Hadahid (Figure 16).

Sediment transport direction data from the Nukhul Formation at Gebel Hammam Faraun indicate a direction of transport downdip to the present-day attitude of the Hammam Faraun structural block.

Figure 14. Volcanic boulder beds in a channel at the base of the Nukhul Formation on the northeastern flank of Gebel Hammam Faraun. The volcanic boulder indicated by the arrow has a vertical dimension of 0.55 m (1.8 ft).

The longitudinal propagation of rifts is frequently cited as a method by which rift segments grow and link with other rift segments (MacDonald et al., 1984; Martin, 1984; Pollard and Aydin, 1984; Bonatti, 1985). Such diachronous behavior can have significant impact on exploration approaches in rifts. A plausible interpretation of the northward decrease in extension, documented from the southern Red Sea to the northern end of the Gulf of Suez, is that this system is a northward-propagating rift system. An alternate interpretation is that the differences in the amounts of extension along the system are related to the distance from a pole of rotational opening somewhere in the current eastern Mediterranean. The opening of a rift about a proximal pole of rotation results in a coeval event, but one demonstrating rapid changes in the amount and/or rate of extension along the axis of the rift. Such an interpretation is favored by Joffe and Garfunkel (1987) and Le Pichon and Gaulier (1988). Both models can generate diachronous major rift-related features. As to which model is appropriate for the Gulf of Suez is beyond the scope of this chapter. However, given the documented changes in the amount of extension along the Gulf of Suez–Red Sea system, it is worthwhile examining the earliest rift-related sequences for systematic variation in the timing of events along the axis of the gulf.

Graphic correlation (Shaw, 1964; Miller, 1977) of off-structure wells along the axis of the Gulf of Suez rift does not show a systematic variation in the age of the base of the Nukhul Formation (F. Miller, W. Hosny, and B. Pierce, personal communication, 1988). Subsidence studies by Moretti and Colletta (1987) and Richardson and Arthur (1988) likewise find no evidence of systematic younging of the Nukhul toward the north. However, all of these approaches require precise paleontologic control within the Nukhul, yet the base of this unit commonly is noted for its lack of abundant, age-diagnostic fauna (e.g., Saoudi and Khalil, 1986; Richardson and Arthur, 1988; Evans, 1988).

The Nukhul Formation lies on pre-Miocene units as old as basement in the southern portion of the gulf (Evans and Moxon, 1986; Richardson and Arthur, 1988). Evans (1990) argued from provenance studies that the quartzose sands documented within the Nukhul Formation in the southern gulf were derived from the northern Red Sea area where "Nubia

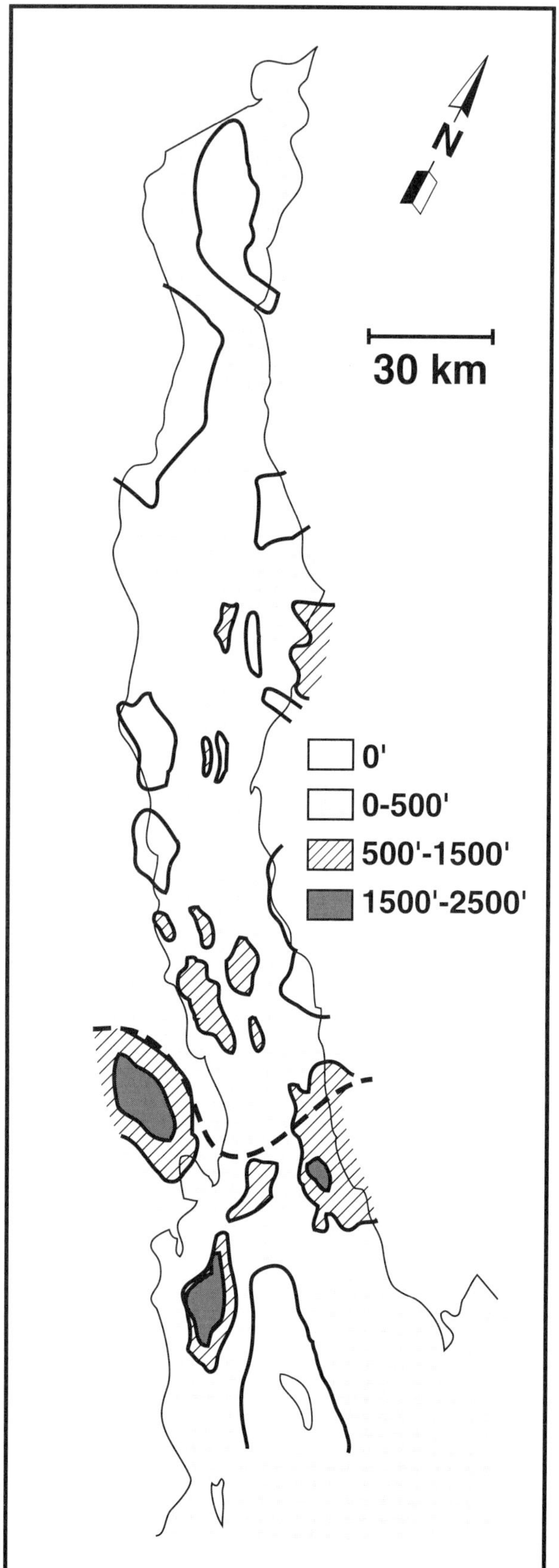

Figure 15. Nukhul isopach map after Saoudi and Khalil (1986). The dashed line crossing the southern gulf represents the northern limit of Shoab Ali Member.

Sandstones" and basement were exposed by uplift in the late Oligocene. These relationships contrast with those documented along the eastern shoulder of the rift where Garfunkel and Bartov (1977) demonstrated that the earliest Miocene sediments consistently overlie the Eocene units. Given comparable pre-Miocene thicknesses between the central and southern Gulf of Suez, these relationships would argue for greater uplift and erosion earlier in the south than in the north. However, as indicated in Figure 5, the pre-Miocene section thins considerably to the south. Reconstructed isopach values for the pre-Miocene section in the southern Gebel Zeit area ranges between 500 and 670 m (1600 and 2200 ft), and in the southernmost gulf the pre-Miocene section is likely less than 305 m (1000 ft) thick. These thicknesses contrast sharply with values for the St. Paul–Abu Zenima area of the central gulf for the Eocene alone, where thicknesses approach 520 m (1700 ft). Therefore, the erosional relief at the base of the Nukhul Formation does not necessarily demand greater structural relief in the southern gulf than in the central or northern gulf.

The best arguments for northward younging of the Red Sea rift into the gulf area derive from isopach and facies data of the Nukhul Formation. Saoudi and Khalil (1986) subdivided the Nukhul Formation into four members: a lower Shoab Ali Member (Figure 12), composed of nonfossiliferous fluvial sands, and three coeval fluvial to shallow-marine members (Ghara, Gharamul, and October; Figure 12). Although they have no direct paleontologic evidence, Saoudi and Khalil (1986) argued that the lower fluvial sandstones (Shoab Ali Member) of the southern gulf are the oldest rocks of the Nukhul Formation. This age assignment is based on stratigraphic position and the similarities in depositional environments of the marine affinities of the three younger members and the total lack of marine influences in the Shoab Ali Member (B. Khalil, personal communication, 1988). The northern extent of the Shoab Ali Member (Figure 15) corresponds roughly with a line on Saoudi and Khalil's (1986) Nukhul isopach map; this line subdivides the Nukhul Basin into two provinces: a southern province with depocenters containing sediment thicknesses significantly greater than 500 ft (150 m) and a northern province with depocenters containing sediment thicknesses generally equal to or less than 500 ft (150 m). A similar trend for deeper Nukhul basins in the southern Gulf of Suez is depicted on the isopach maps of Richardson and Arthur (1988). The Shoab Ali Member may be coeval with the Sharik Formation (Purser and Hotzl, 1988) in the Midyan Peninsula of Saudi Arabia, possibly making it Oligocene in age.

In summary, the evidence in support of the idea of systematic changes in the timing of events along the early axis of the gulf is inconclusive. The best arguments in support of this idea from data within the gulf derive from stratigraphic relationships and isopach data of poorly age-constrained early Nukhul units. Palinspastic restoration of the Musayr Formation documented in the Midyan Peninsula of Saudi Arabia (Purser and Hotzl, 1988) indicate that

Figure 16. Onlap of Nukhul Formation onto the eroded surface of the middle Eocene Mokhatam Formation at Gebel Hadahid.

marine facies existed immediately south of the gulf in the Oligocene, while coeval(?) facies to the north in the gulf were continental in origin. As proposed by Saoudi and Khalil (1986), these data are consistent with an early structural depocenter restricted to the southern gulf. We envision this basin as the northern termination of the earlier forming (Hempton, 1987; Purser and Hotzl, 1988) Red Sea extensional basin. This basin was followed by the more widespread Nukhul Basin, in which the Ghara, Gharamul, and October members were deposited.

Major Rift Development

A significant, gulf-wide depositional hiatus (post-Nukhul event; Beleity, 1984) is documented at the top of the Nukhul Formation (Beleity, 1984; Evans, 1988; Evans and Moxon, 1986) separating the relatively shallow-water deposits of the Nukhul Formation from the deeper-water, marine deposits of the overlying Rudeis Formation (Figure 11). Evans and Moxon (1986) argued that the hiatus marks the initiation of accelerated rifting. Tectonic subsidence curves in the gulf (Scott and Govean, 1986; Moretti and Colletta, 1987; Richardson and Arthur, 1988; Evans, 1988, Steckler et al., 1988) demonstrate an increase in subsidence rate corresponding approximately to this Nukhul–Rudeis boundary.

The increased tectonic subsidence rates immediately following the deposition of the Nukhul Formation resulted in the development of a deep-marine (550 m [1800 ft] water depth, Beleity, 1984; 200 m [670 ft] water depth, Scott and Govean, 1986; 1000 m [3300 ft] water depth, Evans, 1988) environment throughout much of the early gulf. The dominant lithology of the lower Rudeis Formation is Globigerinal marls with secondary development of sands and limestones. Garfunkel and Bartov (1977) stated that marls of the lower Rudeis Formation are found immediately adjacent to the eastern shoulder of the rift without any significant development of coarse clastics. These authors cited this as evidence of an absence of active shoulder uplift during the lower Rudeis. However, in the southern portion of the rift, significant local sand development in the lower Rudeis (e.g., productive sands in July field; Brown, 1980; Abdine, 1981; Evans, 1990) argues for active siliciclastic source terranes within or marginal to the early rift basin. Evans (1990) recorded basement rock fragments in lower Rudeis sandstones and conglomerates that he argued were derived from the Gebel Gharamul area.

The distinction between the lower and upper Rudeis has figured prominently in discussions of the tectonic history of the Gulf of Suez (e.g., Garfunkel and Bartov, 1977; Webster and Ritson, 1984; Beleity, 1984; Chenet et al., 1986; Sellwood and Netherwood, 1984; Steckler, 1985; Richardson and Arthur, 1988; Smale et al., 1988; Evans, 1988; Steckler et al., 1988). The boundary between the upper and lower Rudeis is commonly portrayed as an unconformity surface. Indeed, composite standard graphing of off-structure wells along the axis of the gulf demonstrates a widespread hiatus (Figure 11) within the Rudeis Formation (F. Miller, W. Hosny, and B. Pierce, personal communication, 1988). The unconformity does not correspond to a consistent lithostratigraphic boundary in the subsurface. Investigations of the unconformity in the subsurface and outcrop along the border structures (structurally intermediate position between the shoulders of the rift and the axial trough) of the rift find good stratigraphic evidence to place this surface near the base of the Asl Formation (Figure 12) (Garfunkel and Bartov, 1977; Sellwood and Netherwood, 1984; Smale et al., 1988; Evans, 1988).

This unconformity surface is thought to have occurred as a result of a regional tectonic event commonly referred to as the "mid-clysmic" or "mid-Rudeis" event. In addition to the time breaks documented from biostratigraphic data (Beleity, 1984; Evans, 1988), several lines of physical stratigraphic evidence lead to its recognition in the rock record.

Significant facies variations exist between the lower and upper Rudeis. The widespread, uniformly distributed Globigerinal marls of the lower Rudeis (Garfunkel and Bartov, 1977; Sellwood and Netherwood, 1984) give way to variable facies patterns (Garfunkel and Bartov, 1977; Montenat et al., 1986) and more clastic-dominated sedimentary sequences (Sellwood and Netherwood, 1984; Evans, 1988; Smale et al., 1988). Alluvial fans along the shoulders of the rift sourced submarine-fan systems, which spread out into the basin (Smale et al., 1988; Evans, 1988) and marginal, shallow-marine carbonate environments sourced sediment gravity flows into the basin (Smale et al., 1988). Local source areas developed along the fronts of blocks, such as those seen off the front of the October block where the clastics of the Asl Formation (Figure 12) are conglomeratic. Concomitant with the influx of clastic debris, a marked change in provenance occurred from eroded, dominantly Mesozoic sedimentary cover to more deeply incised sediment-source terranes down to and including basement (Garfunkel and Bartov, 1977; Hagras and Slocki, 1984; Steckler, 1985; Moretti and Colletta, 1987; Smale et al., 1988). An inverse stratigraphy is preserved in the coarse clastics in some fan systems along the eastern margin of the rift (Garfunkel and Bartov, 1977; Sellwood and Netherwood, 1984).

Downlapping sequences developed in lows are found in the upper Rudeis as argued from seismic data in the southwest gulf (Smale et al., 1988). Large-scale foreset beds composed of coarse clastics prograding from the margins of the rift over marls of the lower(?) Rudeis Formation are seen along the margins of the rift in outcrop at Wadi Sudr (Figure 17) and west of Gebel Abu Alaqa (Garfunkel and Bartov, 1977) in the northeast gulf. Elsewhere in the gulf, onlapping relationships are found on structurally high blocks (Garfunkel and Bartov, 1977; Webster and Ritson, 1984; Montenat et al., 1986; Richardson and Arthur, 1988; Smale et al., 1988). Indeed, the angularity found between units internal to the upper Rudeis and between the upper Rudeis and older units commonly is cited as evidence for the unconformity (Garfunkel and Bartov, 1977; Webster and Ritson, 1984; Thiriet et al., 1986; Vigano and Patton, 1988).

Although a structural reorganization of the rift generally is recognized to have taken place during the mid-Rudeis event, the nature of this reorganization and how it relates to the evolution of the basin is variably interpreted. Garfunkel and Bartov (1977) drew attention to the fact that many faults that offset pre-Miocene units die out or show a significant decrease in displacement as they reach the base of the Asl Formation. From such data these authors argued that the reorganization in the rift resulted in establishment of larger structural units containing numerous smaller blocks, and that these units dominated the character of the rift throughout the rest of its history. Subsequent deformation in the rift was localized along the master faults controlling the large blocks with reduced activity along the previously active blocks within the larger units. Three major subdivisions of the rift are proposed by Garfunkel and Bartov (1977) to have evolved during the mid-Rudeis event—the centrally subsiding trough, border structures (occupying an intermediate position between the central trough and the rift shoulder), and the rift shoulder. Other investigators (Montenat et al., 1986; Prat et al., 1986; Richardson and Arthur, 1988; Smale et al., 1988) interpreted the mid-Rudeis event to have resulted in a more highly fragmented rift basin than before the event. Lyberis (1988) and Steckler et al. (1988), using detailed studies of fault kinematic data, argued that the orientation of σ_3 (minimum principal stress) rotated from a bearing of 30 to 60° at the time of the mid-Rudeis event.

Subsidence data throughout the gulf generally record abrupt rate changes (both increasing and decreasing) in tectonic subsidence rates (Figure 11) or sediment accumulation rates around the mid-Rudeis event (Beleity, 1984; Scott and Govean, 1986; Richardson and Arthur, 1988; Evans, 1988; Steckler et al., 1988). Furthermore, the mid-Rudeis event always occurred at or immediately prior to the transition between the phase of rapid subsidence found in the lower Rudeis and the phase of relatively slow subsidence characteristic of the post-middle Miocene history of the Gulf of Suez rift.

Several regional events appear to correspond in timing to the mid-Rudeis event. Bayer et al. (1988) and Voggenreitter et al. (1988) proposed that the

Figure 17. Large-scale foreset strata in the Rudeis Formation along the northeastern margin of the rift at Wadi Sudr.

Arabian plate motion changed from a northeasterly direction to a more north-northeast direction at about the time of the mid-Rudeis event in the Gulf of Suez. However, other investigators argue for only mild variations in plate motion direction. Hempton (1987) used a dominantly northerly travel direction for the Arabian plate since the late Eocene, and Steckler et al. (1988) and Le Pichon and Gaulier (1988) used a northeast (30° in the northern Red Sea and 47° in the southern Red Sea) motion direction since the onset of rifting in the Red Sea. Although Lyberis (1988) and Steckler et al. (1988) keep the Arabian plate motion relatively constant, they proposed significant changes in the velocity and direction of the Sinai microplate around the time of the mid-Rudeis event. Le Pichon and Gaulier (1988) cited a four-fold increase in rotation rate in the late Serravallian, immediately following the mid-Rudeis event in the Gulf of Suez. Despite these differences in the kinematic histories of the plates in the Red Sea–Gulf of Suez area, most investigators acknowledge the coincidence in timing of the mid-Rudeis event with a change in activity along the Dead Sea wrench system and, as proposed by Steckler and ten Brink (1986), it would appear that motion along the Aqaba–Dead Sea wrench system supplanted extension in the Gulf of Suez.

The facies distribution patterns and the irregular subsidence history established after the mid-Rudeis event continued during the deposition of the Kareem Formation. The base of the Kareem commonly is delineated by the presence of the Markha anhydrite (Figure 12). However, the distribution and character of the evaporite are variable in that the Markha may be absent or, alternatively, represented by several anhydrite beds separated by shales and marls. Composite standard graphing of off-structure wells in the gulf demonstrates a gulf-wide hiatus at the beginning of deposition of the Kareem Formation (F. Miller, W. Hosny, and B. Pierce, personal communication, 1988). Although the Markha is closely associated in time with this hiatus, it does not show a consistent relationship to the hiatus, sometimes occurring above and sometimes below the hiatus. Indeed, Beleity (1984) demonstrated that the Markha is diachronous in the gulf. Richardson and Arthur (1988) interpreted the Markha as being laid down in a sabkha environment. Both Beleity (1984) and Richardson and Arthur (1988) noted reworking of the Markha into younger parts of the Kareem Formation. Certainly the reworking of the anhydrite and the diachronous and irregular character of the Markha anhydrite suggest a picture of active structural blocks with subdued relief in a relatively shallow environment. However, the Markha may be closely associated in time with a significant, short-term fall in eustatic sea level and separating these two effects (eustasy and tectonism) is difficult.

Waning tectonic subsidence continued throughout deposition of the later part of the Kareem Formation (Richardson and Arthur, 1988; Evans, 1988) as recorded by the Shagar Member (Figure 12), which shows outer shelf to upper bathyal environments with water depths ranging between 150 and 500 m (500 and 1640 ft) (Scott and Govean, 1986; Evans, 1988). In the central and southern gulf, thick sequences of sand, which were increasingly dominated by basement-derived debris (Evans, 1990), were shed into the gulf from the actively rising rift shoulders.

Rift Abandonment

The top of the Kareem Formation is marked by a pronounced unconformity (Beleity, 1984; Evans and Moxon, 1986) and is succeeded by an increase in the amount of evaporite deposition in the gulf. This, in conjunction with a marked decrease in the subsidence rates (Moretti and Colletta, 1987; Richardson and Arthur, 1988; Evans, 1988; Steckler et al., 1988), records a period of quiescence and the onset of abandonment of the gulf as a site of active extension. The Belayim Formation is composed of an alternating sequence of evaporites (containing both anhydrite and subordinate halite as in the Baba and Feiran members; Figure 12), shales, sandstones, and carbonates (Sidri and Hammam Faraun members; Figure 12). These lithologies show rapidly changing and complex facies relationships (e.g., Heybroek, 1965; Thiebaud and Robson, 1979) reflecting a structurally related topography associated with the preceding period of tectonism.

Mediterranean fauna are documented in the marls of the Belayim Formation (Thiebaud and Robson, 1979) and the development of the evaporite basin appears to have occurred through the restriction of the gulf from the Mediterranean by a structural high in the northern gulf. The effects of the high may have been brought about by a fall in eustatic sea level (Figure 11). Said and Basiouni (1958) and Said (1961) proposed that the Ayun Musa structural block caused the restriction. However, the northern limit of significant Belayim evaporite deposition lies south of the structural expression of this block, following the southern and eastern edge of the Wadi Araba structure in the south and then trending northward toward the Ayun Musa block along the updip edge of the main structural block that forms the floor of the Darag Basin. The restriction of the gulf appears to have been through the combined effects of residual topographic expression on the Wadi Araba structure and an elongate clysmic-trending block. This limit of evaporite deposition is essentially the same throughout the deposition of the South Gharib and Zeit formations.

A significant fall in eustatic sea level (Figure 11) in the earliest Tortonian (Haq et al., 1987), coupled with the restriction between the gulf and the Mediterranean, likely was responsible for the change from interbedded evaporites and marls to the dominantly halite deposition of the South Gharib Formation (Richardson and Arthur, 1988). Subsidence in the central gulf during this time (Moretti and Colletta, 1987; Richardson and Arthur, 1988; Steckler et al., 1988) (Figure 11) has been attributed by Richardson and Arthur (1988) primarily to sediment loading effects, although some mild extension and subsidence may have persisted (e.g., Steckler et al., 1988). Wells in the southernmost portion of the gulf demonstrate an active period of subsidence starting at about 10 Ma and extending to the present (Moretti and Colletta, 1987; Richardson and Arthur, 1988). Although salt diapirism complicates tectonic subsidence interpretations (Steckler et al., 1988), this period of subsidence in the south probably reflects continued extension along clysmic-trending faults in response to the left slip on the Aqaba–Dead Sea wrench system. The absence of subsidence during this period in wells farther to the north of the gulf argues for confinement of tectonism associated with the Dead Sea wrench system to the southernmost portions of the gulf.

Subsidence curves (Moretti and Colletta, 1987; Richardson and Arthur, 1988) document renewed tectonism at the onset of deposition of the Zeit Formation (Figure 11), which has been limited in age by Richardson and Arthur (1988) to the Messinian (at 5 Ma). Although the axial portions of the gulf demonstrate subsidence, the subsidence data in other portions of the gulf do not record uniform positive (downdropping) subsidence. Areas of negative subsidence (uplift) are recorded along the central and southern onshore (Moretti and Colletta, 1987; Richardson and Arthur, 1988; Evans, 1988), northeastern onshore (Evans, 1988), and northern offshore (Moretti and Colletta, 1987; Richardson and Arthur, 1988) portions of the gulf (Figure 11). The coincidence of renewed tectonism and the return to less-restricted environmental conditions of the Zeit Formation led Richardson and Arthur (1988) to propose that this event records the opening of a seaway that allowed faunal communication with the Indian Ocean. Such an idea receives further support from regional tectonic considerations in that this period records readjustments in the relative motions about the Sinai triple junction (Joffe and Garfunkel, 1987) and renewed motion on the Dead Sea wrench system (Garfunkel, 1981). Indeed, Joffe and Garfunkel (1987) depicted a two-phase displacement for the Dead Sea system with the second phase (5 Ma to Holocene) accomplishing 35 km (22 mi) of the 105 km (65 mi) total strike-slip displacement. Likewise, Voggenreitter et al. (1988) discussed a similarly timed, two-phase displacement for the Red Sea.

The subsidence trends recorded in the latest Miocene locally continue to the present (e.g., Moretti and Colletta, 1987; Richardson and Arthur, 1988). Thick sequences of Pliocene–Pleistocene sediments are found in the gulf, but their tectonic significance commonly is obscured due to the growth of salt structures, especially in the south. Present-day seismic activity is restricted to the southern third of the gulf (e.g., Ben Menahem et al., 1976; Daggett et al., 1986; Boulos et al., 1987). Outcrop evidence of Holocene fault movements in Quaternary alluvial sediments, along with examples of uplifted young reefs, is cited by Garfunkel and Bartov (1977) as evidence of continued tectonism located primarily in the southern gulf.

RIFT STRUCTURE

In our discussion of the Gulf of Suez rift structure, we have chosen to focus primarily upon the plan-view expression of the rift faults. We restrict our comments and observations to the regional scale, addressing fault patterns and changes in these patterns as a function of location in the rift. We do this in an attempt to provide a framework that explorationists may use as a starting point to build upon or modify as they see fit. We initiate the discussion by summarizing the orientations and relative abundances of the four fault populations that dominate the gulf. In the next section, we discuss how these faults interrelate in the rift process and propose that the model of displacement transfer is an effective tool in understanding poorly imaged fault configurations in the subsurface. In the final section, we generalize about regional changes in structure along the rift.

Major Fault Populations

Excellent exposures of rift-related structures are found along the margins of the Gulf of Suez. The most obvious surface features expressed are large, rotated fault blocks such as the Abu Durba and Araba blocks (Figures 18, 19). The degree of internal deformation of the larger blocks is highly variable, ranging from very little, as in the case of the Esh El Mellaha (Figure 20, southwest end of section F-F′) and Araba blocks (Figure 20, northeast end of section E-E′), to intensely faulted, as in the case of the Hadahid block (Figure 20, northeast end of section D-D′). A variety of structural configurations have been recorded at block edges. Robson (1971) drew attention to the role of ramps in the lateral terminations of blocks as the frontal faults lose throw along strike. Thiebaud and Robson (1981) demonstrated that in the Miocene section, block terminations may be expressed as monoclines. The role of drape folding in the pre-Miocene sedimentary units associated with the development of block-bounding faults has been documented by Patton (1984), Moustafa (1987), and Coffield and Schamel (1989); however, with large fault displacements, and at the basement level, faulting is the dominant mechanism responsible for block termination in the gulf.

The faults that determine the shapes of the rotated blocks in the rift can be grouped into four major populations on the basis of orientation. Statistically, the most significant of the trends is the rift-parallel trend or the clysmic trend (Figure 21). The clysmic trend lies between 310 and 340° azimuth, with its maximum occurring in the range of 320–330°. It is consistently found throughout the gulf and has been active over the entire phase of rift development (Lyberis, 1988). The clysmic faults demonstrate a normal sense of separation, with both pure dip-slip (Robson, 1971; Angelier, 1985) and oblique-slip (Lyberis, 1988) motion recorded from outcrop data. In the northern Gulf of Suez, a subpopulation lying between 310 and 315° is present (Robson, 1971; Moustafa and El Shaarawy, 1987) (Figure 21). The trend is clearly expressed where faults form the coastlines of the gulf along the southwestern side of Gebel Hammam Faraun and the northeastern edge of the North Galala

Figure 18. Northeast-dipping basement blocks of gebels Araba and Abu Durba. View in photo looks southeast along the eastern shore of the Gulf of Suez. (See Figure 1 for location.)

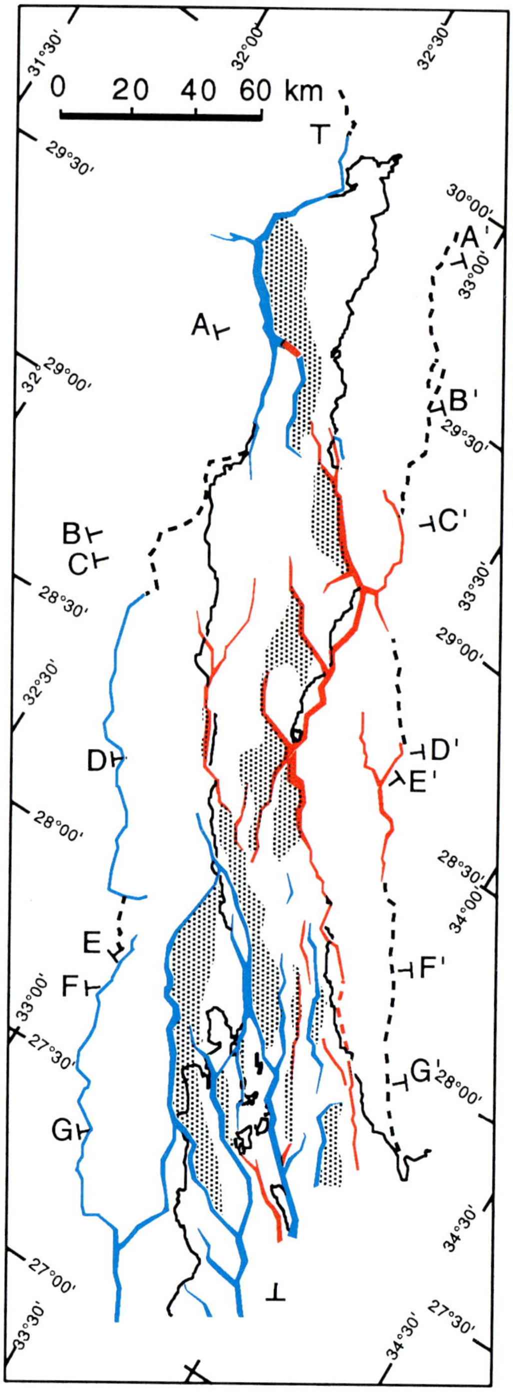

Figure 19. Simplified fault map of the Gulf of Suez showing faults possessing in excess of 1 km (0.6 mi) of throw; blue faults are downthrown to the northeast and red faults are downthrown to the southwest. Stippled areas represent the pre-Miocene surface below 4 km (2.5 mi). The thin dashed line represents the locations of the rift-bounding faults. The locations of the serial sections (A–G) of Figure 20 are shown along the margins of the rift; the vertical line, marked by the tabs at the north end of the gulf and south of Shadwan Island, is the tie line for the sections shown in Figure 20. The map is grossly simplified after 1:50,000 scale structure maps constructed by H.M. Proctor and T.L. Patton.

Plateau. The trend is similarly expressed along the southern portion of Gebel Araba (Figures 1, 19).

In the case of the Galala and Hammam Faraun fault segments, note that removal of the extension experienced by the rift in the north results in the alignment of these two fault segments. Furthermore, Moustafa and El Shaarawy (1987) stated that these faults align with a similarly oriented, major basement lineament expressed in Sinai basement exposures on the shoulder of the rift, and argued that these two faults have exploited a preexisting zone of weakness in the basement during the rift opening. Other surface data exist that support the concept that a clysmic trend in the gulf has great antiquity. Schurmann (1966) mapped basement dikes with a clysmic trend in the Wadi Dib area. Abu Zeid (1985) assigned a Precambrian age to clysmically oriented fabrics in the Eastern Desert, as far south as Gebel Duwi. The clysmic trend was active during the Paleozoic as documented through the intrusion of Gharib and Zeit granites and their associated dikes along a 320–330° azimuth trend (Montenat et al., 1986, 1988; Ott d'Estevou et al., 1986a; Prat et al., 1986). El Tarabili and Adawy (1972) suggested that the northwest strike of the foliations in the metamorphic rocks of the Wadi Feiran area may have influenced the development of clysmic-trending faults in the Wadi Baba area. Furthermore, these authors stated that faults with north-northwest–south-southeast strikes, which correspond to the presently expressed rift boundary, were active in the Carboniferous in the Baba area.

The second most significant fault trend is oblique to the clysmic trend and spans a wide area between 350 and 30° (Figure 21). Two subpopulations exist within this trend occurring between 350 and 360° and between 15 and 30° (Figure 21). The former trend has been previously referred to as the Erythrean or African (Said, 1962) or Nubia trend, and the latter has been referred to as the Aualitic (Said, 1962) or Aqaba trend. For the purposes of this discussion, we have lumped both trends under the single name of North-oblique (N-oblique) trend, describing its orientation relative to the dominant clysmic fabric, and avoiding any genetic association with the parallel and longer-lived wrench faulting of the Gulf of Aqaba. The N-oblique faults were active during the initial stages of rifting, as evidenced by the presence of Oligocene–Miocene dikes with an N-oblique orientation (Robson, 1971; Moustafa and Abdeen, in press) and unpublished isopach data of early rift sediments (Nukhul and Mheiherrat formations; Figure 12). Major faults of this orientation do not occur uniformly throughout the gulf, but appear to be most frequently expressed on the surface at the terminations of major rift blocks (Montenat et al., 1988), such as the southern end of Gebel Zeit, the northern end of Esh El Mellaha, and the northern termination of the Abu Durba and Hadahid blocks (Figure 19). Although dominantly dip-slip motions have been recorded from slikensides on faults of the N-oblique orientation, these faults frequently demonstrate a component of left-lateral,

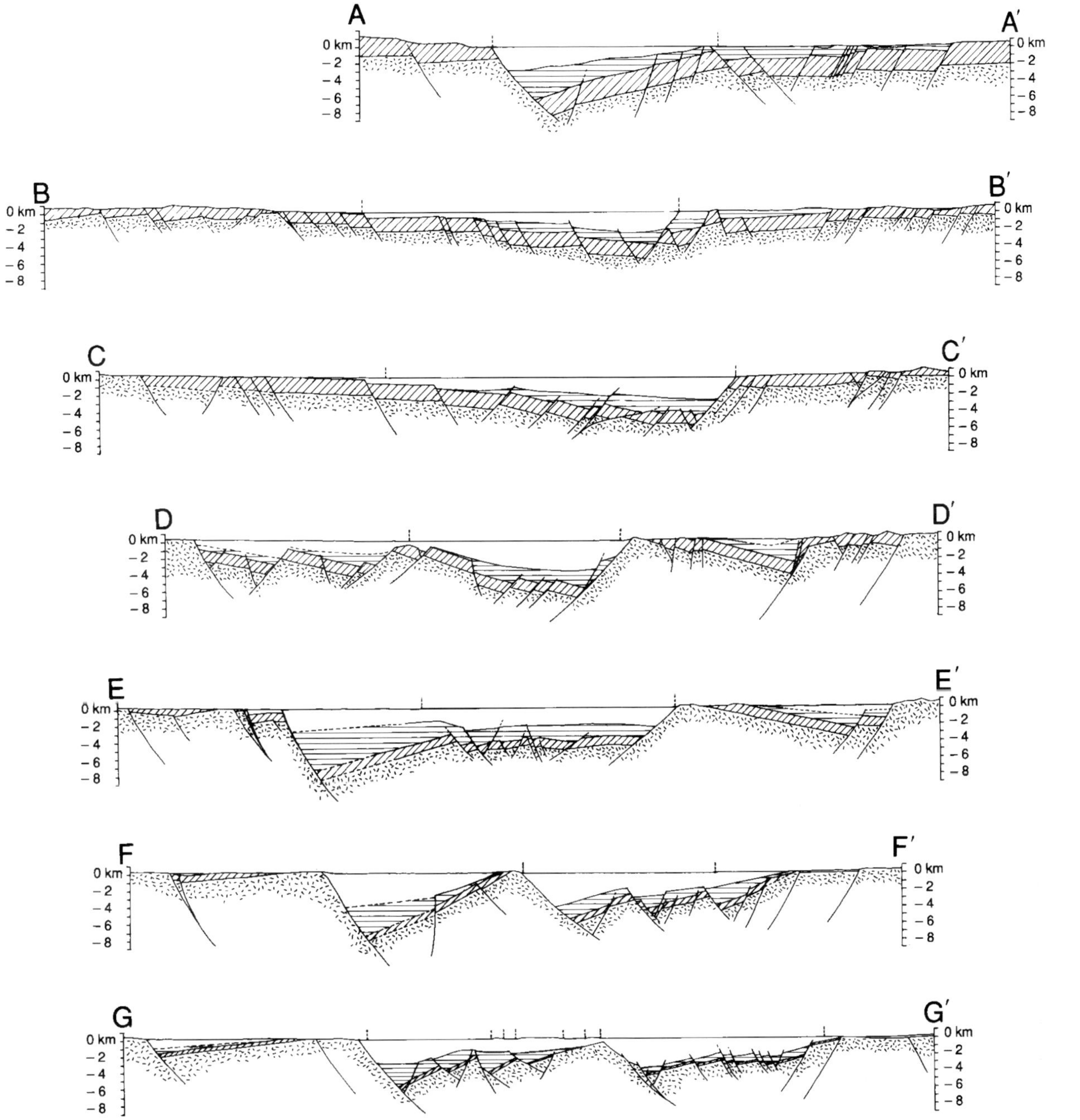

Figure 20. Serial cross sections along the Gulf of Suez. Section locations are shown in Figure 19. Random stipples = basement; sloping parallel lines = pre-Miocene; horizontal parallel lines = Miocene clastics; unpatterned = Miocene evaporites and post-Miocene.

strike-slip motion (Angelier, 1985; Ott d'Estevou et al., 1986a, b; Prat et al., 1986; Montenat et al, 1986, 1988; Richert et al., 1986; Lyberis, 1988; Moustafa and Abdeen, 1992) and an infrequent component of right-lateral, strike-slip motion (Moustafa and Abdeen, 1992).

Again, arguments can be marshaled supporting the antiquity of this trend in the gulf area. Schurmann (1966) recorded the presence of north-south– to north-northeast–oriented dikes in the Precambrian southwest of the Gharib area along the west-central portion of the gulf and southwest of Hurgada in the southern gulf. Robson (1971) mapped a pronounced basement fabric with a maximum between 20 and 25° azimuth in the Precambrian of the Sinai Peninsula. He argued for the antiquity of this trend along the Suez side of the peninsula based on the absence of the trend in the overlying Phanerozoic cover. Unpublished air-photo–based geologic maps of the basement along the northeastern margin of the gulf document dike swarms with orientations ranging between north-south and northeast-southwest. Beyth (1981) proposed that the north-south–oriented, rift-bounding fault in the Gebel Nukhul area was active during

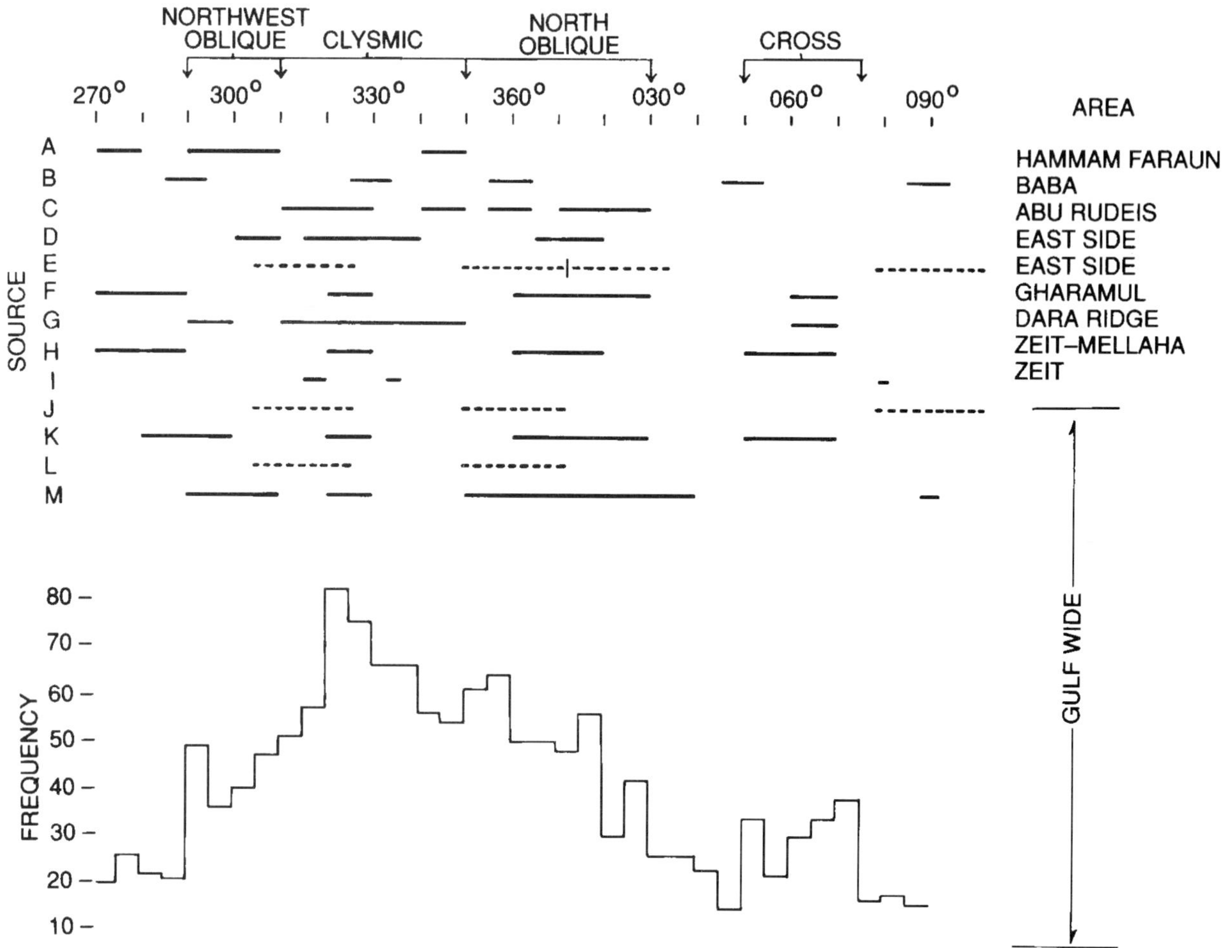

Figure 21. Summary of fault orientation data in the Gulf of Suez showing the four main trends: northwest oblique, clysmic, north oblique, and cross. Solid lines represent data from specified bearing ranges; dashed lines represent data from quadrants only. The histogram at the base of the diagram shows the distribution of linear elements with lengths in excess of 2 mi (3.2 km) extracted from air photos of the Gulf of Suez basin. For the histogram n = 1463. Sources: A— Moustafa and Abdeen, in press; B—El Tarabili and Adawy, 1972; C—Ott d'Estevou et al., 1986b; D—Robson, 1971; E—Garfunkel and Bartov, 1977; F—Ott d'Estevou et al., 1986a; G—Moustafa and Fouda, 1988; H—Prat et al., 1986; I—Angelier, 1985; J—Richert et al., 1986; K—Montenat et al., 1986, 1988; L—Lyberis, 1988; M—Colletta et al., 1988.

Paleozoic uplift and subsidence cycles. Other workers proposed that the N-oblique fabric was exploited immediately preceding Miocene extension during the Syrian Arc, north-northwest–directed compressive events (Garfunkel and Bartov, 1977), with the N-oblique trend being the left-lateral shear member of an array of conjugate strike-slip faults.

A third trend, also obliquely oriented to the clysmic trend, lies between 280 and 310° azimuth, with its maximum lying between 290 and 300°. This trend is weakly expressed relative to the clysmic and the N-oblique trends, and has been referred to as the Duwi trend (Montenat et al., 1986, 1988). In the southern gulf, Prat et al. (1986) extended the range of faults assigned to the Duwi trend to nearly an east-west direction (Figure 21). For the purpose of this discussion, we refer to faults with an orientation lying between 280 and 310° azimuth as belonging to the Northwest-oblique (NW-oblique) trend. Faults of this trend are most commonly expressed in surface data from the southwestern margins of the gulf between the Dara ridge and Gebel Esh El Mellaha (Montenat et al., 1986, 1988; Ott d'Estevou et al., 1986a; Prat et al., 1986; Moustafa and Fouda, 1988). However, El Tarabili and Adawy (1972) documented faults of this orientation in the Wadi Baba area. Kinematic data on the slip of the NW-oblique faults, like the N-oblique faults, record a significant component of strike-slip motion. However, the NW-oblique faults show a preference for right-lateral strike-slip motion (Angelier, 1985; Prat et al., 1986; Montenat et al., 1986, 1988; Richert et al., 1986; Lyberis, 1988). There are no data strongly supporting the presence of NW-oblique trend prior to the opening of the Suez rift. Robson (1971), in his basement fabric data, showed a weakly expressed trend lying approximately between an azimuth of 305 and 310°. However, the age and location of this trend within the Sinai Peninsula are not presented. Unlike the clysmic and N-oblique trends, the NW-oblique trend is only weakly expressed in the basement dike swarms mapped by Schurmann (1966).

Finally, a distinctive, although relatively infrequent fault population occurs between 50 and 75° (Figure 21); we refer to this fault population as the cross trend (Figure 21). Within the structural geology literature of the Gulf of Suez, the terms "cross" (Robson, 1971; Ott d'Estevou et al., 1986a; Montenat et al., 1986; Sultan and Schutz, 1986; Argenton and Maccagni, 1988; Colletta et al., 1988) and "transverse" (Angelier, 1985; Moretti and Colleta, 1988) have been indiscriminately used by workers to describe rift faults that do not parallel the clysmic orientation. We continue the use of the term "cross" here, but restrict its meaning to faults having near-orthogonal orientations to the clysmic trend.

Cross faults have been most frequently documented in surface data from the southwestern portion of the gulf in the Dara ridge, Gebel Zeit, and Gebel Esh El Mellaha areas (Angelier, 1985; Montenat et al., 1986; Ott d'Estevou et al., 1986a; Prat et al., 1986; Moustafa and Fouda, 1988). Montenat et al. (1988) drew attention to the spatial coincidence between the area of activation of this fabric (i.e., Dara ridge) and the location of a possible Upper Cretaceous Syrian Arc structure. Indeed, several other localities at which the cross-fault fabric becomes important also are located proximal to Syrian Arc structures. Sultan and Schutz (1986) seismically documented the presence of cross faults in the offshore area along the southern margin of the Wadi Araba structure. Surface examples of cross faults of significant length (on the order of 25 km [15.5 mi]) run parallel to the southern side of the Wadi Araba structure in the St. Paul Monastery area and form the coastal fault along the northern termination of the Darag Basin. This latter fault parallels the subsurface-defined Ayn Musa feature, a Syrian Arc structure (Renolds, 1979). The fault possesses significant throw in the southwest, but loses throw rapidly to the northeast. Both right-lateral (Angelier, 1985) and left-lateral (Moustafa and Fouda, 1988) strike-slip motions have been proposed for the cross-trending faults of the Dara ridge. However, despite these occurrences, the cross faults are less frequent (Figure 21) and show considerably less throw relative to the oblique- and clysmic-trending fault populations in the gulf (Montenat et al., 1986, 1988; Colletta et al., 1988).

The cross trend parallels the most pronounced of the preexisting basement fabrics. Northeast-trending dike swarms represent the most pervasive of the late Precambrian–early Paleozoic basement fabrics on both sides of the gulf. The cross trend is clearly expressed on magnetic data in the gulf (Meshref et al., 1988; Argenton and Maccagni, 1988). Furthermore, the Syrian Arc features expressed in the northern gulf have orientations parallel to this basement fabric and both the Neo-Tethyan extensional event and the Late Cretaceous–early Eocene Syrian Arc compressional event likely exploited this basement fabric.

The shapes of the rotated rift blocks in the gulf are defined through the interaction of faults from these four major fault populations. The faults have worked in unison to accommodate the extension the upper crust underwent during Miocene rifting. In discussing the interrelationships of these faults, we assume that the principal direction of extension has been nearly perpendicular to the major axis of the rift (Chenet and Letouzey, 1983; Angelier, 1985) or slightly oblique to this direction (Lyberis, 1988; Steckler et al., 1988) throughout the deformation history of the rift. Surface and subsurface data collected in our studies of the rift structures strongly support such an interpretation. However, such a view is not universally held among investigators of the rift. The Gulf of Suez has run a gamut of kinematic models for its origin. The early interpretations of the gulf proposed crustal shortening orthogonal to the rift axis (Ball, 1910, 1911; Hume, 1916; all in Gregory, 1921). More recent interpretations have suggested varying degrees of near axial-parallel wrenching or marked oblique extension for different stages of rift development (cf. Youssef, 1968; Abdel-Gawad, 1969; Heath, 1975, in Riad, 1981; Riad, 1977, 1981; Montenat et al., 1986, 1988; Ott d'Estevou, 1986a, b; Prat et al., 1986; Courtillot et al., 1987; Richardson and Arthur, 1988).

The observations we present on the four major fault trends in the gulf derive primarily from outcrop-scale observations of individual fault segments. These fault segments have played several different roles in the development of the major fault trends depicted in Figure 19. First, although many of the faults at the scale of observation of Figure 19 possess an overall clysmic trend, they display a pronounced zigzag fault pattern (e.g., Abdel-Gawad, 1969; Garfunkel and Bartov, 1977; Richert et al., 1986; Moretti and Colletta, 1988; Moustafa and Abdeen, 1992) consisting of fault segments of the orientations described in the preceding section. This pattern is expressed at a variety of scales ranging from individual outcrops to fault patterns mapped from satellite imagery. The pattern is most clearly expressed along faults exposed in basement outcrops, and, although present, becomes less pronounced in exposures of younger portions of the stratigraphic section. Because, as proposed previously, these fault segments are likely to have developed through the reactivation of preexisting fabrics, it would appear that the major clysmic faults in the basin developed through the early exploitation of a variety of trends that linked and worked in unison to form the major, extension-accommodating faults of the rift. The low mean stress of rifting favors exploitation of preexisting fabrics (Nelson, 1987). This can result in anomalous structural configurations that are restricted especially to the edges of the rift blocks. For instance, significant promontories and embayments in the fault surface may develop within the central portion of the fault, such as is seen west of Zeit Bay at the edge of Esh El Mellaha block.

Second, the major clysmic faults in the basin possess arcuate traces with both a concave and convex sense of curvature toward the hanging wall, as is expressed in the larger faults portrayed in Figure 19. Toward the ends of the clysmic faults, reactivated oblique elements, which contribute to the overall arcuate trace of the major faults, appear to play an active role in terminating major clysmic faults.

Displacement Transfer

Another aspect of the faults portrayed in Figure 19 is that they frequently bifurcate or splay into two segments. Significantly large splays also may initiate along the main trunk of the fault, commonly, although not exclusively, occurring at steps or bends in the fault trace. The splays involve both clysmic and oblique faults, with the oblique faults generally linking between clysmic faults. Examples are found at the northern end of Gebel Abu Durba, at the southern end of Gebel Hadahid, and along the eastern margin of the Wadi Araba Syrian Arc structure (Figure 19). The oblique faults in these situations act as transfer faults (Gibbs, 1984), along which the displacement or extension experienced by one clysmic fault is relayed to another en echelon clysmic fault. Robson (1971) noted in the gulf that cross faults (Robson's terminology) would link between the major faults of the rift. Colletta et al. (1988) more fully developed the concept of oblique fault trends acting as transfer faults, and proposed major rift-scale transfer faults for the north end of Gebel Zeit and at the north end of Abu Durba. These authors stated that the transfer faults have a limited extent and interact with only one or two blocks, a contention supported by our work. More detailed maps of the transfer fault geometries and the resulting rhombic block patterns at these localities can be found for the Gebel Zeit area in Colletta et al. (1986), Prat et al. (1986), and Moretti and Colletta (1988), and for the Abu Durba area in Chenet and Letouzey (1983), Ott d'Estevou et al. (1986b), and Moustafa and Khalil (1987b). Cross faults also may act as transfer faults, as seen at the northern termination of the western boundary fault immediately south of the Wadi Araba structure, and in the termination of the Darag fault along the northern margin of Darag Basin (Figure 19).

Morley et al. (1990) defined extensional transfer zones as coordinated systems of faults that conserve regional extensional strain (or displacement) from one area of the rift to another. The process of displacement transfer is most easily visualized when it occurs between two isolated, en echelon normal faults (Figure 22). Two major types of transfer zones, as defined by Morley et al. (1990), are conjugate and synthetic (Figure 22). The former type of transfer zone exists where the transfer process occurs between faults of opposite dip direction, and the latter type occurs when transfer takes place between faults possessing similar dip direction (e.g., the examples of Gebel Zeit and Gebel Abu Durba). In the case of conjugate transfer systems, two additional subcategories exist: convergent systems, in which faults of opposing dip face toward each other, and divergent systems in which faults of opposing dip face away from one another. In the former case, a graben develops between the two oppositely dipping faults, whereas in the latter case a horst develops between the two oppositely dipping faults (Figure 22b, c). The transfer of displacement between the two faults can be accomplished by one or both of two deformation processes, flexure (Figure 22a–d) and faulting (Figure 22e, f). The flexure case corresponds to ramps (e.g., Griffiths, 1980), twist zones (Colletta et al., 1988) or flexure troughs (Rosendahl, 1987). The faulting case occurs when either a single, intervening transfer fault (Gibbs, 1984), also known as a transform fault in the terminology of Bally (Bally et al., 1981; Bally, 1982), or a network of transfer faults (Morley et al., 1990) links the two rift faults.

The transfer geometries schematically portrayed in Figure 22 are scale independent and may be applied to shear fracture relationships in hand specimen at one end of the scale continuum, or to the linking of major rift segments composed of systems of similarly dipping faults at the other end of the scale continuum. Because of this self-similar relationship, different transfer geometries may be nested within one another at different scales of observation.

The major faults internal to the boundaries of the Gulf of Suez rift can be grouped into three major fault-dip domains (Figure 19). The northern and southern domains are dominated by faults that dip to the northeast, while a central domain is dominated by faults that dip to the southwest. Faults of lesser throw within each of these domains tend to have similar dip directions as the major faults (Figure 20), but lesser-throw faults with opposite dip direction are also common. The asymmetry of the faults, in turn, produces a marked asymmetry to the tilt direction of the fault blocks within these domains, producing the frequently cited dip provinces described by Moustafa (1976) and Abdine (1981). The dominant direction of block dip in the northern and southern basins is to the southwest, but structural dip of blocks in the central province is to the northeast (Figure 20). Note the distinction made here between block-dip domains and fault-dip domains. The latter remains the focus of the following discussion.

The structural transition from one fault-dip domain to another is accomplished through displacement transfer between fault systems. In the Gulf of Suez, changes between the fault-dip domains described occur primarily through the process of conjugate transfer. The transfer zone between the northern and central fault-dip domains develops between the northeast-dipping Darag fault (located along the northeastern edge of the north Galala Plateau, Figure 20) and the southwest-dipping Abu Zenima fault (forming the coastline between Abu Zenima and Gebel Hammam Faraun, Figure 20). Throw on the Darag fault at the top of the pre-Miocene surface probably is in excess of 6 km (20,000 ft) and generates southwest-block dips approaching 20° (Figure 20, cross section A-A′). Projected throw on the Abu Zenima fault just south of cross section C-C′ (Figure 20) is in excess of 4.5 km (15,000 ft) with pronounced northeast block dips. In the conjugate, convergent transfer zone between these two major

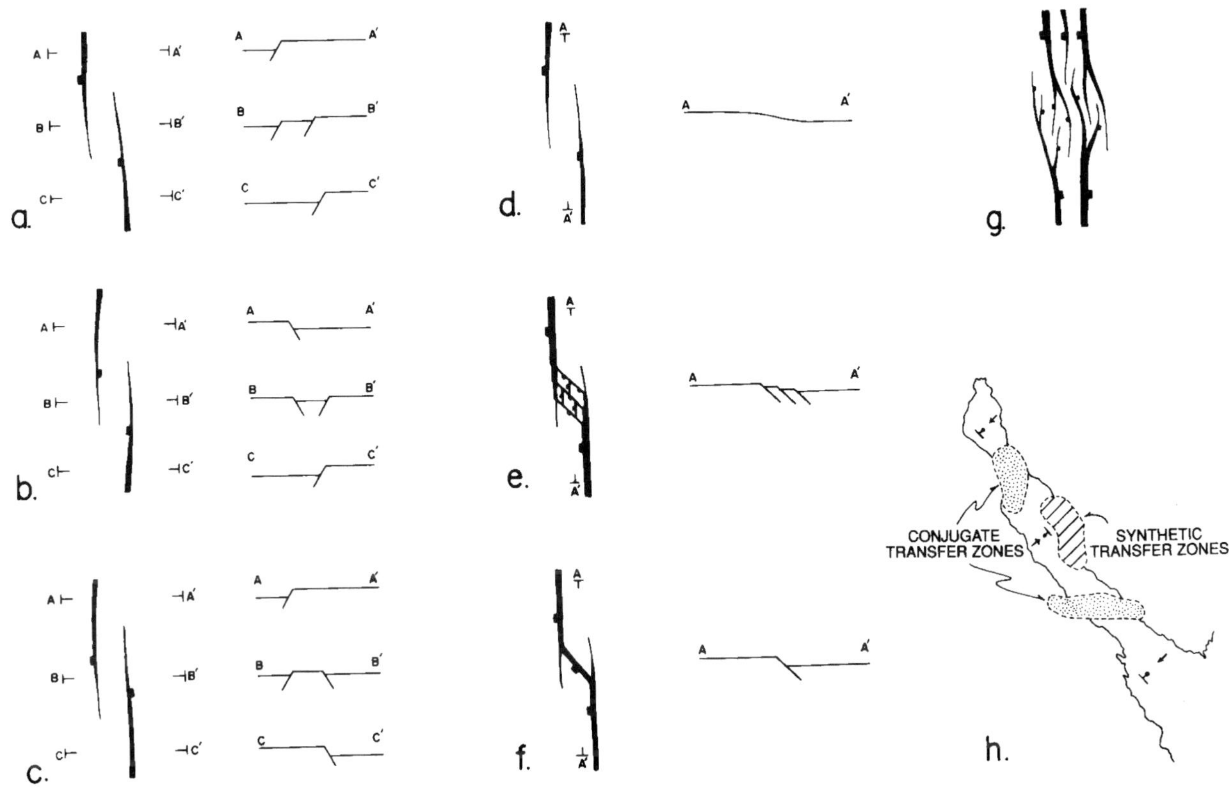

Figure 22. Idealized examples of displacement transfer between en echelon, overlapping normal faults (after Morley et al., 1990). (a) Synthetic transfer; (b) Convergent conjugate transfer; (c) Divergent conjugate transfer; (d–f) Synthetic transfer zone showing (d) unfaulted flexure or ramp (Griffiths, 1980); (e) Diffuse network of clysmic- and oblique-oriented transfer faults; (f) single, obliquely-oriented transfer fault; (g) Systems of similarly dipping faults transferring displacement to a system of oppositely-dipping faults; the simplified configuration in g supplemented with networks of oblique, transfer faults, by our interpretation, is the mechanism by which displacement transfer between the major fault-dip provinces occurs; (h) The locations and orientations of three major transfer zones in the Gulf of Suez; block dip directions are shown by arrows and fault dip directions are indicated by the dots on the downthrown side of the fault symbols.

faults, a full graben develops with the pre-Miocene surface generally at subsea depths of less than 4 km (13,000 ft) and possessing gentle to flat dips. Colletta et al. (1988) described an identical geometry for this transfer zone and termed it a graben twist zone. These authors stated that there is an absence of transverse faulting accompanying this transfer zone; however, geophysical mapping at the Belayim horizon (T. Romero, personal communication, 1988) demonstrates both clysmic and NW-oblique to nearly east-west–oriented trends within this zone. These fabrics are portrayed on the pre-Miocene map (Figure 19) as segments near the termination of the Darag and Abu Zenima faults, and suggest that small throw (<1 km or <0.6 mi) faults of this orientation may link these two Clysmic faults at the pre-Miocene level.

Although the transfer zone between the central and southern fault-dip domains is more complex and dominated by several large faults, a cross section through the transfer zone (Figure 20, cross section E-E′) shows an overall conjugate, convergent transfer geometry. The northeastern side of the section shows the northeast-dipping Gebel Araba complex of blocks, and the southwest portion of the section shows the southwest-dipping blocks of Esh El Mellaha and Gebel Zeit. Between these oppositely dipping blocks is a full graben with its floor composed of gently dipping to flat-lying blocks. The horst and graben configuration in the central portion of the section suggests that both convergent and divergent conjugate transfer is taking place in the graben. A seismic section displaying a productive example of one of these horsts is shown in Figure 23.

Our subsurface mapping of the pre-Miocene surface in this area suggests that the fault fabric in the transfer zone is composed of a diffuse network of faults of both oblique and clysmic orientations forming numerous, rhombic-shaped, fault-bound blocks. This interpretation differs from that proposed for the area by Colletta et al. (1988), who call upon a major oblique transfer fault to accommodate an abrupt change in block dip for the area. Onshore in the area of the Dara ridge, cross faults are active in the transfer process as well (Moustafa and Fouda, 1988; Coffield and Schamel, 1989). By our interpretation, the orientation of the transfer zone between the central and

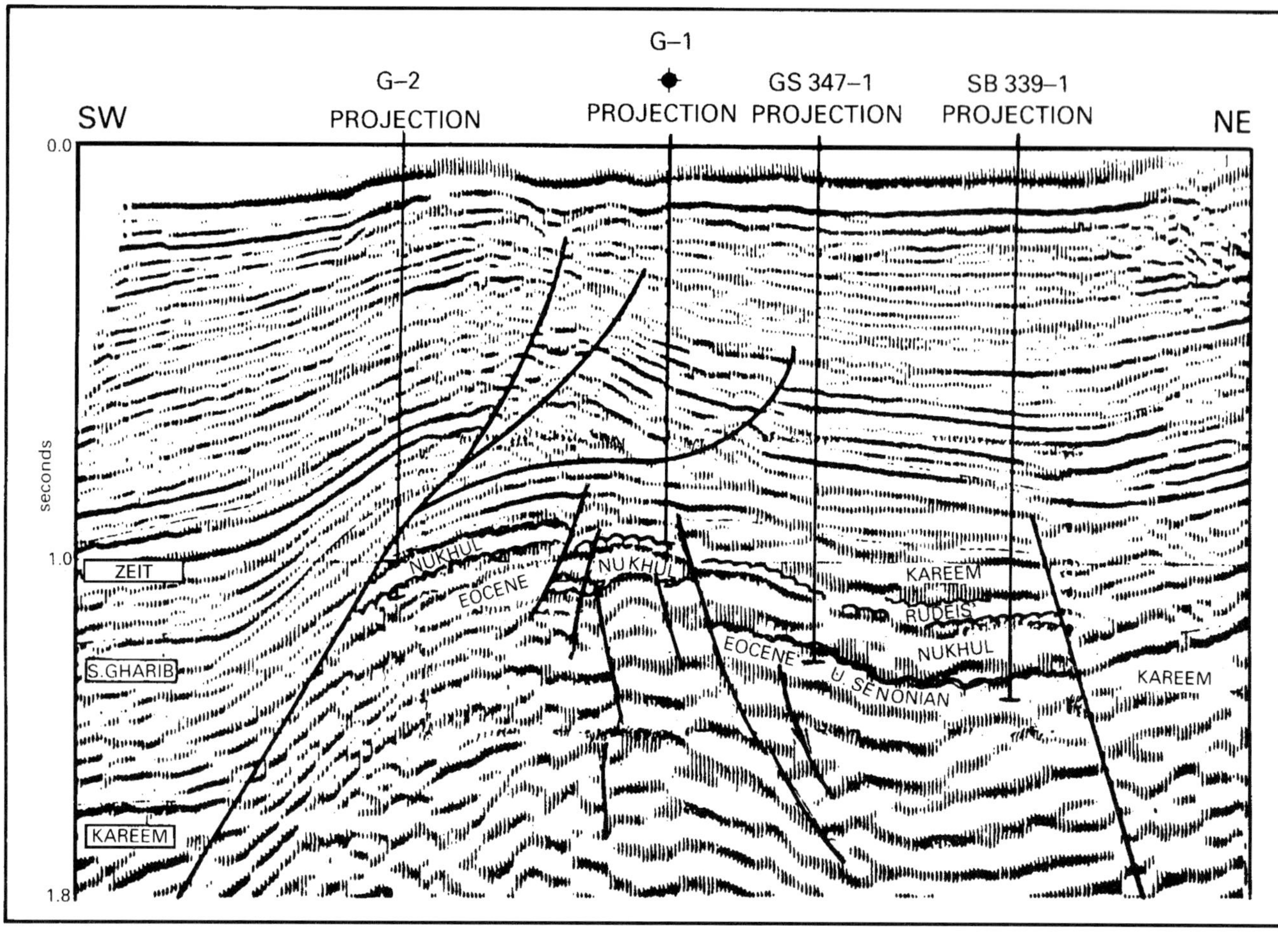

Figure 23. Seismic line across the GS-347 horst immediately south of El Tor (see Figure 1 for location). The interpretation was generated by M. Sattar and P. Pangman of Gupco during 1982. The figure is courtesy of ERICO (London).

southern provinces, as characterized by a zone of interfingering, oppositely dipping faults, has an east-west orientation (Figure 22h) similar to that proposed by Meshref et al. (1988), rather than an orientation orthogonal to the rift as proposed by other investigators (e.g., Moustafa, 1976; Steckler et al., 1988; Coffield and Schamel, 1989).

An example of a major synthetic transfer system also exists in the gulf in the area between Abu Durba and Abu Zenima (Figures 1, 19). Three major, en echelon clysmic faults are involved, including the coastal fault bounding Gebel Araba, the subsurface-defined fault immediately west of Abu Rudeis, and the coastal fault between Abu Zenima and Hammam Faraun (Figure 20). A major N-oblique transfer fault linking these three clysmic faults is exposed onshore and extends from immediately north of Abu Durba to the northern side of the Baba Plain. The northward extension of this fault is likely the rift-bounding fault along the western margin of the El Tih Plateau; however, its throw is much diminished north of the Baba Plain. The southern end of the transfer zone is marked by a distinctive change in structure, specifically from the network of oblique and clysmic faults (see Figure 19) characteristic of the area from Abu Durba to the Baba Plain (e.g., see geologic maps in Chenet and Letouzey, 1983; Chenet et al., 1986; Ott d'Estevou et al., 1986b) to the long, continuous and unfaulted block forming Gebel Araba. The onshore structure suggests an increased role of N-oblique faults in the deformation in the transfer zone, generating a number of rhombic-shaped blocks. In our subsurface mapping of the area between these three clysmic faults, we have interpreted a similar fabric.

We interpret this network of oblique and clysmic faults as transferring the major locus of extension from in front of Gebel Araba in the central gulf (at the southern end of the synthetic transfer zone) around the Wadi Araba structure, to the more easterly located Darag Basin in the north. Note in Figure 19 how the structural lows associated with these three faults step in an en echelon manner to the northeast. A similar shift in isopach thicks of synrift deposits is documented over the same area by Fawzy and Abdel Aal (1986) and Zahran and Meshref (1988). Patton and Nelson (1988) used this shift to argue for a rift offset around the end of the Wadi Araba structure.

These three areas represent the most significant examples of the transfer process taking place on the regional scale in the Gulf of Suez; however, these areas by no means represent the only examples. Numerous other localities in the gulf demonstrate the same phenomenon over a variety of different scales.

Extension Estimates

In addition to the changes in asymmetry, several other significant structural variations occur from northwest to southeast, longitudinally along the rift. The number of faults possessing greater than 1 km (0.6 mi) of throw increases from north to south (Figure 19), as do both the number of structural lows and the total area of the pre-Miocene surface below 4 km (2.5 mi) subsea depth. In addition, the width of the rift increases abruptly south of the Wadi Araba structure from an average of 50 km (31 mi) to 80 km (50 mi), and then more gradually to 100 km (62 mi) in the southernmost rift. Colletta et al. (1988) stated that the average tilt angle for the northern gulf is 5–10° and increases to 20–25° toward the south in the Gebel Zeit area. They also stated that the average size of blocks bounded by faults with greater than 1 km (3300 ft) of throw decreases from 50 km (31 mi) in the north to 15 km (9 mi) in the south, a feature consistent with our observation of a greater density of large faults in the southern gulf. All of these observations suggest that the amount of extension in the gulf increases from northwest to southeast along the rift axis.

Numerous investigators have attempted to quantify the amount of extension at various locations along the rift using several different approaches (Table 1). However, the estimates derived from these approaches have been far from consistent. As pointed out by Chenet and Letouzey (1983), Angelier (1985), and Gaulier et al. (1988), extension estimates from surface data in the gulf are consistently lower than the amount of extension determined from subsidence and deep-seismic data. Quantification of the true strain from surface sections is difficult, even under the best of conditions. The task is particularly difficult in the gulf in that the structural configuration below the Belayim (the last reliable industry seismic reflector) is significantly obscured on seismic profiles due to reflection multiples brought about by the overlying salt section. Geologic cross sections and maps of the pre-Belayim horizons in the gulf are highly interpretive, being constrained principally by the shape of the Belayim surface and industry wells. Extension values derived from these sources thus should be considered as minimum estimates. To estimate the amount of unseen extension in our structural profiles, we have employed fractal population statistics.

Studies of size distributions of brittle fault populations argue that they are fractally distributed (Kakimi, 1980; Villemin and Sunwoo, 1987; Childs et al., 1990; Scholz and Cowie, 1990; Walsh et al., 1991). Marrett and Allmendinger (1991) and Walsh et al. (1991) argued that small faults account for a significant amount of strain in the brittle faulting process. Marrett and Allmendinger (1992) used the fractal nature of faults to estimate the contribution of small faults (those beyond the resolution of the sampling technique) to the total strain in the Viking Graben of the North Sea. These authors explicitly treat the problem of extracting an extension estimate through one-dimensionally sampled fault population documented in cross section data. Walsh et al. (1991) did a similar treatment, also using North Sea data. We apply here the technique used by Marrett and Allmendinger (1992) in an effort to establish a more complete estimate of the extension in the Gulf of Suez, as derived from fault mapping in structural profiles of the shallow crust.

Figure 24 shows the fractal nature of the stratigraphic heaves (the component of displacement measured parallel to bedding) extracted from faults on structural profiles A-A´, C-C´, D-D´, F-F´, and G-G´ (Figure 20). We have not used profiles B-B´ and E-E´ due to their more oblique orientation to the axis of the rift (therefore also oblique to the assumed extension direction). Use of the stratigraphic heave to estimate the measurable extension ignores the contribution of block rotation to the total extension. Because the dominant fault fabric dips in the opposite direction to that of the dominant dip direction of the fault blocks, using the stratigraphic heave as an estimate of the extension will underestimate the total extension. The underestimate will be relatively minor for gently dipping blocks, but will become significant for more steeply dipping blocks, such as those found in the southern portions of the gulf. When differential rotation has occurred between two blocks, the reference horizon for the stratigraphic heave measurement is the shallower of the two dipping blocks.

The fractal dimension (C_1) calculated from all faults on these sections is 0.80 (Figure 24). From outcrop observations, Angelier (1985) recognized the significance of the contribution of small faults with throws of less than 100 m (330 ft). R. Marrett (1991) used Angelier's (1985) estimates of extension due to small faults to calculate a fractal dimension. Marrett obtained a value of C_1 between 0.71 and 0.83, a range in good agreement with the 0.80 estimate from our structural profile data.

To estimate the contribution of unobserved faults to the total extension in each profile, we empirically fit a line with slope of –0.80 to the log-log, fault-heave vs. cumulative-frequency plots for each of the five sections (Figure 25). Fitting of the curve identifies two populations of faults: large faults that lie along the fractal trend and have been effectively sampled, and smaller faults that drift away from the upper portion of the line. The latter population has been interpreted to indicate undersampling (e.g., Walsh et al., 1991; Marrett and Allmendinger, 1992). The last fault intersected by the fractal trend (i.e., the fault with the smallest stratigraphic heave along the trend) is designated as the Nth fault in the population with stratigraphic heave h_N. These values are listed for each profile in Table 2. We estimated the measurable extension by summing the stratigraphic heaves of the first N faults of the profile. The summed heaves are listed in the fifth column in Table 2. The extension due to faults

Table 1. Published Extension Estimates, Gulf of Suez

Area	Extension Ratios		Extension (km)	Source
	Profiles	Subsidence		
North	1.33		16.3	Present Study (A–A′)
	1.10		4–5	Colletta et al., 1988
		1.2	5–10	Richardson and Arthur, 1988
Central	1.21		17.7	Present Study (Profile C–C′)
	1.62		32.6	Present Study (Profile D–D′)
	1.12–1.18			Colletta et al., 1988
	1.13	1.7		Chenet and Letouzey, 1983
		1.65	25–27	Steckler, 1985
		2.1(*)	25	Gaulier et al., 1988
		2.5(**)		Chenet and Letouzey, 1983
		1.25–1.5	16–27	Richardson and Arthur, 1988
South			15–20	Garfunkel and Bartov, 1977
	1.2			Jackson et al., 1988
	1.26		20	Colletta et al., 1988
	1.35			Chenet and Letouzy, 1983
	1.23			Moretti and Colletta, 1988
	1.37		21.1	Present Study (Profile F–F′; see Figure 20)
	1.51		29.3	Present Study (Profile G–G′; see Figure 20)
	1.2–1.4	1.65–1.7	25	Angelier, 1985
		1.70	30–40	Richardson and Arthur, 1988
		1.7	27–31	Steckler et al., 1988
Junction Red Sea			32 ± 4 (†)	Steckler et al., 1988

(*) Deep seismic data.
(**) Unspecified locality.
(†) Extrapolated estimate.

with smaller stratigraphic heaves than fault N is determined by numerically summing the heave contributions of all the postulated faults along fractal trends that lie beyond fault N. This approach assumes that N is not the fractal limit of the fault population. These values are listed in Table 2 as h_e. The sum of the extension measured along the first N faults of the profile and the extension determined from the fractal extrapolation (h_e) are added, yielding an estimate of the total extension along the profile. Comparison of the extension measured from structural profiles and the calculated total extension (Table 2) suggests that the total extension along these profiles may be 40–90% greater than that directly measurable from structural profiles.

The empirical fit of the line (–0.80 slope) to the log-log plots of fault heaves from profiles A-A′, D-D′, and G-G′ is relatively straightforward (Figure 25). However, less obvious is where to position the line for data points in profiles C-C′ and F-F′. Note that a significant number of points lie to the right of the fractal trend line for these profiles. We have chosen conservative locations for the placement of the fractal trend line in these two plots, which has resulted in reduced estimates of the amount of total extension. Profile C-C′ lies just south of the Wadi Araba, Syrian Arc structure, where the rift narrows significantly. The poorer quality heave-frequency data pattern here may reflect our inability to correctly portray the structural complexities of this area. None of the profiles have been rigorously balanced. In comparing extension estimates calculated from the top of the pre-Miocene and the basement horizons, profile F-F′ showed the greatest inconsistency of any of the profiles. The difficulty of fitting the fractal trend to data derived from this profile may be an artifact of the construction geometry used in generating the profile. In our attempt to account for the unobserved extension, we emphasize those profiles in which the data show a more consistent relationship to the fractal trend line (i.e., profiles A-A′, D-D′, and G-G′).

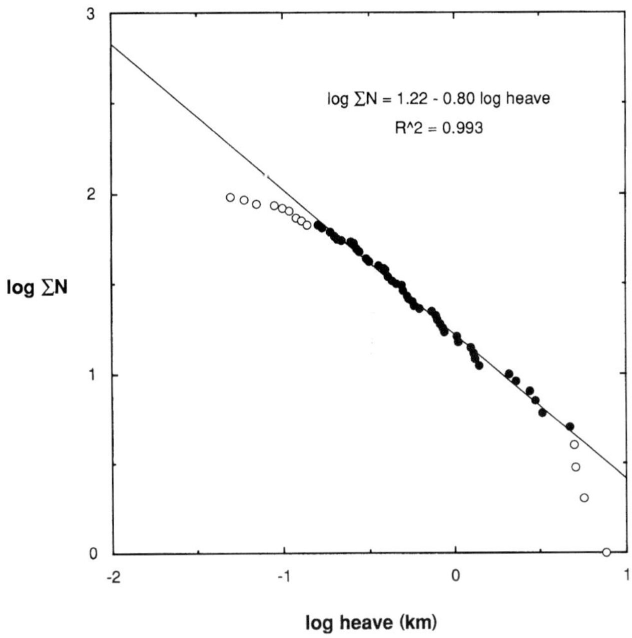

Figure 24. Size distribution of Gulf of Suez stratigraphic heaves at the basement horizon of faults from five structural profiles (A, C, D, F, and G; see Figure 20) in the Gulf of Suez. The fractal dimension is 0.80. R^2 is the correlation coefficient for the fit of the line to the data represented by solid circles. Analysis courtesy of R.A. Marrett.

We collected estimates of total extension and extension ratios from the literature, and compiled them by their approximate locations along the axis of the gulf (Table 1). We subdivided the estimates into three domains based on the locations of the two major conjugate transfer zones. The estimate for the amount of extension (in kilometers), as well as the extension ratio in the northernmost gulf (profile A-A´), is significantly greater than that previously proposed by Colletta et al. (1988) and Richardson and Arthur (1988). Likewise, the amount of extension derived from profile D-D´ is nearly 6 km (3.7 mi) greater than the closest estimate in the central gulf. However, the extension ratio derived from this profile is similar to subsidence extension ratio estimates proposed by Chenet and Letouzey (1983) and Steckler (1985). In the southern gulf, the total extension documented in profile G-G´ compares well with estimates of Angelier (1985) and Steckler et al. (1988). However, the extension ratio of this profile lies between estimates from published structural profiles and estimates derived from subsidence studies.

The data in Table 1 support the often cited increase of extension in the gulf from the northwest to the southeast. Colletta et al. (1988) used block dip data and extension estimates to suggest that the southeastward increase in extension is a continuous process resulting from a "scissor-like" opening of the gulf. Although we concur with the increase in extension to the southeast, we prefer a model of discontinuous extension domains in the gulf.

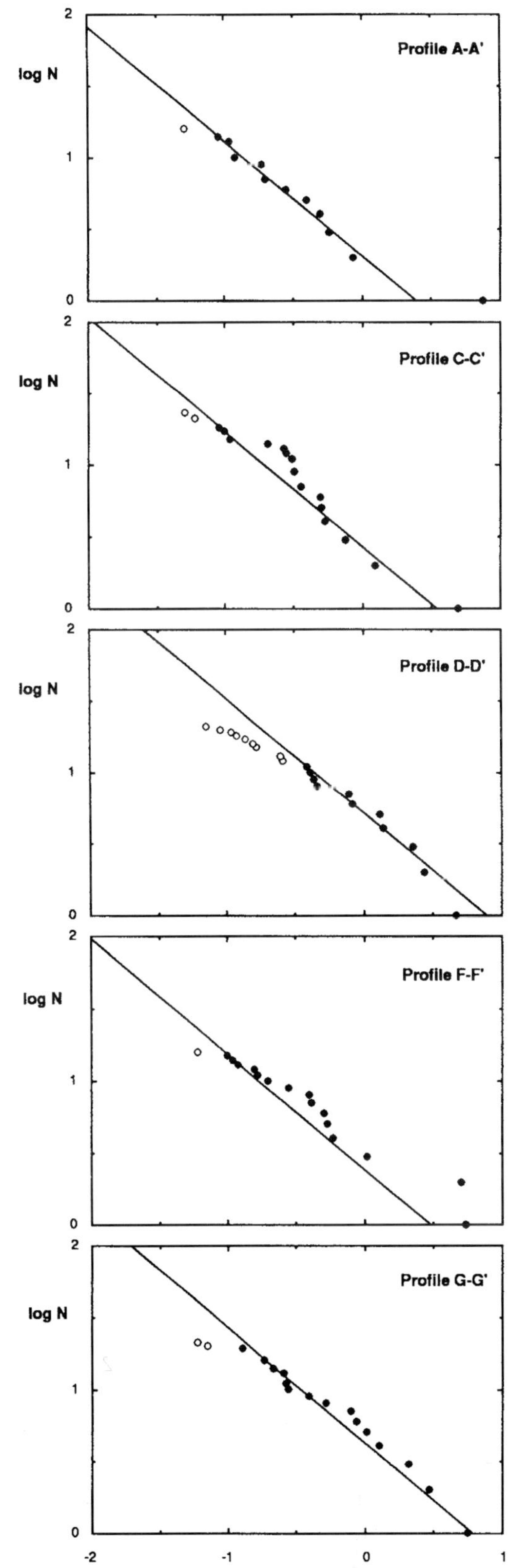

Figure 25. Size distributions of stratigraphic heaves at the basement horizon of faults from individual profiles (see Figure 20). The fractal trend line is taken from Figure 24 and has been empirically fit to the data represented by solid circles. Analysis courtesy of R.A. Marrett.

Table 2. Extension Estimates from Structural Profiles and Fractal Extrapolation

Profile	Measured Extension (km)	N	h_N (km)	$\sum_{i=1}^{N} h_i$ (km)	Extrapolated h_e (km)	Total Extension (km)
A–A′	11.37	14	0.09	11.3	5.0	16.3
C–C′	11.59	18	0.09	11.3	6.4	17.7
D–D′	17.32	11	0.39	15.8	16.8	32.6
F–F′	15.26	15	0.10	15.2	5.9	21.1
G–G′	17.87	16	0.19	17.3	12.0	29.3

Throughout our discussion of the rift, we emphasized the impact that the Mesozoic hinge zone has had on partitioning the gulf into domains demonstrating different geologic behavior. In the absence of more reproducible extension estimates, we suggest that the Mesozoic hinge zone has partitioned the gulf into two major extensional domains. The model emphasizes the role of cross-gulf features in defining the extension domains. The northern domain is bound by the Cairo-Suez trough and the northeast-striking fault bounding the north margin of the Darag Basin. North of this zone, shallow crustal extension has been substantially reduced relative to the 16 km (10 mi) proposed from profile A-A′. The southern termination of this extension domain is marked by the southern edge of the Wadi Araba structure, or equivalently, the southern limit of the Mesozoic hinge zone. A zone of increasing extension occurs between the southern limit of the Wadi Araba structure and the Gebel Araba block, a zone corresponding to the rift offset proposed by Patton and Nelson (1988) and Nelson et al. (1992). The second extension domain runs from south of the north end of the Araba block to the southern end of the gulf, where it is terminated by another cross-gulf feature, the Aqaba wrench zone, which separates the domain from an area of greater extension immediately south in the northern Red Sea. Extension values in the southern domain cluster around 30 km (18.5 mi), nearly twice that experienced in the northern domain of the Darag Basin.

Finally, although our inability to accurately account for the total extension in the gulf is in some ways frustrating, it also provides an element of optimism for exploration. The statement that there may be 40–90% more extension than our current technology can document suggests that economically attractive, undetected fault closures may yet exist in the Gulf of Suez.

SUMMARY

The structural fabrics imposed on the basement rocks of the Gulf of Suez prior to the early Paleozoic transgression appear to have had significant influence upon the structural development of the gulf throughout the Phanerozoic. Either of two roughly orthogonal, recurring structural trends have been expressed in each of the major tectonic episodes of the gulf area; the trends lie within the northwest to north and northeast to east quadrants. Immediately following the Pan-African tectono-thermal event, northeast-southwest–oriented dikes (Schurmann, 1966; Stern et al., 1984) were injected and similarly oriented depositional troughs formed in response to a Cambrian extensional event. An Early Cambrian marine transgression occurred throughout the Gulf of Suez area as a possible response to subsidence associated with the extensional event. The general northward tilt of the margin persisted during a northward-directed fluvial progradation sourced by east-west–trending structural highs.

The northeast-southwest to east-west structural trend of the margin persisted throughout the later part of the Paleozoic through the early Tertiary. The Hercynian orogeny reactivated generally east-west–trending fabrics, generating widely spaced highs and lows in the gulf area (Beleity et al., 1986). Although some evidence of north-south and north-northwest–south-southeast elements were expressed during this phase, they are of secondary importance relative to the orientation of the highs and lows. Features with orientations falling within the northeast to east quadrant continued to be active in the Paleo-Tethyan and Neo-Tethyan margin development for both extensional and compressional phases. This is particularly true in the northern gulf where recurring deformation along the unstable shelf prevailed. Trending orthogonally to these features was the Suez ridge expressed in the Early Cretaceous (Van Houten et al., 1984).

Finally, during the development of the Gulf of Suez rift in the Tertiary, the north-northwest–south-southeast structural trend again became dominant. The principal direction of extension during the opening of the rift was orthogonal to slightly oblique to this trend. Faults orthogonal to the extension direction primarily experienced dip-slip motion; faults obliquely oriented to the extension direction experienced a net oblique-slip motion, which may be manifest in both dip-slip and strike-slip kinematic indicators. Estimates of extension in the gulf show an increase from approximately 10–16 km (6–10 mi) in the north to as much as 30 km (18.5 mi) in the southeast. Corresponding increases in the magnitude of block rotation (Colletta et al., 1988), the number of faults possessing large throws (Colletta et al, 1988), and the geothermal gradient (Shahin, 1988) accompanied this increase in extension, whereas the average size of the structural blocks (Colletta et al., 1988) decreased with increasing extension. Subsidence phases associated with extension correlate well with regional motions and kinematic adjustments of the surrounding plates.

Along the axis of the rift, three major fault-dip (or block-dip; Moustafa, 1976; Abdine, 1981) domains exist. The domains terminate in major, rift-scale transfer zones in which displacement on one system of normal faults is transferred to an along-strike system of oppositely dipping normal faults. Both conjugate and synthetic, normal-fault transfer systems are functional at a variety of scales in the rift.

Initial rifting in the gulf does not appear to have been preceded by local uplift. However, shoulder uplift is documented subsequent to the onset of rifting.

The salient points of the development of the Gulf of Suez rift are summarized in the following paragraphs, starting in the late Eocene and ending with the Holocene.

Late Eocene

By the latest Eocene, growth on the Syrian Arc structures had ceased. Most of these features are restricted to the northernmost portion of the rift (Figure 26a). However, Montenat et al. (1986) and Ott d'Estevou et al. (1986a) proposed that the Dara ridge was expressed during the time of development of the Syrian Arc features (Figure 26a). Undeformed middle Eocene carbonates overlie large portions of these features (Said, 1962; Soliman et al., 1965; El Akkad and Abdallah, 1971; Robson, 1971), suggesting that a period of tectonic quiescence existed between the Syrian Arc deformation and the initiation of rifting. However, locally some of these folds may have remained exposed (Figure 5).

Oligocene

Following the deposition of Eocene units and preceding the Oligocene–Miocene volcanism, a period of right-lateral wrenching took place in the Cairo-Suez trough, intersecting the northern end of the present-day Gulf of Suez structural basin (Figure 26b). Continental sands and conglomerates were deposited in the resulting trough. Continental rifting in the Red Sea area had also initiated in the south by this time (Bayer et al., 1988; Purser and Hotzl, 1988; Voggenreitter et al., 1988). Khalil and Meshref (1988) proposed that the Shoab Ali Member of the Nukhul Formation was deposited in an early basin in the southernmost gulf, and that it preceded the initiation of rifting in the rest of the gulf as recorded by stratigraphically higher members of the Nukhul Formation. We propose, in the absence of paleontologic data, that these continental clastics record early rift deformation that represents the northern termination of the Red Sea rift, and that these sediments in the southernmost gulf (Figure 26b) may be as old as Oligocene. We assume an Oligocene age for the red beds of the Abu Zenima Formation in the central gulf.

We speculate that the Shoab Ali depression, the Abu Zenima depression, and the Cairo-Suez trough were structurally linked in the Oligocene(?), and that the continental clastics of these units represent deposition in the initial rift sag characterized by isolated depocenters. The transition to Oligocene marine units occurs immediately south of the gulf (Purser and Hotzl, 1988). The right-lateral wrenching in the Cairo-Suez trough and the opening of the Red Sea are kinematically compatible.

Oligocene–Miocene

Minor Oligocene–Miocene volcanic activity (Figure 26c) is recorded throughout the gulf, as well as in other surrounding areas. The greatest abundance of volcanism is found in the north-central area of the gulf where basalt flows, sills, and dikes are common, both in outcrop and in the subsurface. Dikes were frequently intruded along rift-parallel fissures.

Aquitanian(?)–Burdigalian

Fluvial and shallow-marine sediments of the Nukhul Formation were deposited during the Burdigalian in a well defined rift depression (Figure 26d). The depression extended the present day length of the gulf, and was 60–70 km (37–43 mi) wide in the south and central portions and 30 km (19 mi) in the north (non-restored widths). The Nukhul Formation is found adjacent to the present-day, rift-bounding faults as well as in the deeper portions of the basin, attesting to its early and widespread distribution. We envision a depression during Nukhul deposition characterized by diffuse faulting throughout the basin, with normal faults of multiple orientations and dip directions active (Vigano and Patton, 1988). Faulting was sufficiently advanced to result in pronounced local angular unconformities between the base of the Nukhul Formation and the pre-Miocene section. The palinspastically restored rift basin shows a slight rift offset (Figure 26d; Patton and Nelson, 1988) in the north as the area of maximum extension stepped around the pre-existing Wadi Araba structure. During this period of the rift-basin development, slow subsidence rates dominated (Moretti and Colletta 1987; Richardson and Arthur, 1988; Steckler et al., 1988).

Burdigalian

Subsidence rates accelerated and remained high throughout the rest of the Burdigalian (Moretti and Colletta, 1987; Richardson and Arthur, 1988; Steckler et al., 1988), during which time the deep-marine marls and sands of the lower Rudeis Formation were deposited. A majority of the extension experienced by the basin occurred during this phase of development, widening the rift depression to near its present-day width (Figure 26e). Logically, this period

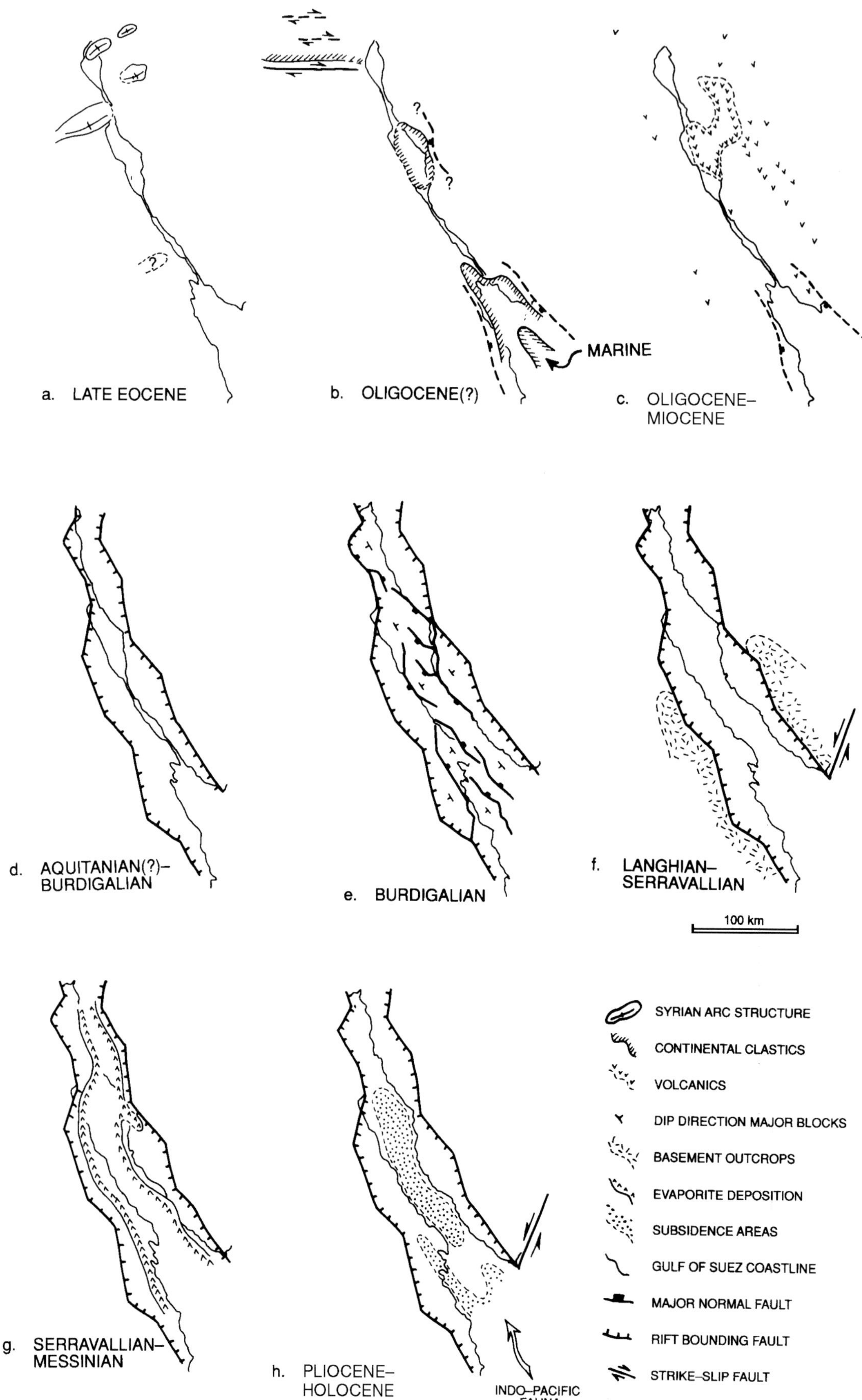

Figure 26. Sequential development of the Gulf of Suez from the late Eocene to the present.

also represents the time of maximum structural development in which the larger faults (Figure 26e) and the associated basin asymmetries and transfer zones were established.

The end of this phase of the rifting is marked by a major structural readjustment in the rift, which is manifest in the stratigraphic record by the mid-Rudeis event. Subsidence rates became variable and showed a general trend toward slowing (Moretti and Colletta, 1987; Richardson and Arthur, 1988; Steckler et al., 1988). These events are roughly synchronous with the increased activity of the Dead Sea wrench system, which is thought to have supplanted the Gulf of Suez as a site of rifting.

Langhian–Serravallian

The Kareem Formation of this time period records the unroofing of the basement (Evans, 1990) along the rift shoulders (Figure 26f). The basin began to fill as subsidence waned and a greater portion of the deformation that the Gulf of Suez had experienced was accommodated along the Dead Sea wrench system. Toward the end of this period, evaporite deposition marked the onset of basin restriction.

Serravallian–Messinian

The thick evaporite sequences of the South Gharib and Zeit formations were deposited (Figure 26g) during the later part of the Miocene. Basin restriction, possibly occurring in the northern gulf, isolated both the Gulf of Suez and the Red Sea. The Dead Sea wrench system experienced a period of relative inactivity.

Pliocene–Holocene

During the Pliocene, the Dead Sea wrench system again experienced a surge of activity (Garfunkel, 1981). Further opening of the Red Sea and communication of the Indian Ocean (Said, 1962; Richardson and Arthur, 1988) with the Gulf of Suez occurred as indicated by the presence of open-marine facies and the abundance of Indo-Pacific fauna (Figure 26h). Renewed subsidence also occurred in the central and southern offshore portions of the gulf during this time (Moretti and Colletta, 1987; Richardson and Arthur, 1988). The present-day subsidence in the southernmost gulf is likely in response to continued motion of the Dead Sea wrench system.

ACKNOWLEDGMENTS

This study would not have been possible without the support of the Gulf of Suez Petroleum Company, the Egyptian General Petroleum Corporation, and Amoco Production Company. We gratefully extend our appreciation to these organizations and, in particular, to Wafik Meshref, Gupco Exploration General Manager. Compilation of this work has taken place over the period of a year, during which time we have had numerous discussions with many geologists and geophysicists in the exploration and exploitation departments of Gupco. These discussions have provided a vehicle through which we have been able (or at least attempted) to synthesize the numerous and often contradictory ideas expressed in the literature addressing nearly 800 m.y. of geologic history. We wish to sincerely thank these individuals for their free and frank exchange of ideas. Those individuals who had a particularly significant impact on this work include Harry Proctor, P.L. Vigano, Amr El Homossany, A.H. Beleity, Safi Wasfi, Badr Khalil, Mosbah Khalil, Hussein G. Fouda, Glen Steen, and Wafik Meshref. Waseem Hosny, Ihab Gaafar, Raouf Rateb, F.X. Miller, and Bob Pierce provided valuable discussions on the biostratigraphy and age of the Miocene synrift and postrift section in the gulf. Robert Stern provided helpful comments and insights into the complexities of the Pan-African terranes. Randall Marrett provided analytical and interpretive assistance in generating fractal-derived estimates of basin extension. Susan Landon, Mike Steckler, and Jim Lowell made many helpful suggestions on a previous version of this manuscript.

REFERENCES CITED

Abdallah, A. M., and A. El Adindani, 1963, Stratigraphy of upper Paleozoic rocks, western side of the Gulf of Suez: Geological Survey of Egypt, paper no. 25, 18 p.

Abdallah, A.M., A. El Adindani, and N. Fahmy, 1963, Stratigraphy of the lower Mesozoic rocks, western side of Gulf of Suez: Geological Survey Egypt, paper no. 27, 28 p.

Abdel-Gawad, M., 1969, New evidence of transcurrent movements in Red Sea area and petroleum implications: AAPG Bulletin, v. 53, p. 1466–1479.

Abdel-Rahman, A.M., and R. Doig, 1987, The Rb-Sr geochronological evolution of the Ras Gharib segment of the northern Nubian shield: Journal of the Geological Society of London, v. 144, p. 577–586.

Abdel-Rahman, A.M., and R.F. Martin, 1987, Late Pan-African magmatism and crustal development in northeastern Egypt: Geological Journal, v. 22, p. 281–301.

Abdine, A.S., 1981, Egypt's petroleum geology: good grounds for optimism: World Oil, December, v. 193, p. 99–112.

Abu Zied, H.T., 1985, Latest Pan-African tectonic event in W. Hamrawin, A possible extension of Najd wrench faulting event of Saudi Arabia into the Eastern Desert of Egypt (abs.): 23rd Annual Meeting, Geological Society of Egypt, p. 15–16.

Abul-Nasr, R.A., 1986, Syntectonic slope deposit at Gebel Hammam Faraun, Sinai (Egypt) (abs.): Proceedings of the 59th Annual Meeting of the South Carolina Academy of Science (April 16–19, 1986; Columbia, SC), Bulletin South Carolina Academy of Science, v. 48, p. 100–101.

Abul-Nasr, R.A., 1987, Biostratigraphy and facies analysis of some Eocene exposures in west-central Sinai, Egypt: Ph.D. dissertation, University of South Carolina, Columbia, South Carolina, 161 p.

Abul-Nasr, R.A., 1990, Re-evaluation of the upper Eocene rock units in west-central Sinai, Egypt: Middle East Research Center, Ain Shams University, Earth Science Series, v. 4, p. 234–247.

Abul-Nasr, R.A., and R.C. Thunell, 1987, Eocene eustatic sea level changes: evidence from western Sinai, Egypt: Palaeogeography, Palaeoclimatology, Palaeoecology, v. 58, p. 1–9.

Angelier, J., 1985, Extension and rifting: the Zeit region, Gulf of Suez: Journal of Structural Geology, v. 7, p. 605–612.

Argenton, A., and A. Maccagni, 1988, Possible axial extension in the Gulf of Suez rift (work hypothesis): Tectonophysics, v. 153, p. 297–306.

Argyriadis, I., P.C. De Garciansky, J. Marcoux, and L.E. Ricou, 1980, The opening of the Mesozoic Tethys between Eurasia and Arabia-Africa: Memoire du Bureau de Recherches Geologiques et Minieres, v. 115, p. 199–214.

Ball, J., 1910, On the origin of the Nile Valley and the Gulf of Suez: Geological Magazine, v. VII, p. 71–76.

Ball, J., 1911, The Gulf of Suez: Geological Magazine, v. VIII, p. 1–10.

Bally, A.W., 1982, Musings over sedimentary basin evolution: Philosophical Transactions of the Royal Society of London, A 305, p. 325–338.

Bally, A.W., D. Bernoulli, G.A. Davis, and L. Montadert, 1981, Listric normal faults: Oceanologica Acta, Proceedings of the 26th International Geologic Congress, Geology of Continental Margins Symposium, p. 87–101.

Barakat, M.G., M. Darwish, and A.N. El Barkooky, 1986, Lithostratigraphy of the post Carboniferous–pre-Cenomanian clastics in west-central Sinai and Gulf of Suez, Egypt: (preprint) Egyptian General Petroleum Corporation, Eighth Exploration Conference, 15 p.

Bartov, Y., Z. Lewy, G. Steinitz, and I. Zak, 1980a, Mesozoic and Tertiary stratigraphy, paleogeography and structural history of the Gebel Areif En Naqa area, eastern Sinai: Israel Journal of Earth-Sciences, v. 29, p. 114–139.

Bartov, Y., M. Eyal, A.E. Shimron, and Y.K. Bentor, 1980b, Sinai—geological photomap: Survey of Israel, scale 1:500,000.

Bayer, H.J., H. Hotzl, A.R. Jado, B. Roscher, and W. Voggenreiter, 1988, Sedimentary and structural evolution of the northwest Arabian Red Sea margin: Tectonophysics, v. 153, p. 137–151.

Beleity, A.M., 1984, The composite standard and definition of paleoevents in the Gulf of Suez: Egyptian General Petroleum Corporation, Sixth Exploration Conference, 1982, p. 181–198.

Beleity, A.M., M. Ghoneim, M. Hinawi, M. Fathi, G. Gebali, and M. Kamel, 1986, Paleozoic stratigraphy, paleogeography and paleotectonics in the Gulf of Suez: (preprint) Egyptian General Petroleum Corporation, Eighth Exploration Conference, 21 p.

Ben Menahem, A., A. Nur, and M. Vered, 1976, Tectonics, seismicity, and structure of the Afro-Eurasian junction—the breaking of an incoherent plate: Physics Earth Planetary Interiors, v. 12, p. 1–50.

Bentor, Y.K., 1985, The crustal evolution of the Arabo-Nubian massif with special reference to the Sinai Peninsula: Precambrian Research, v. 28, p. 1–74.

Bernoulli, D., and H.C. Jenkins, 1974, Alpine, Mediterranean and North Atlantic Mesozoic facies in relation to the early evolution of the Tethys, *in* R.H. Dott and R.H. Shaver, eds., Symposium on modern and ancient geosynclinal sedimentation: Special Publication Society of Economic Paleontologists and Mineralogists, no. 19, p. 129–160.

Beyth, M., 1981, Paleozoic vertical movements in Umm Bogma area, southwestern Sinai: AAPG Bulletin, v. 65, p. 160–165.

Bhattacharyya, D.P., and L.G. Dunn, 1986, Sedimentologic evidence for repeated pre-Cenozoic vertical movements along the northeast margin of the Nubian Craton: Journal of African Earth Sciences, v. 5, p. 147–153.

Bielski, M., 1982, Stages in the evolution of the Arabian-Nubian massif in Sinai: Ph.D. dissertation, Hebrew University, Jerusalem, Israel, 155 p.

Biju-Duval, B., J. Letouzey, and L. Montadert, 1979, Variety of margins and deep basins in the Mediterranean: AAPG Bulletin, v. 29, p. 293–317.

Bonatti, E., 1985, Punctiform initiation of seafloor spreading in the Red Sea during transition from a continental to an oceanic rift: Nature, v. 316, p. 33–37.

Boukhary, M., and W. Abdelmalik, 1983, Revision of the stratigraphy of the Eocene deposits in Egypt: Neues Jahrbuch für Mineralogie Monatshefte, v. 6, p. 321–337.

Boulos, F.K., P. Morgan, and T.R. Toppozada, 1987, Microearthquake studies in Egypt carried out by the Geological Survey of Egypt: Journal of Geodynamics, v. 7, p. 227–249.

Braun, M., A. Flexer, A. Honigstein, A. Rosenfeld, and U. Baida, 1987, Sedimentary setting of the Turonian, mainly Nezer Formation in northern Negev (Israel) structures—a precursor of the Syrian Arc folding phase: Newsletters on Stratigraphy, v. 17, p. 57–70.

Brenckle, P.L., and T.R. Marchant, 1984, Calcareous microfossils from the Um Bogma Formation (Lower Carboniferous) of the lower Nubia Group at Gebel Nukhul, Sinai, Egypt: Egyptian General Petroleum Corporation, Sixth Exploration Seminar, 1982, p. 302–309.

Brown, G.F., and R.G. Coleman, 1972, Tectonic framework of the Arabian Peninsula: 24th International Geological Congress, Montreal, v. 3, p. 300–305.

Brown, G.F., and R.O. Jackson, 1960, The Arabian shield: 21st International Geological Congress Proceedings, Copenhagen, v. 9, p. 69–77.

Brown, R.N., 1980, History of exploration and discovery of Morgan, Ramadan and July oilfields, Gulf of Suez, Egypt: Canadian Society of Petroleum Geology, Memoir 6, p. 733–764.

Buchbinder, B., C. Benjamin, Y. Mimran, and G. Gvirtzman, 1988, Mass transport in Eocene pelagic chalk on the northwestern edge of the Arabian platform, Shefela area, Israel: Sedimentology, v. 35, p. 257–274.

Buck, W.R., 1986, Small-scale convection induced by passive rifting: the cause for uplift of rift shoulders: Earth and Planetary Science Letters, v. 77, p. 362–372.

Burke, K., and C. Sengor, 1986, Tectonic escape in the evolution of the continental crust, *in* M. Barazangi and L. Brown, eds., Reflection seismology: the continental crust: American Geophysical Union, Geodynamic Series 14, p. 41–53.

Burollet, P.F., J. Bolze, and P. Ott d'Estevou, 1982, Sedimentology and tectonics of the Gharamul area (preprint): Egyptian General Petroleum Corporation, Sixth Exploration Seminar, 1982, Cairo, p. 1–17.

Chenet, P.Y., and J. Letouzey, 1983, Tectonique de la zone comprise entre Abu Durba et Gebel Mezzazat (Sinai Egypte) dans le contexte de evolution du rift de Suez: Bulletin Centres Recherche Exploration-Production, Elf-Aquitaine, v. 7, p. 201–215.

Chenet, P.Y., J. Letouzey, and E. S. Zaghloul, 1986, Some observations in the rift tectonics in the eastern part of the Suez rift: Egyptian General Petroleum Corporation, Seventh Exploration Seminar, 1984, p. 18–36.

Childs, C., J.J. Walsh, and J. Watterson, 1990, A method for estimation of the density of fault displacements below the limits of seismic resolution in reservoir formations, *in* A.T. Buller et al., eds., North Sea oil and gas reservoirs II: Norwegian Institute of Technology, p. 309–318.

Chowdhary, L.R., and S. Taha, 1987, Geology and habitat of oil in Ras Budran field, Gulf of Suez, Egypt: AAPG Bulletin, v. 71, p. 1274–1293.

Coffield, D.Q., and S. Schamel, 1989, Surface expression of an accommodation zone within the Gulf of Suez rift, Egypt: Geology, v. 17, p. 76–79.

Colletta, B., I. Moretti, P.Y. Chenet, C. Muller, and P. Gerard, 1986, The structure of the Gebel Zeit area: a field example of tilted block crest in the Suez rift: Institut Français du Pétrole, Ref. I.F.P. 34 547, 14 p.

Colletta, B., P. Le Quellec, J. Letouzey, and I. Moretti, 1988, Longitudinal variation of the Suez rift structure (Egypt): Tectonophysics, v. 153, p. 221–233.

Courtillot, V., R. Armijo, and P. Tapponnier, 1987, The Sinai triple junction revisited: Tectonophysics, v. 141, p. 181–190.

Daggett, P.H., P. Morgan, F.K. Boulos, S.F. Hennin, A.A. El-Sherif, A.A. El-Sayed, N.Z. Basta, and Y.S. Melek, 1986, Seismicity and active tectonics of the Egyptian Red Sea margin and the northern Red Sea: Tectonophysics, v. 125, p. 313–324.

Davies, F.B., 1984, Strain analysis of wrench faults and collision tectonics of the Arabian-Nubian shield: Journal of Geology, v. 82, p. 37-53.

Dercourt, J., et al., 1986, Geologic evolution of the Tethys belt from the Atlantic to the Pamirs since the Lias: Tectonophysics, v. 123, p. 241–315.

Dewey, J.F., W.C. Pitman III, W.B.F. Ryan, and J. Bonnin, 1973, Plate tectonics and the evolution of the Alpine system: Geological Society of America Bulletin, v. 84, p. 3137–3180.

Dixon, T.H., and M.P. Golombek, 1988, Late Precambrian crustal accretion rates in northeast Africa and Arabia: Geology, v. 16, p. 991–994.

Donovan, S.K., 1987, The fit of the continents in the late Precambrian: Nature, v. 327, p. 139–141.

Druckman, Y., 1984, Evidence for Early-Middle Triassic faulting and possible rifting from the Helez deep borehole in the coastal plain of Israel, *in* J.E. Dixon and A.H.F. Robertson, eds., The geologic evolution of the eastern Mediterranean: Geological Society of London, Special Publication no. 17, p. 203–212.

Eames, L., 1986, Palynologic definition of Paleozoic unconformity-bounded sequences, Gulf of Suez region: Egypt, Egyptian General Petroleum Corporation, Seventh Exploration Seminar, 1984, p. 117–125.

Egyptian General Petroleum Corporation, 1964, Oligocene and Miocene rock-stratigraphy of the Gulf of Suez region, report of the Stratigraphic Committee: Egyptian General Petroleum Corporation, 142 p.

El Akkad, S., and A.M. Abdallah, 1971, Contribution to geology of Gebel Ataqa area: Annals of the Geological Survey of Egypt, v. 1, p. 21–42.

El Heiny, I., and S. Morsi, 1986, Review of the upper Eocene deposits in the Gulf of Suez, Egypt: (preprint) Egyptian General Petroleum Corporation, Eighth Exploration Conference, 18 p.

El Shazly, E.M., 1977, Geology of the Egyptian region, *in* A.E.M. Narin, W.H. Kanes, and F.G. Stehli, eds., The ocean basins and margins, the eastern Mediterranean: New York, Plenum Press, 4A, p. 379–344.

El Tarabili, E., and N. Adawy, 1972, Geologic history of the Nukhul-Baba area, Gulf of Suez, Egypt: AAPG Bulletin, v. 56, p. 882–902.

Engel, A.E.J., T.H. Dixon, and R.J. Stern, 1980, Late Precambrian evolution of Afro-Arabian crust from ocean arc to craton: Geological Society of America Bulletin, v. 91, p. 699–706.

Evans, A.L., 1988, Neogene tectonic and stratigraphic events in the Gulf of Suez rift area, Egypt: Tectonophysics, v. 153, p. 235–247.

Evans, A.L., 1990, Miocene sandstone provenance relations in the Gulf of Suez: insights into synrift unroofing and uplift history: AAPG Bulletin, v. 74, p. 1386–1400.

Evans, A.L., and I.W. Moxon, 1986, Gebel Zeit chronostratigraphy: Neogene syn-rift sedimentation atop a long-lived paleohigh: (preprint) Egyptian General Petroleum Corporation, Eighth Exploration Seminar, 1986, 23 p.

Eyal, Y., and Z. Reches, 1983, Tectonic analysis of the Dead Sea rift region since the Late Cretaceous based on mesostructures: Tectonics, v. 2, p. 167–185.

Fawzy, H., and A. Abdel Aal, 1986, Regional study of Miocene evaporites and Pliocene–Recent sediments in the Gulf of Suez: Egyptian General Petroleum Corporation, Seventh Exploration Conference, 1984, p. 49–74.

Freund, R., M. Goldberg, T. Weissbrod, Y. Druckman, and B. Derin, 1975, The Triassic–Jurassic structure of Israel and its relation to the origin of the Eastern Mediterranean: Geological Survey of Israel Bulletin, no. 65, 26 p.

Garfunkel, Z., 1981, Internal structure of the Dead Sea leaky transform (rift) in relation to plate kinematics: Tectonophysics, v. 80, p. 81–108.

Garfunkel, Z., and Y. Bartov, 1977, The tectonics of the Suez rift: Geological Survey of Israel Bulletin, no. 71, 45 p.

Garfunkel, Z., and B. Derin, 1984, Permian–early Mesozoic tectonism and continental margin formation in Israel and its implications for the history of the Eastern Mediterranean, *in* J.E. Dixon and A.H.F. Robertson, eds., The geologic evolution of the eastern Mediterranean: Geological Society of London, Special Publication no. 17, p. 187–201.

Gass, I.G., 1981, Pan-African (upper Proterozoic) plate tectonics of the Arabian-Nubian shield, *in* A. Kroner, ed., Precambrian plate tectonics: Elsevier Scientific Publishing Company, Amsterdam, p. 387–405.

Gaulier, J. M., X. Le Pichon, N. Lyberis, F. Avedik, L. Geli, I. Moretti, A. Deschamps, and S. Hafez, 1988, Seismic study of the crust of the northern Red Sea and Gulf of Suez: Tectonophysics, v. 153, p. 55–88.

Gibbs, A.D., 1984, Structural evolution of extensional basin margins: Journal of the Geological Society of London, v. 141, p. 609–620.

Glennie, K.W., M.G.A. Boeuf, M.W. Hughes Clarke, M. Moody-Stuart, W.F.H. Pilaar, and B.M. Reinhardt, 1973, Late Cretaceous nappes in Oman Mountains and their geologic evolution: AAPG Bulletin, v. 57, p. 5–27.

Gorur, N., F.Y. Oktay, I. Seymen, and A.M.C. Sengor, 1984, Palaeotectonic evolution of the Tuzgolu basin complex, central Turkey: sedimentary record of a Neo-Tethyan closure, in J.E. Dixon and A.H.F. Robertson, eds., The geologic evolution of the eastern Mediterranean: Geological Society of London, Special Publication no. 17, p. 467–482.

Greenberg, J.K., 1981, Characteristics and origin of Egyptian Younger Granites: summary: Geological Society of America Bulletin, v. 20, p. 63–77.

Gregory, J.W., 1921, The rift valleys and geology of east Africa: Seeley, Service and Company Limited, London, 479 p.

Griffiths, P.S., 1980, Box-fault systems and ramps: a typical association of structures from the eastern shoulder of the Kenya rift: Geological Magazine, v. 117, p. 579–586.

Gvirtzman, G., and T. Weissbrod, 1984, The Hercynian geanticline of Helez and the late Paleozoic history of the Levant, *in* J.E. Dixon and A.H.F. Robertson, eds., The geologic evolution of the eastern Mediterranean: Geological Society of London, Special Publication no. 17, p. 177–186.

Hagras, M., and S. Slocki, 1984, Sand distribution of the Miocene clastics in the Gulf of Suez: Egyptian General Petroleum Corporation, Sixth Exploration Seminar, 1982, p. 201–212.

Hamza, F.H., 1988, Miocene litho- and biostratigraphy in west-central Sinai, Egypt: Middle East Research Center, Ain Shams University, Earth Science Series, v. 2, p. 91–103.

Hanter, G., 1965, Remarks on the distribution of Miocene sediments in the Gulf of Suez region: Fifth Arab Petroleum Congress, Cairo, 13 p.

Haq, B.U., J. Hardenbol, and P.R. Vail, 1987, Chronology of fluctuating sea levels since the Triassic: Science, v. 235, p. 1156–1167.

Hassan, A.A., 1967, A new Carboniferous occurrence in Abu Durba, Sinai, Egypt: Sixth Arab Petroleum Congress, v. 2, p. 1–8.

Hempton, M.R., 1985, Structure and deformation history of the Bitlis suture near Lake Hazar, southeastern Turkey: Geological Society of America Bulletin, v. 96, p. 233–243.

Hempton, M.R., 1987, Constraints on Arabian plate motion and extensional history of the Red Sea: Tectonics, v. 6, p. 687–705.

Hermina, M., E. Klitzsch, and F.K. List, 1989, Stratigraphic lexicon and explanatory notes to the geologic map of Egypt, scale 1:500,000: Conoco Incorporated, Cairo, Egypt, 264 p.

Henry, C., J. Chorowicz, F. Bernard, G. Gess, and C. Lallemand, 1986, Region du Sinai, carte geologique et geomorphologique teleanalytique au 1/500,000, d'apres les images Landsat: Laboratoire de Teledetection: Institut Français du Pétrole.

Heybroek, F., 1965, The Red Sea Miocene evaporite basin, *in* Salt basins around Africa: Institute of Petroleum, London, p. 17–40.

Hume, W.F., 1916, Report on the oilfields region of Egypt: Ministry of Finance Egypt, Cairo, 103 p.

Hume, W.F., J.C. Magdwick, F.W. Moon, and H. Sadek, 1920, Preliminary geological report on Gebel Tanka area: Petroleum Research Bulletin, Cairo, v. 4, 15 p.

Husseini, M.I., 1988, The Arabian Infracambrian extensional system: Tectonophysics, v. 148, p. 93–103.

Ismail, M., and A.M. Abdallah, 1966, Contribution to the stratigraphy of St. Paul's Monastery area by microfacies: Bulletin of the Faculty of Science, Alexandria University, Egypt, v. 2, p. 325–334.

Issawi, B., and U. Jux, 1982, Contributions to the stratigraphy of the Paleozoic rocks in Egypt: Geological Survey of Egypt, paper no. 64, 28 p.

Jackson, J.A., N.J. White, Z. Garfunkel, and H. Anderson, 1988, Relations between normal-fault geometry, tilting and vertical motions in extensional terrains: an example from the southern Gulf of Suez: Journal of Structural Geology, v. 10, p. 155–170.

Joffe, S., and Garfunkel, Z., 1987, Plate kinematics of the circum Red Sea—a re-evaluation: Tectonophysics, v. 141, p. 5–22.

Kakimi, T., 1980, Magnitude-frequency relation for displacement of minor faults and its significance in crustal deformation: Bulletin of the Geological Survey of Japan, v. 31, p. 467–487.

Keeley, M.L., 1989, The Paleozoic history of the Western Desert of Egypt: Basin Research, v. 2, p. 35–48.

Khalil, B., and W. Meshref, 1988, Hydrocarbon occurrences and structural style of the southern Suez rift basin: (preprint) Egyptian General Petroleum Corporation, Ninth Exploration and Production Conference, 1988, 13 p.

Klitzsch, E., 1986, Plate tectonics and craton geology in Northeast Africa (Egypt, Sudan): Geologisch Rundschau, v. 75, p. 755–768.

Klitzsch, E., and C.H. Squyres, 1990, Paleozoic and Mesozoic geologic history of Northeastern Africa based upon new interpretation of Nubia strata: AAPG Bulletin, v. 74, p. 1203–1211.

Klitzsch, E., and P. Wycisk, 1987, Geology of the sedimentary basins of northern Sudan and bordering areas: Berliner Geowissenschaftliche Abhandlungen, (A) 75.

Kroner, A., et al., 1987, Pan-African crustal evolution in the Nubian segment of northeast Africa, *in* A. Kroner, ed., Proterozoic lithospheric evolution: American Geophysical Union Geodynamic Series, v. 17, p. 235–257.

Le Pichon, X., and J.M. Gaulier, 1988, Rotation of the Arabian and the Levant fault system: Tectonophysics, v. 153, p. 271–294.

Lovelock, P.E.R., 1984, A review of the tectonics of the northern Middle East region: Geological Magazine, v. 121, p. 577–587.

Lyberis, N., 1988, Tectonic evolution of the Gulf of Suez and the Gulf of Aqaba: Tectonophysics, v. 153, p. 209–220.

MacDonald, K.C., J.C. Sempere, and P.J. Fox, 1984, East Pacific Rise from Sigueiros to Orozco fracture zone: along-strike continuity of axial neovolcanic zone and structure and evolution of overlapping spreading centers: Journal of Geophysical Research, v. 89, p. 6049–6069.

Marrett, R.A., and R.W. Allmendinger, 1991, Estimates of strain due to brittle faulting: sampling of fault populations: Journal of Structural Geology, v. 13, p. 735–738.

Marrett, R.A., and R.W. Allmendinger, 1992, Amount of extension on "small" faults: An example from the Viking graben: Geology, v. 20, p. 47–50.

Martin, A.K., 1984, Propagating rifts: crustal extension during continental rifting: Tectonics, v. 3, p. 611–617.

Meneisy, M.Y., and A.Y. Abdel Aal, 1984, Geochronology of Phanerozoic volcanic rocks in Egypt: Bulletin of the Faculty of Science, Ain Shams University, v. 24B, p. 163–176.

Meneisy, M.Y., and H. Kreuzer, 1974, Potassium-argon ages of Egyptian basaltic rocks: Geologisches Jahrbuch, v. 9, p. 33–39.

Meshref, W.M., 1984, Regional structural setting of northern Egypt: Egyptian General Petroleum Corporation, Sixth Exploration Seminar, 1982, p. 17–34.

Meshref, W.M., M.S. Abu El Karamat, and M.K. El Gindi, 1988, Exploration concepts for oil in the Gulf of Suez: (preprint) Egyptian General Petroleum Corporation, Ninth Exploration and Production Conference, 1988, 28 p.

Miller, F.X., 1977, The graphic correlation method in biostratigraphy, *in* E.G. Kauffman and J.E. Hazel, eds., Concepts and methods of biostratigraphy: Stroudsburg, Pennsylvania, Dowden, Hutchinson and Ross, p. 165–186.

Montenat, C., P. Ott d'Estevou, and B. Purser, 1986, Tectonic and sedimentary evolution of the Gulf of Suez and the northwestern Red Sea: a review, *in* C. Montenat, ed., Geological studies of the Gulf of Suez, the northwestern Red Sea coasts, tectonic and sedimentary evolution of a Neogene rift: Documents et Travaux Institut Geologique Albert de Lapparent, no. 10, p. 7–18.

Montenat, C., et al., 1988, Tectonic and sedimentary evolution of the Gulf of Suez and the northwestern Red Sea: Tectonophysics, v. 153, p. 161–177.

Moretti, I., and B. Colletta, 1987, Spatial and temporal evolution of the Suez rift subsidence: Journal of Geodynamics, v. 7, p. 151–168.

Moretti, I., and B. Colletta, 1988, Fault-block tilting: the Gebel Zeit example, Gulf of Suez: Journal of Structural Geology, v. 10, p. 9–19.

Moretti, I., B. Colletta, and R. Vially, 1988, Theoretical model of block rotation along circular faults: Tectonophysics, v. 153, p. 313–320.

Morley, C.K., R.A. Nelson, T.L. Patton, and S.G. Munn, 1990, Transfer zones in the East African rift system and their relevance to hydrocarbon exploration in rifts: AAPG Bulletin, v. 74, p. 1234–1253.

Moussa, H.E., 1987, Geologic studies and genetic correlation of basaltic rocks in west central Sinai: Ph.D. dissertation, Ain Shams University, Egypt, 308 p.

Moustafa, A.M., 1976, Block faulting in the Gulf of Suez: (preprint) Egyptian General Petroleum Corporation, Fifth Exploration Seminar, 1976, 36 p.

Moustafa, A.R., 1987, Drape folding in the Baba-Sidri area, eastern side of the Suez rift: Egyptian Journal of Geology, v. 31, p. 15–27.

Moustafa, A.R., 1988, Wrench tectonics in the north Western Desert of Egypt (Abu Roash area, southwest of Cairo): Middle East Research Center, Ain Shams University, Earth Science Series, v. 2, p. 1–16.

Moustafa, A.R., and M. Abdeen, 1992, Structural setting of the Hammam Faraun block, Eastern side of the Suez rift: Journal of the Faculty of Science, University of Kuwait, v. 19, p. 291-310.

Moustafa, A.R., and A.M. Abd-Allah, in manuscript a, Transfer zones with en echelon faulting at the northern end of the Suez rift: Tectonics, v. 11, p. 499–506.

Moustafa, A.R., and A.M. Abd-Allah, in manuscript b, Structural setting of the central part of the Cairo-Suez District: Middle East Research Center, Earth Science Series, Ain Shams University, v. 5, p. 133–145.

Moustafa, A.R., and H. G. Fouda, 1988, Gebel Sufr El Dara accommodation zone, southwestern part of the Suez rift: Middle East Research Center Earth Science Series, Ain Shams University, v. 2, p. 227–239.

Moustafa, A.R., and M.H. Khalil, 1987a, North Sinai structures and tectonic evolution: Middle East Research Center, Ain Shams University, Earth Science Series, v. 3, p. 215–231.

Moustafa, A.R., and M.H. Khalil, 1987b, The Durba-Araba fault, southwest Sinai: Egyptian Journal of Geology, v. 31, p. 1–13.

Moustafa, A.R., and D.A. El Shaarawy, 1987, Tectonic setting of the northern Gulf of Suez: Proceedings of the 5th Annual Meeting, Egyptian Geophysical Society, p. 339–368.

Moustafa, A.R., M.A. Yehia, and S. Abdel-Tawab, 1985, Structural setting of the area east of Cairo, Maadi, and Helwan: Middle East Research Center, Ain Shams University, Science Research Series, v. 5, p. 40–64.

Murris, R.J., 1980, Middle East: stratigraphic evolution and oil habitat: AAPG Bulletin, v. 64, p. 597–618.

Nelson, R.A., 1987, Notes on the northern termination of the Gulf of Suez rift into a pre-existing east-west fault system: Al-Petrol, November–December, Cairo, p. 57–62.

Nelson, R.A., T.L. Patton, and C.K. Morley, 1992, Rift-segment interaction and its relation to hydrocarbon exploration in continental rift systems: AAPG Bulletin, v. 76, p. 1153–1169.

Omar, G.I., 1989, Fission-track analysis of basement apatites at the western margin of the Gulf of Suez rift, Egypt: evidence for synchroneity of uplift and subsidence: Earth and Planetary Science Letters, v. 94, p. 316–328.

Omara, S., 1972, An Early Cambrian outcrop in southwestern Sinai, Egypt: Neues Jahrbuch für Geologie und Paläontologie Monatshefte, v. 1972, p. 306–314.

Ott d'Estevou, P., J. Bolze, and C. Montenat, 1986a, Etude geologique de la marge occidentale du Golfe de Suez: le Massif des Gharamul et le Gebel Dara: *in* C. Montenat, ed., Geological studies of the Gulf of Suez, the northwestern Red Sea coasts, tectonic and sedimentary evolution of a Neogene rift: Documents et Travaux Institut Geologique Albert de Lapparent, no. 10, p. 19–43.

Ott d'Estevou, P., J.J. Jarrige, J.C. Icart, C. Montenat, C. Henry, and J. Cravatte, 1986b, Observations structurales sur le secteur d'Abu Rudeis, Marge orientale du Golfe de Suez, *in* C. Montenat, ed., Geological studies of the Gulf of Suez, the northwestern Red Sea coasts, tectonic and sedimentary evolution of a Neogene rift: Documents et Travaux Institut Geologique Albert de Lapparent, no. 10, p. 75–92.

Patton, T.L., 1984, Surface studies of normal-fault geometries in the pre-Miocene stratigraphy, west central Sinai Peninsula: Egyptian General Petroleum Corporation, Sixth Exploration Seminar, 1982, p. 437–452.

Patton, T.L., and R.A. Nelson, 1988, Rift offsets, Gulf of Suez, Egypt (abs.): AAPG Bulletin, v. 72, p. 233.

Patton, T.L., and S.J. O'Connor, 1988, Cretaceous flexural history of northern Oman Mountain foredeep, United Arab Emirates: AAPG Bulletin, v. 72, p. 797–809.

Piper, J.D.A., 1983, Dynamics of the continental crust in Proterozoic times, *in* L.G. Medaris, Jr., et al., eds., Proterozoic geology: selected papers from an international Proterozoic symposium: Geological Society of America, Memoir 161, p. 11–34.

Piper, J.D.A., 1985, Continental movements and break-up in late Precambrian–Cambrian times: prelude to Caledonian orogenesis, *in* D.G. Gee and B.A. Sturt, eds., The Caledonide orogen: Scandinavia and related areas: John Wiley, London, p. 19–34.

Piper, J.D.A., 1987, Palaeomagnetism and the continental crust: John Wiley, New York, 434 p.

Pollard, D.D., and A. Aydin, 1984, Propagation and linkage of oceanic ridge segments: Journal of Geophysical Research, v. 89, p. 10,017–10,028.

Prat, P., C. Montenat, P. Ott d'Estevou, and J. Bolze, 1986, La marge occidentale du Golfe de Suez d'apres l'etude des Gebels Zeit et Mellaha, *in* C. Montenat, ed., Geological studies of the Gulf of Suez, the northwestern Red Sea coasts, tectonic and sedimentary evolution of a Neogene rift: Documents et Travaux Institut Geologique Albert de Lapparent, no. 10, p. 45–74.

Purser, B.H., and H. Hotzl, 1988, Sedimentary history of the Red Sea rift: a comparison of the northwest (Egyptian) and northeast (Saudi Arabian) margins: Tectonophysics, v. 153, p. 193–208.

Reches, Z., D.F. Hoexter, and F. Hirsch, 1981, The structure of a monocline in the Syrian Arc system, Middle East—surface and subsurface analysis: Journal of Petroleum Geology, v. 3–4, p. 413–425.

Renolds, M.L., 1979, Geology of the northern Gulf of Suez: Annals of the Geological Survey of Egypt, v. 9, p. 322–343.

Riad, S., 1977, Shear zones in north Egypt interpreted from gravity data: Geophysics, v. 42, p. 1207–1214.

Riad, S., 1981, Shear zones in north Egypt interpreted from gravity data, reply: Geophysics, v. 46, p. 1471–1473.

Richardson, M., and M.A. Arthur, 1988, The Gulf of Suez–northern Red Sea Neogene rift: a quantitative basin analysis: Marine and Petroleum Geology, v. 5, p. 247–270.

Richert, J.P., J.J. Jarrige, C. Montenat, and P. Ott d'Estevout, 1986, Compression and transpression mechanism during the first stage of opening of Gulf of Suez and northern part of the Red Sea: (preprint) Egyptian General Petroleum Corporation, Eighth Exploration Conference, 1986, 10 p.

Robertson, A.H.F., and J.E. Dixon, 1984, Introduction: aspects of the geological evolution of the eastern Mediterranean: *in* J.E. Dixon and A.H.F. Robertson, eds., The geologic evolution of the eastern Mediterranean: Geological Society of London, Special Publication no. 17, p. 1–74.

Robertson, A.H.F., C.D. Blome, D.J.W. Cooper, A.E.S. Kemp, and M.P. Searle, 1990, Evolution of the Arabian continental margin in the Dibba zone, northern Oman Mountains *in* A.H.F. Robertson, M.P. Searle and A.C. Ries, eds., The geology and tectonics of the Oman region, Geological Society of London, Special Publication no. 49, p. 251–284.

Robson, D.A., 1971, The structure of the Gulf of Suez (Clysmic) rift, with special reference to the eastern side: Journal of the Geological Society, v. 127, p. 247–276.

Rosendahl, B.R., 1987, Architecture of continental rifts with special reference to east Africa: Annual Review of Earth and Planetary Science, v. 15, p. 445–503.

Sadek, H., 1959, The Miocene in the Gulf of Suez region (Egypt): Egyptian Geological Survey and Mineral Research Department, 118 p.

Said, R., 1961, Tectonic framework of Egypt and its influence on distribution of foraminifera: AAPG Bulletin, v. 45, p. 198–218.

Said, R., 1962, Geology of Egypt: Elsevier, Amsterdam, 377 p.

Said, R., and M.A. Basiouni, 1958, Miocene foraminifera of Gulf of Suez region: AAPG Bulletin, v. 42, p. 1958–1977.

Samuel, M.D., N.N. Yanni, and A.I.M. Akarish, 1988, Critical remarks on the lithostratigraphic classification of the Miocene in the Gulf of Suez region: Middle East Research Center, Ain Shams University, Earth Science Series, v. 2, p. 49–54.

Saoudi, A., and B. Khalil, 1986, Distribution and hydrocarbon potential of Nukhul sediments in the Gulf of Suez: Egyptian General Petroleum Corporation, Seventh Exploration Seminar, 1984, p. 75–96.

Schmidt, D.L., D.G. Hadley, and D.B. Stoesser, 1979, Late Proterozoic crustal history of the Arabian shield, southern Najd Province, Kingdom of Saudi Arabia, *in* S. Talhoun, ed., Evolution and mineralization of the Arabian-Nubian shield, v. 2: Pergamon, New York, p. 41–58.

Scholz, C.H., and P.A. Cowie, 1990, Determination of total strain from faulting using slope measurements: Nature, v. 346, p. 837–839.

Schurmann, H.M.E., 1966, The Pre-Cambrian along the Gulf of Suez and the northern part of the Red Sea: Leiden, E.V. Brill, 405 p.

Scott, R., and F. Govean, 1986, Early depositional history of a rift basin: Miocene in the western Sinai: Egyptian General Petroleum Corporation, Seventh Exploration Seminar, 1984, p. 37–48.

Seilacher, A., 1990, Paleozoic trace fossils, *in* R. Said, ed., The geology of Egypt: Rotterdam, A.A. Balkema, p. 649–670.

Sellwood, B.W., and R.E. Netherwood, 1984, Facies evolution in the Gulf of Suez area: sedimentation history as an indicator of rift initiation and development: Modern Geology, v. 9, p. 43–69.

Sengor, A.M.C., 1984, The Cimmeride orogenic system and the tectonics of Eurasia: Geological Society of America Special Paper 195, 82 p.

Sengor, A.M.C., 1990, A new model for the late Paleozoic–Mesozoic tectonic evolution of Iran and implications for Oman, *in* A.H.F. Robertson, M.P. Searle, and A.C. Ries, eds., The geology and tectonics of the Oman region: Geological Society of London Special Publication no. 49, p. 797–831.

Sengor, A.M.C., and Y. Yilmaz, 1981, Tethyan evolution of Turkey: a plate tectonic approach: Tectonophysics, v. 75, p. 181–241.

Sengor, A.M.C., M. Satir, and R. Akkok, 1984a, Timing of tectonic events in the Menderes massif, western Turkey: implications for tectonic evolution and evidence for pan-African basement in Turkey: Tectonics, v. 3, p. 693–707.

Sengor, A.M.C., Y. Yilmaz, and O. Sungurlu, 1984b, Tectonics of the Mediterranean Cimmerides: nature and evolution of the western termination of Palaeo-Tethys, *in* J.E. Dixon and A.H.F. Robertson, eds., The geologic evolution of the eastern Mediterranean: Geological Society of London, Special Publication no. 17, p. 77–112.

Sestini, G., 1984, Tectonic and sedimentary history of the NE African margin (Egypt-Libya), *in* J.E. Dixon and A.H.F. Robertson, eds., The geologic evolution of the eastern Mediterranean: Geological Society of London, Special Publication no. 17, p. 161–175.

Shahin, A.N., 1988, Oil window in the Gulf of Suez basin, Egypt (abs.): AAPG Bulletin, v. 72, p. 1024–1025.

Shaw, A.B., 1964, Time and stratigraphy: New York, McGraw-Hill, 365 p.

Smale, J.L., R.C. Thunell, and S. Schamel, 1988, Sedimentological evidence for early Miocene fault reactivation in the Gulf of Suez: Geology, v. 16, p. 113–116.

Soliman, S.M., and K.M. Amer, 1972, Cretaceous sedimentation in Eastern Desert, Egypt, and evaluation for petroleum prospecting: Eighth Arab Petroleum Congress, Algiers, p. 1–33.

Soliman, S.M., and M.A. El-Fetouh, 1969, Petrology of the Carboniferous sandstones in west central Sinai: Journal of Geology United Arab Republic, v. 13, p. 61–143.

Soliman S.M., M.I. Faris, and M.F. Hassan, 1965, Geologic setting of the Gulf of Suez province during the Eocene Period: Fifth Arab Petroleum Congress, Cairo, p. 1–29.

Steckler, M.S., 1985, Uplift and extension at the Gulf of Suez: indications of induced mantle convection: Nature, v. 317, p. 135–139.

Steckler, M.S., and U.S. ten Brink, 1986, Lithospheric strength variations as a control on new plate boundaries: examples from the northern Red Sea region: Earth and Planetary Science Letters, v. 79, p. 120–132.

Steckler, M.S., and A.B. Watts, 1980, The Gulf of Lion: subsidence of a young continental margin: Nature, v. 287, p. 425–429.

Steckler, M.S., F. Berthelot, N. Lyberis, and X. LePichon, 1988, Subsidence in the Gulf of Suez: implications for rifting and plate kinematics: Tectonophysics, v. 153, p. 249–270.

Steen, G., 1984, Radiometric age dating and tectonic significance of some Gulf of Suez igneous rocks: Egyptian General Petroleum Corporation, Sixth Exploration Seminar, 1982, p. 199–211.

Stern, R.J., 1985, The Najd Fault system, Saudi Arabia and Egypt: a late Precambrian rift-related transform system?: Tectonics, v. 4, p. 497–511.

Stern, R.J., and C.E. Hedge, 1985, Geochronological and isotopic constraints on late Precambrian crustal evolution in the Eastern Desert of Egypt: American Journal of Science, v. 285, p. 97–127.

Stern, R.J., and W.I. Manton, 1987, Age of Feiran basement rocks, Sinai: implications for late Precambrian crustal evolution in the northern Arabian-Nubian shield: Journal of the Geological Society of London, v. 144, p. 569–575.

Stern, R.J., et al., 1989, Discussion of the paper "Late Pan-African magmatism and crustal development in northeastern Egypt": Geological Journal, v. 24, p. 281–301.

Stern, R.J., D. Gottfried, and C.E. Hedge, 1984, Late Precambrian rifting and crustal evolution in the Northeastern Desert of Egypt: Geology, v. 12, p. 168–172.

Stern, R.J., G. Sellers, and D. Gottfried, 1988, Bimodal dike swarms in the north Eastern Desert of Egypt: significance for the origin of late Precambrian "A-type" granites in northern Afro-Arabia, *in* S. El Gaby and R.O. Greiling, eds., The Pan-African belt of northeast Africa and adjacent areas, tectonic evolution and economic aspects of a Late Proterozoic orogen: Braunschweig/Wiesbaden, Friedr. Vieweg, p. 147–179.

Stocklin, J., 1968, Structural history and tectonics of Iran: a review: AAPG Bulletin, v. 52, p. 1229–1258.

Stoneley, R., 1990, Arabian continental margin in Iran during the Late Cretaceous, *in* A.H.F. Robertson, M.P. Searle, and A.C. Ries, eds., The geology and tectonics of the Oman region: Geological Society of London, Special Publication no. 49, p. 797–831.

Stratigraphic Subcommittee of the National Committee of Geological Sciences, 1976, Miocene rock stratigraphy of Egypt: Egyptian Journal of Geology, v. 18, p. 1–69.

Strougo, A., 1986, The Velascoensis event: a significant episode of tectonic activity in the Egyptian Paleogene: Neues Jahrbuch für Geologie und Paläontologie Monatshefte, v. 173, p. 253–269.

Sultan, N., and K. Schutz, 1986, Cross faults in the Gulf of Suez area: Egyptian General Petroleum Corporation, Seventh Exploration Seminar, 1984, p. 5–17.

Sultan, M., R.E. Arvidson, I.J. Duncan, R.J. Stern, and B. El Kaliouby, 1988, Extension of the Najd shear system from Saudi Arabia to the central Eastern Desert of Egypt based on integrated field and Landsat observations: Tectonics, v. 7, p. 1291–1306.

Tewfik, N., and Z. Ebeid, 1975, On the stratigraphy of the Upper Cretaceous in the Gulf of Suez region, Egypt: Revista Española de Micropaleontologia, v. 7, no. 3, p. 659–701.

Thiebaud, C.E., and D.A. Robson, 1979, The geology of the area between Wadi Wardan and Wadi Gharandal, East Clysmic rift, Sinai, Egypt: Journal of Petroleum Geology, v. 1, p. 63–75.

Thiebaud, C.E., and D.A. Robson, 1981, The geology of the Asl oilfield, western Sinai, Egypt: Journal of Petroleum Geology, v. 4, p. 77–87.

Thiriet, J., P.F. Burollet, C. Montenat, and P. Ott d'Estevou, 1986, Evolution tectonique et sedimentaire Neogene a la transition du Golfe de Suez et de la Mer Rouge: Le secteur de Port-Safaga (Egypte), *in* C. Montenat, ed., Geological studies of the Gulf of Suez, the northwestern Red Sea coasts, tectonic and sedimentary evolution of a Neogene rift: Documents et Travaux Institut Geologique Albert de Lapparent, no. 10, p. 45–74.

Travis, C.J., 1984, Upper Cretaceous stratigraphy of eastern Galala, Egypt: M.S. thesis, University of South Carolina, Columbia, South Carolina, 89 p.

Vail, J.R., 1983, Pan-African crustal accretion in northeast Africa: Journal of African Earth Sciences, v. 1, p. 285–294.

Vail, J.R., 1985, Pan-African (late Precambrian) tectonic terrains and the reconstruction of the Arabian-Nubian shield: Geology, v. 13, p. 839–842.

Van Houten, F.B., 1980, Latest Jurassic–Early Cretaceous regressive facies, northeast Africa craton: AAPG Bulletin, v. 64, p. 857–867.

Van Houten, F.B., D.P. Bhattacharyya, and S.E.I. Mansour, 1984, Cretaceous Nubia Formation and correlative deposits, eastern Egypt: major regressive-transgressive complex: Geological Society of America Bulletin, v. 95, p. 397–405.

Vigano, P.L., and T.L. Patton, 1988, Mid-Clysmic event, Gulf of Suez rotational deformation associated with a deep crustal detachment fault (abs.): AAPG Bulletin, v. 72, p. 1029.

Villemin, T., and C. Sunwoo, 1987, Distribution logarithmique self-similaire des rejets et longueurs de failles: exemple du Bassin Houiller Lorrain: Comptes Rendus, Académie des Sciences, Paris, v. 305, p. 1309–1312.

Viotti, C., and G. El Demerdash, 1969, Studies in Eocene sediments of Wadi Nukhul area, east coast—Gulf of Suez: Proceedings of the Third African Micropaleontological Colloquium, Cairo, p. 403–423.

Voggenreitter, W., H. Hotzl, and A.R. Jado, 1988, Red Sea related history of extension and magmatism in the Jizan area (SW Saudi Arabia): indication for simple-shear during early Red Sea rifting: Geologische Rundschau, v. 77, p. 257–274.

Walsh, J., J. Watterson, and G. Yielding, 1991, The importance of small-scale faulting in regional extension: Nature, v. 351, p. 391–393.

Wasfi, S., and G. Azazi, 1979, Stratigraphy of the northern Gulf of Suez: Annals of the Geological Survey of Egypt, v. 9, p. 308–321.

Webster, D.J., and N. Ritson, 1984, Post-Eocene stratigraphy of the Suez rift in the southwest Sinai, Egypt: Egyptian General Petroleum Corporation, Sixth Exploration Seminar, 1982, Cairo, p. 276–288.

Weissbrod, T., 1969a, The Paleozoic outcrops of Israel and adjacent countries, part I: Geological Survey of Israel Bulletin, no. 48, 31 p.

Weissbrod, T., 1969b, The Paleozoic outcrops of Israel and adjacent countries, part II: Geological Survey of Israel Bulletin, no. 48, 25 p.

Weissbrod, T., and J. Nachmias, 1986, Stratigraphic significance of heavy minerals in the late Precambrian–Mesozoic clastic sequence ("Nubia Sandstone") in the Near East: Sedimentary Geology, v. 47, p. 263–291.

Willis, K.M., R.J. Stern, and N. Clauer, 1988, Age and geochemistry of late Precambrian sediments of the Hammamat series from the northeastern desert of Egypt: Precambrian Research, v. 42, p. 173–187.

Youseff, M.I., 1968, Structural pattern of Egypt and its interpretation: AAPG Bulletin, v. 52, p. 601–614.

Zahran, M.E., and W. Meshref, 1988, The northern Gulf of Suez basin evolution, stratigraphy and facies relationship: (preprint) Egyptian General Petroleum Corporation, Ninth Exploration and Production Conference, Cairo, 30 p.

Ziegler, P.A., 1987, Evolution of the Arctic–North Atlantic borderlands, *in* J. Brooks and K.W. Glennie, eds., Petroleum geology of north-west Europe: London, Graham and Trotman, v. 2, p. 1201–1204.

Ziegler, P.A., 1989, Evolution of Laurussia, a study in late Paleozoic plate tectonics: Boston, Kluwer Academic Publishers, 102 p.

Zilberman, E., 1981, Tectonic activity along the Sa'ad-Nafha lineament in the central Negev: Geological Survey of Israel, Current Research 1980, p. 86–88.

Chapter 2

Structure and Stratigraphy of the Gulf of Suez, Egypt

Karsten I. Schütz
Conoco Inc.
Houston, Texas, U.S.A.

INTRODUCTION

Since 1886 when the first oil was discovered and subsequently produced in the Middle East on Gemsa Peninsula, the Gulf of Suez has challenged the imaginations of geologists and continues to do so. The Gulf of Suez stretches over 300 km between the city of Suez and Ras Mohamed (Figures 1–3). The modern sea inundates only one-third of the geologic graben feature between the Sinai basement uplift and the Eastern Desert mountains. For simplicity, the whole graben system onshore and offshore is referred to as Gulf of Suez or "Clysmic Gulf" ("Clysma" being the Roman name of Suez) (Hume, 1921; Heybroek, 1965).

The water body of the Gulf of Suez is rather shallow with a water depth of meters. Only in the southern part off Ras Mohamed does a deeper trench enter the gulf from the Red Sea. We define the southern end of the gulf as the line from Ras Mohamed to Shadwan Island and Hurghada. This line follows approximately a 200-m water-depth contour, where the Red Sea begins. The Gulf of Suez is the northern extension of the 2000-km-long Red Sea rift system, which is a Cenozoic structure slicing through the once-continuous Arabo-African craton. Geologically, the Gulf of Suez is a complex graben/half-graben system located between two basement uplifts—the Sinai and the Eastern Desert mountains. The graben cuts through older structural trends having ages from Precambrian to Eocene. The structure of the Gulf of Suez is dominated by normal faults and tilted fault blocks, which were formed after the Oligocene, mainly during Miocene time. The axial portion of the gulf underwent the most subsidence and contains the thickest deposits of lower Miocene sediments and middle/upper Miocene evaporites. In contrast to the Red Sea graben system, where motion is presently occurring along its entire length, the Gulf of Suez experienced extension primarily during Miocene time.

"The origins of rifts, whether by lithospheric stretching (passive) or as grabens over sites of mantle upwarps (active) is currently controversial, particularly in the interpretation of ancient 'failed' rift systems. . . . Although it is now generally accepted that rifts reflect tension, there are divergent opinions as to what causes the tensional regime. Until relatively recently rift grabens were viewed as fault-bounded basins initiated at the crest of lithospheric domes or upwarps . . . sited over mantle plumes. . . . Therefore, they have often been interpreted in plate tectonic terms . . . even though many such rifts are wholly confined to continental areas. . . ." (Sellwood and Netherwood, 1984, p. 43, 44). This is particularly true for the Gulf of Suez because, similar to other African grabens, it is located entirely inside a cratonic area.

Toward the north, the African continent, including the Sinai, extends far into the eastern Mediterranean Sea, which is underlain by a thick section of Cenozoic, Mesozoic and, in part at least, Paleozoic sediments on a continental crust (Orwig, 1982; Hinz, 1974; Morelli, 1978; Harsch et al., 1981). Toward the south, the Gulf of Suez meets the Red Sea, which is clearly a nascent ocean basin. However, the northern part of the Red Sea is still underlain by tilted basement blocks. "An axial trough is not present north of about 25°N" (Drake and Girdler, 1964) ". . . and the large, linear magnetic anomalies associated with the axial trough [in the southern Red Sea] are not present" (Cochran, 1983, p. 55). Only south of 21°N can oceanic crust with magnetic anomalies parallel to the central trough be found (Cochran, 1983, p. 41, 55).

The Gulf of Suez graben was formed after the Eocene, as shown by structural trends and sediment distribution of the Cretaceous and lower and middle Eocene. Pre-Eocene or even Carboniferous origin is not supported by the geological evidence. Pre-Miocene sediments were deposited on a craton or in epicontinental seas. Sedimentation was controlled, at least during the Jurassic, Cretaceous, and Neogene, by east-west–trending faults (Figures 4, 5). Rifting began in the Gulf of Suez after the Oligocene, and is documented primarily by an upper Oligocene planation surface on both sides of the Gulf of Aqaba and the southern Gulf of Suez (Quennell, 1984). In addition, basalt intrusions and flows in the Gulf of Suez

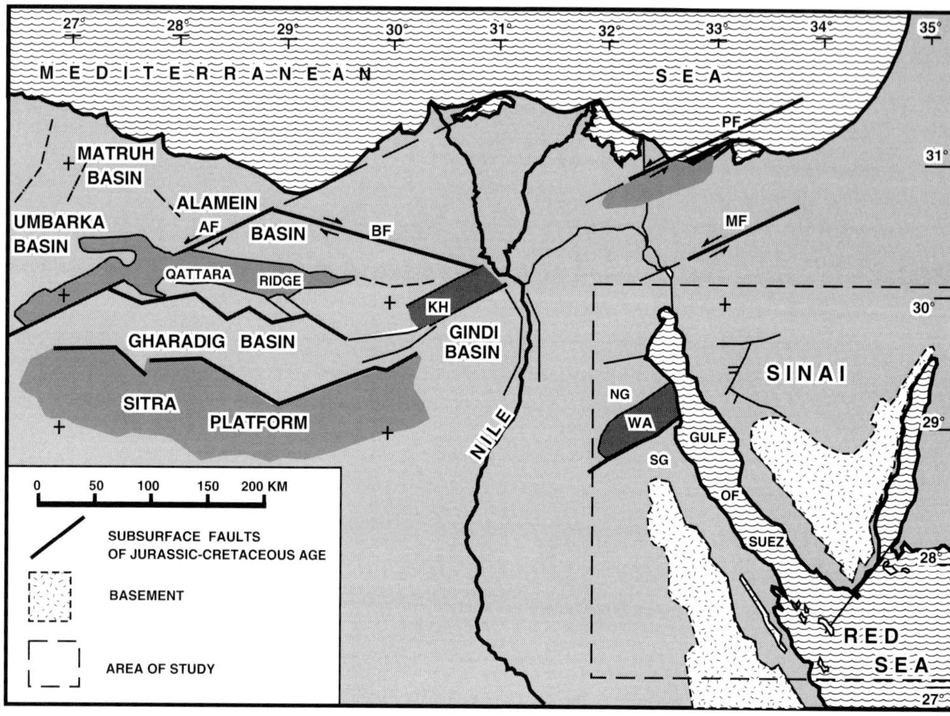

Figure 1. Major tectonic elements of northern Egypt and Sinai. AF = Alamein fault, BF = Alam el Bueib fault, KH = Kattanyia Horst, PF = Pelusium fault, MF = Gebel Maghara fault, NG = Northern Galala Plateau, WA = Wadi Araba, SG = Southern Galala Plateau.

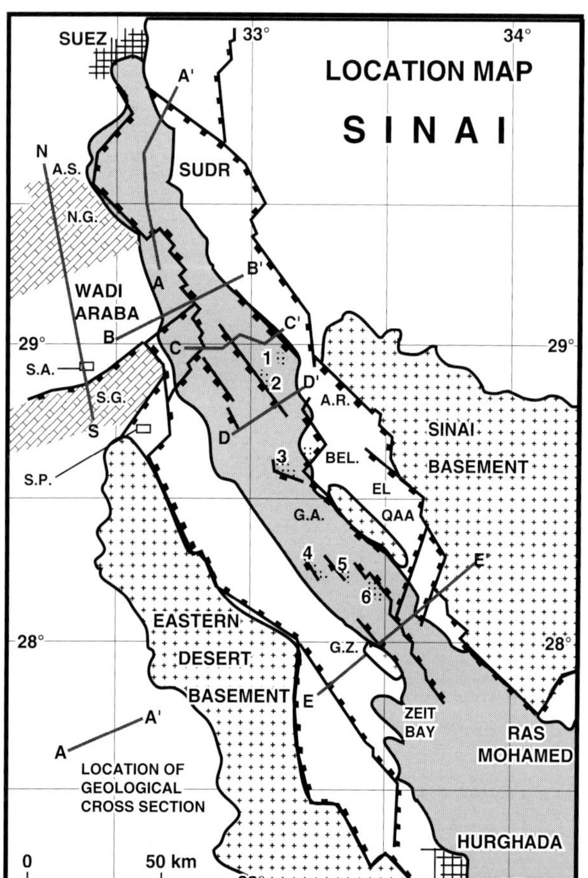

are all dated as earliest Miocene (20 to 23 Ma; Garfunkel and Bartov, 1977). The basalts at Gebel Qatrani and in the Western Desert (Williams and Small, 1984) are age equivalent.

The sea invaded the Gulf of Suez area again during the early Miocene. The basal Nukhul Formation contains conglomerates, limestones, and evaporites. Differential erosion is limited to the crest of fault blocks, indicating that initial phases of rifting occurred near sea level. Initial faulting, therefore, was small scale, with displacement in the 10-m range. Subsidence and sedimentation, however, accelerated shortly after Nukhul and contemporaneous with Rudeis Formation (Burdigalian) deposition, concurrent with uplift of the graben shoulders. Erosion reached basement at Gebel Zeit during the Burdigalian, as shown by *Globigerina* limestones directly on basement. These *Globigerina* limestones were deposited on a submerged basement ridge, which originally had experienced subaerial erosion.

Figure 2. Location map. A.R. = Abu Rudeis, A.S. = Ain Sukna, BEL = Belayim, G.A. = Gebel Araba, G.Z. = Gebel Zeit, N.G. = Northern Galala Plateau, S.G. Southern Galala Plateau, S.A. = Saint Anthony's Monastery, S.P. = Saint Paul's Monastery, 1 = Ras Budran oil field, 2 = October oil field, 3 = Belayim Marine oil field, 4 = July oil field, 5 = Ramadan oil field, 6 = Morgan oil field.

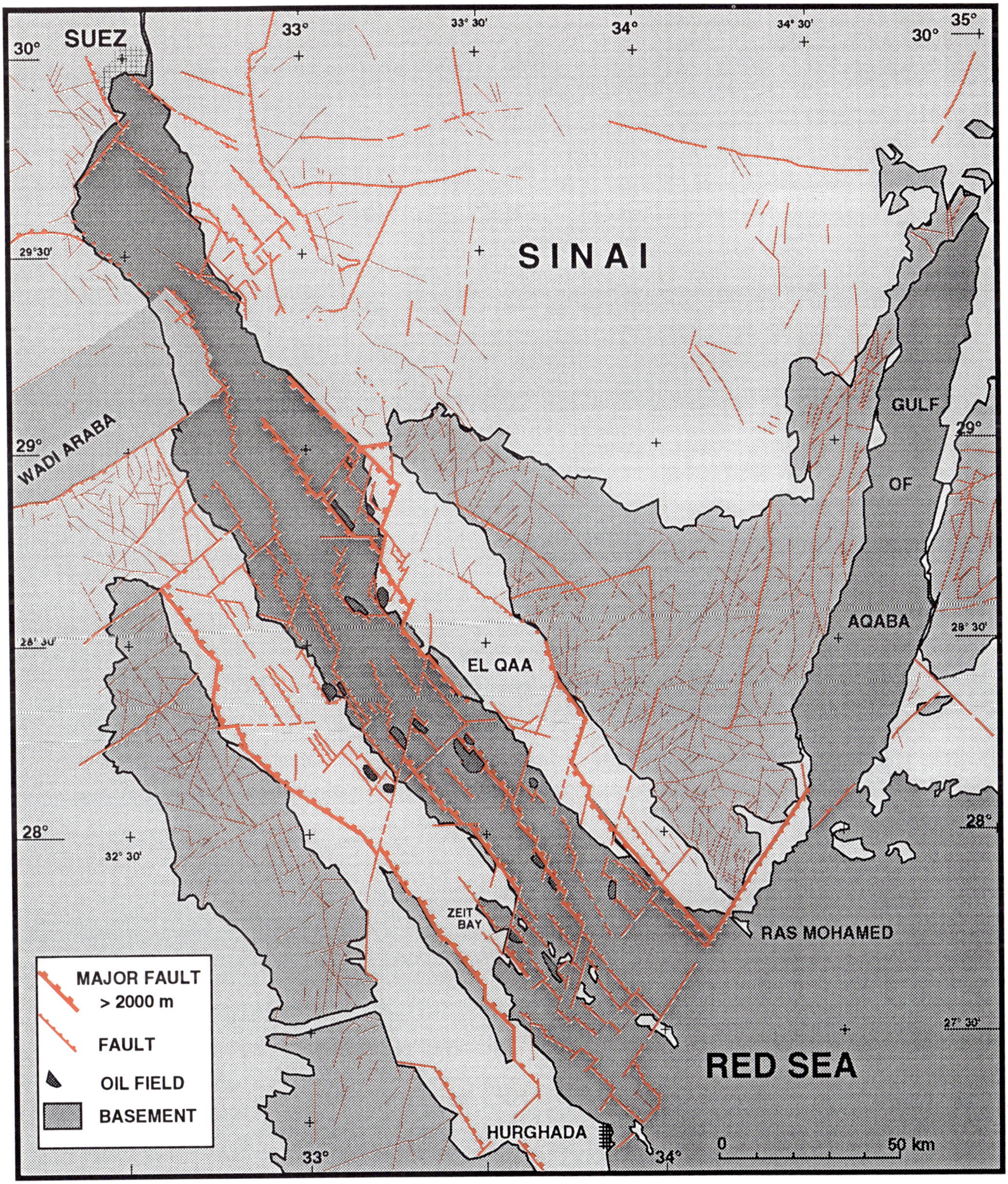

Figure 3. Structural features map.

Tectonic activity slowed several times during the Miocene. Major quiescent events occurred during deposition of the mid-Rudeis (marking the transition from synrift to postrift sedimentation), at the base of the Kareem, and throughout the Belayim. After the Langhian, block-faulting decreased and subsidence occurred on a more regional scale, predominantly along the major graben-shoulder faults.

In summary, after continental erosion and minor deposition of red beds in the Oligocene, the first

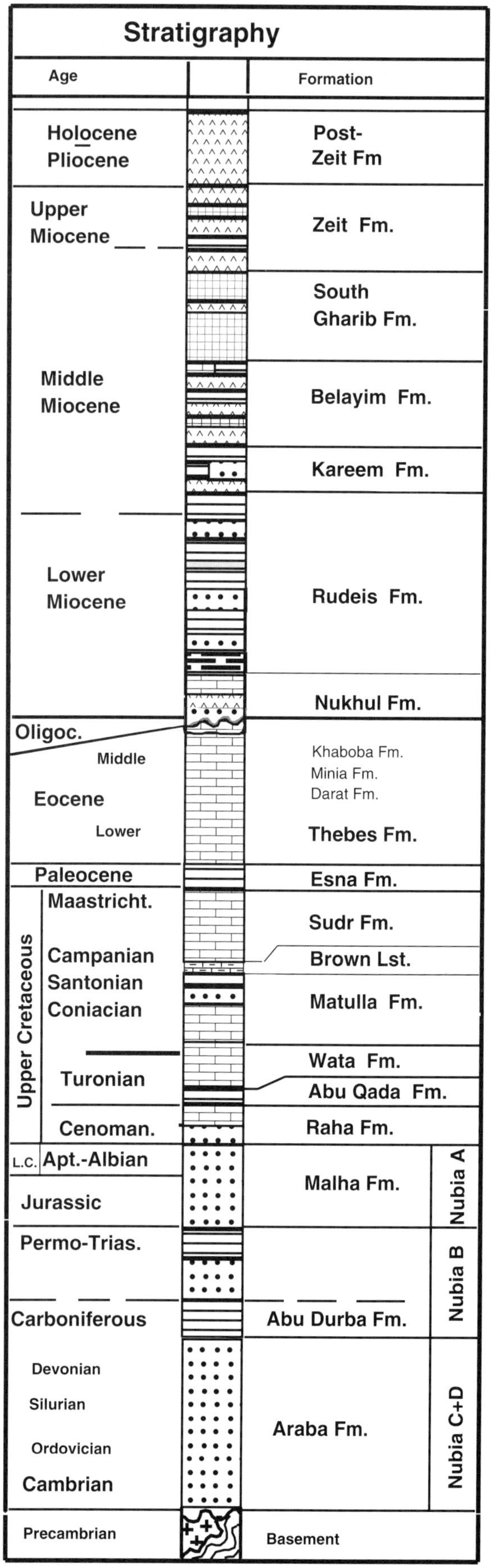

Figure 4. Stratigraphic column of the study area.

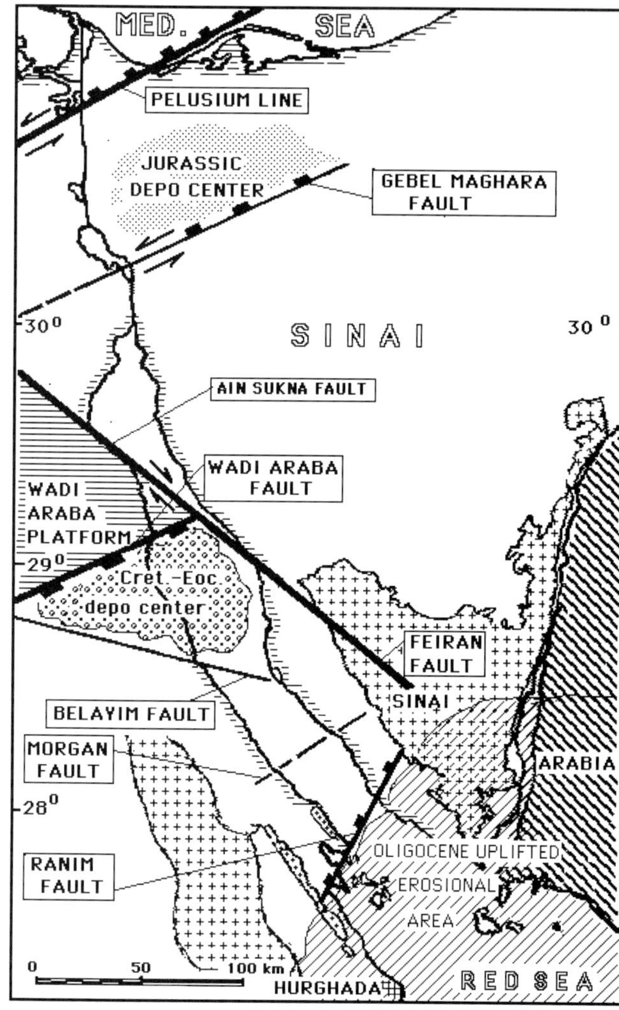

Figure 5. Postulated pre-Miocene fault pattern.

breakup started in the earliest Miocene, accelerated during the early Miocene, quieted in middle Miocene (late Langhian and Serravallian), resumed deposition during the late Serravallian, and continued during the Tortonian; however, on a broader scale, the whole Gulf of Suez subsided as one graben along the main graben faults. This indicates that extension had decoupled the two sides of the graben. In the Gulf of Suez, major uplift and rifting clearly occurred simultaneously. Because the uplift is believed to indicate heating and expansion of the lithosphere, timing relative to the faulting and subsidence is crucial to understanding the rifting process (Le Pichon et al., 1982).

Since 1962, when Said's *Geology of Egypt* was published, many kilometers of seismic new exploration wells, LANDSAT, and several mapping projects have increased our geological knowledge. The best way to proceed seems to be to present a regional compilation of generally agreed upon facts, add our new oil exploration data, and document, via stratigraphic fine-tuning, the geologic history of the Gulf of Suez. The aim of the first part of this chapter is to present the state-of-the-art of the structural interpretation of the Gulf of Suez; the second part of this chapter shows that, prior to the Eocene, the stratigraphic units in the Gulf of Suez correlate well with those in the Western Desert and the Sinai, because they were deposited on the same part of the African craton. From the Miocene, a marine environment prevails and extensional tectonics controlled sedimentation in the Gulf of Suez.

The figures in this chapter were prepared using logs of more than 100 exploration wells and more than 2000 km of mainly marine seismic data.

STRUCTURE

Regional Tectonic Framework and Pre-Miocene Fault Pattern

A tectonic interpretation of the Gulf of Suez has to include the regional structural configuration including the Western Desert, the eastern Mediterranean Sea, the Sinai, the Gulf of Aqaba, and the Red Sea. This interpretation also should include the underlying pre-Oligocene fault pattern, which can be studied much easier outside the Gulf of Suez, for example in the Western Desert, where it is not overprinted by Miocene faulting.

I consider the northern part of Africa, the eastern Mediterranean, and Sinai, including the Gulf of Suez, as one major block that, since the Cretaceous, has changed relatively little in comparison to Arabia, where motion presently is occurring along the entire length of the Red Sea. Geophysical investigations (seismic refraction profiling) in the Herodotus Basin and in the Levantine Basin in the eastern Mediterranean area (Morelli, 1978) show that the Moho discontinuity lies at 26 to 27 km. "The results clearly demonstrate the continental nature of the entire region south of the allochthonous nappes and thrust sheets of the Alpine orogenic zone (i.e., south of Cyprus and Turkey)" (Orwig, 1982, p. 4). "Comparable data were reported by Ben-Avraham et al. (1976) for the Levantine Basin, although these authors maintain that the crust is a composite of oceanic and continental fragments. Granted that this may be so, the age of the crust is nevertheless pre-Jurassic and probably pre-Carboniferous. For all practical purposes, since the time of the formation of the crust, the region has evolved structurally and stratigraphically upon a rigid crustal block whose properties and thicknesses are within limits defined as continental. The significance of this conclusion is vital for the understanding of structural evolution and style, and especially in evolution of hydrocarbon prospectiveness" (Orwig, 1982). In addition, it now seems that the Turkish–African plate boundary is not related to a subduction zone, but to an accretional continental margin. "The nature of the assumed Turkish–African plate boundary is investigated in the light of new seismological and geologic-tectonic data. . . . From the observed relationship between extensional, compressional and strike-slip deformation, it is concluded that the Turkish–African convergence results in a sort of complex crustal welding within a broad and diffuse zone rather than in active consumption of African lithosphere beneath the Cypriote–Tauride arc" (Harsch et al., 1981).

"The crust of the Levantine Sea, according to new interpretations of geophysical data, is continental rather than oceanic. Thickness of the crust is about 20 km to the north of the Nile Cone and increases up to 30 km beneath the Mediterranean Ridge" (Harsch et al., 1981, p. 374).

Taking these interpretations into account, one should consider the eastern Mediterranean, northern Egypt, and Sinai as part of one plate (in the sense of plate tectonics), that is, as part of Africa.

The opening of the Gulf of Suez is certainly related to the initial opening of the Red Sea. Based on the sediment and evaporite record in both basins, this is probably one event. During most of early and middle Miocene, all subbasins of the gulf were in full marine connection with each other and with the major basin to the south, the Red Sea. Exploration wells drilled on both sides of the northern Red Sea document similar Miocene sediment and evaporite sequences as occur in the Gulf of Suez, and, hence, a comparable tectonic history. Wells (Quseir B-1X, RSO B′′95-1, RSO Z′95-1) south of Hurghada and on the Saudi Arabian side (Barqan-2) encountered equivalents to the Kareem and Rudeis formations under a thick Miocene evaporite cover. Likewise, lower(?) and middle Miocene lithothamnium carbonates bordering the western scarp of the Gulf of Suez and the Red Sea prove a uniform, long, coastline, which indicates that the Gulf of Suez and the early Red Sea were flooded at essentially the same time.

The Aqaba/Dead Sea rift is considered a transform plate boundary between the Arabian plate and the Sinai/Levant plate without uniformity along its length (Quennell, 1984). The origin of the Aqaba/Dead Sea rift generally has been linked with that of the Red Sea. The widening of the Red Sea caused strike-slip in the Gulf of Aqaba, ceating a rhomb-shaped pull-apart graben very different from the Gulf of Suez.

"The Red Sea occupies an elongate escarpment-bounded depression 250 to 450 km wide between the uplifted Arabian and African shields. Morphologically, the Red Sea consists of shallow continental shelves, a wide 'main trough' which extends from about 15°N to the tip of the Sinai peninsula (28°N) at a depth of 600 to 1000 m, and a narrow 'axial trough' found from 15°N to 24°N which is about 2000 m deep, usually less than 50 km wide, and is characterized by steep walls and irregular bottom topography. . . . The axial trough is associated with large-amplitude magnetic anomalies, which have been identified as sea floor spreading anomalies . . . and have led to a general acceptance of the fact that the axial trough has been generated by sea floor spreading over the past 3 to 5 m.y. . . . The origin and nature of the crust underlying the main trough and shelves is much less certain. Seismic reflection studies showed a strong reflector (reflector S) at about 0.5 s below the surface of the main trough. . . . This reflector was later demonstrated by DSDP drilling to be the top Miocene evaporite deposit, which underlies all of the main trough" (Cochran, 1983, p. 41). Cochran further proves that in the northern Red Sea, north of 24°N, where there is no evidence of mid-ocean ridge tectonics, the main trough is underlain in general by tilted fault blocks of continental crust hidden under the extremely thick salt. The magnetic anomalies north of 24°N coincide with gravity anomalies and thus are interpreted as intrusive bodies and not as sea-floor spreading anomalies.

"The geological and geophysical data of the Red Sea suggest that oceanic crust in the Red Sea is limited to the axial trough and the main trough is underlain by faulted and intruded continental crust. This, however, does not necessarily mean that only a minimal amount of motion has occurred between Africa and Arabia. In fact, significant non-sea floor spreading extension is required by the documented motion on the Dead Sea transform" (Cochran, 1983, p. 42). The geological results from the Gulf of Suez, and especially from the Gebel Zeit and the Hurghada/Islands areas, strongly support this view.

Africa moved from the Triassic to the Cretaceous some 2000 km eastward in relation to Europe. Wrenching due to an unknown resistance (probably collision and accretion of microcontinents) at its northern margin occurred, and major blocks on the continent were sheared off along regional wrench faults. These wrench faults are well documented in the Western Desert and near the Gulf of Suez (Figures 1, 5) and are as follows (fault direction or position is in parentheses): Alamein fault (northeast-southwest), Pelusium line continuing in the Kattanyia fault, corresponding northwest-southeast Alam el Bueib/Wadi Natrun fault, Gebel Maghara fault (northeast-southwest), Wadi Araba faults (parallel to the Pelusium line), Ain Sukna fault (possible ancient continuation of the Natrun fault), and Belayim, Morgan, and Ranim faults (Figure 5).

Considerable changes of sediment thickness in the Jurassic accompany these major fault lines, e.g., at the Kattanyia fault west of Cairo, in the Abu Gharadig basin in the Western Desert, and in the Darag Basin of the Gulf of Suez.

The Pelusium line is an important limit of the Sinai block, especially during the Oligocene as a major hinge line downthrown to the north, and as a boundary between continental sedimentation in the south and marine sedimentation in the north. It has a modern left-lateral component of more than 1 km.

In the Gulf of Suez, only the Ain Sukna fault and the Wadi Araba structure are well known enough to show the Jurassic/Cretaceous origin. However, similar tectonic elements exist in the Belayim Marine high, north of Morgan field and farther south at Gebel Zeit, proven by thickness and facies changes in Cretaceous sediments (Figure 5).

Determining the time of active faulting is difficult. The Ain Sukna fault and the Gebel Maghara fault certainly were active during the Jurassic, as is indicated by the enormous increase in the thickness of the Jurassic toward the north. The Wadi Araba fault was reactivated several times because it controlled deposition of the Paleozoic Nubia sandstones and the Sudr, Esna, and Thebes formations. In the Western Desert, strike-slip motion is documented in the subsurface by flower structures in post-Cretaceous horizons, for example at the Alam el Bueib fault. Horizontal offsets of 1–10 km at these major faults exist (Figure 1). In comparison, no strike-slip motion is evident on seismic in the Gulf of Suez. On the surface, the Wadi Araba fault, with right-lateral offset, seems to be the only possible site of strike-slip motion.

Gulf of Suez Structural Features

Gulf of Suez Subdivision

The Gulf of Suez can be subdivided into a series of subbasins and platforms that, to some degree, have their own structural and sedimentary history (Figures 6, 7).

The onshore subbasins are the West Bakr basin, Zeit Bay basin, Sinai onshore area with Wadi Gharandal and the El Qaa basin. The offshore subbasins are the Darag Basin, Issaran–October–Abu Zenima basins, South Belayim Basin, Morgan Basin, Gebel Zeit Basin, and Hurghada/Islands area.

These basins are generally large half-grabens and are separated by structural highs. They are generally bounded on one side by a major fault with a displacement of 1 km to a maximum of 5 km. Complications occur when the regional dip of the half-grabens changes from southwest to northeast. Previous workers have recognized the dip changes and have accordingly subdivided the gulf into dip regimes; however, not all workers have realized the importance of cross faults, which separate these dip regimes.

Basement and Onshore Plateaus

The Gulf of Suez is bounded by two major basement massifs, the Eastern Desert basement and the Sinai basement (Figures 3, 6, 7). These two massifs are plunging to the north and are bounded by major faults on the gulf side. The Eastern Desert basement is slightly tilted to the west. The Sinai basement is triangular and dips northward. Basement structures generally trend east-west or, like many of the dikes, northeast-southwest. Some possible Najd-trending faults (north-northwest–south-southeast) are present in the Wadi Feiran area.

Three smaller elongate basement blocks inside the Gulf of Suez acted as relatively stable highs during the Miocene. These are the Esh el Melaha range and Gebel Zeit on the western side, and the Gebel Araba on the Sinai side. They are tilted fault blocks with major faults bounding the blocks toward the gulf. Fault displacements range between 1000 m at Esh el Melaha and 5 km at Gebel Zeit. The pre-Miocene sedimentary cover dips with the same angle as basement and reaches 45° at Gebel Zeit (Figures 8, 9, 10). These surface fault blocks are used as analogs for the subsurface, about which we have much less information.

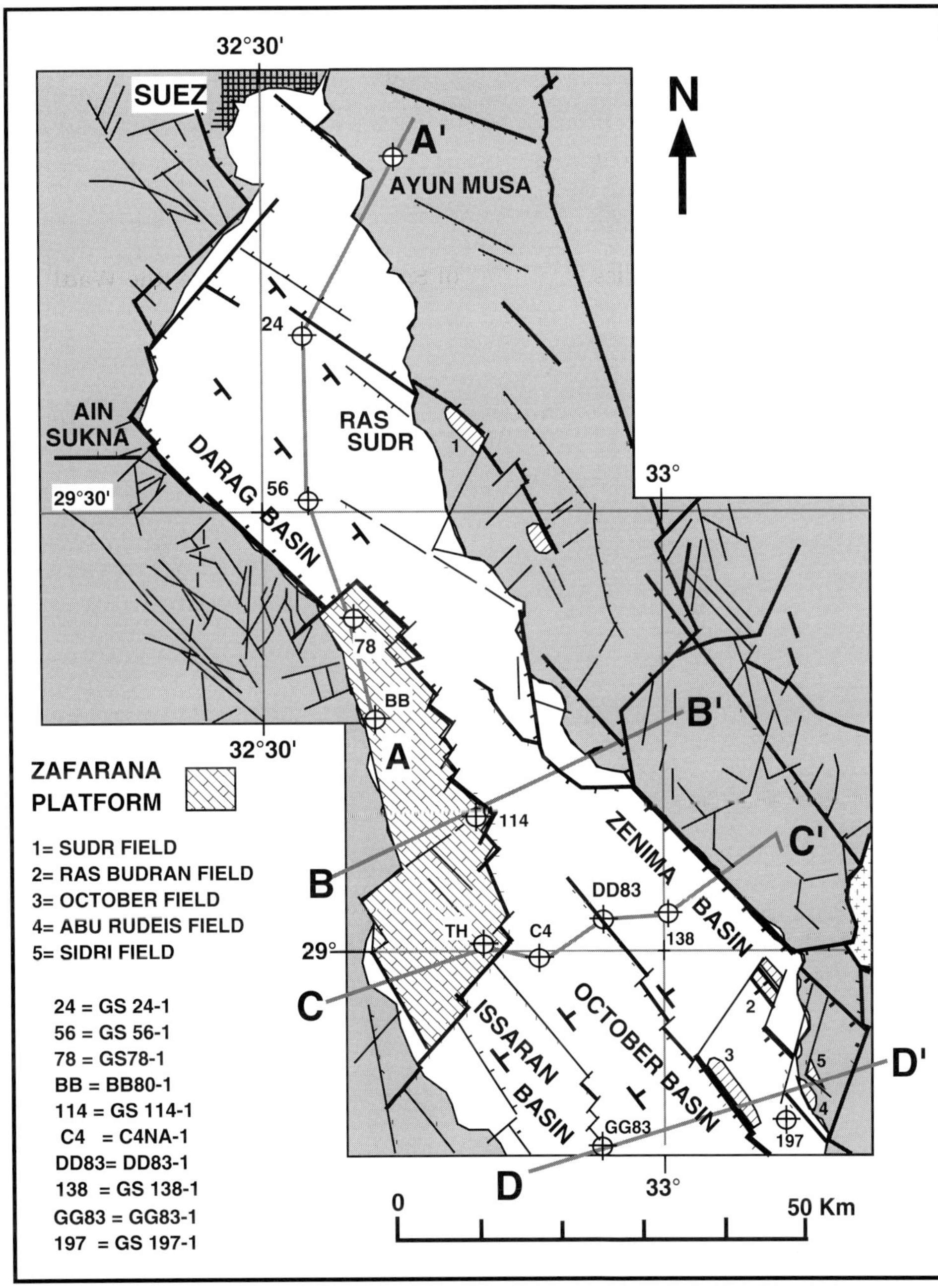

Figure 6. Structure of the northern Gulf of Suez.

The main faults at these basement ridges generally are linear, but are cut and offset by transverse faults, which dissect main faults into segments that are not necessarily aligned (Figures 3, 6). This misalignment is very obvious at the Gebel Araba block, where a major Gulf of Aqaba parallel fault offsets the basement in a left lateral direction. Farther north at Gebel Abu Durba and Gebel Nazzazat, other Aqaba parallel transverse faults have displacements of about 2000 m down to the north (Figures 6, 7).

Similar transverse faults are present at the southern end of Gebel Zeit and at Esh el Melaha; however, the offsets are not so obvious and probably smaller scale. The Aqaba trend seems to have less influence on the west side of the gulf, which is the more stable African block. In general, very little lateral displacement or shear is visible in the Gulf of Suez. The observation of transverse faults (or cross faults) in outcrops and on LANDSAT has been used to help interpret the subsurface structure (Sultan and Schütz, 1984) (Figures 3, 6, 7).

Gebel Zeit

Gebel Zeit (Figures 2, 7–10) is the crest of the steeply dipping pre-Miocene flank of the Zeit Bay basin and is one of the prominent basement outcrops along the Gulf of Suez coast. The main Gebel Zeit has a maximum elevation of 456 m above sea level. Excellent outcrops of all strata allow a detailed interpretation of the tectonic

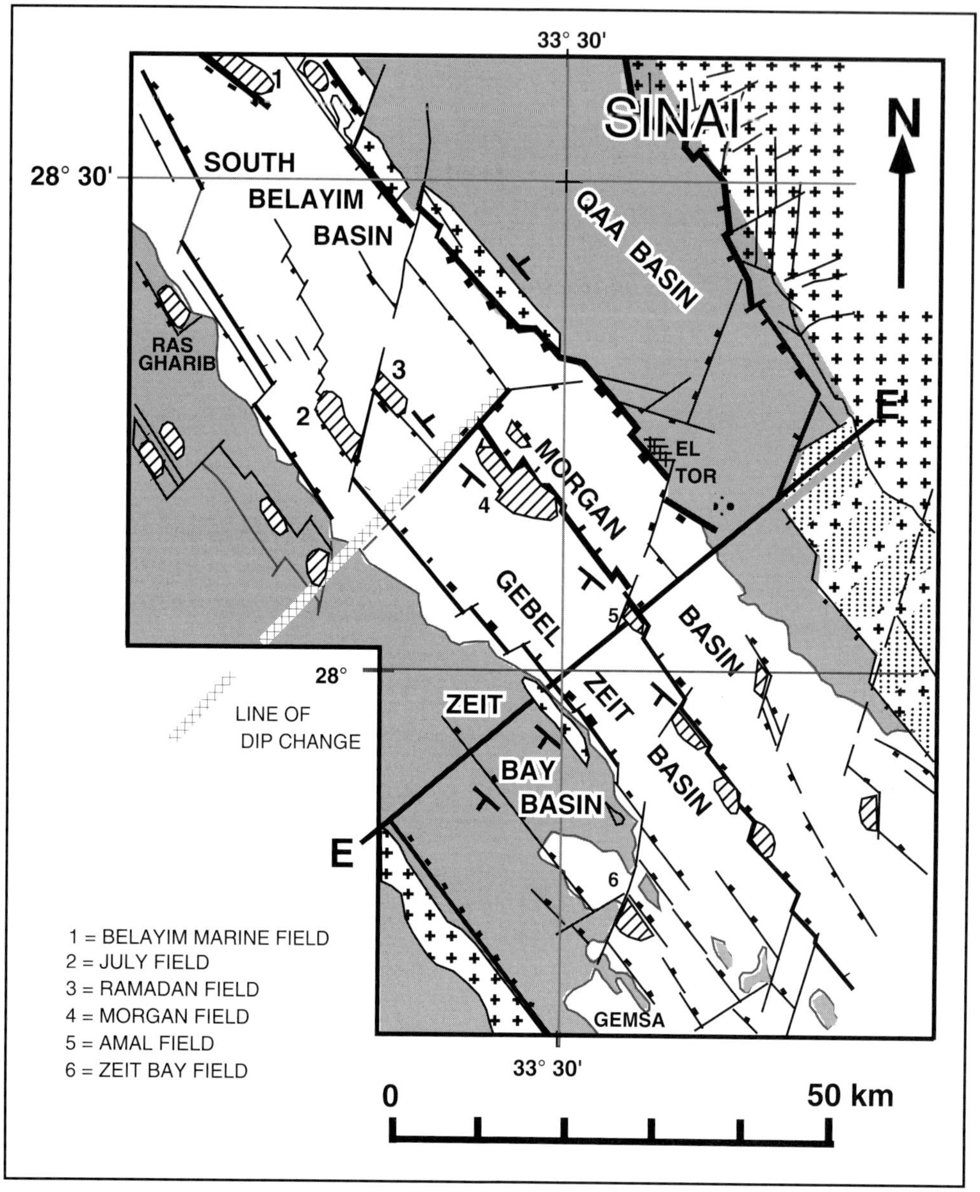

Figure 7. Structure of the southern Gulf of Suez.

history (Zankl and Klitzsch, 1985). A large area, probably bounded by old Jurassic east-west–trending faults, was gently tilted northward during the pre-Miocene and the southern elevated areas were eroded. "Encrustation of an irregular pre-Miocene relief by silica indicates a long-lasting influence of ascending solutions during an arid period of time. Depressions are filled with residual material, mainly cherty conglomerates and boulders" (Zankl and Klitzsch, 1985, p. 32).

During the early Miocene, the formation of elongated, narrow, gulf-parallel blocks was initiated. In the Aquitanian, a shallow-marine sea transgressed the initially gently tilted blocks. Shallow-water carbonates cover the remaining relief in the northwestern part of the main Gebel Zeit and little Gebel Zeit. The block of the little Gebel Zeit was down-faulted in relation to the southern edge of the main Gebel Zeit. In the lower Miocene, the tectonic activity increased and the Gebel Zeit block was further rotated to the west. Its crest, for a short time, was uplifted, subaerially exposed, and extensively eroded down to basement. During Rudeis deposition, the block was rapidly down-faulted and submerged. Submarine canyons developed along fractures perpendicular to the elongation of the blocks. On submarine relief directly on the basement, thin pelagic *Globigerina* limestones were deposited indicating relatively deep water. Pelagic marls ("*Globigerina* Marl") fill the relief of the Rudeis sea. Over the submerged high of the Gebel Zeit a relatively thin Rudeis Formation (several meters to 20 m) was deposited in contrast to adjacent lows, where the Rudeis Formation reaches over 1000 m. The Gebel Zeit high remained submerged, but received only reduced sedimentation during deposition of the

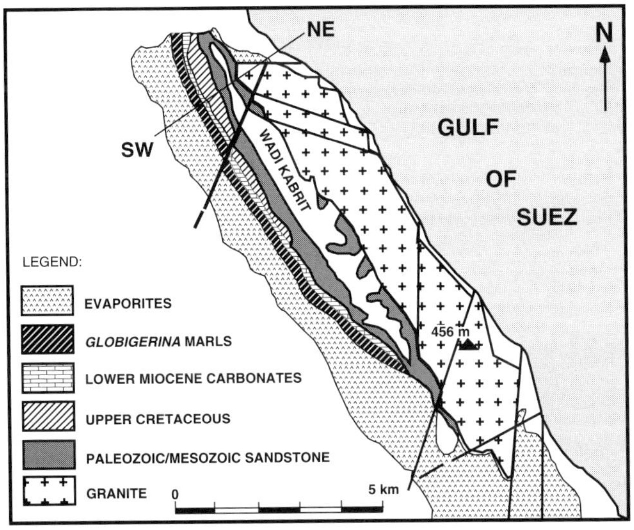

Figure 8. Gebel Zeit geological map.

Kareem and Belayim, and the evaporite deposition. Uplift and subsequent erosion of Miocene strata to basement occurred only after the Pliocene, as Pliocene-age shallow-water carbonates on the western flank prove.

During the Miocene, Gebel Zeit was continuously tilted westward, as the dip increase in the various formations demonstrate. Pre-Miocene strata attain dips of up to 45°. The evaporites strata dip approximately 15–20° and Pliocene sediments dip at least 5°. The enormous uplift of Gebel Zeit is due to isostatic movement. Displacement on the east-bounding fault is in the range of 5 km from the top basement on Gebel Zeit down to basement in the Gebel Zeit Basin (Figures 8–10).

Esh El Melaha Range

The Esh el Melaha Range is a smaller, but longer, edition of Gebel Zeit. Esh el Melaha is not uplifted as much as Gebel Zeit and possibly stayed with its crest

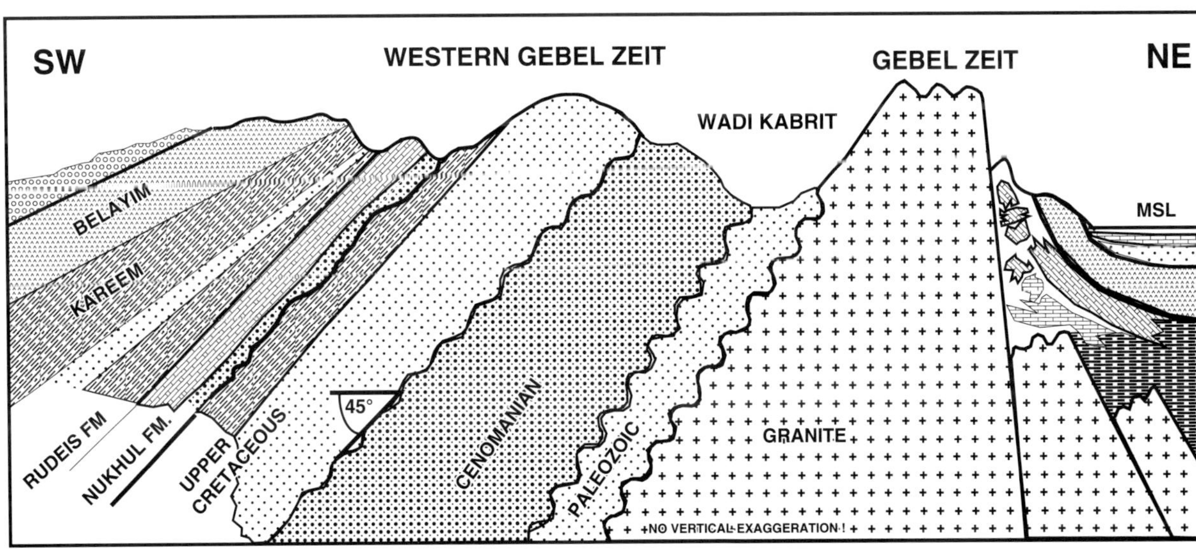

Figure 9. Geological cross section Gebel Zeit (from Zankl and Klitzsch, 1985).

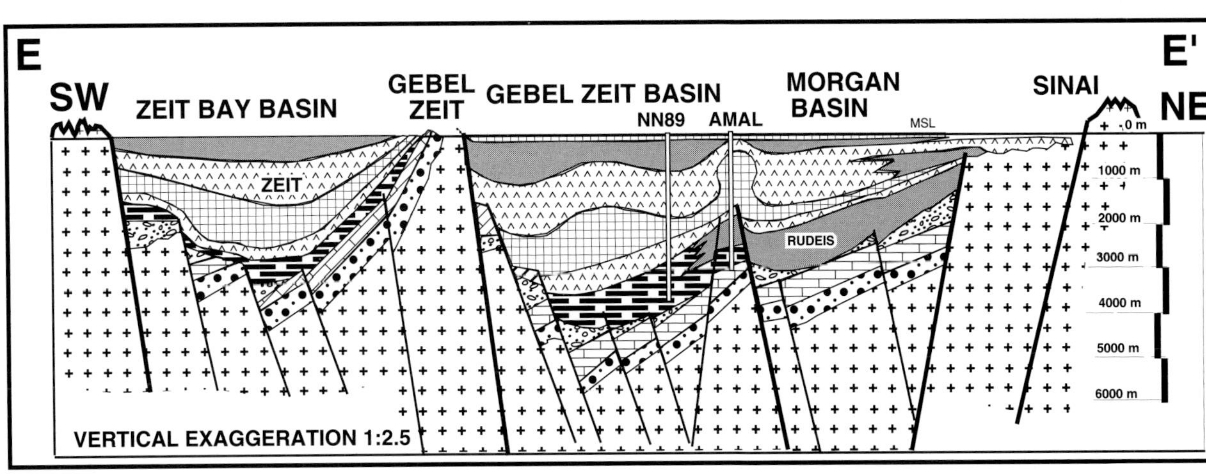

Figure 10. Geological cross section E-E′.

near sea level for long time intervals as the fringing Miocene reefs on its eastern side prove. Parts of the crest of Esh el Melaha were exposed subaerially during the early Miocene as the onlapping Miocene reefal limestones and granite boulder conglomerates within the limestones demonstrate. The recent uplift ranges between 50 and 100 m. On the western side, a thin layer of Paleozoic Nubia sandstones and Cretaceous sediments is present.

Gebel Araba/Gebel Abu Durba (Sinai)

Gebel Araba/Gebel Abu Durba (Figures 2, 7) is the largest, exposed, tilted fault block inside the Gulf of Suez graben. Basement is exposed along the coast over a distance of nearly 40 km. Well-exposed easterly dipping Nubia and Cretaceous sediments trail the high over 60 km as far north as Gebel Nazzazat. The basement is segmented by Aqaba-trending faults, which are the best-exposed transverse faults in outcrop. These faults, however, show very little horizontal displacement. The fault north of Gebel Abu Durba, in fact, dies out in the Upper Cretaceous sediment section south of Gebel Ekma.

The amount of exposure of this block during the Miocene is difficult to judge. The block probably was not eroded down to basement during the Miocene; however, it acted as major barrier for coarse clastic sediment transport to the west. The block kept most sands in the Qaa basin and sheltered the South Belayim Basin from sand input. Only at Belayim field was the barrier breached and sands deposited in upper Rudeis strata and later in Kareem strata at Belayim Marine field, providing additional hydrocarbon reservoirs. South of the Gebel Araba block near El Tor, a high amount of clastics was transported out into the Morgan Basin during Rudeis and Kareem deposition (Figures 11, 12). At its eastern flank and southern end, lithothamnion limestones, probably deposited at the same time as the upper Rudeis, Kareem, and Belayim formations, were deposited in a shallow-marine environment.

Onshore Plateaus

On the African side of the gulf, north of the Eastern Desert basement, are four major structural units (Figures 2, 6): Southern Galala Plateau, Wadi Araba, Northern Galala Plateau, and the Zafarana Platform.

The most striking character of these units are their east-northeast–trending boundaries, perpendicular to the gulf axis. The four units appear to be shifted toward the east, one in respect to the other. This apparent right-lateral shifting corresponds with the east-northeast–trending boundaries, thereby indicating an important control of the larger overall structure by an east-northeast trend, perpendicular to the clysmic trend (Figure 5).

Southern and Northern Galala Plateaus: The Northern and Southern Galala plateaus are mostly covered by Eocene limestone. Both plateaus are slightly tilted to the south as the modern drainage pattern suggests. Originally, these blocks were flat or even tilted to the north as an older drainage pattern proves. Most of the faults visible on the plateaus are of Gulf of Suez trend, but have no apparent throw. Only the faults closer to the gulf show displacements down to the east.

The Northern Galala Plateau is bounded on the north by a series of faults trending east-west and east-northeast, which intersect the clysmic trend. The eastern boundary is the major Ain Sukna fault.

Wadi Araba: The Wadi Araba east-northeast–west-southwest–trending "horst" (Figures 3, 13) crosses the Miocene clysmic gulf at nearly a right angle. The Wadi Araba is bounded to the north and south by the two Galala escarpments. The base of the Wadi is Nubian sandstones mainly belonging to the Carboniferous. The wadi, although strongly modified by erosion, appears flat and the dips of the Nubia sandstones are always very gentle. In the westernmost part, the Nubia shows steep dips and small folds, which appear related to local faulting. The southern boundary of the horst is marked by a fault that separates the Southern Galala Plateau from the valley. Along this fault, especially close to the St. Anthony Monastery, near-vertical dips in Cretaceous sediments south of the fault indicate drape over the fault, which is downthrown to the south (Figure 13).

The origin of this cross trend is obvious when examining the Western Desert structural trends, such as the North Qattara ridge, the Abu Gharadig basin, and Bahariya, but it is best expressed in the Kattaniya horst west of Cairo. These Western Desert structural features are clearly of Jurassic age, which also is true for the Wadi Araba (Figure 1).

Wadi Araba, similar to the Kattaniya horst, was a high block throughout the Jurassic, Cretaceous, and Neogene that shed sediments to the north and south. This shedding is best documented south of Gebel Galala near Thelemet and St. Paul Monastery by the Eocene limestone conglomerates and reworked shallow-water nummulitic sediments in a basinal limestone; these sediments document a paleoslope away from Wadi Araba. These sediments were slump masses that accumulated south of the major fault bounding the Wadi Araba to the south. On top of the Wadi Araba during the Eocene, a shallow-water environment must have existed, as the large *Nummulites* in the slumps prove. In some layers, large *Nummulites* are mixed with small *Nummulites* or occasionally are sorted (only large species float in a lime-mud matrix).

Similar slump masses with synsedimentary folds are present at the base of the Eocene and within the Sudr Chalk, proving a strong relief during the Late Cretaceous, also. The Wadi Araba seemed to act as a transport barrier for Nubia sediments as well. South of Wadi Araba, thick Nubia "A" sandstones occur and are present only in reduced thickness north of it.

Today, the Wadi Araba horst is a geomorphological depression. The core of the horst is eroded down to the Carboniferous Nubia Formation (Figure 13). The bounding fault south of Wadi Araba is well defined. Steeply dipping Cretaceous marls and

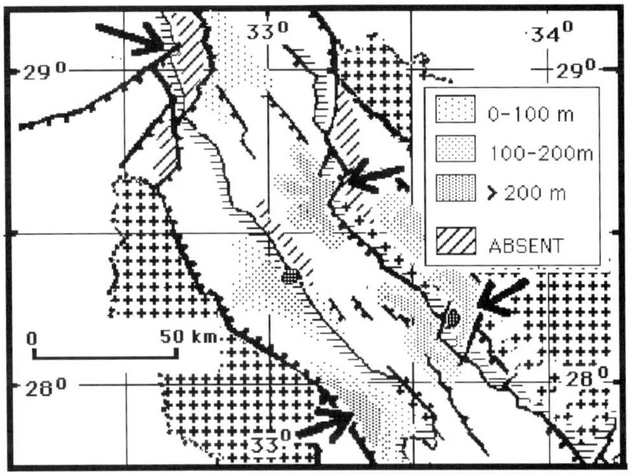

Figure 11. Upper Rudeis sand isopach map.

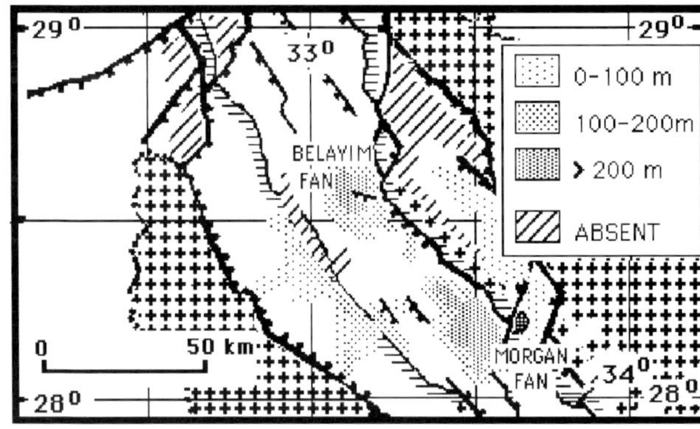

Figure 12. Kareem sand isopach map.

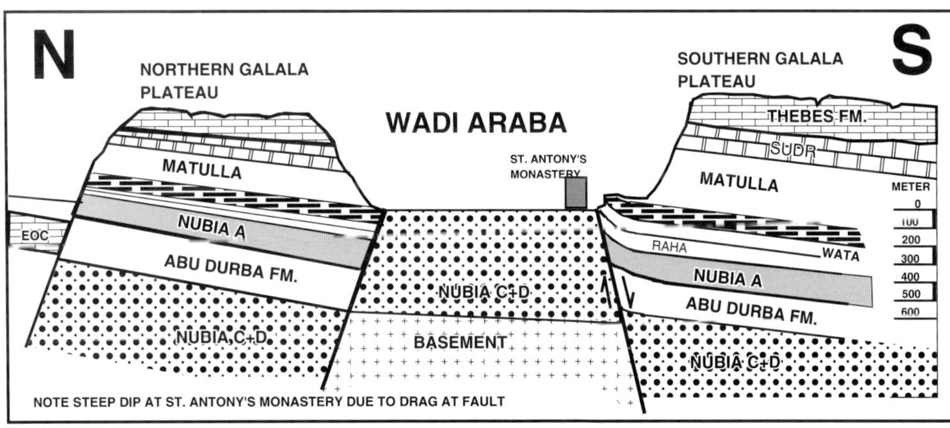

Figure 13. Geological north-south cross section Wadi Araba.

limestones mark the fault at the outcrops near the St. Anthony Monastery. Vertical displacement today seems to be 500 m. During geological time, this displacement was possibly greater, but was readjusted after the Miocene(?). The fault probably has a left-lateral wrench component of several hundred meters.

Zafarana Platform: The Wadi Araba possesses a slightly eastward-dipping axis and extends into the offshore. There, it is called the Zafarana Platform, which is relatively well known now as a result of several exploration wells. The Zafarana Platform has a very thin Miocene and post-Miocene cover without Miocene evaporites; however, nullipore-type limestone/dolomite facies or sands are present.

The pre-Miocene on the Zafarana Platform is generally not deeply eroded, but we assume deep karst erosion of Eocene carbonates because this area must have been exposed for long periods during the Miocene. Nubia sandstones are generally preserved and provide an excellent migration path for oil out of the basin and for freshwater inflow from Wadi Araba.

Nondeposition or erosion took place during the Miocene. Near the rim of the Zafarana Platform, thick sands occur in the Rudeis and Kareem formations as observed in well GS 114-1A (Figure 6). These sands originated in the Wadi Araba and were shed from the west to the east and southeast (Figure 11). During the Miocene, the Zafarana Platform extended farther eastward to the southern continuation of the Ain Sukna fault. After the middle Miocene, this fault line was set back westward in a zigzag pattern near well GS 114-1A. The southern limit of the Zafarana Platform is a N60°E cross fault, which extends from St. Paul Monastery northeastward near well Thelemet-1. This fault has a displacement of 250–500 m down to the south and, due to its significant throw, is one of the most visible cross faults on seismic in the Gulf of Suez (Sultan and Schütz, 1984). Strong Miocene facies changes (shales to nullipore limestones/dolomites) accompany this fault line (Figure 14).

In conclusion, the Wadi Araba/Zafarana structure is a positive feature throughout time, and played a major role in blocking the southern part of the Gulf of Suez and the Red Sea from the eastern Mediterranean during the middle and late Miocene.

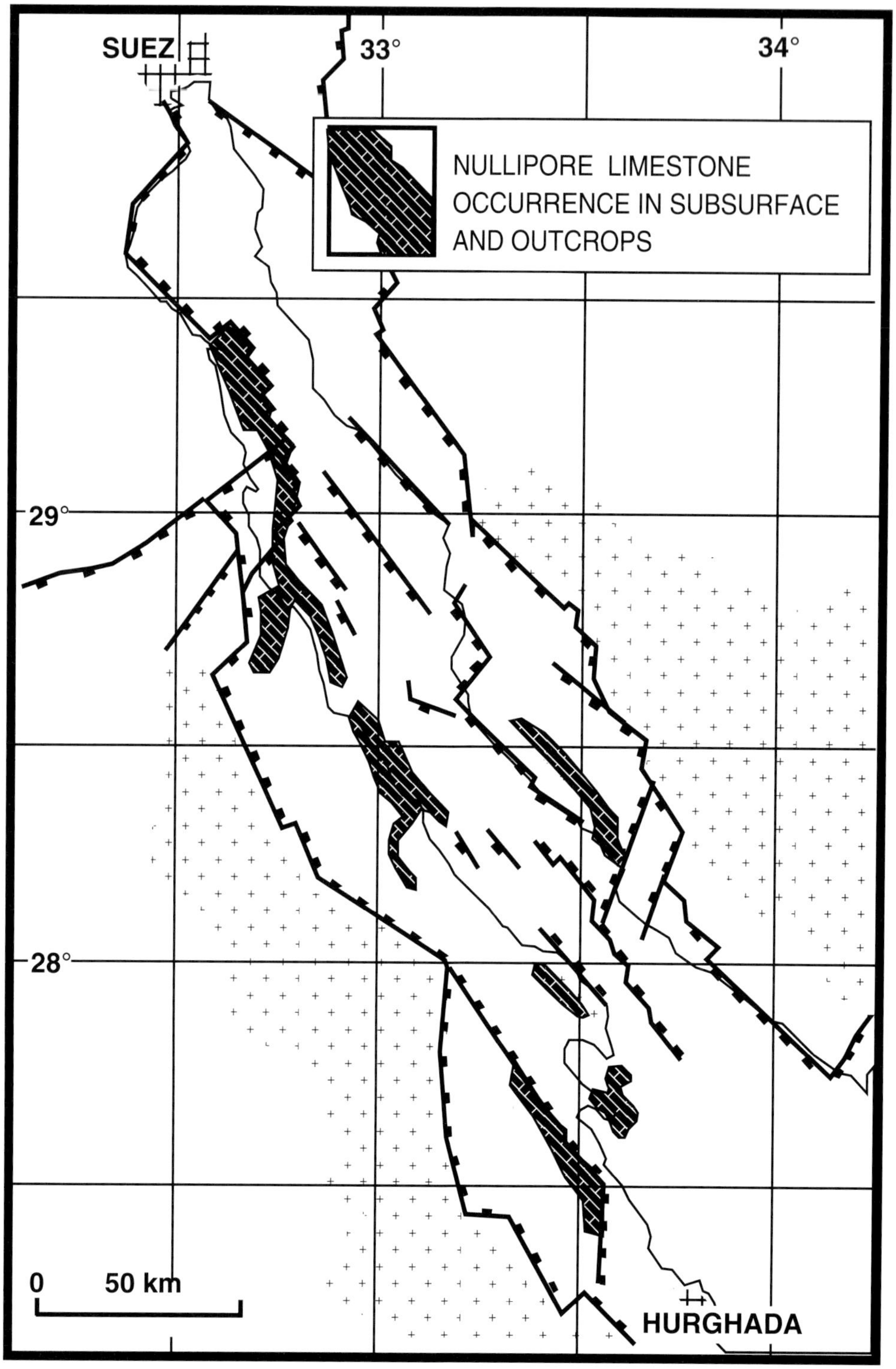

Figure 14. Nullipore limestone.

Miocene Subbasins

Northern Gulf

Darag Basin: The Darag Basin is a large, southwest-dipping half-graben that covers the whole northern part of the Gulf of Suez. The basin is as deep as the other Gulf of Suez half-grabens, but has an independent history and simple structural style. It is bounded to the southwest by a major fault, the Ain Sukna fault, which has a displacement in the range of 3600 m. The long southwest-dipping flank is broken by the Sudr-GS 9-1 trend (Figure 15).

Unique to the Darag Basin is the presence of Jurassic sediments that thicken from several meters in

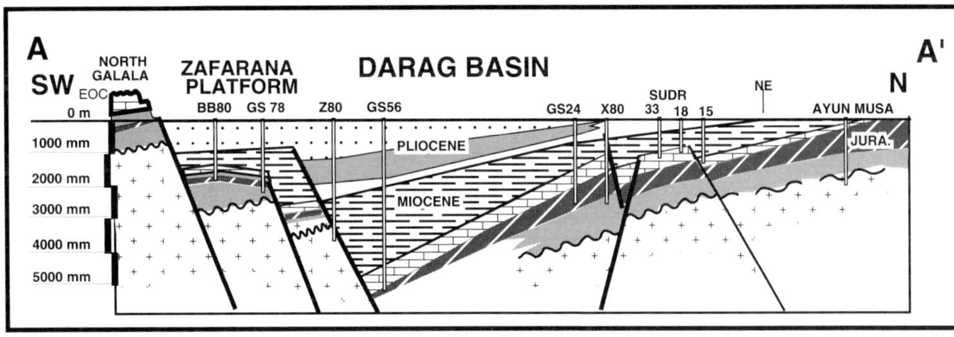

Figure 15. Geological cross section A-A´.

the southern part near Ras Abu Darag to several hundred meters in the north near well GS 24-1, Ayun Musa-2, and North Sinai (Figures 5, 6). South of the Darag Basin, Jurassic marine sediments are absent and continental sediments are very thin and generally not well documented.

Coarse clastic material is scarce in the lower Miocene Rudeis section (Figures 15, 16). The extremely thin Kareem/Belayim/South Gharib section and the lack of thick evaporites indicate a time of nonsubsidence and low deposition in the middle Miocene (Figure 15). During this interval, the Darag Basin was probably cut off from the rest of the Gulf of Suez by the Zafarana Platform, which acted as a barrier between the Mediterranean and the Red Sea. A narrow strait probably existed during the entire Miocene between Ras Sudr and Lagia, connecting the Mediterranean with the Gulf of Suez and the Red Sea (Mitla Strait, Garfunkel and Bartov, 1977). During evaporite deposition in the Gulf of Suez and the Red Sea this marine connection was cut off.

After the Miocene, the Darag Basin subsided again with high sedimentation rates, but was cut off again during the Pliocene, as the Pliocene evaporites in the Darag Basin confirm.

Ain Sukna Fault: The Ain Sukna fault is a major, important, old structural element in the Gulf of Suez. It is only a part of a much longer fault system, which cuts through the Gulf of Suez to the southeast along the west coast of Sinai near Lagia and farther into the Sinai basement up into Wadi Feiran (Figures 5, 6). This fault is probably an older Jurassic fault, but reactivation occurred during the early Miocene, resulting in major displacement. The Ain Sukna remained relatively inactive during middle and late Miocene, but subsided again 2000 m after the Pliocene. An element important in understanding this fault is the complicated fault system offshore Lagia, where the fault reverses its throw (Figure 17). The original depocenter was located northeast of Lagia with Rudeis sediment having a thickness of 1300 m, whereas southwest of Lagia (wells GS 101-1, CLB-1) relatively thin Rudeis sediments with a thickness of less than 300 m were laid down. The area southwest of Lagia was part of the Zafarana Platform during the early Miocene. The throw of the fault was reversed during the middle Miocene and has continued up to the Holocene, downdropping sediments to the southwest, with a displacement of up to 1800 m at base evaporites. This point near Lagia was a critical point during the middle Miocene (starting in the Langhian, important during Belayim and South Gharib deposition) for the marine conditions of Gulf of Suez because it provided a barrier between the Darag Basin (and Mediterranean Sea) and the southern part of the Gulf of Suez and the Red Sea so that the large evaporite basins could develop.

This fault is remarkable for the volcanic dikes associated with it that were penetrated by well Galala-1 and were found in outcrops near Abu Zenima, as well as hot springs at Ain Sukna on the west coast and at Hamman Faraun, south of Lagia on the east coast.

Central Gulf

The central Gulf of Suez includes the Issaran–October and Abu Zenima basins, the West Bakr Basin, the South Belayim Basin, and the El Qaa basin (Figures 6, 7). These half-grabens show northeasterly dips at the pre-Miocene level and the Miocene sequence developed similarly. The stratigraphy (Figure 4) is representative for most of the central and southern parts of the Gulf of Suez.

The pre-Miocene section is without anomalies as described in the stratigraphy, except that the Eocene has a depocenter in this area exceeding a thickness of 300 m (Figure 18).

The Miocene section starts with the generally thin conglomeratic Nukhul formation with some sand streaks, but without an anhydrite member. The Rudeis Formation is generally thick (300–600 m) and contains nearly no sand except along the margin of the basin. The Kareem Formation consists of a thin anhydrite member (Markha Member) and a 60- to 300-m-thick Shagar Shale Member.

The Belayim Formation is typical for its four-fold subdivision: Hamman Faraun Member, Feiran Member, Sidri Member, and Baba Member.

During Belayim deposition, the Gulf of Suez remained tectonically relatively quiet and thicknesses range from 150 to 300 m with very little lateral lithological variations.

The South Gharib Formation in the Issaran–October–Abu Zenima basins is a series of intercala-

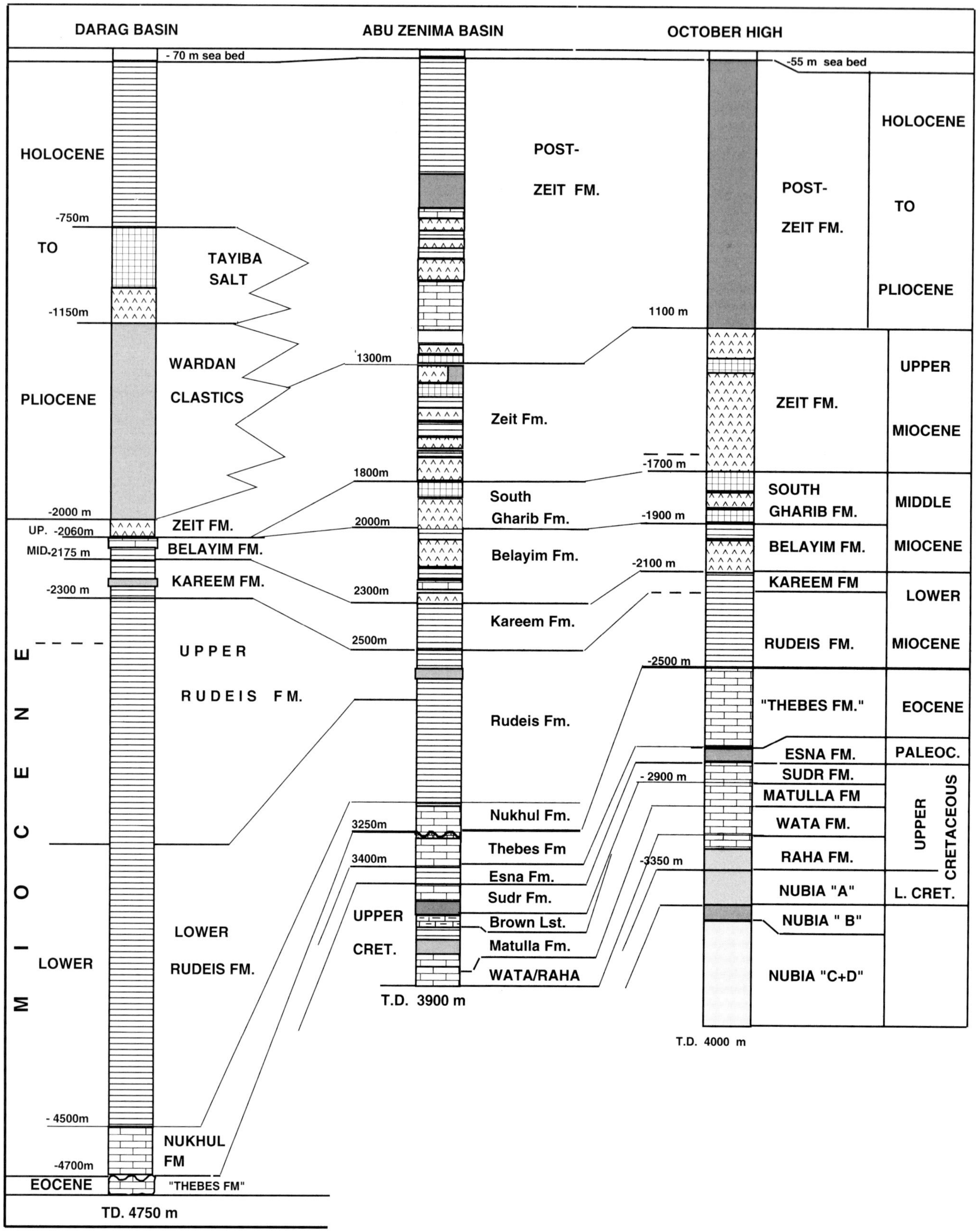

Figure 16. Type log correlation of Darag Basin, Abu Zenima Basin, October high.

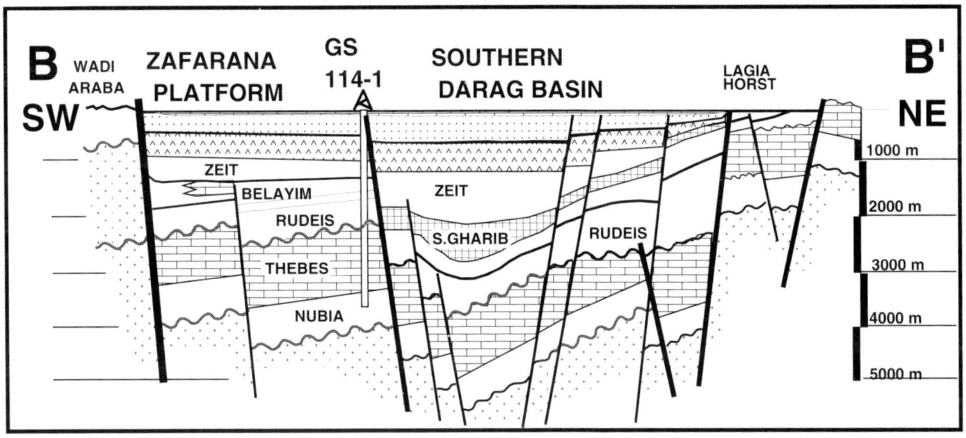

Figure 17. Geological cross section B-B′.

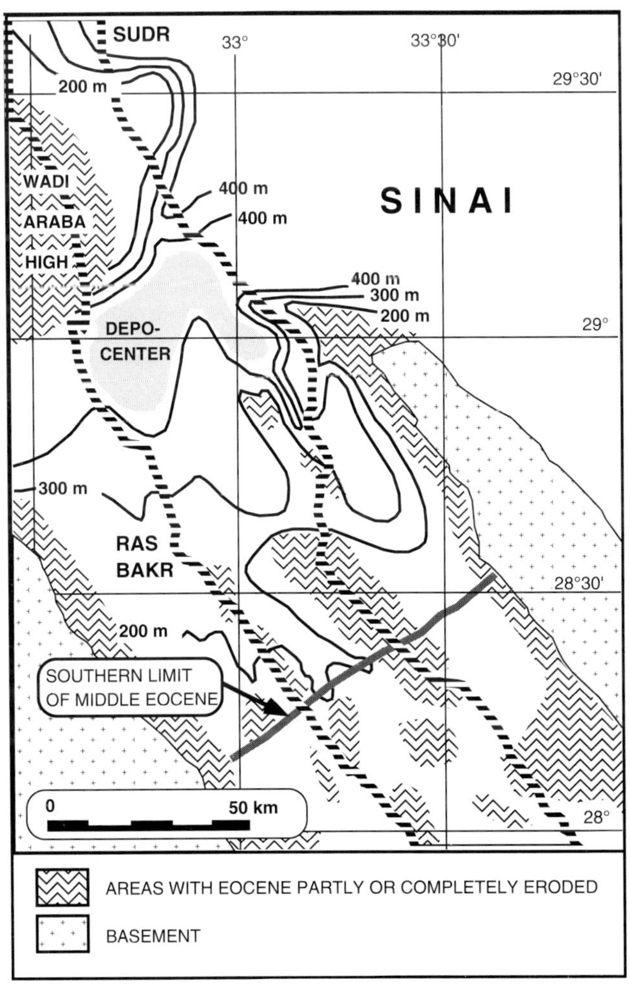

Figure 18. Total Eocene isopach map (after Breman, 1984).

tions of anhydrite, salt, and thin shales with a thickness averaging 300 m.

In the Zeit Formation, the evaporite depositional environment continues, however, with much less salt deposition. The topmost salt layer (the "first" salt) contains a potassium salt layer of 1–3 m, which allows for good log correlation. The Zeit Formation thickness ranges from 400 to over 600 m.

Issaran–October–Abu Zenima Basins: These three parallel adjoining basins are located in the broadest part of the Gulf of Suez and provide the most typical type of deep basins structurally and sedimentologically in the Gulf of Suez. All three basins are half-grabens with northeasterly-dipping long flanks (Figure 6). The crests of the flanks, the October high and the northern extension of the Amer-Bakr-Gharib high, are penetrated by many exploration wells and are geologically and seismically well known. On the flanks, only limited geological information is available, and from the deeper parts of the basin only the Abu Zenima Basin possesses two deep exploration wells, GS 138-1 and C3A-1. The deepest part of the October Basin is untouched by drilling and, hence, practically unknown.

The Issaran Basin is bounded to the west by a fault and to the east by the GG-83 trend. To the south, the basin terminates in the GG-83 area along a major cross fault (Figure 6). Only a narrow graben leads south into the West Bakr basin. To the north, the basin is limited by the Zafarana Platform. The Issaran Basin is relatively shallow, with its deepest part west of the GG-83 area where the top Eocene reaches 3000 m thick.

The October Basin is the largest and probably the deepest of all basins in the Gulf. This basin adjoins the Issaran Basin to the east, where the 20-km-broad flank plunges to the east down to more than 5000 m at top Eocene (or 6000 m at top basement). The October Basin is bounded to the east by the DD 83-1-October trend. A structural high at Belayim Marine field forms the southern limit of the basin; however, during the Miocene, this crossing high block did not exist in these proportions. To the north, the basin narrows and ends south of the Zafarana Platform near GS 114-1A. The deeper portion of the October Basin is the least well known because it does not have any deep wells. A magnetic anomaly is present in the deepest part, west of Feiran, possibly as a result of the presence of a basalt. The nearest basalt dikes are known

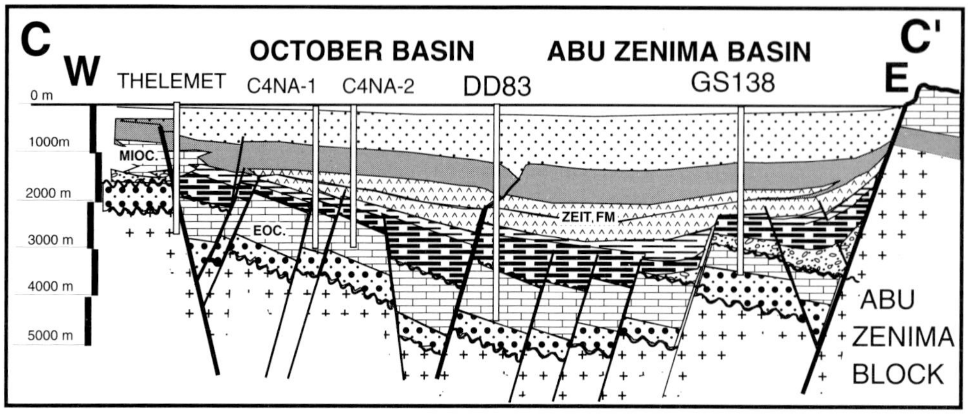

Figure 19. Geological cross section C-C′.

from Ras Budran field and outcrops at Abu Zenima. The low axis west of the October high is offset at several points indicating cross elements (Sultan and Schütz, 1984).

The Abu Zenima Basin is the eastern deep basin with northeast-dipping fault blocks. Top Eocene is located as deep as 4500–5000 m at the eastern limit near the Abu Zenima fault. Seismic mapping of the pre-Miocene horizons in the basin north of the Ras Budran high shows right-lateral offset of the pre-Miocene fault blocks and the Miocene low axis. Aqaba-parallel faults delineate and create the northeast-dipping fault blocks of the Ras Budran field.

October High: The October high trend is similar to the "B"-trend in the southern part of the Gulf of Suez, one of the best-known subsurface high trends. The October high plunges from a depth of 3000 m at the top of the Eocene near the October field to more than 4000 m north of well DD 83-1 (Figures 3, 6, 19). The high is bounded to the west by a major clysmic fault, which is downthrown to the west with a maximum displacement of 2000 m. Several intermediate fault blocks, 100–250 m wide, accompany the main fault.

The trend is cut by several major transverse faults. One major east-west–trending transverse fault delineates the October trend to the south and offsets the main fault eastward to the Feiran area. This offset is not associated with a strike-slip component, but probably represents an older (Jurassic?) structural direction. North of the October field, several Gulf of Aqaba–parallel faults cut the high and are left-laterally offset. The same faults can be traced into the Ras Budran field. Other N60°E–trending faults cross the trend and displace the southern block downward, forming traps for oil migrating along the trend.

West Bakr Basin: The West Bakr basin is located between the Western Desert basement outcrops and the coast (Figure 3). Toward the north, the basin shallows and is bounded by the South Galala mountains. To the south, the West Bakr continues west of Gebel Zeit into the Zeit Bay basin. The boundary between these two basins is defined by the dip change from northeasterly dips in the West Bakr basin to southwesterly dips in the Zeit Bay basin. This line passes from north of Morgan field through Shukheir Bay, onshore toward Gebel Gharamul (Figure 7).

Structurally, the basin is complex and consists of narrow horsts and half-grabens with northeast-dipping flanks. Remarkably, this onshore basin is still actively subsiding, as sabkha depressions west of Ras Amer and south of Ras Gharib show. As a basement-bordering basin, it contains large amounts of coarse clastic material, mainly sand throughout the Miocene section. In the Rudeis Formation, gross sand thicknesses in wells west of Ras Gharib field reach 500 m.

Amer-Bakr-Gharib High: A broad, high, horst-type block, the Amer-Bakr-Gharib high separates the West Bakr basin from the October and South Belayim basins and acted as an important barrier for sediment transport from the west during the Miocene. At its crest, the top pre-Miocene rises to 600 m near Ras Gharib (Figures 3, 7), 1000 m near Ras Bakr, and 750–1000 m at Amer field. The high was severely eroded, exposing Nubia sandstones during the early Miocene at Ras Gharib and Bakr fields, and it stayed high throughout the Miocene. Nukhul, Rudeis, and Kareem formations are very thin and commonly contain limestones. During Belayim deposition, nullipore carbonates were deposited on the highest parts east of the Ras Gharib field at Ras Fanar and at Bakr field. Over the high, only anhydrite was laid down during Belayim, South Gharib, and Zeit formation deposition, and strong facies changes occur surrounding the Amer-Bakr-Gharib high.

EE 82-FF/GG 83 High: The crestal part of the long October Basin flank is bounded to the west by a double fault system, the faults being separated by approximately 1 km. The western fault is the older fault and has a displacement down to the west of about 300–600 m (Figures 6, 20). Erosion capped the 1-km-wide block-tilted fault block at its crest near the western fault. In one of the FF/GG 83 exploration wells, erosion cut down to the Sudr Chalk.

Unlike the Amer-Bakr-Gharib high, the erosion of the pre-Miocene on this high trend is less severe. The eastern fault is younger, has less displacement, and

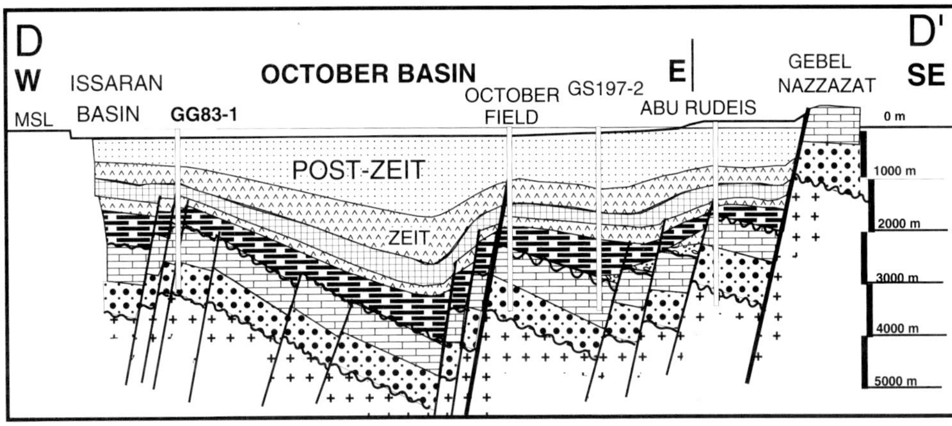

Figure 20. Geological cross section D-D´.

was probably formed when the tilt of the long flank exceeded a certain amount and the older fault became too flat to release tension. So, a new set of faults with a steeper fault hade formed. (This phenomenon can be observed along many major faults in the Gulf of Suez as blocks commonly are tilted more than 30° and thereby older faults reach fault hades of less than 55°. When more than 55° is reached, the overburden forces merge both sides of the fault and a new fault is formed. Depending on the amount of throw, a wider or smaller fault block is formed.)

The axis of the EE 82-FF/GG-83 high plunges from the north near well Thelemet-1 to the south to well FF 83-3. From there, it rises toward the GG-83 area and farther to the HH-82 high, which can be taken as the southern continuation of this trend. Associated, and typical for the EE 82-FF/GG-83 high, is a listric fault in the post-Belayim section. This listric fault was initiated by the deeper fault system and because of increasing tension with the widening gulf, the more brittle anhydrites broke and slid downward on the basal salt layer. This growth fault masks deeper seismic events and creates a problem for seismic interpretation.

South Belayim Basin: Originally the October Basin extended farther south into the Belayim Basin. The South Belayim's western flank is still unbroken; however, during Kareem deposition, the structure of the Belayim Marine field was uplifted and now separates the two basins (Figure 7). This is documented by significant facies and thickness changes from the Kareem through the Zeit Formation.

The Belayim marine high bounds the South Belayim Basin to the north. To the east, the basin is delimited by the Gebel Abu Durba fault. To the west, the South Belayim Basin is bounded by the Amer-Bakr-Gharib high. To the south, the basin is open toward the Morgan Basin and the Gebel Zeit Basin. However, the prevailing northeasterly dip of the central Gulf of Suez changes to southwesterly dips south of the Ramadan and July fields. The dividing line of this important dip change runs north of the Morgan field across the LL 87-1 area toward Shukheir and Gebel Gharamul, where synsedimentary slump masses in Cretaceous sediments indicate an old tectonic heritage.

A unique feature of the South Belayim Basin is the bending of its western flank near the exploration wells East Bakr M-1 and 2. The northeast-dipping strata change direction to southerly dips and then turn back into northeasterly dips south of the wells. Thus, the flank is offset approximately 2 km. South of the bend, the flank gradually steepens and is replaced by a major fault southeast of Ras Fanar.

El Qaa Basin: The El Qaa basin is a relatively unexplored basin; however, seismic and the few bore holes indicate a simple half-graben with a long flank rising up to the Gebel Araba/Abu Durba and Gebel Nazzazat (Figures 7, 21). The pre-Miocene is well known along the outcrops west and east of the basin. The Miocene sequence differs from most of the Gulf of Suez basins by its extremely high sand content due to its proximity to the Sinai basement uplift. The Gebel Araba high remained a high throughout the Miocene and trapped sand input from the east, preventing sand transport farther west. Only near El Tor was sand shed to the west into the Morgan area (Figures 11, 12) and marine ingressions occurred. Another gap, however much shorter lived, existed near Belayim Bay (Figures 11, 12). Structurally, the basin is well defined by faults to the east and southeast.

The main fault bounding the basin along the Sinai basement has a displacement down to the west of 4000–5000 m and is comparable in size to the Gebel Zeit fault. This large displacement results from the large uplift of the southern part of Sinai. Toward Ras Mohamed (Figure 3) the onshore area is underlain by a shallow basement block, which extends as far west as the coast.

The basin axis of the Qaa basin runs gulf-parallel close to the Sinai basement, and plunges from east of Gebel Nazzazat to its deepest part east of El Tor. There the low turns into a Gulf of Aqaba direction and joins the Morgan Basin south of El Tor. The break of direction is due to a major Aqaba-parallel fault that bounds the basin to the southeast (Figures 3, 7).

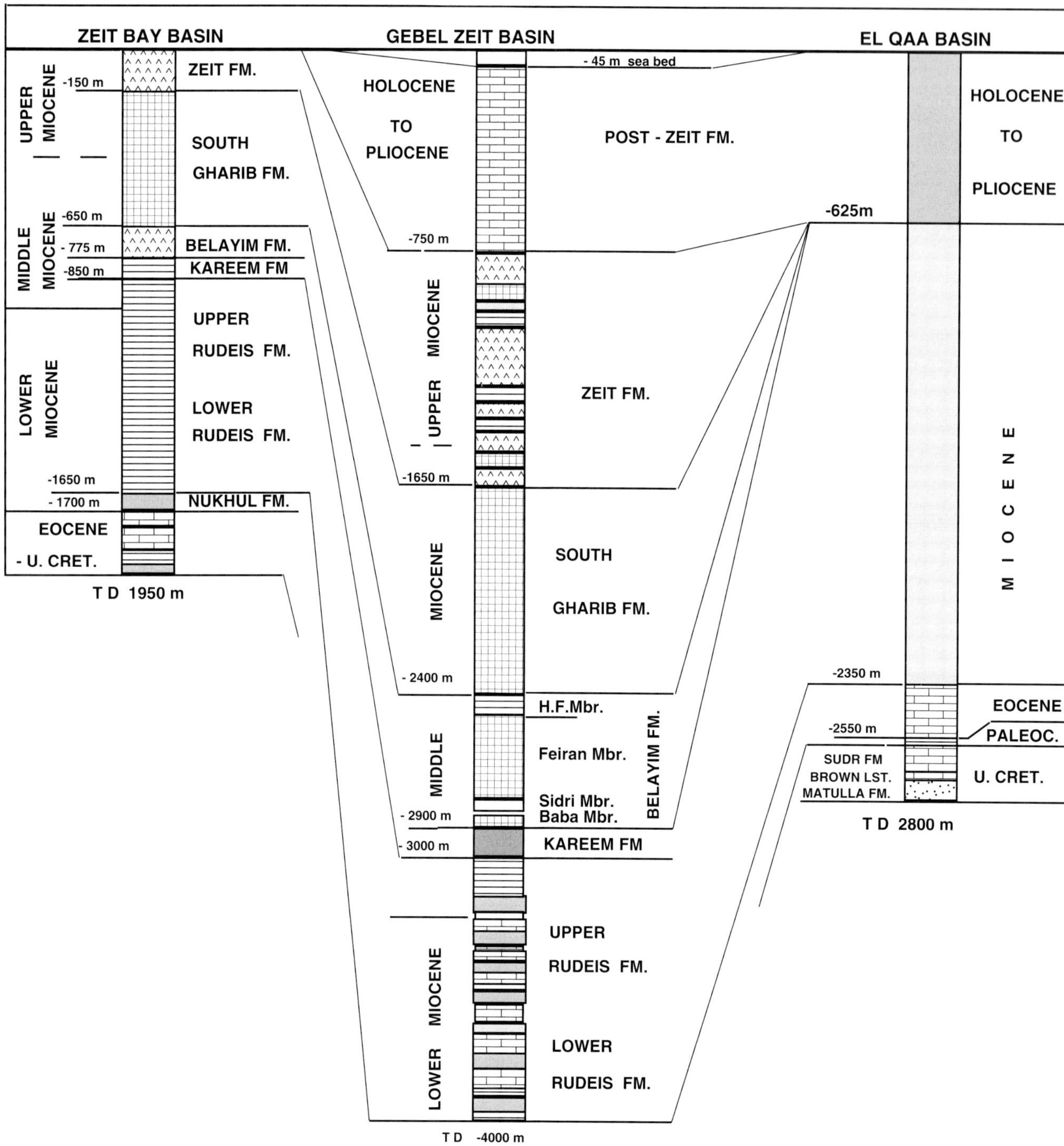

Figure 21. Type log correlation of Zeit Bay basin, Gebel Zeit Basin, El Qaa basin.

Southern Gulf

The southern Gulf of Suez comprises the Morgan Basin, the Gebel Zeit Basin with the "B"-trend, the Zeit Bay Basin, Gebel Zeit, Esh el Melaha range, and the Hurghada/Islands area (Figures 3, 7). To the south, the Miocene basins continue into the Red Sea. The predominant dip of the pre-Miocene fault blocks in this area is to the southwest. The northern limit of the southwesterly dips crosses the gulf from north of the Morgan field to Shukheir on the west coast and continues toward Gebel Gharamul (Figure 7). This structural boundary is probably of Jurassic origin.

Morgan Basin: The Morgan Basin is located between the Sinai coast and the "B"-trend (Figure 7). The northern part of the basin may be the only full graben with two bounding faults in the Gulf of Suez. This graben has been an exceptionally fast subsiding depocenter east of the Morgan field, which was rapidly filled with sands derived from the El Tor point source. The bounding faults are the Gebel Araba–El

Tor fault to the east and, to the west, the "B"-trend fault. Crossing faults of Aqaba orientation created a cutoff for the subsiding compartment to the south and aided the subsidence. To the north, the possible Jurassic-aged fault, where the dip change occurred, might have acted the same way.

Gebel Zeit Basin: The Gebel Zeit Basin is a major, long half-graben with steep dips to the southwest. It is bounded to the west by a major fault against Gebel Zeit, which displaces basement approximately 5000 m (Figures 7, 10). To the east, the long flank rises to the "B"-trend. To the north the basin continues into the south Belayim basin, and to the south it is connected to the Hurghada/Islands area and the Red Sea.

The basin contains the thickest Miocene sediments and evaporites in the Gulf of Suez and salt diapirs could develop only here. However, only few wells have been drilled in the Gebel Zeit Basin, so geological knowledge is sparse. The deep well NN 89-1 (Figures 10, 21) proves extremely thick Rudeis Formation (1000 m) with many sand streaks, but the base of the Miocene was not reached at total depth.

The salt in the Belayim Formation (Feiran Member, 390 m of salt) is thickened due to salt pillowing along the steep flank of the "B"-trend. Although this salt does not represent the original salt thickness, it proves that salt in the 100-m-thick range was laid down in the depocenter of this basin. A much thicker salt pillow within the Zeit Formation developed in the same basin axis, documented only on seismic lines (Figure 10). Different from the half-grabens to the north, the pre-Miocene strata of Gebel Zeit (as well as the Zeit Bay Basin and the Hurghada/Islands area) are tilted from over 25 up to 45° (Figure 9). These high dips cannot be recognized on seismic, and it was a hard and expensive lesson for explorationists to accept these steep dips. Outcrops of steeply dipping Kareem/Belayim formations on the southern flank of Gebel Zeit were obvious evidence of the steep pre-Miocene dips (Figure 9).

In the southern part of the Gebel Zeit Basin near Ashrafi (well Ashrafi East-1) a narrow, high basement block covered with Rudeis Formation was drilled. Pre-Miocene strata are missing and basement is deeply weathered, indicating that some basement blocks were uplifted more than 1500 m prior to the Burdigalian.

"B"-Trend: The crest of the southwest-dipping long flank of the Gebel Zeit Basin is called "B"-trend. It is probably the best known area in the gulf, as many exploration wells and oil fields are located on it. The crest of the pre-Miocene blocks rise to 3000 m and locally as high as 2600 m (e.g., Amal, GS 382-1). The pre-Miocene crest is lined up with South Gharib salt pillows and diapirs (e.g., Amal), which were initiated by the continuous tilting and faulting after deposition of the salt. During the Burdigalian and Langhian (Rudeis and Kareem formations) the east-bounding fault acted as a barrier for sand transport from the east and only few sands spilled over the crest (except at Morgan field). The "B"-trend is segmented by cross faults and the individual blocks acted as traps for the migrating oil. The whole trend seems to plunge slightly to the south, so the oil generated in the Gebel Zeit Basin migrated north in the Nubian sandstone conduit.

Zeit Bay Basin: The first exploration wells in Egypt were drilled onshore in the Zeit Bay basin at the end of the last century and the beginning of this century, with little success except for the Gemsa oil field (Figure 7). The discovery of the Zeit Bay field in 1980 proved the old explorationists right.

The Zeit Bay basin extends onshore parallel to the Gebel Zeit mountain. To the north, the basin continues to the West Bakr basin. To the west and the east, basement outcrops of the Esh el Melaha mountains and Gebel Zeit delineate the basin. To the south, the Zeit Bay basin extends into the Hurghada/Island area.

The main features of the Zeit Bay basin are (1) the steep flank—up to 45°—west of Gebel Zeit; this flank plunges into the deep central low (Figure 10); (2) a narrow horst trend west of Zeit Bay with its highest point at Gemsa peninsula; and (3) the Ras el Bahar/Zeit Bay field high, which was probably created by transverse faults (Figure 7).

All units of the pre-Miocene sediment package decrease in thickness toward the south indicating a paleohigh in the Hurghada/Islands area. During the early Miocene, extremely thick Nukhul was deposited in the lows proving early subsidence. On highs, "nullipore" carbonates and reefs were deposited marking the Miocene coastline and islands, as at Gemsa and Zeit Bay field. Similar to the El Qaa basin, high input of coarse clastics from the basement to the west occurred throughout the lower and middle Miocene (Figures 7, 22, 23); however, the clastic wedges do not extend over the whole basin. The Rudeis Formation is predominantly marly in the basin center.

From the Eastern Desert basement, coarse clastic sediments were shed into the lagoonal area of the Esh el Melaha Basin, but sands were blocked by the Esh el Melaha range. Only locally were sediments transported through channel-type breaches into the Zeit Bay basin. One Miocene channel is exposed at Gebel Abu Gerfan.

The high subsidence rate or the tilting occurred continuously throughout the Miocene as the thickness in the basin axis (e.g., in well Gebel Zeit-1) demonstrates (Figure 21).

Hurghada/Islands Area: The Hurghada/Islands area is intermediate between the Gulf of Suez and the Red Sea. The structural style and pre-Miocene geology is very different from most of the Gulf of Suez basins and belongs very much to the Red Sea province. This area includes the offshore region south of Zeit Bay and the Gemsa peninsula. To the east, it is bounded by the "B"-trend and the Morgan Basin. To the west, it is separated from the Esh el Melaha Basin by the Gebel Esh el Melaha. Separation from the Zeit Bay basin is justified because the pre-Miocene sequence in both basins is different and, south of Zeit Bay, the salt is considerably thicker and forms pillows and diapirs.

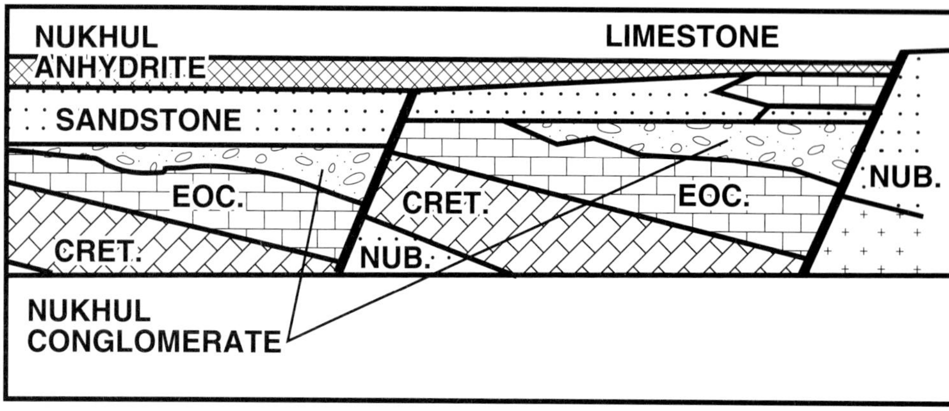

Figure 22. Nukhul Formation facies.

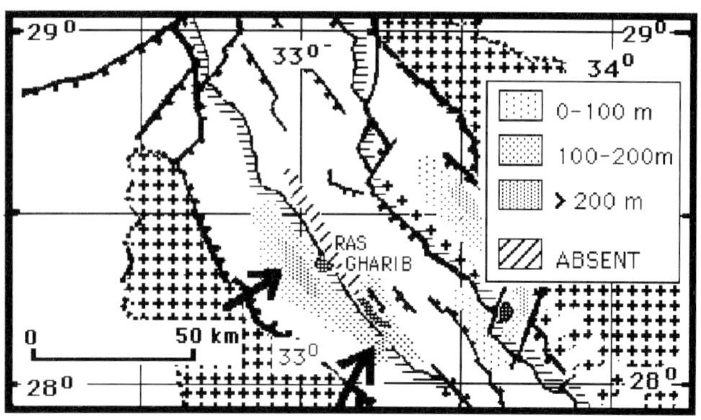

Figure 23. Lower Rudeis sand isopach map.

The northern part of the islands is relatively well known as a result of exploration wells and oil fields. The southern part, west of Shadwan, is relatively unexplored. Seismic is unreliable due to steeply dipping fault blocks and a very disturbed evaporite section. Therefore, the fault pattern on Figure 3 is, to some degree, simplified.

The pre-Miocene section in this area is thin or nonexistent due to originally thin deposition or subsequent erosion in Oligocene/Aquitanian.

Most wells drilled on highs found "bald" structures covered with evaporites as was already known from outcrops. Wells drilled on the western flanks found thin pre-Miocene sediments with a few meters of Nubia sandstones, usually of Early Cretaceous age.

The fault blocks are tilted 45° to the west and, therefore, structures are very narrow. Generally, they are interpreted as horsts; however, wells drilled on the east side of structures encountered basement after passing through 45° faults. An indisputable example with measured dips exists in the Geisum field. The fault pattern of the area, as well as the individual fault block geometry, leaves much room for speculation. However, I consider the fault tectonics of this area as essentially the same as in deeper parts of the Northern Red Sea, where 3000-m-thick salt disguises the pre-Miocene/basement structures.

We should consider the Gulf of Suez as a set of smaller basins having their own structural style and, to some degree, their independent structural history. The distribution of the mature source rocks (Figure 24) and the geothermal gradient map (Figure 25) contribute evidence for this conclusion.

Structural Evolution

The Gulf of Suez is a good example of a model extensional regime because all stages of extension are preserved in the different subbasins. In the northernmost basin, the Darag Basin, a simple tectonic style with a single, long, relatively flat-dipping flank is present, showing the least amount of extension in the gulf. Nevertheless, the basin is as deep as the some of the basins to the south. Neither border basins nor high uplifted shoulders accompany this basin. The shoulders are plateau style and not tilted, therefore not of isostatic origin. The basins farther south, the Issaran–October–Zenima basins and the Belayim Basin, show more extension and a more complicated tectonic style. Dips in this basin average 25° but do not exceed 35°. These basins are accompanied by border basins, the West Bakr and the El Qaa basins, which are bounded by high, uplifted basement shoulders.

The basins in the southern gulf, especially in the Gebel Zeit and Zeit Bay basins, display a large amount of extension, which is mainly illustrated by their high dips exceeding 35° and reaching 45°. These steep dips, together with high basement flanks, are typical for the Hurghada/Islands area. These steep dips already occurred during the early Miocene, very early in the basin formation, but continued after evaporite deposition. (This tectonic style can be predicted in the Red Sea also.)

From north to south in the Gulf of Suez, we can observe an increase in extension, possibly caused by a counterclockwise rotation of the Sinai plate. Larger amounts of extension produced higher dips and more complicated tectonics in the southern basins.

A regional differentiation of the structural evolution of the basins exists. The basins to the south (and the Red Sea) probably opened first, subsided more

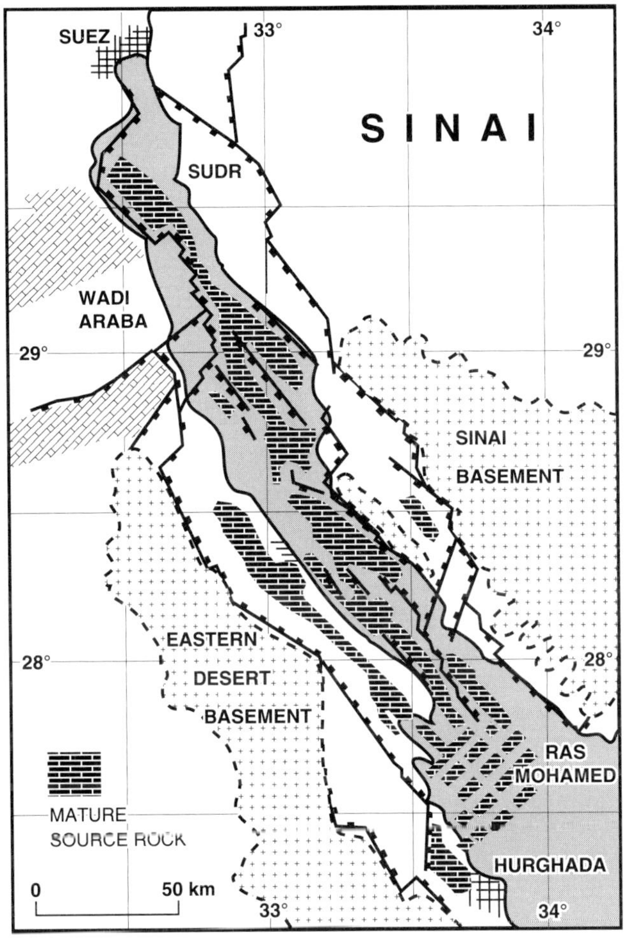

Figure 24. Mature source rocks.

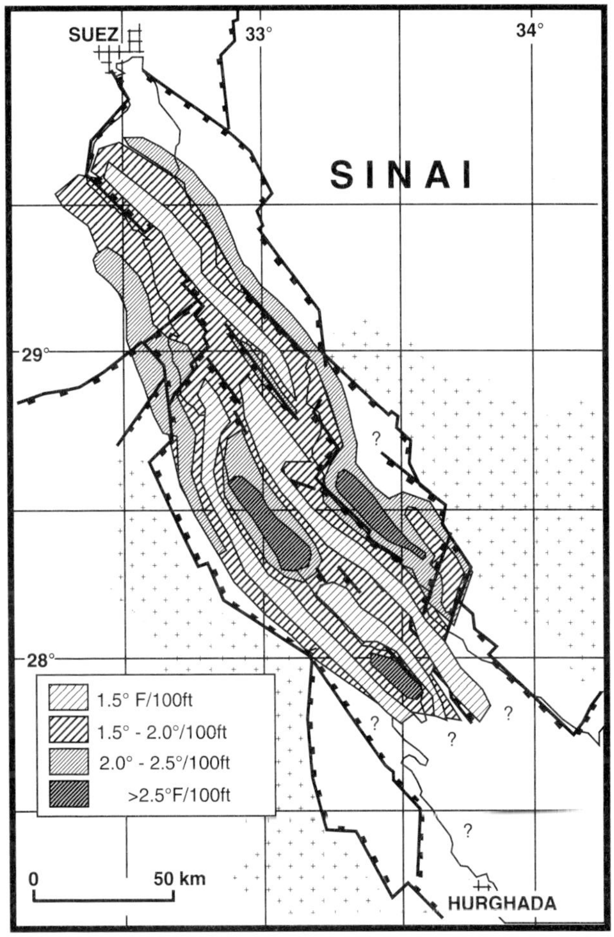

Figure 25. Geothermal gradient map.

deeply, and, during the early Miocene, experienced a similar extensional style, which the basins to the north reached only in middle Miocene.

STRATIGRAPHY

Prerift

Basement

A complex variety of Precambrian metamorphic and crystalline rocks with abundant sills and dikes crop out on both sides of the Gulf of Suez and form the basement of the Gulf of Suez. These rock types determine, to a certain degree, the fault and fracture patterns, depending on the homogeneity of the material. On the Sinai, older metamorphic rocks are intruded over wide areas by granites and other plutonic rocks of various Precambrian ages. Several large, circular, sometimes dome-shaped features of these plutonic rocks are clearly visible on LANDSAT images. Still younger alkaline granites and the northeast-southwest–trending dike swarms of various compositions cross these features. On the west side of the gulf, a similar basement configuration is recognizable on LANDSAT images and wide areas of Dokhan metamorphosed volcanics, which are intensely fractured, exist.

Basalts of Oligocene/earliest Miocene age are relatively rare in the Gulf of Suez, but are known from outcrops on both sides as dikes (e.g., in Wadi Araba and on North Sinai) and as sills (e.g., near Abu Zenima). Basalt dikes were encountered in some wells, especially in the Ras Budran field. More widespread basalt flows may occur in the southern and deepest part of the October Basin as the magnetic anomaly indicates.

Pre-Cenomanian

The sedimentary sequence overlying the Precambrian basement embraces almost all stratigraphic stages from Paleozoic to Holocene (Figure 4). Essentially three units can be differentiated: (1) A pre-Cenomanian, predominantly sandy, continental unit up to 600 m thick, with few, thin, marine ingressions in the Cambrian, Carboniferous, Permian–Triassic, and Jurassic. (2) A Cretaceous to Eocene marine unit predominantly consisting of limestones 500–1000 m thick and minor sands and shales. (3) A Miocene to Holocene predominantly marine unit up to 4000 m thick with thick shales, several evaporite cycles, and sandstones.

Generally, thicknesses increase from south to north for the pre-Miocene section and for short distances from west to east for the Miocene units. In this chapter, I make no attempt to revise the stratigraphy of the Gulf of Suez. I do attempt to clarify some facies relationships within some rock units, because these were often not recognized previously. I also emphasize the major tectonic events as documented in the sediments. At present, the best attempt to simplify and better define the rock stratigraphy is a study from Robertson Research (preliminary edition, April 1985); however, this study is available to subscribers only.

The generally red, thick sandstone unit overlying basement is commonly called the "Nubia Formation." These sandstones are from different marine environments, different continental environments, are of different ages and should be differentiated. The term "Nubia" is very handy for the exploration geologist, but not for the stratigrapher, and many papers about the correct usage of this name have been written (Klitzsch et al., 1979). In Gulf of Suez geology, this name is not applicable because, for explorationists, the age of the pre-Cretaceous sandstones is of relatively little significance, and nowhere is the terminology and stratigraphy of the Nubia more difficult than in the area of the Gulf of Suez. In this paper, the "Nubia" unit is subdivided into a Paleozoic and Mesozoic Nubia (Harms, 1985).

An interesting model for the distribution of the Paleozoic strata was proposed by Gvirtzman and Weissbrod (1985). According to their interpretation, Cambrian, Ordovician, Silurian, and Devonian strata were laid down concordantly over a wide area in northeastern Egypt, Sinai, and the Levant, but were truncated by a Carboniferous unconformity. Marine Carboniferous sediments transgressed a major Hercynian geanticline. A second pre-Permian transgression capped some of the Carboniferous strata.

Paleozoic

Cambrian sandstones crop out in western Sinai (Um Bogma and east of Abu Durba/Gebel Araba and near Wadi Feiran). In eastern Sinai, the Cambrian is described by Israeli authors (Weissbrod, 1969, 1970, 1981; Gvirtzman and Weissbrod, 1985).

The Um Bogma area, east of Abu Zenima, has perhaps the best known Paleozoic sequence in Western Sinai. It consists at Wadi Shellal, Um Bogma, and Sinai, of a basal Cambrian(?) section as much as 80 m thick. The basal 13 m of arkosic sandstones are followed by a middle unit (19 m thick) of shale, arkosic sandstone and dolomite carrying *Cruziana* sp. and problematica. These are overlain by a massive cross-bedded, variegated, arkosic sandstone (the lower part is 18 m thick, and the upper part is 30 m thick).

In Abu Durba, east of Belayim Bay, Hassan (1967) and Omara (1972) described a second important Nubia outcrop. There, Said (1980) described a limestone outcrop packed with laminated algal stromatolites that have been identified as belonging to the species *Vetella ushbasica,* known from the Lower Cambrian from Kasakhstan. Besides these stromatolites, there are organic remains that belong to Archeocyathidae.

To date, no Cambrian sediments have been identified on the western side of the gulf. In offshore wells, however, these strata likely exist.

Between the Cambrian and Carboniferous strata lies a 30- to several-100-m-thick cross-bedded fluvial unit that is poorly dated. This unit was named the Nubia "C" or Araba Formation by Hassan (1967). The type section lies in Wadi Araba, but this unit is also often encountered in wells. It is productive in the July, Ramadan, October, and Ras Budran fields and consists of fine to coarse sandstones with porosities of less than 16%, generally with siliceous cement. If the assumptions of Gvirtzman and Weissbrod (1985) are correct, this unit could belong to the Ordovician, Silurian, and Devonian. Devonian microflora are known from the Belayim field; however, stratigraphic information on this part of the Paleozoic is very incomplete.

Carboniferous

Marine Carboniferous sediments with black shales and *"Productus"* fossils are well known in extent and thickness distribution in the Gulf (Figure 26). Generally called the Nubia "B," the name "Abu Durba Formation" probably is more appropriate (Hassan, 1967) for the Lower and Upper Carboniferous.

Stratigraphically, the Abu Durba Formation is well defined. It spans the whole Carboniferous and Permian. The best dated outcrops on the Sinai are east of Gebel Abu Durba and at Um Bogma (see Hassan, 1967).

The Abu Durba Formation is a 40- to 250-m-thick unit composed of dark-gray shales and interbedded fine-grained sandstone layers. The shale is generally noncalcareous in the central part of gulf and becomes carbonaceous toward the east and the north. Dolomite beds are known from outcrops on the Sinai (Um Bogma, up to 40 m thick) and are relatively fossiliferous (foraminifera present are Tetrataxidae, Archaediscidae, and Endothyridae; solitary and colonial corals: brachiopoda, crinoids, lamellibranchata) indicating a Tournaisian–Visean age. The greatest thickness of 250 m of the Abu Durba Formation is recorded from the East Gharib M-1 well. There, the section consists of sandstone and shale intercalations with a few coal layers, indicating a shallow-marine environment with coastal-plain deposits. In this environment, lateral facies changes can be expected. To the north at Wadi Araba, the Carboniferous beds are younger than those at Um Bogma and range from Westfalian to Permian. At the eastern flank of the Northern Galala Plateau north of Abu Darag, the section consists of 400-m-thick shales, sandstone, and fossiliferous limestone. A similar 162-m section was drilled in a borehole at Wadi Araba. Still younger strata are exposed around Wadi Aheimer, where 260 m of sandstone, shale, and thin fossiliferous carbonates are dated Stephanian to early Permian.

The black radioactive Carboniferous shales are a common marker horizon in most exploration wells in the Gulf of Suez north of the Ramadan and July fields. Especially well developed, however less well dated, are the black shales in the Ras Budran field. They are not present in the southern Gulf of Suez, probably due to erosion prior to the Early Cretaceous (Figure 26).

Permian

Permian marine shales and limestones with crinoids and bryozoa are recorded locally east of Gebel Abu Durba (Hassan, 1967) and west of the gulf, north of 29°N along the cliff of the Northern Galala plateau (Abdallah and Adindani, 1963; Abdallah et al., 1963). These strata should be included in the Abu Durba Formation. At Wadi Rod el Hamal in Wadi Araba, more than 100 m of interbedded shales and sandstones top the Carboniferous beds. The Permian is characterized by a thick shale unit at the base, a *Rhynchouella* sp. horizon in the middle, and a coral horizon at the top. Thick sandstone beds separate the coral and the *Rhynchoullea* sp. horizon.

Jurassic

Jurassic marine sediments are known only in the northern Gulf of Suez from outcrops south of Ain Sukna and from wells in the Darag Basin. Thicknesses increase over a short distance from 0 m north of Wadi Araba to 955 m in well Ayun Musa-2 on the east side of Suez Bay. North of the Gebel Maghara fault (Figure 5), the Jurassic attains a thickness of 2000 m.

Lithologies and stratigraphy allow comparison of the Jurassic of the northern Gulf of Suez with the Western Desert and outcrops on North Sinai, e.g., Gebel Maghara. In this respect, it is reasonable to subdivide the marine Jurassic into an upper clastic unit, the Alam el Bueib Member of the Burg el Arab Formation, overlying a predominantly carbonate sequence, the Masajid Formation of Callovian/Bathonian age. The Masajid Formation is a widespread unit recognizable along the northern African shelf from the Western Desert to Northern Sinai and far into Arabia and represents uniform sedimentation and transgression on the southern margin of the Tethys.

Below the Masajid Formation a lower clastic unit, the Khatatba Formation, was found in the Ayun Musa-2 well (Figure 15).

The distribution of the Jurassic follows tectonic lineations as outlined by the Jurassic/Cretaceous paleofault pattern (Figures 1, 5).

Nubia "A" (Malha Formation) (Aptian–Albian)

The Nubia "A" (Figure 27) includes all clastic sediments between the Abu Durba Formation and the Cenomanian carbonates. Palynomorphs have been used to date these sandstones as essentially Albian to Aptian in age. For Nubia "A" the equivalent term "Malha Formation" was proposed by Abdallah and Adindani (1963) and by Robertson Research (1985) with reference to the type locality at Wadi Malha at the Northern Galala Plateau.

The Nubia "A" is drilled by more than 80 wells and crops out on both sides of the Gulf of Suez. It is

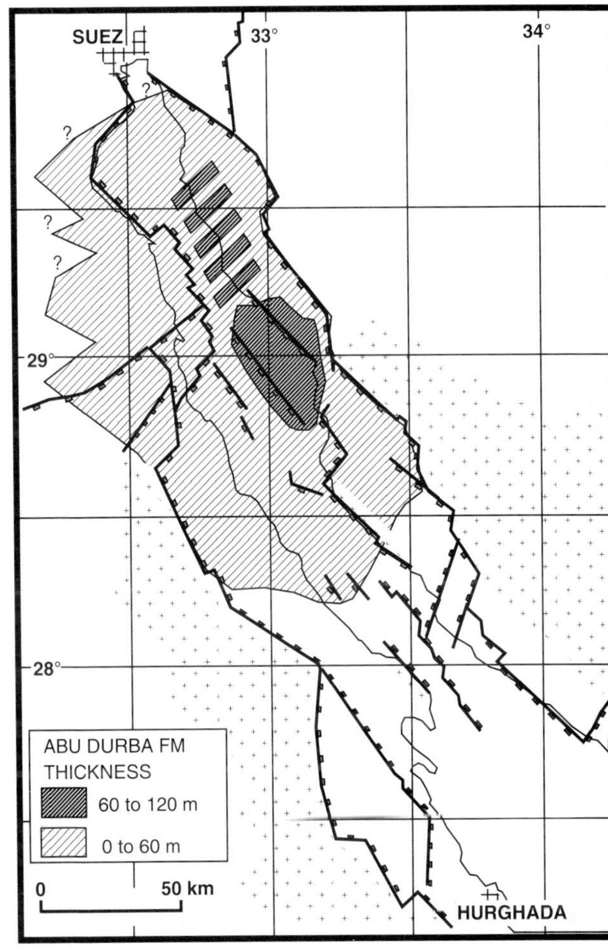

Figure 26. Abu Durba Formation (Nubia "B") net sand isopach map.

mainly composed of medium to coarse, cross-bedded red sandstones of a braided-river environment, with occasional shale and rare limestone layers. In offshore wells, its thickness is relatively consistent in the 30–100 m range. However, onshore, west of the Eastern Desert mountains, a considerably thicker sequence, probably up to 1000 m, is exposed. (At Wadi Dakhal, however, these onshore sandstones could reach into the Cenomanian as ammonite limestone beds indicate.) On the Sinai, 880 m are reported at Wadi Feiran and 560 m from Baba. The Nubia "A" overlies the Abu Durba Formation unconformably. In the southern Gulf, Nubia "A" rests on pre-Carboniferous strata. In the central Gulf, it transgressed Permian or Carboniferous sediments and in the northern Gulf it was deposited on Jurassic sediments. Toward the south, the Nubia "A" thins over a relatively short distance as can be seen along Gebel Zeit. South of Gebel Zeit we have to assume a zero line (Figures 5, 27).

The pre-Cenomanian unit consists mainly of continental sandstones with several marine incursions on the African continent. A significant pre-Carboniferous tectonic phase allowed erosion and the marine Carboniferous then discordantly transgressed wide

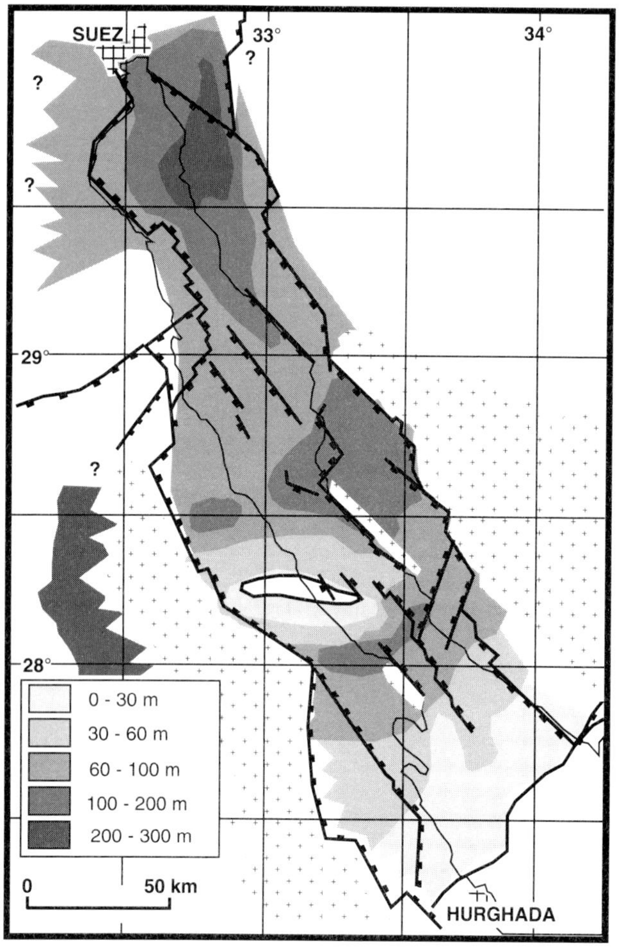

Figure 27. Nubia "A" net sand isopach map.

areas from the north. Thickest Carboniferous strata are reported to the northwest of the Gulf of Suez. A Carboniferous embayment in the area of the Gulf of Suez is unlikely, because Carboniferous strata were deposited extensively all over northern Egypt and Sinai. Relatively quiet tectonic conditions remained until the Jurassic. Only thin Permian strata and little or no Triassic strata were deposited, indicating a period of slight uplift and erosion.

In the Jurassic, due to the eastward movement of Africa in relation to Eurasia, northern Egypt and the Sinai were subjected to tectonic stresses, which resulted in formation of several deep fault-bounded basins (e.g., Abu Gharadig Basin, North Kattania Basin with thick Khatatba Formation, Maghara Basin on North Sinai, Figure 1). These basins were filled mainly with coarse clastics derived from the south. Northward, the clastic content decreases and marine conditions prevail. Simultaneous with the basin formation several shear zones were generated on the craton. These remained active, to varying degrees, from the Jurassic to the Eocene, controlling sedimentation.

During the Early Cretaceous, the craton was further uplifted and coarse clastics were again shed northward and covered the continent over wide areas, including the Gulf of Suez.

Cretaceous–Eocene Limestone Unit

A major break in sedimentation occurred in the Cenomanian when the sea transgressed large areas of the African craton from the north. From the Cenomanian through the Turonian, deposition took place largely in a restricted, shallow-marine, inner sublittoral environment with more marginal marine, littoral environments increasing to the south, marking the proximity of the shoreline. During the early Campanian, the sea invaded far onto northern Africa. Marine Campanian sediments crop out as far south as Quseir at the Red Sea.

The Cretaceous is subdivided as follows: Sudr (Chalk) Formation (Maastrichtian to upper Campanian), Brown Limestone Member (lower Campanian), Nazzazat Group, Matulla Formation (Coniacian–Santonian), Wata Formation (Turonian), Abu Qada Formation (Turonian), and Raha Formation (Cenomanian).

Raha Formation

The Raha Formation contains the first marine Cretaceous sediments. In many places (e.g., at St. Paul Monastery) several meters of glauconite beds overlie the Nubia "A" sandstones unconformably. Overlying the glauconite beds are marls and fossiliferous massive limestones, which, due to their content of ammonites, echinids, and lamelibranchiata (mainly ostrea), allow proper dating. Thick massive ostrea marl and limestone banks emphasize the shallow-water depositional environment. Some sandstones are locally intercalated. The thickness increases from south to north in a remarkably uniform manner from 20 m at Gebel Zeit to 170 m in the central gulf to 280 m in the northern gulf (Nebwi, Ras Sudr, X 80-1). On well logs, the Raha Formation correlates peak to peak over long distances.

On the Sinai (e.g., at Gebel Nazzazat) above the first several meters of limestone, another 20- to 30-m-thick red sandstone/red shale unit occurs, which in wells probably is commonly mistaken as the Nubia "A." These red beds and other sandstones higher in the section indicate proximity of the shoreline, which for most of the Cretaceous was at the southern part of the Gulf of Suez and on the southern Sinai. These sands probably were derived from pre-Carboniferous continental sandstones from uplifted Sinai or Saudi Arabia. The Raha Formation is the time and lithological equivalent of the Bahariya Formation of the Western Desert, proving the widespread Cenomanian transgression probably due to a global sea level rise.

Abu Qada Formation (Turonian)

The Raha Formation is topped by a 12- to 25-m-thick, forminifera-rich, green-gray to dark-gray shale bed. This shale marker occurs in most of the well logs in the subsurface and is relatively easily identifiable.

Robertson Research (1985) assigned *Heterohelix globulosa* in the Abu Qada Formation to the early Turonian. "The location of the Cenomanian/Turonian boundary is directly below the planktonic rich horizon equivalent to Ghorab's Abu Qada Formation . . ."

(Robertson Research, 1985). Correlating with the Western Desert, the organic-rich limestone of the Abu Roash "F" or Mansour Member is equivalent to the Abu Qada Formation.

The Abu Qada probably represents the most extensive marine transgression over northern Africa during this first-marine phase in the Upper Cretaceous.

Wata Formation (Turonian–Lower Coniacian)

The Wata Formation conformably overlies the Abu Qada Formation. It is generally 150 m thick and consists of macrofossil-rich, massive limestones, dolomites, and marls, although it wedges out toward the south. Log correlation and paleontological evidence (Robertson Research, 1985) indicates an erosional unconformity at its upper boundary. It is difficult to judge how much section is missing due to the lateral facies changes in short distances in these warm-water, shallow-marine sediments. The most thoroughly studied sections on the Sinai, e.g., at Gebel Nazzazat, were published by Levy (1975). Age dating with ammonites indicated that the upper parts of the Wata Formation may reach into the early Coniacian.

The Wata Formation is generally recognized in the field by its resistant massive limestone cliffs with excellent outcrops on the Sinai. Correlating with Western Desert stratigraphy, the Wata Formation is the equivalent of the Abu Roash Formation. In Israel, it is represented by the Gerofit and Zihor formations. This widespread occurrence in northern Egypt and Sinai indicates a uniform sedimentation phase during the Late Cretaceous.

Matulla Formation (Coniacian–Early Santonian)

The Matulla Formation unconformably overlies the Wata Formation. The Matulla, on average, is 140 m thick and contains marls, limestones, and shales, as well as several sandstone units (Figure 28). In its type locality at Wadi Matulla, Sinai, it consists of 170 m of variegated sandstone, clays, and a few fossiliferous marl layers. Some of the sandstones are cross-bedded and contain plant remains. Oyster beds are common. An excellent onshore correlation of Coniacian outcrops along the western Sinai was given by Levy (1975) (Figure 8). In the subsurface, log correlation of the Matulla Formation shows thickening and decreasing sand content to the north of the Gulf. Average sand content of the gross Matulla interval is 20% except in wells in the southern part of the Gulf, e.g., in the Ramadan field the net/gross ratio increases to 40%.

The upper part of the Matulla Formation is generally represented by sandy shales that grade into shales toward the top. A sharp break marks the onset of a different depositional environment of radioactive, dark brown, organic-rich limestones of the overlying Brown Limestone Member of the Sudr Formation.

Significant lithofacies and biofacies changes occur between Turonian and Senonian strata. These changes are associated with erosional unconformities and probably tectonic uplift of the region to the south and east. According to Levy (1975, p.19), shallowest water conditions in the Matulla Formation were reached "... in early late Coniacian time, when silt, glauconitic

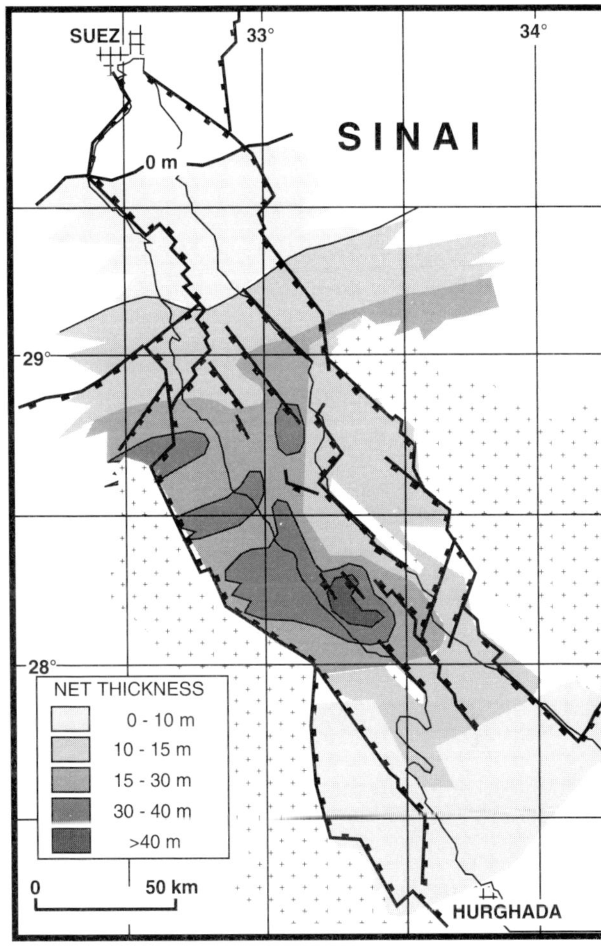

Figure 28. Matulla Formation net sand isopach map.

marls occurred throughout Sinai. In late Coniacian times Sinai and Southern Israel (at least) were slightly tilted southwestwards, and the uplifted north-eastern part of the platform was eroded rapidly, exposing the Coniacian and even Turonian strata. Late Coniacian folding and faulting caused further erosion or nondeposition on uplifted areas (e.g., Northern Sinai [e.g., Gebel Maghara], Gebel Magmar in Central Sinai) during latest Coniacian, whereas in the low regions (e.g., Wadi Sudr) chalks . . . accumulated. At the end of Coniacian time the whole area was quite flat and partly exposed for a short time. . . . At the beginning of the Early Santonian [now corrected to lower Campanian] a shallow sea prevailed over southern Israel and most Sinai, in which phosphoric, chalky marls containing Pycnodonte vesicularis (Lam) and Spinaptychus spinosus (Cox) . . ." were deposited. The late Coniacian compression of northern Sinai did not involve the Gulf of Suez. In the Western Desert, the equivalent formation is the Khoman Formation. There, however, this time interval is often not recorded in the rocks, probably due to erosion, as in central and northern Israel.

Sudr Formation (Upper Campanian–Maastrichtian)

The Sudr Formation predominantly consists of uniform massive white chalk beds reaching a thickness of

100–130 m, with a maximum of 220 m at Gebel Tanka and Hamman Faraun. Farther northeast at Wadi el Bir, up to 300 m are recorded. Very thin chalk is present in the southern Gulf, indicating a wedge-out to the south like the older Cretaceous units. No chalk is found northwest of the gulf at the Northern Galala Plateau and Gebel Ataqa. At the south side of the southern Gebel Galala near St. Paul Monastery, slump masses of chalk crop out, indicating a paleoslope dipping toward the southeast. Said (1962) recognized the east-west–trending basin delineation of the chalk basin, which is located at the southern fault of Wadi Araba (Figures 1–3, 13).

Wadi Araba and the Northern Galala Plateau were positive blocks and probably received a reduced sedimentation of chalk only. (A similar situation continues during the Paleocene and Eocene when shallow-water nummulitic sediments were laid down on the Wadi Araba high, and then shed of the high as slumped masses and debris flows down a slope north of St. Paul Monastery.)

In the Sudr Formation, the open-marine conditions of the Brown Limestone Formation continue, however, with less organic-rich sediments. Thickness changes over short distances and slump masses indicate tectonic events, probably along rejuvenated Jurassic faults and the Syrian arc structures.

Brown Limestone Member of the Sudr Formation (Lower Campanian)

The term "Brown Limestone" was introduced to differentiate the lower unit of the Sudr Formation from the upper Chalk unit (Ansary et al., 1962). Ghorab (1961) described the lower Campanian "Phosphate Formation" near Quseir, Red Sea, which is time equivalent to the "Brown Limestone," but lithologically different. The phosphate beds, rich in fish teeth and bone fragments, reach north as far as St. Paul Monastery at the Southern Galala Plateau. On the northern Sinai, the lower Campanian phosphate-bearing beds are named the Rajim Formation. In the Gulf of Suez, these phosphate beds are represented by brown limestones, rich in organic matter, that are recorded on the gamma-ray logs as very highly radioactive due to their uranium content. The Brown Limestone represents special marine conditions of a vast epicontinental sea over Egypt and the Sinai. The relatively uniform lithology and formation thickness, together with the phosphate and organic matter content, point to cool, oxygen-rich surface waters. The conformably overlying chalks confirm open-marine conditions.

This unit is probably one of the primary source rocks for hydrocarbons in the Gulf of Suez.

Esna Formation (Paleocene)

The Esna Formation in the Gulf of Suez is a relatively thin shale unit (15–40 m), with some limestone beds. Sedimentation across the Cretaceous/Tertiary boundary probably was not continuous because one or several fossil zones are missing. In the subsurface a log break from shale in the Esna Formation to clean chalk allows relatively easy differentiation and correlation. As mentioned, Esna shales are missing in the Southern and Northern Galala plateaus, probably due to submarine erosion by slump masses and debris flows in the Eocene. Clear erosional surfaces mark the gap in the field.

Eocene Unit

Eocene rocks crop out over wide areas in Egypt, as well on both sides of the Gulf of Suez (Figure 29). They are decimeter-bedded limestones with interbedded shale layers. The limestones generally are brown to cream in color, rich in organic matter and pelagic microfossils, and contain chert as nodules or layers. *Nummulites* locally occur mainly around Wadi Araba, but also north and east of Belayim.

Salem (1976) (Figures 6, 15) presented the facies distribution of lower and middle Eocene rocks west of the Nile. Salem clearly shows northeast-southwest–trending highs with associated shallow-water sediments. Breman (1985) extended this depositional pattern into the Gulf of Suez area and proved a high block with shallow water over Wadi Araba (Figures 18, 30).

According to Said (personal communication, 1988) the Eocene rocks of the eastern Gulf are divisible into four units (Figure 29): (1) the Thebes Formation (also Egma Limestone, lower Eocene), (2) the Darat Formation (lower middle Eocene), (3) the Khaboba Formation (upper middle Eocene), and (4) the Tanka Formation (upper Eocene; Priabonian).

The lower Eocene Thebes Formation is made of a lower member of nummulitic limestone with flint, a middle member of chalky limestones, and an upper member of *Cardita*-bearing limestones. The thickness of this unit in Wadi Nukhul is 140 m.

The lower middle Eocene Darat Formation (green beds of earlier authors) is made up of alternating green and brown shales and marls with few fossiliferous limestone (occasionally chalky) intercalations. The thickness of this unit in Wadi Nukhul is 98 m.

The upper middle Eocene Khaboba Formation (gypsiferous marls of earlier authors) is made of shale, marl, and chalky limestone interbeds with a few flint bands. The shale and marl interbeds usually are yellow green or yellow brown in color, occasionally pink or violet, and richly gypsiferous. The beds are rich in megafossils, shell casts, and fish remains. The thickness of this unit in Wadi Nukhul is 99 m.

A detailed distribution of the upper Eocene Tanka Formation is given in Said (1962). According to Said, a major unconformity occurred in the middle Eocene and no deposition took place during the Bartonian (Figure 29). Only in the Priabonian in the northern area of the gulf, especially near Wadi Tanka, limestones, chalky marls, and glauconitic marls, more than 200 m thick, were deposited. The subsequent Oligocene–lower Miocene erosion preserved only the deeply subsided areas. However, the existence of thick, upper Eocene beds in the Gulf of Suez area prove the continuity of the depositional and tectonic conditions existing since the Late Cretaceous.

On the western side of the gulf, the Eocene is composed of units of massive limestones that are easily

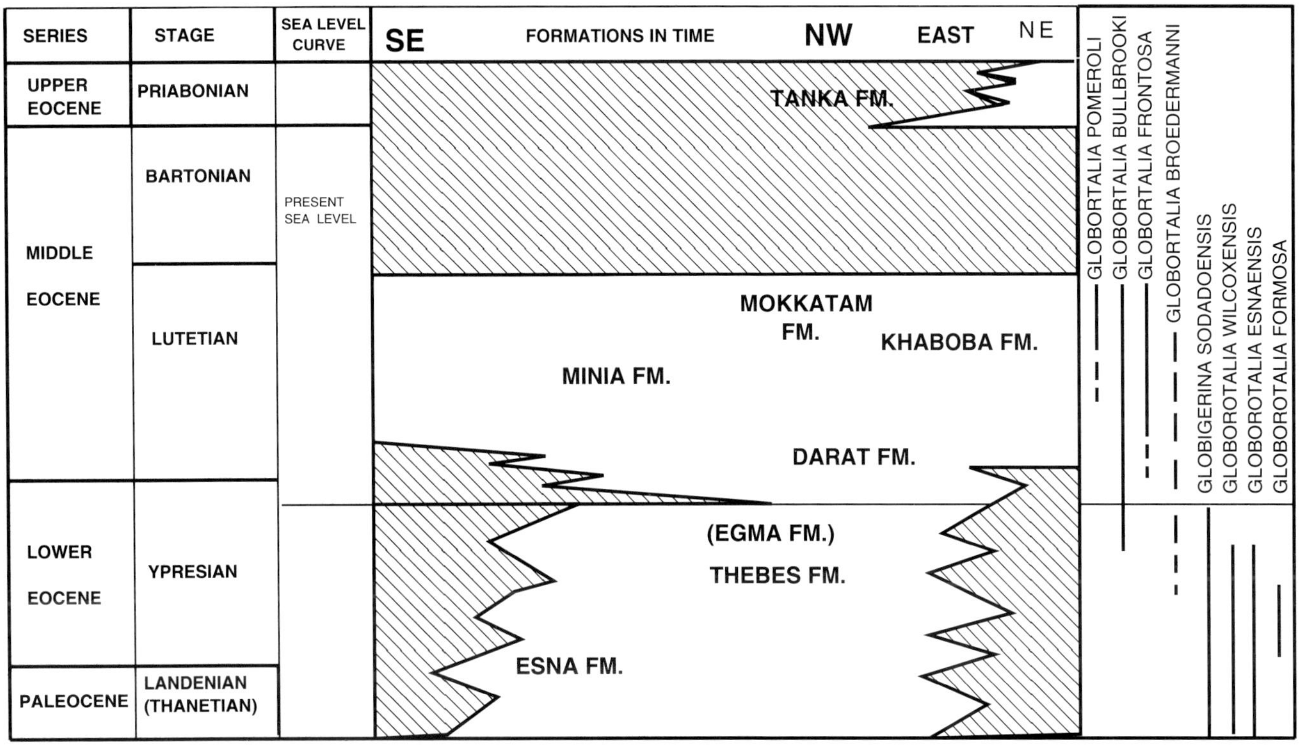

Figure 29. Eocene stratigraphy, ages, unconformities, and hiatuses, Gulf of Suez (after Breman, 1984).

correlatable with those of the Nile valley. The lower Eocene Thebes Formation is absent and the oldest Eocene rocks belong to the Minia Formation (lower middle Eocene) and the Mokattam Formation (upper middle Eocene). In Gebel Ataqa, the Mokattam Formation is followed by a unit of clastics (El-Ramiya), which may be correlatable with the Kaboba Formation of the eastern coastal plain of the gulf.

Oil companies generally have lumped the Eocene limestones under one name, either as "Eocene Limestones" or as "Thebes Formation." This seems sufficient for oil exploration purposes, although certainly at least two units, the lower and middle Eocene, can be distinguished on logs lithologically and with micropaleontology.

Thebes Formation (also Egma Limestone, Yypresian)

Breman (1985) identified four facies in the Thebes Formation (Figure 30). Facies A is composed of pelagic, argillaceous, bituminous, dark brown limestones with abundant foraminifera (*Globigerinoides*). Facies B is pelagic, cream-colored, cherty foraminiferal limestone with abundant chert as nodules and bands, and is locally bituminous. Facies C is a pelagic, cream-colored, chalky foraminiferal limestone (transitional to A and B, and locally near shoals), and *Nummulites* are present in this facies. Facies D is made of tidal or shoal dolomites with sabkha anhydrites. All pelagic facies can be interbedded with each other and lateral facies changes also occur.

In pelagic bituminous limestones, correlation on gamma-ray logs seems to be possible over wide areas of the gulf. On gamma-ray logs, these high-radiation peaks are the result of the high uranium content, especially on spectral gamma-ray logs, and seem to indicate contemporaneous events or at least uniform environments over a wide area.

The Eocene sea transgressed over northern Africa as far south as the Cretaceous sea had. Later erosion on highs (e.g., on the Gebel Zeit ridge), reduced the Eocene distribution to structurally low areas (Figure 18). The thickest total Eocene is 500 m thick and is found in the area surrounding Wadi Araba. To the south, the Eocene wedges out. Average thickness near the Ramadan and July fields is 150 m, and near Zeit Bay only 10 m are recorded. According to Breman (1985) the Thebes Formation presently covers most of the Gulf of Suez as far south as Zeit Bay except on high ridges as shown on Figures 18 and 30.

Minia Formation (Lutetian)

The Minia Formation exists only as far south as a line north of the July/Ramadan fields (Figure 18). Only unconfirmed isolated occurrences seem possible in lows farther south. Lithologically, the Minia Formation can be subdivided into two facies. (1) A pelagic, brownish, argillaceous, bituminous limestone that may grade to marl. It is commonly cherty and contains abundant pelagic *Globigerinoid* foraminifera. Locally, limestones are dolomitized. This facies is predominantly present south of a line from Ras Issaran to Abu Zenima. (2) A white- to cream-colored, chalky limestone with abundant foraminifera and shell beds, locally with *Nummulites*. If chert is present, it is underlain by dark bituminous marls. This facies surrounds the Wadi Araba high.

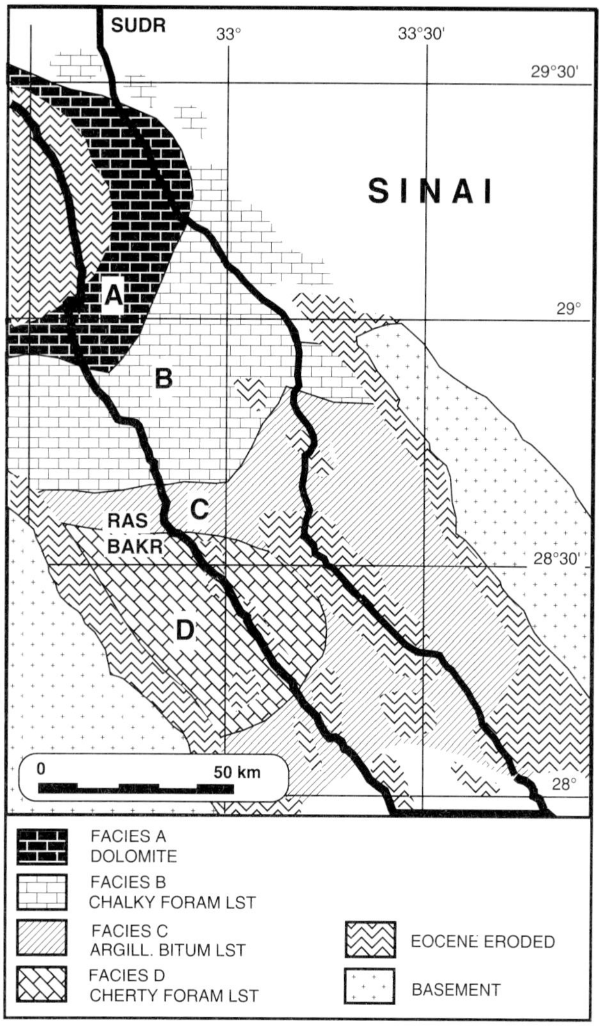

Figure 30. Lower Eocene facies distribution (after Breman, 1984).

On the South Galala Plateau, north of St. Paul's Monastery, limestones with both small and large *Nummulites* floating in a micritic matrix can be found. This indicates mixing of two faunas by sediment transport down a slope. Limestones with small *Nummulites* floating in the lime matrix and limestones with only big *Nummulites* also crop out. In the area north of St. Paul's, Eocene beds commonly are more massive or the bedded strata are slightly folded, but lying within flat strata indicating slumps. Massive "conglomeratic" limestone (marbles), which can be interpreted only as slump masses, are quarried near Thelemet. Similar synsedimentary slumps with Nummulitic limestones are reported from Gebel Gharamul. The facies changes, as well as the thickness distribution, allow one to interpret the depositional environment.

Wadi Araba existed throughout the Eocene as a shoal area where *Nummulites* lived. These accumulated and were transported away from the shoal and deposited in the deeper marine limestones as slump masses and debris flows. Tectonic activity and high subsidence rates, as well as high carbonate production on Wadi Araba, allowed the thick accumulation around the high. The controlling faults are interpreted to follow older Jurassic and Cretaceous (and not clysmic) directions. The most likely fault is the south-bounding Wadi Araba fault, which is downdropped to the south. This is a well-documented case of sedimentation controlled by tectonics.

Eocene limestones, mainly from the more cherty Thebes Formation, are oil reservoirs in several oil fields in the Gulf, for example in the Amer and Bakr fields, as well in the Sudr, Asl, and Matarma fields.

Oligocene Unit

Oligocene sediments are uncommon and often poorly datable in the Gulf of Suez. If present, they are thin red beds (sands, gravel, and shales) of continental origin. Direct dating is not possible. Thus, in the gulf many seemingly Oligocene outcrops probably are younger (see Wadi Tayiba section in Sellwood and Netherwood, 1984).

The middle Oligocene records a major global sea level drop, which could explain the change from carbonate sedimentation in the Eocene to a nondepositional environment in the Oligocene.

From regional geologic evidence, it seems that the Gulf of Suez was located between an uplifted area to the southeast (probably over northwest Arabia) and a depositional area to the west at Gebel Qatrani, west of Cairo. One fast subsiding marine Oligocene basin was located over the Western Desert. Another Oligocene Basin was located north of Sinai with the Pelusium line (old Jurassic element) as the controlling hinge line.

From Arabia and the Sinai, an extensive erosional surface is known. "Before the tectonic disturbance of these surfaces they were coextensive across the site of the (Aqaba) rift system. They are at the same altitude across the Gulf of Aqaba" (Quennell, 1984).

"An oldest age for the surface is upper Oligocene, the cycle having been initiated in the middle or upper Eocene. This is based on stratigraphic evidence from Sinai. . . . The diastrophic episode which ended the erosion cycle was probably coincident with the opening of the Red Sea" (Quennell, 1984, p. 785). In the Gulf of Suez area, only minor erosion occurred during the Oligocene because a maximum of several 100 m of sediment or less are missing from the middle and upper Eocene section. Uplift and tectonic activity were probably very minor as major events would have been recorded in the sediments or in fractures of the Eocene rocks. Good evidence for an uplifted area southeast of the Gulf stems from the Oligocene beds at Gebel Qatrani west of Cairo, where an Oligocene river originating from the east built a delta westward. There, fine-grained fluviatile sands and silts contain excellent fossils, including primates.

Erosion of exclusively Eocene limestone and marl beds in the gulf area would not have produced sandstones at Gebel Qatrani. Basalt flows (Haddadin basalt) overlying the fluviatile Oligocene to lowermost Miocene sequence are dated between 21.7 and 23.3 Ma, i.e., Aquitanian (Williams and Small, 1984).

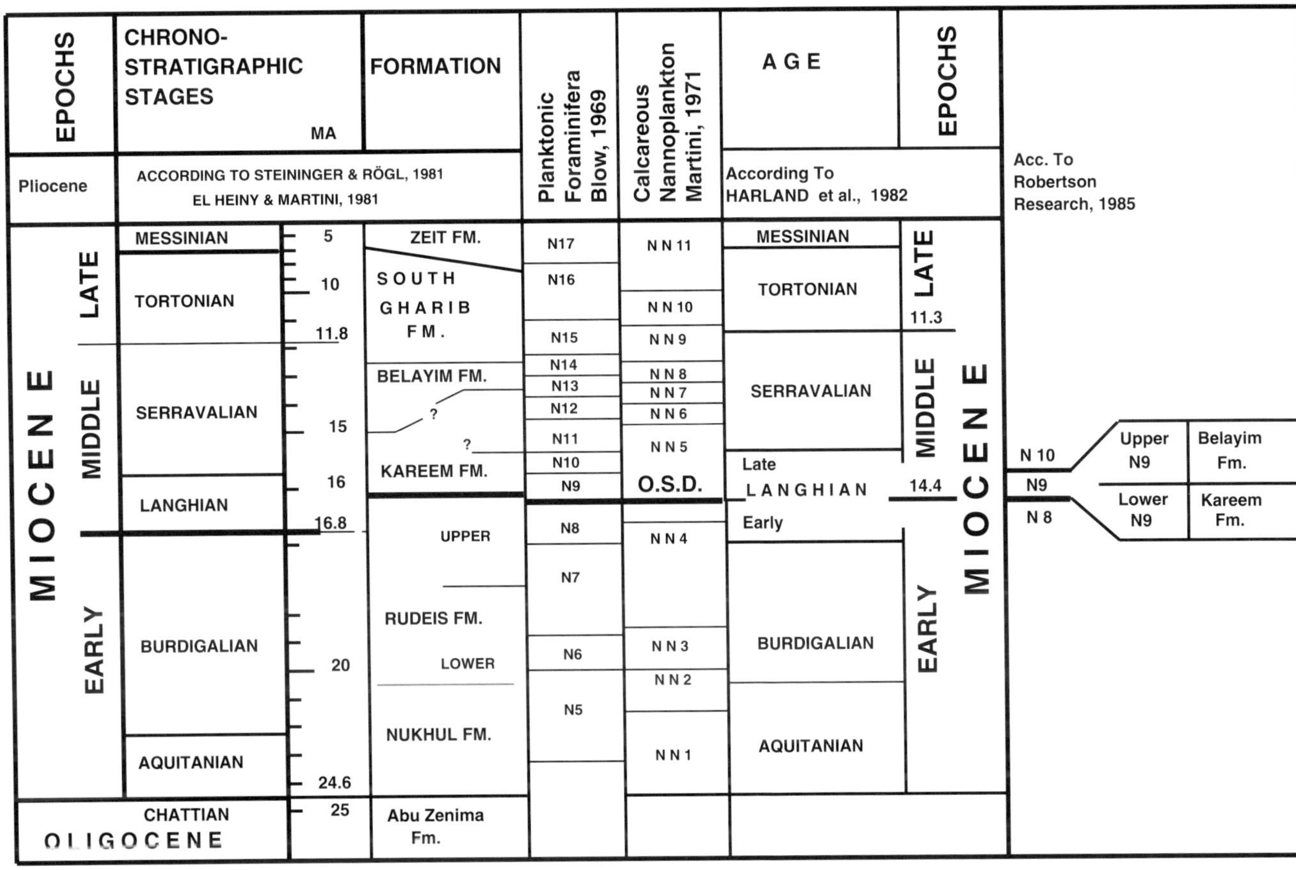

Figure 31. Miocene stratigraphy, Gulf of Suez. Time scale is in Ma, and is not linear to emphasize Langhian. O.S.D. = Orbulina Suturalis datum.

The basalts of the Gulf of Suez area are of the same age (Garfunkel and Bartov, 1977; Steen, 1982; Siedner, 1973).

I assume that the basalts in the Gulf of Suez area and the widespread basalt in the Western Desert at Wadi Natrun and at Gebel Qatrani are essentially of the same age and are the main proof for the tectonic break up of the North African craton in the early Miocene. Until then no clysmic oriented tectonic activity is recorded in the sediments or their depositional patterns.

Synrift

With the onset of the Miocene, a major change in marine sedimentation occurred that was partly due to the change from the epicontinental sea of the Cretaceous–Eocene to fully marine conditions, but also due to a major global change of marine conditions. "By the early Miocene, about 22 m.y. ago, the ocean basins had essentially assumed their modern shapes, if not the same proportions. . . . The early Miocene marks the establishment of a permanent steep temperature gradient between the polar and tropical regions related to the development of polar-to-tropical surface water-mass belts. . . ." (Kennett, 1982, p. 729, 730). In the Gulf of Suez, a deep-water pelagic and bathyal fauna occurred in the Miocene for the first time.

During this time a major change in regional tilt was initiated. The tilt of the African craton during the Paleozoic to early Tertiary was oriented to the north. Since the Miocene, the thicknesses have increased southward and show a close relationship with the Red Sea. Sedimentation is not uniform throughout the Gulf of Suez because it is controlled by the subsidence of the individual subbasins and the proximity to the graben shoulders.

In the Gulf of Suez, the Miocene sequence is divided into the following groups and formations according to the EGPC Stratigraphic Committee (1964) and Stratigraphic Subcommittee of the National Committee of Geological Sciences (1974) (Figures 31, 32). A similar division was followed by the stratigraphic lexicon and explanatory notes to the geological map of Egypt (Hermina et al., 1989); however, no detailed subdivision is given for the Ras Malaab Group, which is critical for the Gulf of Suez. The Miocene groups and formations are as follows. The Ras Malaab Group is divided into the Zeit Formation (Tortonian, Messinian), South Gharib Formation (Serravalian), and the Belayim Formation (Langhian–Serravalian). The Gharandal Group consists of the Kareem Formation (Langhian), Rudeis Formation (Burdigal-

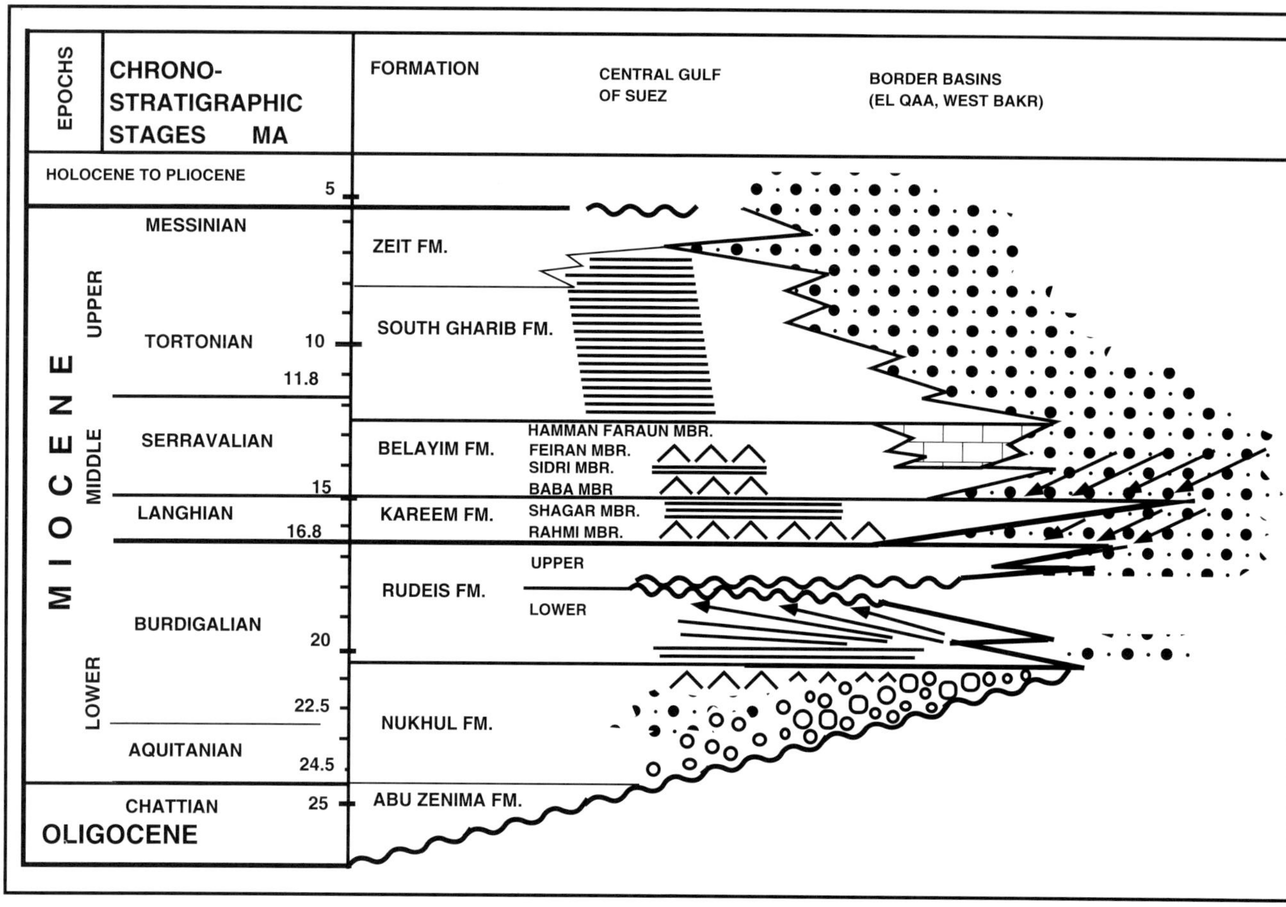

Figure 32. Miocene event stratigraphy, Gulf of Suez.

ian), and the Nukhul Formation (Aquitanian to Burdigalian).

Nukhul Formation

The basal Miocene beds (or the Nukhul Formation) were the first sediments deposited in the clysmic gulf. They are not found elsewhere in Egypt (Figure 12). Nukhul sediments were the result of the first erosional phase after initial tectonic activity, which consisted of uplift and tilting of fault blocks on a relatively small scale. Fault displacement was generally in the 10 m range, only locally up to 100 m. Subsequent subaerial erosion of the crests of these fault blocks resulted in Nukhul sedimentation.

Nukhul sediments consist of limestones, marls, anhydrite and polymictic conglomerates (mainly limestones and chert nodules derived from Eocene rocks) and, increasingly to the south, sandstones. Nukhul sediments overlie, with a sharp unconformity, pre-Miocene strata, generally middle or lower Eocene. Their thickness increases down flanks of fault blocks and is greatest in front of faults (Figure 33). The coarsest sediments (i.e., the polymictic conglomerates) are generally found there (Figure 22). At Ras Gharib field, a thick layer of chert nodules underlying the nullipore carbonates is found over a wider area along the flank, probably due to reworking of Eocene sediments on a beach. On high trends (e.g., on the Amer-Bakr-Gharib high) the Nukhul Formation is represented by shallow-water carbonates. In the southern gulf as far north as Ramadan field, Nukhul evaporites developed, indicating initial desiccation events. The anhydrite layers decrease in number from four in the south to one in the north.

Nukhul sediments are interpreted to be predominantly freshwater (fluviatile to lacustrine) to marginal marine. Deposition occurred near beach or in littoral to inner littoral environments. The carbonates are marine.

The average Nukhul thickness seems to be 10 to less than 30 m over most of the gulf. In general, the Nukhul thickens toward the south. An over 300-m-thick section of the Nukhul is found in the southern part of the gulf (Zeit Bay-1, Shoab Ali-1 wells); however, it is extremely difficult to draw a reliable isopach map for the Nukhul because thicknesses change drastically over short distances (Figure 33). Because most wells in the gulf are drilled on highs in carbonate facies, where the Nukhul tends to be thin, sufficient information is not available. Immense clastic input came from the graben shoulders, which were more severely eroded after the initial breakup. Along the border basins, very thick clastic wedges were built out into the basins (e.g., the El Qaa basin and the Zeit Bay basin). In the Hurghada/Islands area, severe erosion persisted through the early

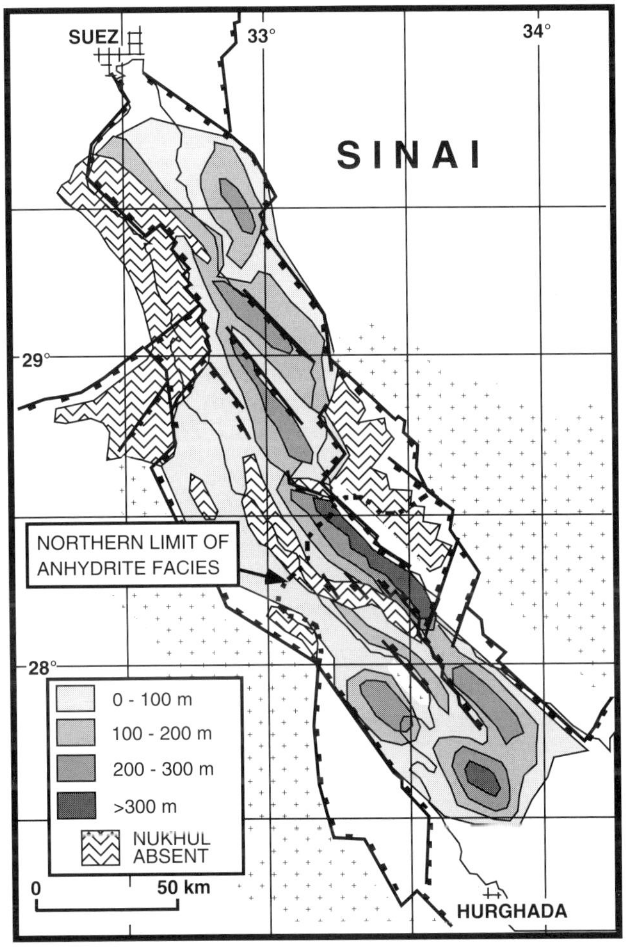

Figure 33. Nukhul Formation isopach map.

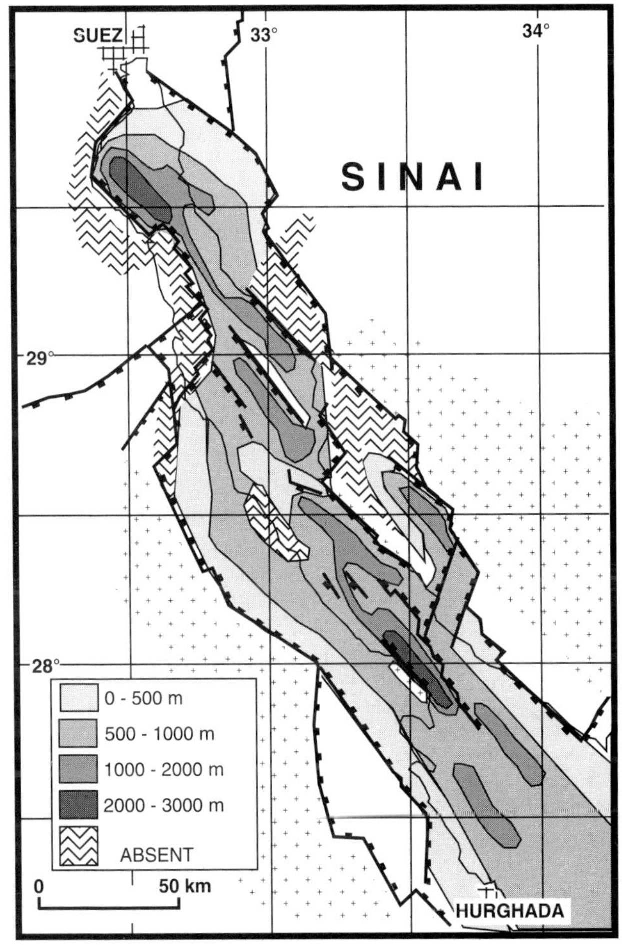

Figure 34. Rudeis Formation isopach map.

Miocene and the extremely thick Nukhul probably includes Burdigalian-aged sediments.

Unfortunately, dating of the Nukhul is difficult because it is poorly fossiliferous. The Nukhul is assigned to Aquitanian (Robertson Research, 1985). Microfossils (*Miogypsina* sp., *Elphidium* sp., and ostracods) indicate a shallow, warm environment.

Several subdivisions were attempted by various authors (e.g., EGPC Stratigraphic Committee, 1964; Saoudi and Khalil, 1984); however, this seemed to be a futile effort because Nukhul sediments change thickness and facies laterally over very short distances. As a mainly clastic sediment, deposition also depended largely on the sediment source, which in most places in the gulf was Eocene limestone. Only in areas with high topographic relief could early Miocene erosion cut deeper into the Upper Cretaceous sediments, especially in the southern part of the gulf, where Eocene and Cretaceous sediments were originally only a few tens of meters thick. The best attempt to subdivide the Nukhul Formation, so far, was made by Saoudi and Khalil (1984) because they consider several subbasins.

Distribution of sediments, especially evaporites and the coarse clastics in the south, indicate that the sea invaded the Gulf of Suez from the south in the early Miocene. (Further investigations are necessary to prove this interpretation.)

The Nukhul Formation produces in 17 oil fields as a result of being the first porosity above Eocene source rocks with Rudeis shales forming a seal.

Rudeis Formation (Burdigalian)

The Rudeis Formation consists mainly of greenish gray to brownish *Globigerina* shales and marls. It is the thickest, most widespread, most uniform and consistent Miocene formation (Figure 34) even though the transition from synrift to postrift sedimentation occurred during deposition. Locally, sands were shed into the deep-marine Rudeis shale basin.

After the shallow-marine conditions of the Nukhul Formation, the Rudeis Formation shows a sharp change to open-marine environment as is indicated by the abundant pelagic foraminifera and the outer shelf and bathyal benthonic fauna (mainly various *Uvigerina* species).

The thickness varies considerably because sedimentation occurred in a quickly subsiding fault-controlled basin (Figure 23). Faults were synsedimentary, and active during the early Miocene. Consequently, the greatest thickness was deposited in front of faults. In general, the Rudeis Formation was deposited everywhere in the gulf and attains an average thickness of 500 m. Over a wider area, it is missing only on the Zafarana platform, which was emergent during the early Miocene (see Hagras and Slocki, 1984). On highs (e.g., the Amer-Bakr-Gharib high), only several tens of meters of the Rudeis was deposited. The Rudeis Formation covers the EE 82-FF/GG-83 high with an average thickness of 200 m. This formation thickens to 600 m a short distance away from the high toward the center of the Issaran Basin. At the October high, the Rudeis averages 400 m thick. A maximum thickness of 2200 m of Rudeis Formation was drilled in the Darag Basin in Gupco's well GS 56-1, which is one of the few wells in the deepest part of a Gulf of Suez subbasin. In the Gebel Zeit basin, the Rudeis Formation probably attains more than 2000 m in thickness. There, more than 1000 m has been penetrated in the DEMINEX well NN 89-1. Nevertheless, these numbers demonstrate the rapid subsidence and sedimentation rates during the early Miocene. On the other hand, the thickness changes over short distance in an east-west direction in the different subbasins show that the individual half-grabens subsided independently during the lower Miocene.

The Rudeis shales, marls, and limestones were deposited in cycles as shown by thick sequences on logs. Rudeis cycles inside the same basin can be relatively reliably correlated peak to peak. However, between two adjacent basins correlation seems to fail. Detailed stratigraphic correlation with foraminifera was attempted, but seems to fail mainly because the foraminifera zones during the early Miocene are too long, especially the N7-zone, the *Globigerina trilobus trilobus* zone (El-Heiny and Martini, 1981). Beleity (1984) attempted a stratigraphic zonation, but succeeded only partially because he included benthonic foraminifera that have value only as environmental indicators. Nevertheless, some foraminiferal events can be used to better subdivide and correlate the Rudeis (and Kareem) Formation as experienced Egyptian paleontologists know (S. Fahmi, personal communication, 1983; Beleity, 1984; El-Heiny and Martini, 1981). The best example is the Orbulina datum or the first occurrence of *Orbulina suturalis* at the top of the Rudeis. Other intra-Rudeis levels can be determined with the "blooming" or extraordinary abundance of foraminifera species in one horizon over the whole gulf, although these are uncommon events.

Various subdivisions of the Rudeis Formation were attempted, but most are valid for one subbasin only (Stratigraphic Subcommittee of the National Committee of the Geological Sciences, 1974; Allen et al., 1984; Webster and Ritson, 1982).

The best attempts so far are those that combine facies biostratigraphically using planktonic foraminifera and calcareous nannofossils (El-Heiny and Martini, 1981) An upgraded version of this biofacies subdivision was presented by Robertson Research (1985). The Rudeis Formation can be subdivided into an upper (postrift) and lower (synrift) Rudeis with two subdivisions.

Lower Rudeis Formation

(1) The upper middle part of the lower Rudeis is the lower part of N8. It is the *Globigerinoides sicanus* zone of El-Heiny and Martini (1981). The depositional environment was deep to shallow outer sublittoral. At the base of the *G. sicanus* zone, relatively shallow water is proven by the sudden occurrence of an arenaceous foraminiferal assemblage. (2) The lower middle part of the Rudeis includes foraminifera zone N7 and was of outer shelf to shallow upper bathyal origin (*Uvigerina*). (3) The lower Rudeis includes the uppermost foraminifera zone N6 and was deposited in upper bathyal conditions.

These biofacies are easily recognized in the central part of the basins. At the basin margins and around high trends the open-marine facies changes into shallower, more calcareous or sandy sediments. The sand content is high especially in the border basins (e.g., El Qaa basin), where predominantly sand was deposited and only limited marine shale reached the Sinai basement. In the West Bakr basin, an enormous amount of sand was shed during lower Rudeis deposition (Figure 23). There, according to Allen et al. (1984, p. 9), ". . . the individual sand deposits tend to be small and confined to structural depression and troughs. . . . The individual sandstones are often of limited extent laterally, i.e., along depositional strike, and abut or onlap onto graben or valley walls. Along depositional dip, sand trends do not extend more than 2 or 3 kilometers. Facies changes and shaling out can occur rapidly in a downdip direction, which is often to the E or SE. Individual sand bodies map as fan or tongue-shaped deposits, sometimes very narrow (< 1 km) and channel-like in the case of narrow valley fills" (Figures 14, 35).

Much sand sourced from Nubia sandstones came from the southwest during lower Rudeis deposition and was trapped up to 300 m thick (Hashem-1 well) in the West Bakr basin by the Amer-Bakr-Gharib high and in the Zeit Bay basin by Gebel Zeit. Only at a narrow pass south of Ras Gharib field, sand spilled over in the central gulf and was trapped again in the Shukheir/July area, where a narrow, very fast subsiding graben was filled. No sand crossed the gulf to the east. The area around the Ramadan field consists mainly of a limestone facies.

On the Sinai side, the El Qaa basin received thick sands derived from Nubian and basement rocks throughout the lower Miocene. These sands were trapped most of the time in the El Qaa basin and could spill over to the Morgan Basin only during deposition of the upper Rudeis and mainly during Kareem deposition (e.g., at Morgan field/GS 316-1 or at the north end of Gebel Durba at Belayim). Only at the Belayim fields did a breach north of Gebel Abu Durba allow sand from the El Qaa basin to be shed for short time intervals (at Belayim Land field during Rudeis Formation deposition and at Belayim Marine field during Kareem Formation deposition, Langhian).

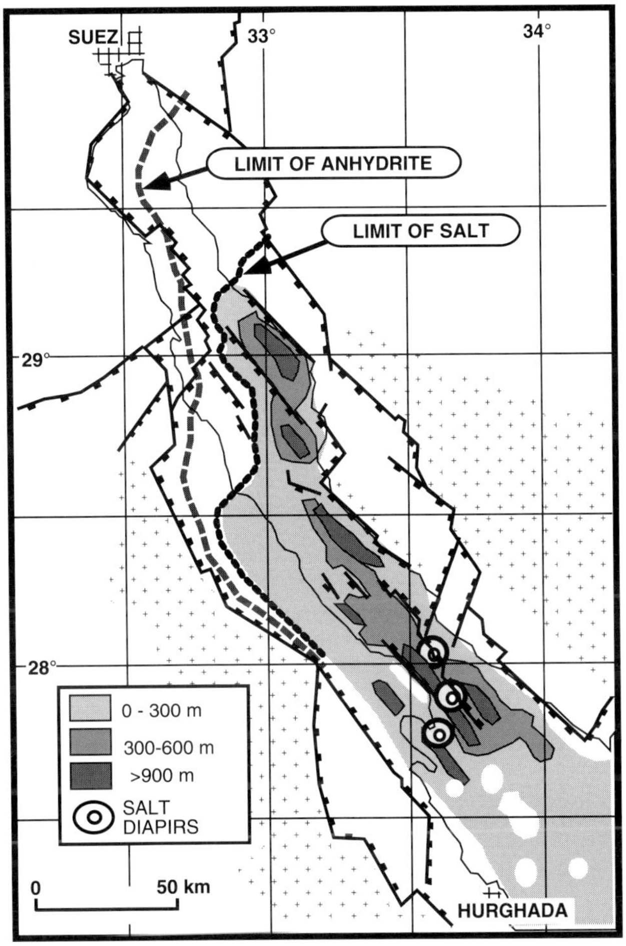

Figure 35. South Gharib Formation isopach map.

The northern part of the basin near Morgan field is an interesting example of an isolated depocenter with sand from a point source (Figure 11). The northern Morgan Basin was blocked off from sand input by the Gebel Araba/El Tor ridge. Only south of El Tor was there an outlet of the El Qaa basin. Through this gate, sands were transported as turbidity flows into the basin axis of the Morgan Basin. Only the basin axis received sands in during Rudeis deposition, as wells prove (gross sand in Rudeis Formation: H-1, 260 m; GS 316, 518 m; GS 326-1, 500 m). On the eastern and western periphery the Rudeis is shaly and marly (compare also Hagras and Slocki, 1984). During Kareem deposition, the basin axis shifted farther west and turbidite sands up to 600 m thick filled the low. The Morgan area was located on a submarine high west of the low axis, but the turbidites had enough energy to travel up the slightly elevated area. There, the flow energy was broken and sands settled. The Kareem sands in the Morgan field attained a gross thickness of 500 m. Only thin sands reached as far west as well LL 87-1. Toward the north, the submarine low was blocked, so only few sands reached GS 305. The entire eastern periphery of the basin was totally bypassed by the Kareem sands. From the breach in the El Tor ridge, a second channel extended toward the southwest and probably supplied the Amal field area with sands.

The "B"-high trend received only very minor amounts of sand (e.g., at Amal field), but some breaches existed to allow the sands from the east to cross into the generally sand-deprived Gebel Zeit Basin (e.g., NN 89-1). No sands from the Sinai side crossed the gulf.

The Rudeis of the northern part of the gulf contains much less sand and it occurs mainly as fans near input areas (e.g., near Wadi Araba). The Darag Basin is essentially devoid of sand.

During the early Miocene, the erosion on the uplifted bordering blocks in the north had not cut deeply enough everywhere to provide sand from the Nubia or from the basement. Near the Darag Basin, thick Eocene and Cretaceous carbonates could not supply sand at all. In the northern part of the Sinai the Nubia sandstones were only recently reached by erosion. Nubia sand is still preserved on basement, e.g., at Wadi Baba east of Abu Rudeis. The southern Sinai as well as Gebel Zeit were uplifted quickly during the early Miocene and at that time erosion cut down to basement, as is proved by *Globigerina* limestones in fractures at Gebel Zeit.

Upper Rudeis sands are more evenly distributed throughout the Gulf. Significant numbers of sand lobes surround the areas of clastic influx.

Postrift

Upper Rudeis Formation

The upper Rudeis coincides with the Praeorbulina glomerosa zone (which is the upper part of the standard planktonic foraminifera zone N8) and the upper limit is well defined by a first occurrence of *Orbulina suturalis*. The environment is mostly shallow outer sublittoral, as benthonic microfossils indicate.

The base of the upper Rudeis coincides with the initiation of significant limestone deposition in the northeast (Asl Member). Because this relatively shallow-water event can be traced all over the gulf, it is interpreted as having been caused by a drop in sea level.

The shoreline of the early Miocene gulf is marked by limestone buildups, which locally contain corals. However, dating of these carbonates is problematic. Although these limestones may not have been deposited throughout the lower Rudeis, they certainly represent the fully marine conditions of the basin (and not the evaporitic cycles). At the Esh el Melaha range, at Gebel Abu Shaar el Gibli, *Globigerina* foraminifera are found in some carbonate layers overlying the reefal buildup. Along the samc paleocoast farther north at the entrance of Wadi Esh, deep-water stromatolites cover steep basement slopes. At the south side of Gebel Zeit, the basement was exposed underwater during the early Miocene, and pelagic micritic limestones containing *Globigerina* were deposited on basement and in basement fractures (Zankl and Klitzsch, 1985).

Kareem Formation (Langhian)

The Kareem Formation has been divided into two members: (1) the upper Shagar Member and (2) the lower Markha Member.

It seems better to introduce a differentiation of three facies: a fully marine facies that coincides with the Shagar Member; an anhydritic, restricted-environment facies (the Markha Member); and a sandy facies, which attains a thickness of 250 m at Morgan and Belayim Marine fields. The average thickness of the Kareem Formation is about 200 m over extensive areas in the gulf. In the basin centers, thicknesses of generally 300 m were attained and thicknesses of up to 500 m were attained at the northern end of the October/Abu Zenima basins (Figure 12).

Markha Member

Several meters above the Orbulina datum in the top section of the Rudeis shales, a break in sedimentation occurs, and anhydrites and shales with little or no fauna indicate a restricted environment. Most of the wells in the gulf encountered this break and the anhydrites; however, some wells drilled in the deeper part of the Kareem basin did not find the anhydrites (e.g., several wells west of the Belayim Marine field). In the deeper parts of the Kareem basin, sedimentation was continuous. The restricted fauna points to less-marine conditions, probably due to a cut off of the marine connection or a sea level drop in the Gulf of Suez. The Markha Member, therefore, was a lateral facies in marginal basinal areas (which cover large areas in the gulf).

Shagar Member

In the Shagar Member, full-marine conditions were established again and sedimentation of *Globigerina*-rich marls continued, as in the Rudeis Formation, with similar high subsidence and sedimentation rates.

All areas of sand provenance were shedding coarse clastics; especially well documented are the fans built out at Belayim and at the southern end of the El Qaa plain toward the Morgan field (Figure 12). At Belayim Marine field, the Kareem sands have scoured deeply into the underlying strata and cores show debris flows and turbidites.

The sandstone geometry in front of the Belayim Marine scarp indicates sharp channeling and submarine fans. Paleontology from interbedded shales supports a deeper water, outer shelf to upper bathyal environment for the fan sands, although fauna in the sands is a mixture of both outer sublittoral to inner sublittoral forms.

Toward the top of the Kareem Formation, the marine conditions seem to persist as the upper shale cycles correlate over large areas in the gulf until another sharp break to anhydrites starts the Belayim Formation.

The Kareem Formation is assigned to the standard foraminifera zone N9-10, which, according to one group of paleontologists specialized in the Mediterranean stratigraphy, belongs to the Langhian, i.e., to the middle Miocene (Steininger and Roegl, 1981; El-Heiny and Martini, 1981). Other authors follow the worldwide stratigraphy of Harland et al. (1982) (e.g., Robertson Research, 1985) and include the Kareem partly in the Burdigalian, i.e., lower Miocene and partly the Langhian.

Belayim Formation (Langhian?/Serravalian)

The Belayim Formation is a four-fold sequence of evaporites and shales with some limestones and sandstone intercalations (Figure 36). "It is the youngest fossiliferous planktonic level in the Gulf of Suez. It is characterized by the presence of *Globigerinoides mitrus, Globorotalia siakensis, Globorotalia* cf. *scitula* and *Orbulina suturalis*. The zone is equivalent to the planktonic foraminiferal zone N11." (El-Heiny and Martini, 1981, p. 104). Other paleontologists place the Belayim in the standard foraminaferal zone upper N9 (Robertson Research, 1985).

In all Gulf subbasins, the Belayim can be subdivided into four members: (1) Hammam Faraun Member, shale, limestones and sand; (2) Feiran Member, evaporites and shale, locally sand streaks; (3) Sidri Member, commonly shale, but locally thick sands; and (4) Baba Member, evaporites, commonly salt.

Facies changes occur at the basin margins, where clastics replaced anhydrite of the Baba and Feiran members or where lithothamnion limestones (nullipore, Figure 14) mark the shoreline of the Hammam Faraun Member. Generally, the boundaries between the underlying Kareem Formation and between the different members are well defined on logs.

The Baba Member consists of a basal anhydrite layer, a middle salt layer in depocenters (southern October Basin and Belayim Marine field up to 180 m thick; in the Gebel Zeit Basin several tens of meters), and an upper anhydrite layer. The average thickness of the Baba Member is 75 m. In areas where the salt is missing, the lower and upper anhydrite are merged and cannot be differentiated.

The Sidri Member is typically a thin shale unit (1–5 m), which is very persistent over all the gulf (except in the Darag Basin). Sands are developed locally and are producing horizons, e.g., Belayim Land and Belayim Marine fields.

The Feiran Member is comprised of several massive anhydrites with intercalating shales. In the southern gulf, in the Morgan and Gebel Zeit basins, thick salt replaces the anhydrite in depocenters, e.g., 400 m of a Feiran salt swell on the flank of the Gebel Zeit Basin in the well NN89-1. Similar salt swells are found in the Morgan field. The salt thickness generally increases southward. Lateral facies variations are common in the Feiran Member. As in other Miocene strata, sandstones increase toward the basin margins. The Feiran Member is the thickest of the Belayim members and reaches an average thickness of 100 m.

The Hammam Faraun Member is composed of a highly variable lithology of limestones, shale, and sandstone. It is the first porous formation under the sealing thick evaporites and generally contains oil. Facies change laterally, especially near the basin margins. Thicknesses average 50 m and rarely exceed 100 m.

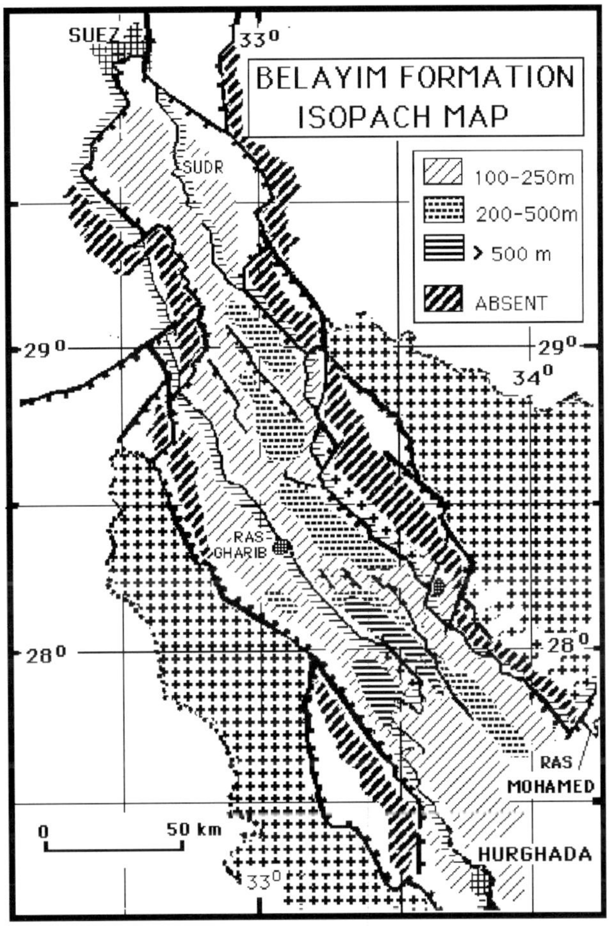

Figure 36. Belayim Formation isopach map.

Along the paleoshore, a unique carbonate facies developed consisting of reefal buildups with corals and lithothamnion algal shoals. The fossiliferous shale intercalations between the nonfossiliferous evaporites contain marine fauna associations of the upper bathyal to outer neritic life zones. This relationship documents a rapid and sudden change between the facies. Each marine clastic intercalation represents a new refill of the basin and a connection with the ocean. The reefal limestones containing corals and lithothamnion are especially important in proving the existence of fully marine conditions. These limestones attain a tremendous thickness up to more than 300 m in several places in the gulf (Gebel Abu Shaar el Gibli, at Zeit Bay and Gemsa fields, at Ras Gharib and Ras Fanar fields, in well EE 81-1).

The relatively uniform thicknesses in all members of the Belayim Formation indicate relatively quiet tectonic conditions with equal subsidence across the gulf (Figure 36). This indicates that the subsidence occurred at the main graben boundaries and the gulf acted as a broad subsiding graben. This was a change in comparison with the Rudeis or Kareem formations, where the individual half-grabens controlled the sedimentation. During Belayim deposition, only the southern part of the gulf shows higher subsidence rates; however, only limited information is available.

Nullipore (Lithothamnion) Limestones

The nullipore or lithothamnion limestones are a unique carbonate facies that marked the shoreline or islands throughout the early and middle Miocene except during deposition of evaporites (Figure 14). The limestones and dolomites consist of biogenic limestones, organoclastic carbonates, and, in some places, coraline algal limestone and laminated algal dolomites. At Gebel Zeit, outcrops with molluskan beds containing oysters and pecten, bryozoan limestones, and foraminiferal limestones with mass occurrence of *Miogypsina* sp. and other foraminifera are found. These represent shallow-marine conditions in the early Miocene (Aquitanian) (Zankl and Klitzsch, 1985).

Superb outcrops of a carbonate sequence exceeding 100 m in thickness and consisting of massive reefs are exposed at Gebel Abu Shaar el Gibli, in Wadi Kharaza. Reef slope, fore reef, reef core, and lagoon with isolated backreef are well exposed. The reef started to grow directly on basement or on a basal conglomerate. At some places, basement boulder conglomerates fill canyons in the carbonates and are cemented with a carbonate matrix (east cliff of Gebel Abu Shaar, south of Bir Abu Shaar). The age of this sequence probably spans early and middle Miocene; however, good stratigraphic proof is lacking (see Cofer et al., 1984). Giant stromatolites, 10 m high, grew in a lagoonal facies at the upper part of Wadi Kharaza (Zankl and Klitzsch, 1984).

Similar carbonate units are documented in the subsurface at Gemsa field (age uncertain), Zeit Bay field (after Kareem deposition, probably Belayim-Hammam Faraun equivalent), Ras Gharib and Fanar fields (top section of Hamman Faraun equivalent, lower section possibly older), the well EE 81-1 (probably spanning lower and middle Miocene), and the Thelemet-1 well.

South of the Gulf of Suez, these carbonates are found on the uplifted shoulders of the Red Sea, commonly on basement, e.g., near Safaga and at many places as far south as Quseir and the Sudanese border.

On the Sinai, nullipore limestones extend along the east flank of Gebel Ekma and to the south at Gebel Abu Suweira.

The nullipore limestone facies seems to have preserved the paleoshore of the early and middle Miocene sea. Due to their excellent porosity, these carbonates are main oil-producing horizons and are a primary exploration target.

South Gharib Formation

The Turonian South Gharib Formation is comprised of anhydrites with shale intercalations at the periphery of the basins and major salt layers in the depocenters (Figure 31). The South Gharib Formation covers most of the gulf with well-correlatable shale units (e.g., "shale III, IV, and V") between the salt. Several major salt layers are recognized by exploration geologists and are called "lower," "main," and "major" salt. The South Gharib Formation attains an average thickness of 200 m (Figure 35); however, in the depocenters and toward the Red

Sea, the thickness increases considerably and reaches 2000 m in salt diapirs. In the gulf, salt diapirs were formed in the Gebel Zeit Basin exclusively due to great salt thickness and continuous tectonic activity during the latest Miocene up to the Holocene. Farther south in the Red Sea, more than 3000 m of Zeit and South Gharib salt were drilled in wells offshore Quseir.

To the north in the Issaran-October and Abu Zenima basins the thickness of the South Gharib Formation thins to 0 m and defines the perimeter of the evaporite basin (Figure 35). In the Darag Basin very thin evaporites were deposited (60 m of South Gharib Formation in Nebwi 81-1).

Fawzy and Aal (1984) compiled a study on Miocene evaporites and subdivided the South Gharib Formation into two informal units: the lower South Gharib and upper South Gharib.

"The lower South Gharib unit represents the informal Gupco rock units: lower, major and main salt, which are traceable only in the Morgan trough. This salt unit contains minor streaks of anhydrite, shale and sands. This member varies in thickness and attains its maximum thickness in the southern part of the gulf (Amal-2). This maximum thickness is attributed in part to salt movement. . . . The upper South Gharib unit is characterized by a sequence of salt and anhydrite interbeds" (Fawzy and Aal, 1984, p. 5).

South Gharib evaporation started abruptly and left the existing fringing lithothamnion carbonate buildups subaerially exposed. Evaporites, layer by layer, filled up the basin, abutting against the reef wall until the sabkha transgressed the carbonate islands. This is well documented in Ras Fanar field where excellent correlation of evaporite and shale layers basinward of, and above, the "nullipore" carbonates is possible. Therefore, anhydrites interfingering with nullipore carbonates are younger than the carbonates and not a lateral facies. The limestones represent fully marine conditions and the evaporites document a later dry environment.

Similar to the Belayim Formation, relatively uniform evaporite deposition continued over large areas in the gulf during deposition of the South Gharib Formation, indicating that the gulf subsided as one broad graben along the main border faults. The general thickness increase toward the south is taken as proof of faster subsidence in the southern part, possibly due to wider extension of the gulf near the Red Sea. The thin South Gharib evaporites in the Darag Basin prove a barrier between the October Basin and the Mediterranean Sea during deposition of the South Gharib. This barrier in the "Mitla Strait" was not very effective because the sea invaded the dried-out basin more than 20 times. The extensive Red Sea was part of the same evaporite basin as the Gulf of Suez.

Zeit Formation (Latest Miocene)

Evaporite sedimentation continued during deposition of the Zeit Formation, but with fewer salt deposits in the Gulf of Suez. Only in the Gebel Zeit Basin are thick salt pillows located in the axis of the basin as seen only on seismic because no well has yet been drilled. In the vertical sequence, several relatively thin salt layers are recognized in many wells, especially in the upper section where a very widespread salt (the "first salt") represents the last complete filling of the basin. This "first salt" contains in the depocenters of the subbasins a potassium salt layer of approximately 1 m thick, which is easily identifiable on gamma-ray logs. Because very few potassium salts occur throughout the South Gharib and Zeit formations, it is an excellent marker horizon for correlation.

Throughout Zeit Formation deposition considerable amounts of sandstones were shed from the uplifted graben shoulders. An exceptionally thick prograding fan was built out from the east around Feiran into the October Basin. The fan extends toward the northern Belayim Marine field. Similar fans are present in all other basins and make it impossible to define the upper boundary of the Zeit Formation with lithological criteria. Only in the Hurghada field have geologists successfully defined the Miocene/Pliocene boundary with the aid of microplankton (Diatomeae) (Bowman, 1926, 1931).

The average thickness of the Zeit Formation is 500 m, but wide parts of the gulf are covered with up to 1000 m of the Zeit (e.g., in the October Basin, wells in the FF/GG-83 area). Subsidence and sedimentation rates, as well as facies trends, continued, as in the underlying South Gharib Formation.

Pliocene to Holocene

Pliocene to Holocene deposits reach a considerable thickness, generally 1000 m, but in some fast-subsiding areas, as in West Feiran, these deposits reach thicknesses of up to 2000 m. This relates to an average subsidence rate of 40 cm/k.y. The post-Miocene strata in the gulf primarily contain coarse clastics, but locally, as in the central axis, also contain limestones, oolitic limestones, and chalk.

The Darag Basin, which after a period of slow subsidence continued to sink rapidly, contains two units: the upper evaporitic unit with the "Zafarana anhydrite" and, locally, with a "Tayiba salt;" and the lower clastic unit, the "Wardan clastics." The evaporitic unit extends southward into the October Basin. There, a Pliocene salt is recorded in well GS 160-1 (Fawzi and Aal, 1984).

In the El Qaa basin, predominantly coarse clastics were deposited and various authors have attempted to name the formation (Webster and Ritson, 1982; Fawzi and Aal, 1984).

In the central axis of the gulf, shale and carbonate facies are represented by thicknesses of up to several 100 m (e.g., wells East Bakr M-1, KK 85-1, Ramadan field, NN 89-1). For this area, outcrops at the flank of Gebel Zeit are typical. According to Said (1962) the Pliocene strata are massively bedded limestone with abundant ostrea, marly chalk, and oolitic limestones with some coarse clastic interbeds.

The thickness and the clastic character of the Pliocene to Holocene strata resemble that of the peripheral Zeit Formation and are taken as proof of the continuity of tectonic subsidence of the gulf with isostatic uplift of the shoulders.

CONCLUSIONS

The Gulf of Suez is an intracratonic rift basin that originated in the early Miocene. Stratigraphic evidence and facies distribution of pre-Miocene rocks do not support rifting prior to the early Miocene. During the Carboniferous, the wide embayment covering the Gulf of Suez area had its depocenter west of Cairo. The carbonate facies in the Cretaceous and Eocene extended mainly east-west, parallel to the Tethyan coast.

During the Oligocene, most of the Gulf of Suez area was subaerially exposed and eroded with the erosion cutting down to basement in the Red Sea area, where only a thin sediment cover existed.

The initial rifting, approximately 22 Ma, was accompanied by basalt intrusions of sills and dikes in the gulf and by basalt flows from fissures several hundred kilometers away from the gulf in the Western Desert. In contrast to the East African rift system, rifting in the Gulf of Suez occurred at sea level. After a short period of fluviatile and lacustrine sedimentation during deposition of the Nukhul Formation, the sea intruded the rift from the very beginning, as evidenced by pillow basalts, forminifera, and evaporites. Marine sedimentation continued with some continental influence along the graben edges.

Extension and subsidence were almost continuous to the present, but there were different amounts of subsidence and different tectonic expression in the different subbasins.

The initial rifting during the Aquitanian created relatively small faults and fault blocks. During the Burdigalian and Langhian (Rudeis and Kareem formation deposition) subsidence and faulting increased, resulting in larger and more tilted fault blocks. In the Serravalian, the extension was such that it decoupled the Sinai from the African stress system and the whole of the gulf subsided, controlled mainly by the faults at the graben shoulders, thus creating the wide evaporite basin of the South Gharib and Zeit formations.

The northernmost basin, the Darag Basin, was not affected at this time by the decoupling and did not subside as much. The rotational opening of the gulf did allow the Darag Basin, after the initial rifting and large subsidence in the Burdigalian, to subside more rapidly again only in the Pliocene to Holocene. The Ain Sukna/Abu Zenima strike-slip fault probably played an obstructing role during the middle Miocene.

Since the beginning of Miocene, the central parts of the rift have subsided between 6 and 7 km, which is about one-third the thickness of the continental crust. The isostatic uplift of the rift shoulders ranges from a few meters in the north to approximately 2–3 km over most of the length of the rift, thereby creating a north-plunging basement. Uplift of the graben shoulders started early and some locally restricted basement outcrops existed relatively early in the Burdigalian/Langhian (e.g., at Gebel Zeit and Sinai basement east of the Belayim and Morgan fields). However, large areas of basement were eroded only during the Pliocene.

The influence of global sea level fluctuations during the Miocene was not studied; however, it would be difficult to detect such influence in the tectonically active Red Sea/Gulf of Suez rift.

Extension is interpreted to be normal to the rift (with only a very small and probably episodic right-lateral component). Generally, Miocene faulting created a new set of fault directions; however, in some cases an older "structural grain" was followed. Older (Jurassic) fault directions are mainly Wadi Araba parallel and often offset clysmic faults, thereby creating "compartmented" fault-block units. Similar long fault-block units with offsets by the Gulf of Aqaba system are obvious on the Sinai.

ACKNOWLEDGMENTS

I wish to thank C.H. Squyres from Conoco Egypt, who initiated this paper in 1985; Professor R. Said, who gave me moral support through his patience; and my colleagues at DEMINEX, S. Shaheen, Abd-El-Aziz El Naggar, Dr. A. Fawaz, Dr. H. El-Hilaly; and at Conoco, N. Sultan, Dr. E. Breman, M. Spangler, P. Van Den Heuvel, H. Bisewski, P. Whitley, I. Taha, Dr. Kerdany, and, last but not least, M. Hermina. The paper was reviewed by J. Davidson, Conoco, and S. Landon.

The fault pattern onshore used in this chapter is derived from a LANDSAT base including field maps from the Technical University Berlin 1:100,000 and from aerophotographic maps from GEOMAP 1:50,000. The fault pattern offshore in Figure 3, etc. is at top pre-Miocene, adopted and simplified for a small scale from unpublished Conoco structural maps (authors, K. Schütz, D. Brown, M. Spangler, E. Breman, P.V.D. Heuvel, et al.).

REFERENCES CITED

Abdallah, A.M., A. Adindani, and S. Fahmi, 1963, Stratigraphy of lower Mesozoic rocks, western side of Gulf of Suez: Egyptian Geological Survey, paper no. 27, Cairo, 23 p.

Abdallah, A.M., and A. Adindani, 1965, Stratigraphy of upper Paleozoic rocks of the western side of the Gulf of Suez: Egyptian Geological Survey, paper no. 25, Cairo.

Allen, G., A. Ayyad, G. Desforges, M. Haddadi, and J. Pizon, 1984, Subsurface sedimentological study of the Rudeis Formation in Kareem, Ayun, Yusr and Shukheir Fields: 6th Egyptian General Petroleum Company Exploration Seminar, Cairo, p. 164–176.

Ansary, S.E., S.F. Andrawis, and S.E. Fahmy, 1963, Biostratigraphic studies of the upper Cretaceous sections in GPC wells, Eastern Desert and Sinai, 4th Arab Petroleum Congress, paper no. 24, Beirut.

Bayoumi, A.I., 1983, Tectonic origin of the Gulf of Suez, Egypt, as deduced from gravity data: Extract, Handbook of Geophysical Exploration at Sea: CRC Press, Florida.

Bayer, H.J., 1985, Kollisions und Kompressionstektonik am Ostsaum des Golfes von Aqaba: Geologisches Rundschau, v. 74, p. 3599–3610.

Beleity, A.M., 1984, The composite standard and definition of paleoevents in the Gulf of Suez: 6th Egyptian General Petroleum Company Exploration Seminar, Cairo, 1982, p. 181–198.

Ben-Avraham, Z., 1978, The structure and tectonic setting of the Levant continental margin, eastern Mediterranean: Tectonophysics, v. 46, p. 313–331.

Ben-Menahem, A., A. Nur, and M. Vered, 1976, Tectonics, seismicity and structure of the Afro-Eurasian junction—the breaking of an incoherent plate: Phys. Earth Planet. Inter., v. 12, p. 1–50.

Biju-Duval, B., I. Dercourt, and X. Le Pichon, 1977, From the Tethys Ocean to the Mediterranean Seas: a plate tectonic model of the evolution of the Western Alpine System: International Symposium on the Structural History of the Mediterranean Basins: Editions Technip, Paris, p. 143–164.

Blow, W.H., 1969, Late middle Eocene to Recent planctonic foraminiferal biostratigraphy, Proc. 1st Intl. Conf. Planct. Microfoss, Geneva, 1967, E.J. Brill, Leiden, v. 1, p. 199–422.

Bowman, T.S., 1926, Report on boring for oil in Egypt, section II: Sinai Mines and Quarries Department, Cairo, 91 p.

Bowman, T.S., 1931, Report on boring for oil in Egypt, section III: Eastern Desert and adjoining islands: Mines and Quarries Department, Cairo, 353 p.

Breman, E., 1985, Facies distribution of the Eocene Thebes Formation, Conoco unpublished internal report, Cairo.

Brown, C., and R.W. Girdler, 1982, Structure of the Red Sea at 20°N from gravity data and its implication for continental margins: Nature, v. 298, p. 51–53.

Chenet, P., J. Letouzey, and E.S. Zaghoul, 1984, Some observations on the rift tectonics in the eastern part of the Suez Rift: 7th Egyptian General Petroleum Company Exploration Seminar, Cairo, March 1984, 33 p.

Cochran, J.R., 1983, A model for development of Red Sea: AAPG Bulletin, v. 67, n. 1, p. 41–69.

Cofer, C.R., K.D. Lee, and J.L. Wray, 1984, Miocene Carbonate microfacies, Esh El Mellaha Range, Gulf of Suez: 7th Egyptian General Petroleum Company Seminar, Cairo, p. 97–115.

Dewey, J.F., W.C. Pitman III, W.B.F. Ryan, and J. Bonnin, 1973, Plate tectonics and the evolution of the Arabian plate system: Geological Society of America Bulletin, v. 84, p. 3137–3180.

Drake, C.L., and R.W. Girdler, 1964, A geophysical study of the Red Sea, Geophysical Journal of the Royal Astronomical Society, v. 8, p. 473–495.

EGPC Stratigraphic Committee, 1964, Oligocene and Miocene rock-stratigraphy of the Gulf of Suez region: Egyptian General Petroleum Company, Cairo, 142 p.

El-Heiny, I., and E. Martini, 1981, Miocene foraminiferal and calcareous nannoplankton assemblages from the Gulf of Suez region and correlations: Geol. Med., Book 8, no. 2, 1981, p. 101–108.

Fawzy, H., and A. Abdel Aal, 1984, Regional study of Miocene evaporites and Pliocene–Recent sediments in the Gulf of Suez: 7th Egyptian General Petroleum Company Seminar, Cairo, p. 49–74.

Garfunkel, Z., 1981, Internal structure of the Dead Sea leaky transform (rift) in relation to plate kinematics: Tectonophysics, v. 80, p. 81–108.

Garfunkel, Z., and Y. Bartov, 1977, The tectonics of the Suez rift: Geological Survey of Israel Bulletin 71, 43 p.

Garfunkel, Z., and B. Derin, 1985, Permian–early Mesozoic tectonism and continental margin formation in Israel and its implications for the history of the Eastern Mediterranean, *in* A. Quennell, ed., Geological evaluation of the eastern Mediterranean Sea: Geological Society of London, Special Publication 17, p. 1871–1201.

Ghorab, M.A., 1961, Abnormal stratigraphic features in Ras Gharib Oil Field, Bull. III Arab Petroleum Congress, Alexandria.

Girdler, R.W., and P. Styles, 1974, Two-stage Red Sea floor spreading: Nature, v. 247, p. 7–11.

Gvirtzman, G., and B. Buchbinder, 1977, The desiccation events in the eastern Mediterranean during Messinian times as compared with other Miocene desiccation events in basins around the Mediterranean: Symposium on the structural history of the Mediterranean basins, Editions Technip, Paris, p. 411–428.

Gvirtzman, G., and T. Weissbrod, 1985, The Hercynian Geanticline of Helez and the late Paleozoic history of the Levant, in A. Quennell, ed., Geological Evaluation of the Eastern Mediterranean Sea: Geological Society of London, Special Publication 17, p. 177–186.

Hagras, M., 1976, The distribution and nature of the Miocene sediments in the Gulf of Suez: 5th Egyptian General Petroleum Company Seminar, Cairo.

Hagras, M., and S. Slocki, 1984, Sand distribution of the Miocene clastics in the Gulf of Suez: 6th Egyptian General Petroleum Company Seminar, Cairo.

Harland, W.P., A.V. Vox, P.C. Llewellyn, C.A.A. Pickton, A.G. Smith, and R. Walters, 1982, A geologic time scale, Cambridge University Press, 131 pp.

Harms, J.C., 1985, An interpretation of the stratigraphy of rock units within the Nubia Group, Gulf of Suez, Egypt: Unpublished Conoco Report.

Harsch, W., T. Kuper, B. Rast, and R. Sagresser, 1981, Seismotectonic considerations of the nature of the Turkish–African plate boundary: Geologisches Rundschau, v. 70, no. 1, p. 368–384.

Hassan, A.A., 1967, A new Carboniferous occurrence in the Abu Durba, Sinai, Egypt: 6th Arabian Petroleum Congress, Baghdad, v. 2, 8 p.

Hermina, M., E. Klitzsch, and F.K. List, eds., 1989, Stratigraphic lexicon and explanatory notes to the geological map of Egypt: Conoco, scale 1:500,000.

Heybroek, F., 1965, The Red Sea Miocene evaporite basin, *in* Salt basins around Africa: Institute of Petroleum, London, pp. 17–40.

Hinz, K., 1974, Results of seismic refraction and seismic reflection measurements in the Ionian Sea: Journal of Geology, v. 2, p. 35–65.

Hume, W.F., 1916, Report on the oil fields region of Egypt: Egypt Survey Department, Cairo, 103 p.

Hume, W.F., 1921, Relations of the northern Red Sea and its associated gulf areas to the "Rift" theory, Q. Jl. Geol. Soc. Lond., v. 77, p. 96–101.

Hume, W.F., 1962, Geology of Egypt, v. 3, part 1: the stratigraphical history of Egypt from the close of the pre-Cambrian episodes to the end of the Cretaceous period: G.S.E., Cairo, 712 p.

Kampschuur, W., et al., 1983, TECTOSAT Northern Sinai Report: Nigel Press Association, Edenbridge, United Kingdom.

Kennett, J.P., 1982, Marine geology: Prentice Hall, Englewood Cliffs, New Jersey.

Klitzsch, H.E., and H.W. Linke, 1983, Gulf of Suez photogeologic interpretation map, prepared for Conoco Coral, Cairo, A.R.E.

Klitzsch, H.E., J.C. Harms, A. Lejal-Nicol, and F.K. List, 1979, Major subdivisions and depositional environment of Nubia strata, southwestern Egypt: AAPG Bulletin, v. 63, n. 6, p. 967–974.

Le Pichon, X., and J. Francheteau, 1978, A plate-tectonic analysis of the Red Sea–Gulf of Aden area: Tectonophysics, v. 46, p. 369–406.

Le Pichon, X., J. Angelier, and J.C. Sibuet, 1982, Subsidence and stretching, *in* J.S. Watkins and C.L. Drake, eds., Studies in continental margin geology: AAPG Memoir 34.

Levy, Z., 1975, The geological history of southern Israel and Sinai during the Coniacian: Israel Journal of Earth Science, v. 24, p. 19–43.

Linke, H.W., 1984, Gulf of Suez—detailed structural work report: Conoco Field Report, 73 p.

Lowman, P.D., 1985, Mechanical obstacles to the movement of continent-bearing plates: Geophysical Research Letters (NASA), v. 12, n. 5, p. 223–225.

Marriott, R.P., 1981, Gulf of Suez, field structure study: Conoco unpublished internal report.

Martini, E., 1971, Standard Tertiary and Quaternary calcareous nannoplancton zonation; Proc. 2nd Intl. Conf. Planct. Microfossils, Rome, v. 2, p. 739–785.

McKenzie, D.P., D. Davies, and P. Molnar, 1970, Plate tectonics of the Red Sea and east Africa: Nature, v. 226, p. 243–248.

Meshref, W.M., 1982, Regional structural setting of northern Egypt: 6th Egyptian General Petroleum Company Exploration Seminar, Cairo, March 1982, 11 p.

Moon, F.W. and H. Sadek, 1923, Preliminary geological report on Wadi Gharandal area, Petrol. Res. Bull. 12, Government Press, Cairo, 42 pp.

Moore, J.M., 1979, Tectonics of the Najd transcurrent fault system, Saudi Arabia: Journal of the Geological Society of London, v. 136, p. 441–454.

Morelli, C., 1978, Eastern Mediterranean geophysical results and implications: Tectonophysics, v. 46, p. 333–346.

Mueller, J.C., S. Agah, and R. Marriott, 1981, Gulf of Suez, Egypt–Sinai: Conoco unpublished internal report.

Neev, D., 1975, Tectonic evolution of the Middle East and the Levantine basin (easternmost Mediterranean): Geology, v. 3, p. 683–686.

Neev, D., G. Almagor, A. Arad, A. Ginzburg, and J.K. Hall, 1976, The geology of the southeastern Mediterranean Sea: Geological Survey of Israel Bulletin, no. 68, p. 1–51.

Neev, D., J.K. Hall, and J.M. Saul, 1982, The Pelusium megashear system across Africa and associated lineament swarms: Journal of Geophysical Research, v. 87, p. 1015–1030.

Nur, A., and Z. Ben-Avraham, 1978, The eastern Mediterranean and the Levant: tectonics of continental collision: Tectonophysics, v. 46, p. 297–311.

Omara, S., 1972, An early Cambrian outcrop in southwestern Sinai, Egypt: N. Jb. Geol. Paleont. Mh., p. 306–314.

Orwig, E.R., 1982, Tectonic framework of northern Egypt and the eastern Mediterranean region: Mobil, 6th Egyptian General Petroleum Company Exploration Seminar, Cairo.

Owen, H.G., 1983, Atlas of continental displacement: Cambridge University Press, Cambridge, 159 p.

Quennell, A.M., 1984, The western Arabian rift system, *in* J. E. Dixon and A.H.F. Robertson, Geological evaluation of the eastern Mediterranean Sea: Geological Society of London, Special Publication 17.

Robertson Research International Ltd., 1985, Gulf of Suez Basin Study, unpublished report by RRI.

Sadek, H., 1959, The Miocene in the Gulf of Suez region: Geological Survey of Egypt, p. 118.

Said, R., 1962, The geology of Egypt: Elsevier, New York, 377 p.

Said, R., 1980, The Paleozoic of the Gulf of Suez: Conoco unpublished internal report.

Salem, R., 1976, Evolution of Eocene–Miocene sedimentation patterns in parts of northern Egypt: AAPG Bulletin, v. 60, n. 1, p. 34–64.

Saoudi, A., and B. Khalil, 1984, Distribution and hydrocarbon potential of Nukhul sediments in the Gulf of Suez: 7th Egyptian General Petroleum Company Exploration Seminar, Cairo, p. 75–96.

Sattar, M.A., and P.G. Pangman, 1982, Gupco block 347 discovery: 6th Egyptian General Petroleum Company Seminar, Cairo.

Sellwood, B.W., and R.E. Netherwood, 1984, Facies evolution in the Gulf of Suez area: sedimentation history as an indicator of rift initiation and development; Modern Geology, v. 9. p. 43–69.

Siedner, G., 1973, K-Ar chronology of Cenozoic volcanics from northern Israel and Sinai: Fortschritte der Mineralogie, v. 50, p. 129–130.

Smith, C.M., ed., 1984, Well evaluation conference, Egypt: Schlumberger.

Steen, G., 1982, Radiometric age dating and tectonic significance of some Gulf of Suez igneous rocks, *in* G. Hantar, ed., 6th Egyptian General Petroleum Company Exploration Seminar.

Steininger, F.F., and F. Rögl, 1981, Paleogeographic and palinspastic reconstruction of the Neogene of the Mediterranean and Paratethys, *in* A. Quennell, ed., Geological evaluation of the eastern Mediterranean

Sea: Geological Society of London, Special Publication 17, p. 659–668.

Stratigraphic Subcommitee of the National Committee of the Geological Sciences, 1974, Miocene rock stratigraphy of Egypt: Egyptian Journal of Geology, v. 18, n. 1.

Sultan, N., and K. Schütz, 1984, Cross faults in the Gulf of Suez area: 7th Egyptian General Petroleum Company Exploration Seminar, Cairo, March 1984, 15 p.

Van Houten, F.B., 1980, Latest Jurassic–Early Cretaceous regressive facies, North Africa: AAPG Bulletin, v. 64, n. 6, p. 857–867.

Van Houten, F.B., D.P. Bhattacharyya, and S.E.I. Mansour, 1984, Cretaceous Nubia Formation and correlative deposits, eastern Egypt: Major regressive-transgressive complex, Bull. Geol. Soc. Am., v. 95, p. 397–405.

Vink, G.E., W.J. Morgan, and P.R. Vogt, 1985, The Earth's hot spots: Scientific American.

Viiotti, C., and G. El Demerdash, 1969, Studies in Eocene sediments of Wadi Nukhul area, Gulf of Suez: 3rd African Micropaleontology Colloquium, Cairo, p. 403–423.

Webster, D.J., and N. Ritson, 1982, Post-Eocene stratigraphy of the Suez rift in SW Sinai, Egypt, *in* G. Hantar, ed., 6th Egyptian General Petroleum Company Exploration Seminar.

Weissbrod, T., 1969, The Paleozoic of Israel and adjacent countries: the Paleozoic outcrops in southwestern Sinai and their correlation with those in southern Israel: Geological Survey of Israel Bulletin 48, 32 p.

Weissbrod, T., 1970, "Nubian Sandstone" discussion, Bull. Am. Assoc. Petrol. Geol., v. 54.

Wernicke, B., and B.C. Burchfiel, 1982, Modes of extensional tectonics: Journal of Structural Geology, v. 4, n. 2, p. 105–115.

Williams, G.A.D., and J.O. Small, 1984, A study of the Oligo-Miocene basalts in the Western Desert: 7th Egyptian General Petroleum Company Exploration Seminar, Cairo, p. 252–268.

Zankl, H., and E. Klitzsch, 1985, Gebel Zeit report and new aspects on the development of the Red Sea–Gulf of Suez Graben: unpublished Conoco report.

Chapter 3

Ramadan Field—Egypt: Gulf of Suez Basin

A. Shawky Abdine
Wafik Meshref
A. Nabil Shahin
Paul Garissino
Salah Shazly
Gulf of Suez Petroleum Company
Cairo, Egypt

LOCATION

Ramadan field covers some 2850 ac (11.5 km^2; 4.5 mi^2) in the offshore, central province of the Gulf of Suez basin, some 174 km (108 mi) to the southeast from Suez City (Figure 1). The field is estimated to contain several hundred million barrels of oil as ultimate recoverable reserves, making it the fourth largest oil field in Egypt. The field is located on a gulf parallel (clysmic trending) tilted fault block system, generated during two pulses of rifting. Within the same area, and situated on similar structures, are the July and El-Morgan oil fields which, together with Ramadan field, account for over 39% of the oil reserves in Egypt. The field is operated by the Gulf of Suez Petroleum Company (GUPCO), which is a joint venture between the Egyptian General Petroleum Corporation (EGPC) and AMOCO Egypt Oil Company (a subsidiary of AMOCO Production Company).

HISTORY

Pre-Discovery

In 1964, aeromagnetic and seismic surveys were conducted over several concessions assigned to the Pan American UAR Oil Company (now AMOCO Egypt Oil Company). Two mappable seismic horizons "M" and "C" were identified (Brown, 1978). These coincide, respectively, with the upper and lower limits of the Miocene evaporite section. Several structural anomalies were mapped using these horizons (Figure 2). In 1965, the well El Tor-1 (later renamed Morgan-1) was spudded on the "C" structure and became the discovery well for El-Morgan field. Later that year, the well Alef-1 was spudded on the crest of the "A" (Cretaceous) sandstones with no hydrocarbon indications in that zone. In 1966, the "J" structure (Figure 2)

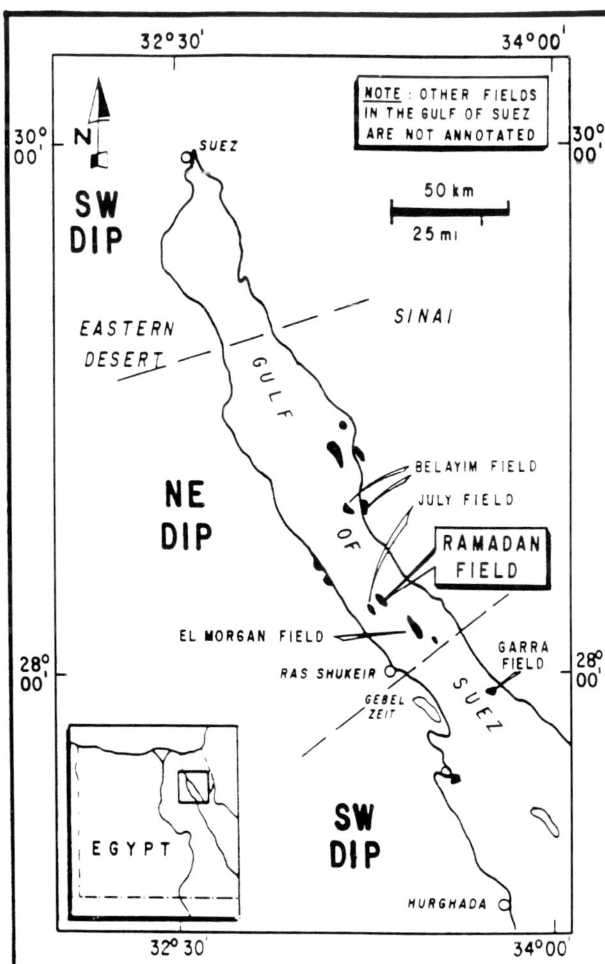

Figure 1. Ramadan field is located in the central Gulf of Suez in a northeasterly dipping province that separated two southwesterly dipping provinces.

Reprinted from *Atlas of Oil and Gas Fields: Structural Traps VI*, published by the AAPG.

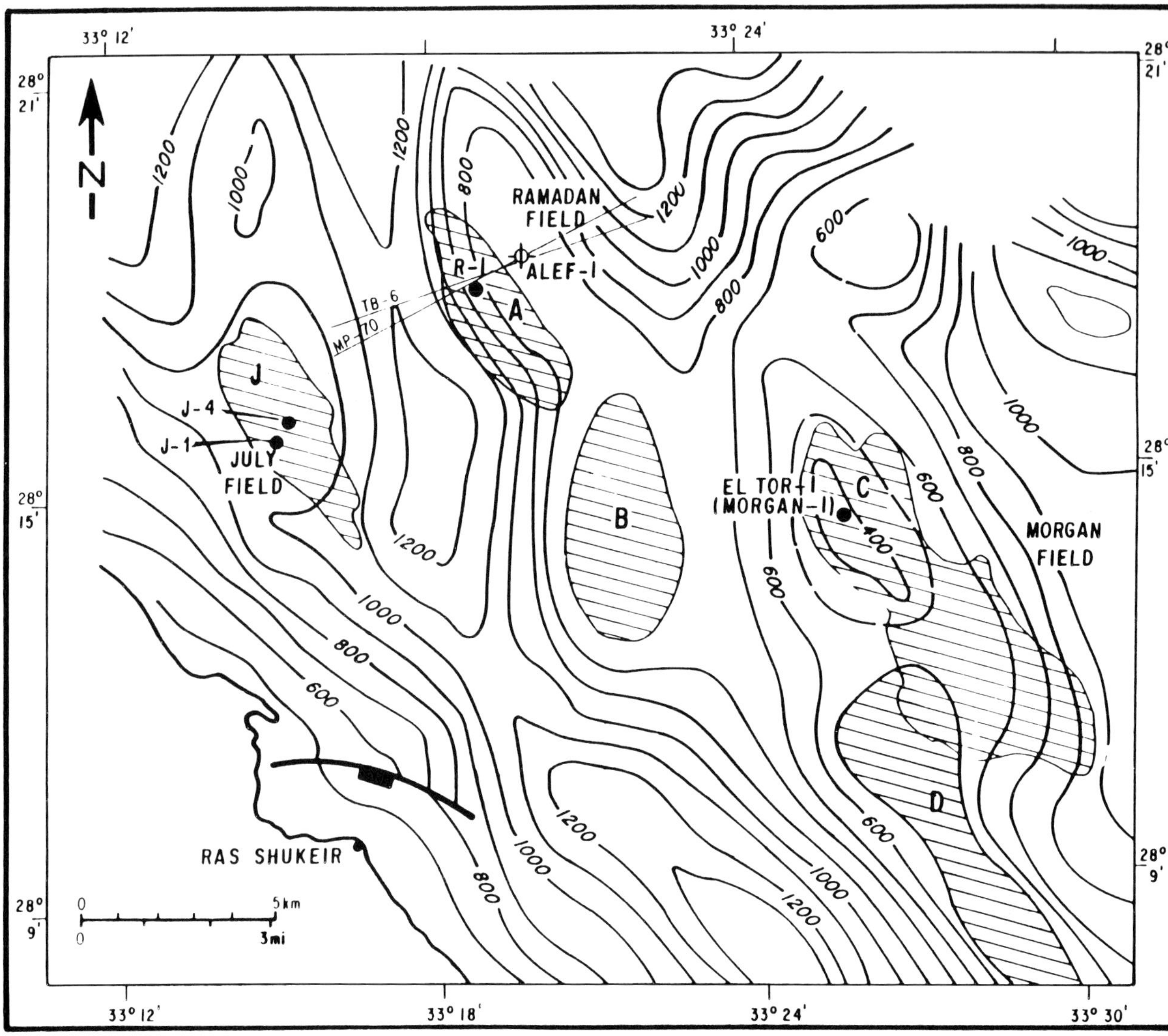

Figure 2. The "M" horizon (see Figure 3B) seismic structure map represents the top of the Miocene evaporites. Several structural anomalies are shown at A, B, C, D, and J structures (hachured). Contours are in meters below sea level. Contour interval is 100 m (328 ft).

was tested by the well J-1, which encountered uncommercial quantities of oil within the lower Miocene Nukhul Formation and the pre-Miocene Cretaceous sands of the Senonian, Turonian, and Cenomanian (Figure 5). Two additional wells were drilled on the structure before the well GS 311-1, later renamed July-4 (J-4), was spudded in May 1973 and encountered the lower Rudeis (lower Miocene) sand reservoir of July field.

Discovery

In May 1974 the well GS 303-1 (Figure 2), renamed Ramadan-1 (R-1), was spudded by GUPCO. The well R-1 ultimately encountered 415 m (1361 ft) of net oil pay extending from 3018 to 3674 m (9900–12,050 ft) across several pre-Miocene formations. Within these, 338 m (1110 ft) of pay was recorded within the Nubia "C" massive sandstones, 55 m (180 ft) within the Senonian, Turonian, and Cenomanian sandstones (Matulla, Wata, and Taha formations, respectively), and 22 m (71 ft) from within the early Cenomanian Nubia "A" sandstone. Ramadan's main pay zone, the Nubia "C" sandstone, was put on production in December 1974 at an initial rate of 25,000 barrels of oil per day (BOPD).

Post-Discovery

In April 1983, production from within the Nezzazat Group (Senonian, Turonian, and Cenomanian age) was tested at a rate of 2422 BOPD. Production from the Nubia "A" sandstone was first tested in November 1986 at a rate of 3200 BOPD.

To date, 51 wells have been drilled over the field area, resulting in a cumulative production of 360 million barrels of oil (MMBO). This drilling has delineated the northern, eastern, and western limits of the field. The southern limit has yet to be reached by the development program, though recent mapping has identified its potential location.

DISCOVERY METHOD

Just to the east of July field, interest was being regenerated in the "A" structure (Figure 2). Although Alef-1 (Figure 2) was a dry hole, dipmeter data indicated a northeast dip in the pre-Miocene section which, if present far enough to the west, could have resulted in a commercial accumulation of hydrocarbons structurally high to the Alef-1 well location. In 1971 and 1972, additional seismic data were obtained which, together with much improved seismic processing techniques, allowed the interpretation of deeper pre-salt structure. Seismic processing also verified that Alef-1 had been drilled downdip on a northeasterly dipping pre-Miocene fault block. The bounding fault of the structure was subsequently mapped at about 3.2 km (2 mi) farther west from the Alef-1 well location (Figures 4A and 4B).

STRUCTURE

Tectonic History

Following the layer cake deposition of the pre-Miocene sediments, the entire area was subjected to the first phase of rifting during the Oligocene, accompanied by uplifting and tilting of faulted blocks and followed by Oligocene-Miocene crestal erosion. This was followed by a transgressive phase that resulted in the deposition of the Miocene clastics and evaporites. A second phase of rifting started during the late Miocene, resulting in the rejuvenation of the older faults.

Geologic Evolution

Initially the Ramadan pre-Miocene structure was shown bounded by only one fault system (Figure 6) extending from south to north of the field. Recently it has been discovered that two fault systems, of very different ages and histories, combine to form the field-bounding fault complex resulting in three fault zones of unique structural style (Figure 7). In the "Northern Zone," the two systems "A" and "B" are separate, with the eastern fault "B" bounding the field. Over the central "Transfer Zone," a portion of the throw from the "B" system was transferred through a series of cross-elements to the "A" system, which trends toward the field from the northwest. In the "Southern Zone," the two fault systems have combined to form the western-bounding fault "AB" of the field. Since the geologic evolution of the northern and southern zones of the field have been so different, a separate analysis has been generated for each.

Evolution of the Northern Zone

Prior to the late Eocene, the entire Ramadan area was covered by a flat-lying blanket of sediments from the Carboniferous Nubia sandstones through the Eocene Thebes Formation (Figure 8-1). The sedimentary record from the upper Eocene to the lower Miocene is missing in Ramadan, making it possible that rifting could have begun over this area as early as late Eocene. Oligocene-age volcanics are present to the north of this area, indicating that the initial phase of rifting occurred when the major clysmic (Gulf of Suez trend) "A" fault system and many associated cross-elements (faults that are normal to clysmic) were formed (Figure 8-2). The "A" fault system incurred in excess of 610 m (2000 ft) of displacement, imparting a 10° dip to the upthrown pre-Miocene section. During the 15 to 20 million years available from late Eocene to early Miocene, crestal erosion occurred along this fault system (Figure 8-3). The cross-elements that formed during the initial rift event tend to have maximum throw near the crests of upthrown rocks, with diminishing throw downdip and terminating in regionally dipping pre-Miocene strata.

The basal Miocene Nukhul Formation onlapped the pre-existing pre-Miocene structure some 3.5 km (2.2 mi) northeast from fault "A" (Figure 8-4). As about 30 m (100 ft) of Nukhul sediments are present in the well R-19 (Figure 7) at the bounding fault on the first block south of this zone, a cross-element (normal to the Gulf trend) must have existed near the crest of fault "A" with a magnitude in excess of 457 m (1500 ft), down to the south. This fault separated the 10° dip of the northern block from the 4° dip over the rest of the field and diminished to the east where the pre-Miocene dip was relatively constant. An incipient "B" fault system may have been present at this time separating the 10° dip regime, upthrown to the bounding fault, from the approximately 8° regional dip to the east. As Miocene sedimentation continued, the first sediments to completely cover the upthrown block were laid down sometime between deposition of the Miocene clastics (Rudeis Formation) and the Markha anhydrites of the Kareem Formation (Figure 5). Sometime after deposition of the Belayim Formation, a second pulse of rifting occurred, resulting in an additional 10° rotation of the block formed between the "B" and "A" systems and allowing for over 610 m (2000 ft) of displacement on the "B" fault. It was during this rotation that the Miocene fault system formed over fault "B" which extended up to but not completely through the Belayim Formation (Figure 8-5). The Belayim surface was seen to drape over the underlying structure.

Evolution of the Southern Zone

As in the Northern Zone, prior to late Eocene the stratigraphic section was basically flat-lying (Figure 9-1). Sometime during late Eocene the first pulse of rifting occurred, resulting in the initiation of about 550 m (1800 ft) displacement on the "A" fault system and probably the initiation of the incipient "C" fault system. The pre-Miocene dip of the upthrown block was in the order of 4° toward the northeast to the west of the "C" fault and as much as 8° to the east (Figure 9-2). In the time

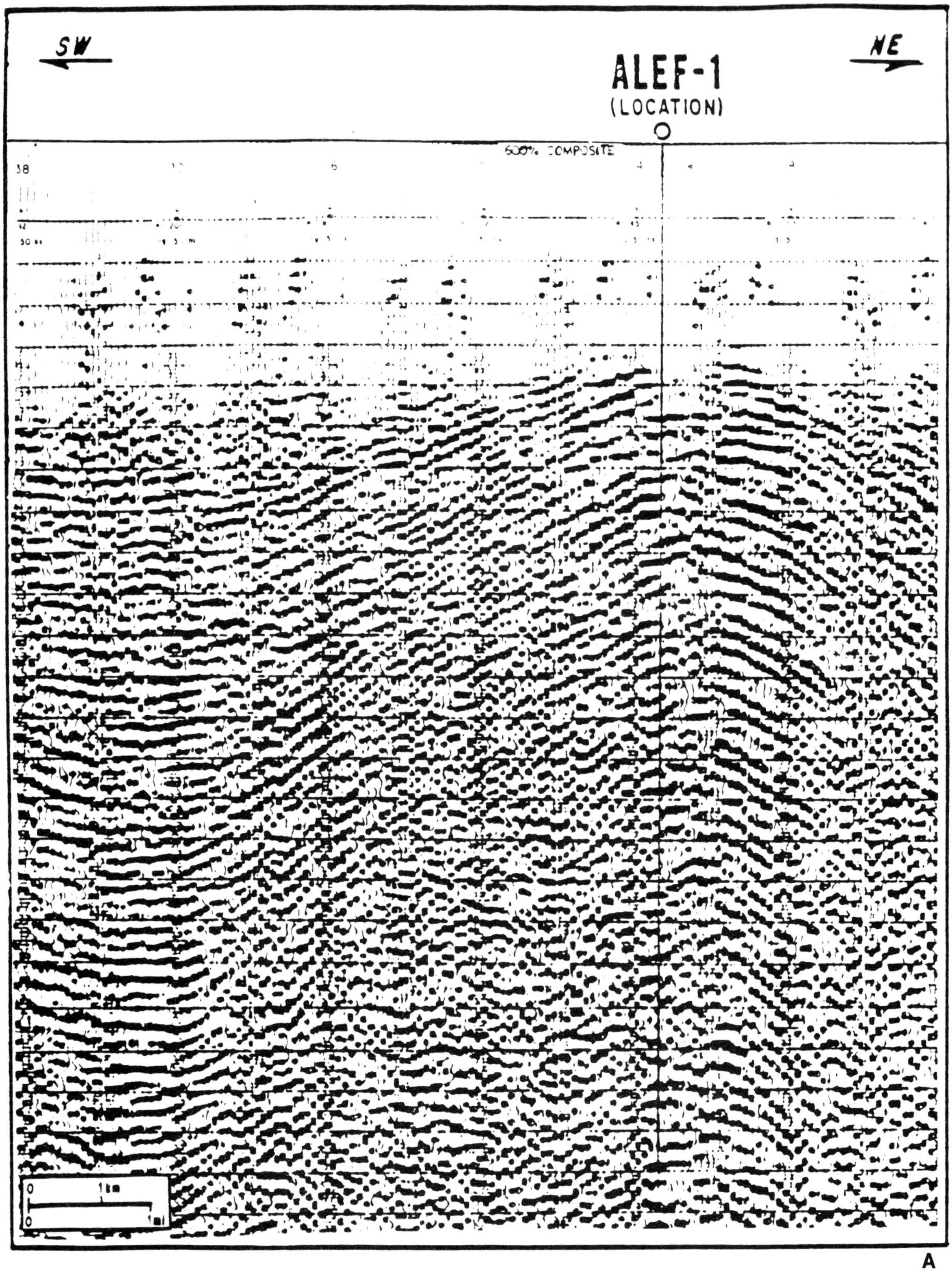

Figure 3. Seismic line TB-6 (see Figure 2 for location), drill line for well Alef-1 to test a crestal position of an upthrown side of a clysmic (Gulf of Suez trend) normal fault on the "A" structural anomaly (Figure 2). (A) Uninterpreted.

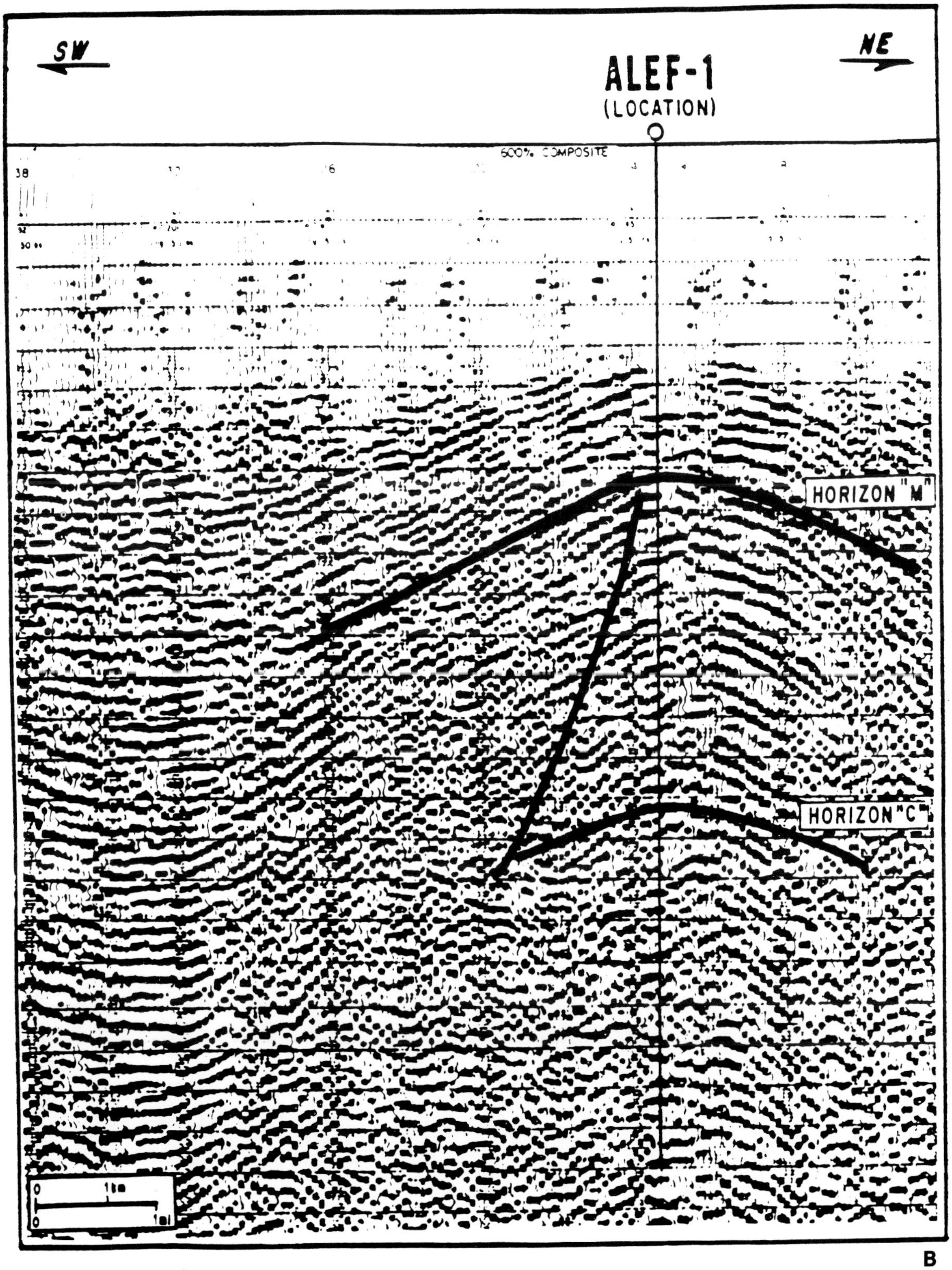

Figure 3. (B) Interpreted.

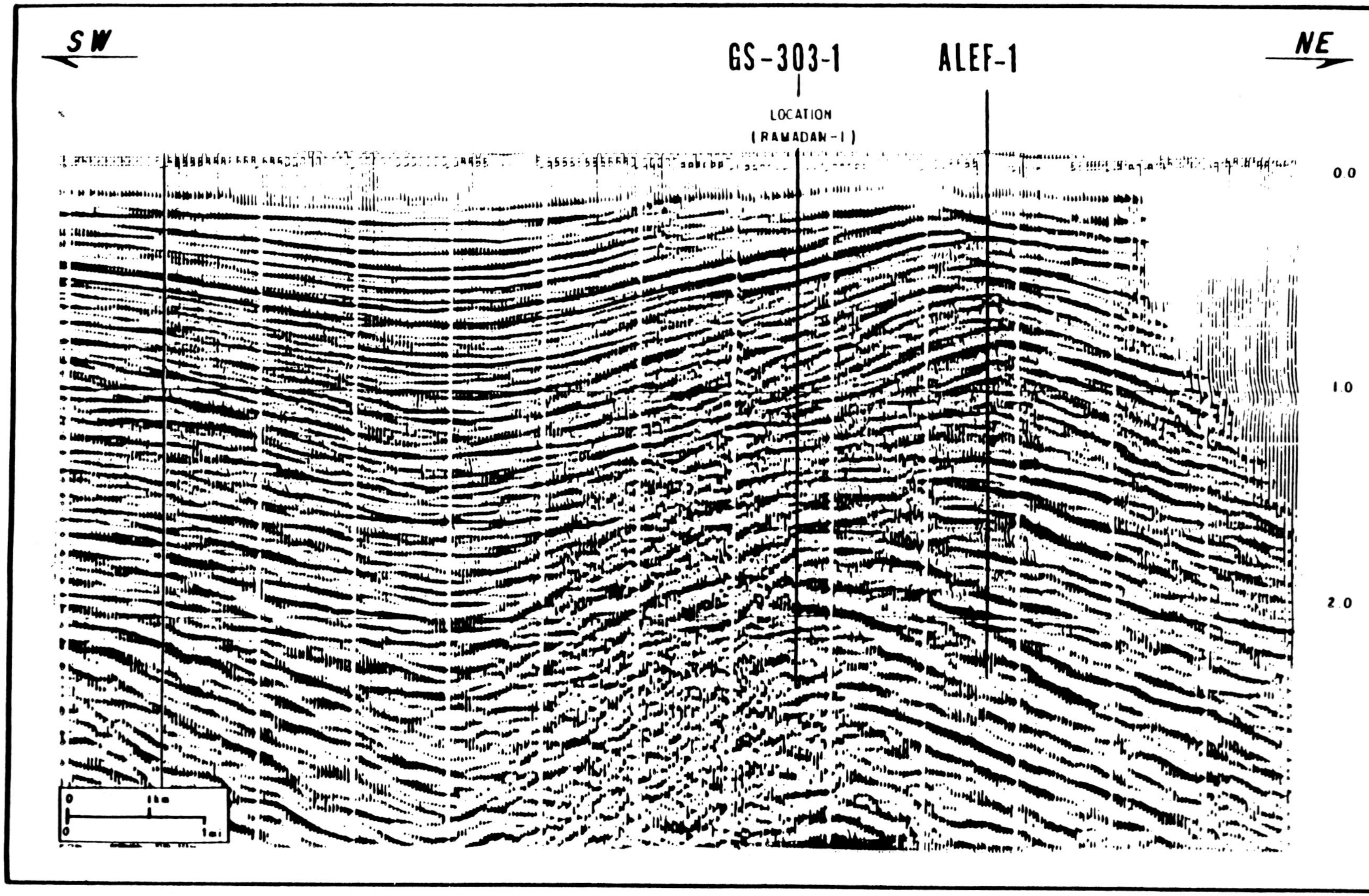

A

Figure 4. Seismic line MP-70 (see Figures 2 and 7 for location), drill line for well GS 303-1 (the discovery well, Ramadan-1), showing that well Alef-1 was drilled downdip on a northeasterly dipping pre-Miocene fault block. (A) Uninterpreted. (B) Interpreted (1974 processing and interpretation).

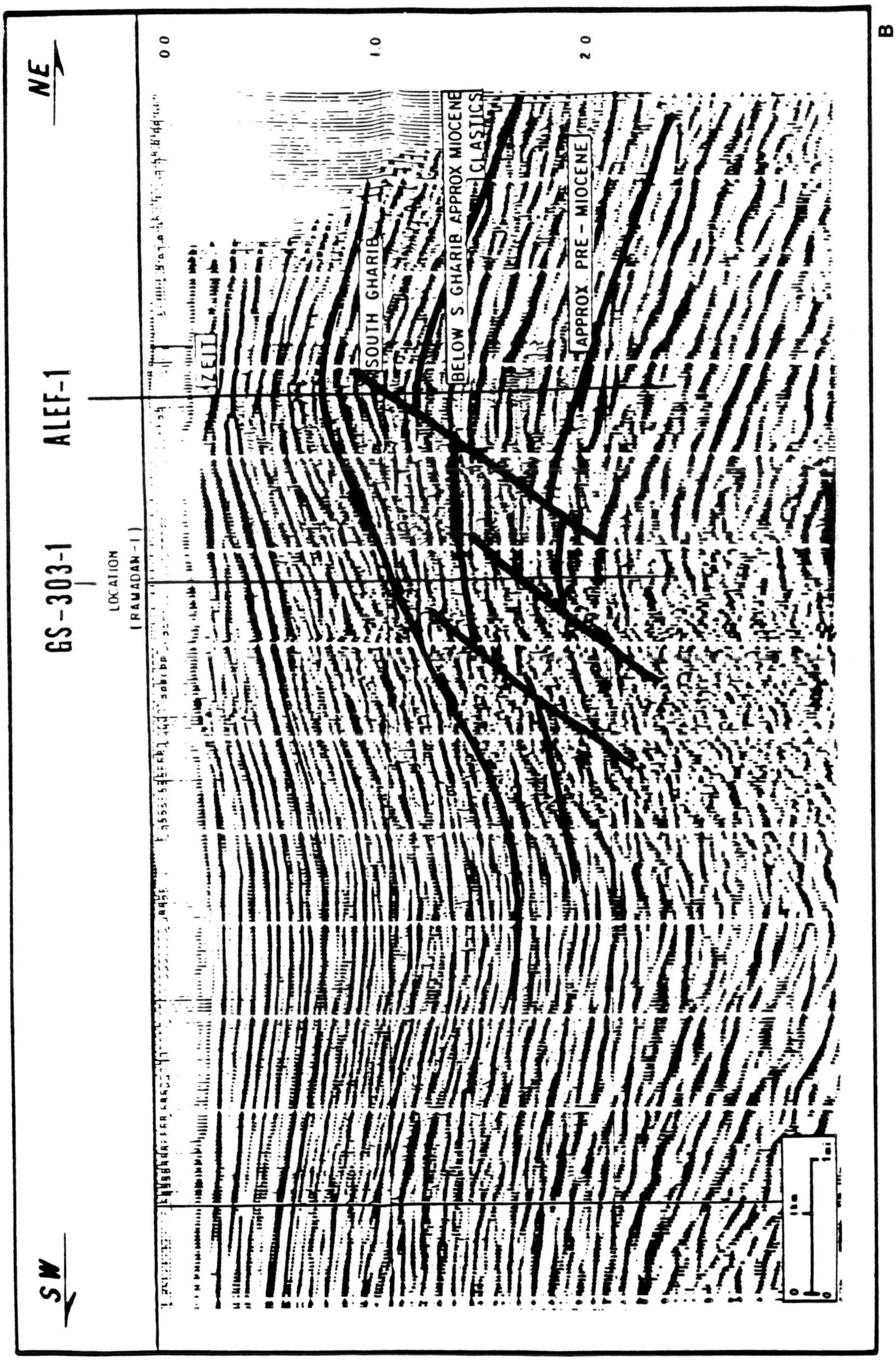

Figure 4 (continued).

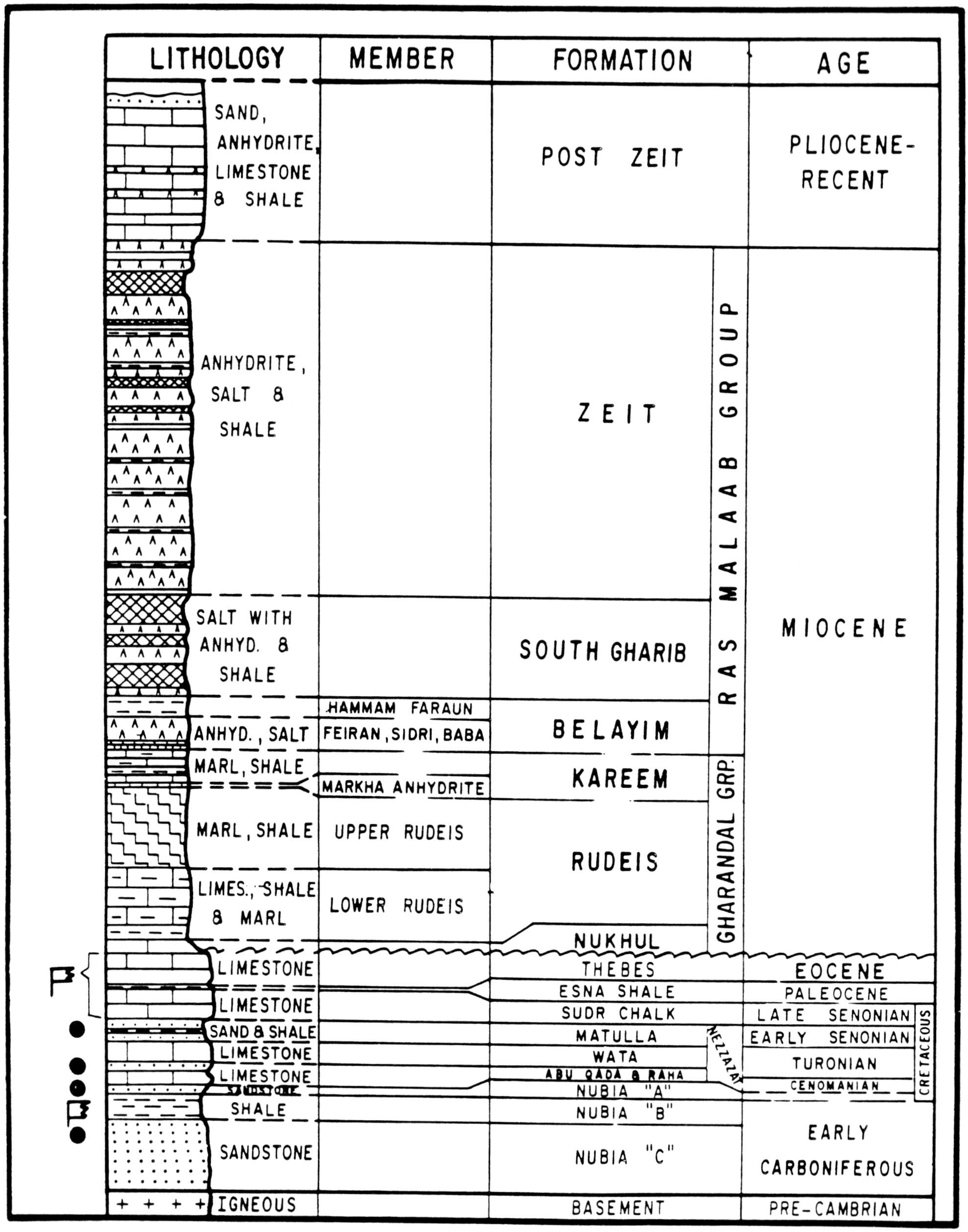

Figure 5. Stratigraphic column of Ramadan field area, typical of the central Gulf of Suez. Reservoir rocks (solid circles) are the limestones and sandstones of the Nezzazat Group and the Nubia "A" and "B." Possible source rocks (flags) are the organically rich limestones and shales of the Eocene Thebes Formation, the Paleocene Esna shales, the Campanian Sudr Formation, and the Carboniferous Nubia "B."

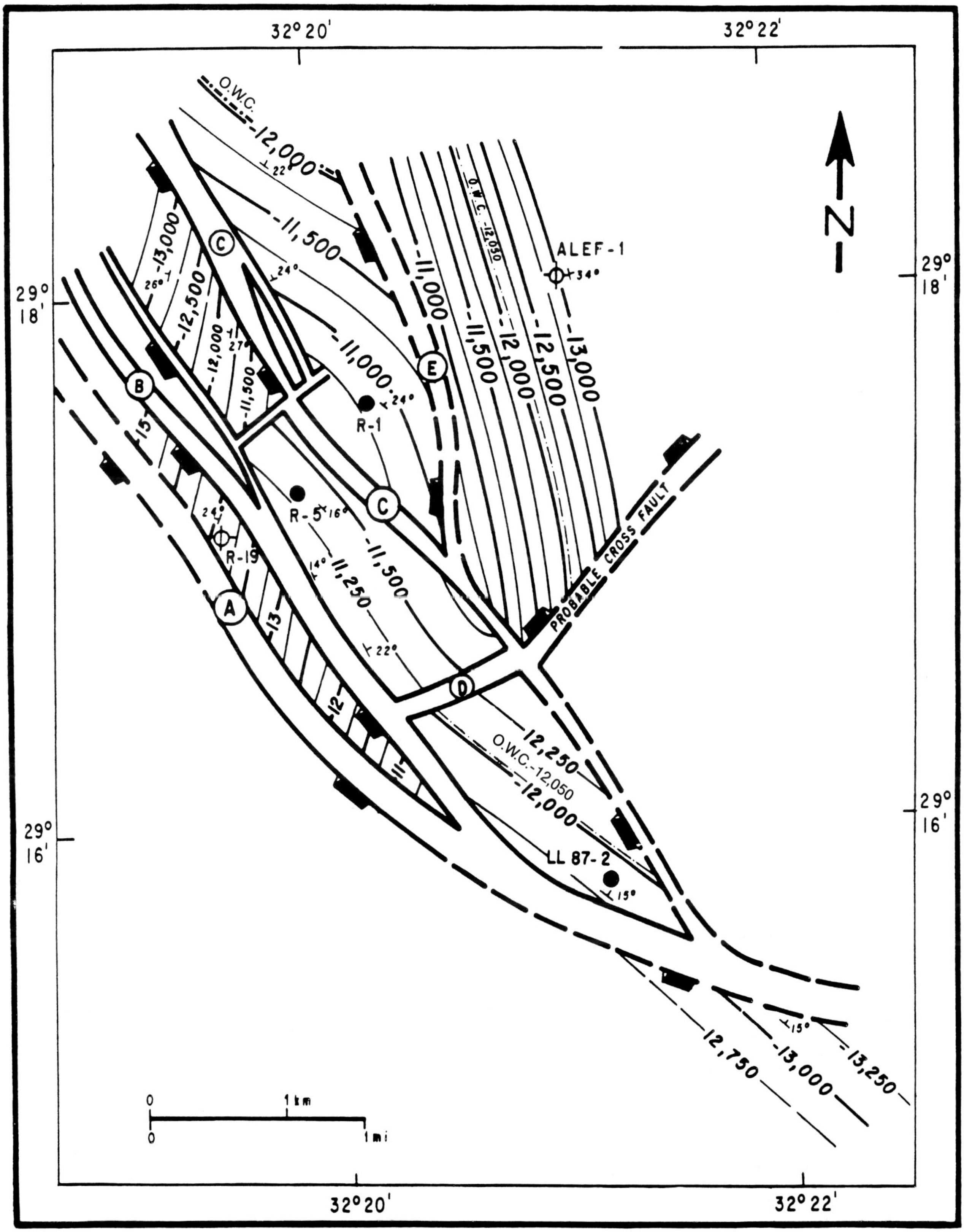

Figure 6. Top of the Nubia "C" structure contour map of Ramadan field (Brown, 1978) showing that the pre-Miocene faulted tilted block is bounded from south to north by one fault system "A." This system was initiated during the main rifting as shown by the fact that it is parallel to the Gulf of Suez trend (clysmic). Oil-water contact (OWC) is –12,050 ft (–3073 m). Contour interval is 250 ft (76.2 m).

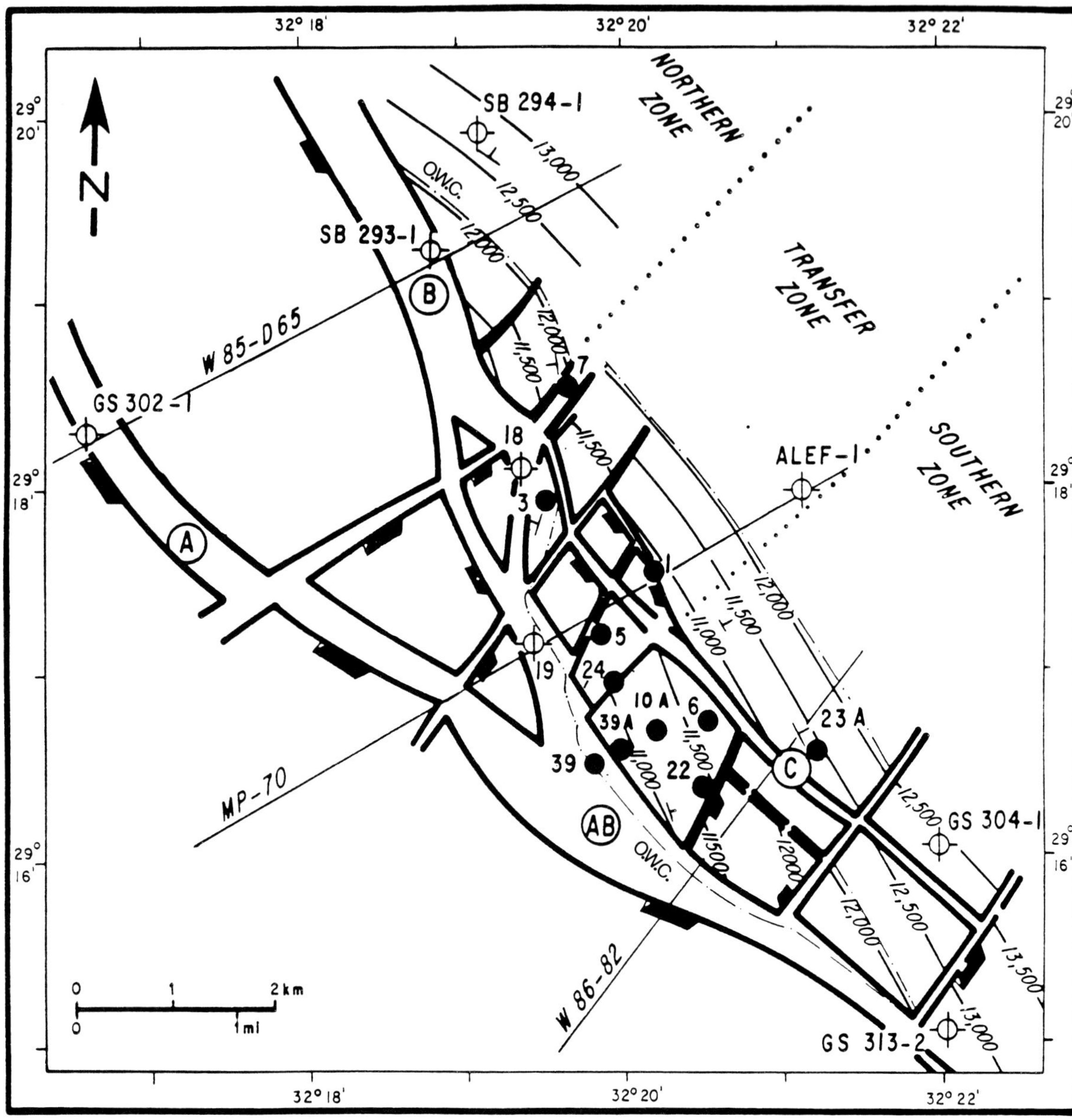

Figure 7. Top of the Nubia "C" structure contour map showing that the pre-Miocene faulted and tilted block of the Ramadan field is bounded by two major fault systems (A and B) of different ages (rifting and pre-rifting, respectively). The two systems combine in one system (AB) toward the south via a Transfer Zone that separates the Northern Zone from the Southern Zone with markedly different histories. Contour interval is 500 ft (152.4 m). OWC is –12,050 ft (–3072 m).

elapsed between the formation of the "A" system and the deposition of the lower Miocene sediments, crestal erosion along the "A" system resulted in the development of a pre-Miocene erosional scarp at that location. The angle of that scarp surface with respect to the horizontal would have been 37° (Figure 9-3).

During the early Miocene, the Nukhul Formation was deposited over the block, thickening over any existing cross-elements. The rest of the Miocene section, through the Belayim and possibly including some of the South Gharib Formation, was laid down (Figure 9-4). During the second phase of rifting, the "A" and "C" fault systems were rejuvenated. These systems each received a portion of the throw attributed to the "B" system of northern Ramadan.

During this phase, an additional 10° of block rotation occurred over the western block of the field, resulting in a final scarp angle of 27° and a bounding fault angle of about 54° with respect to the horizontal (Figure 9-5).

During this second rotational phase, a Miocene fault system formed in much the same fashion as in the Northern Zone. The existence of the underlying pre-Miocene scarp allowed for a detachment surface

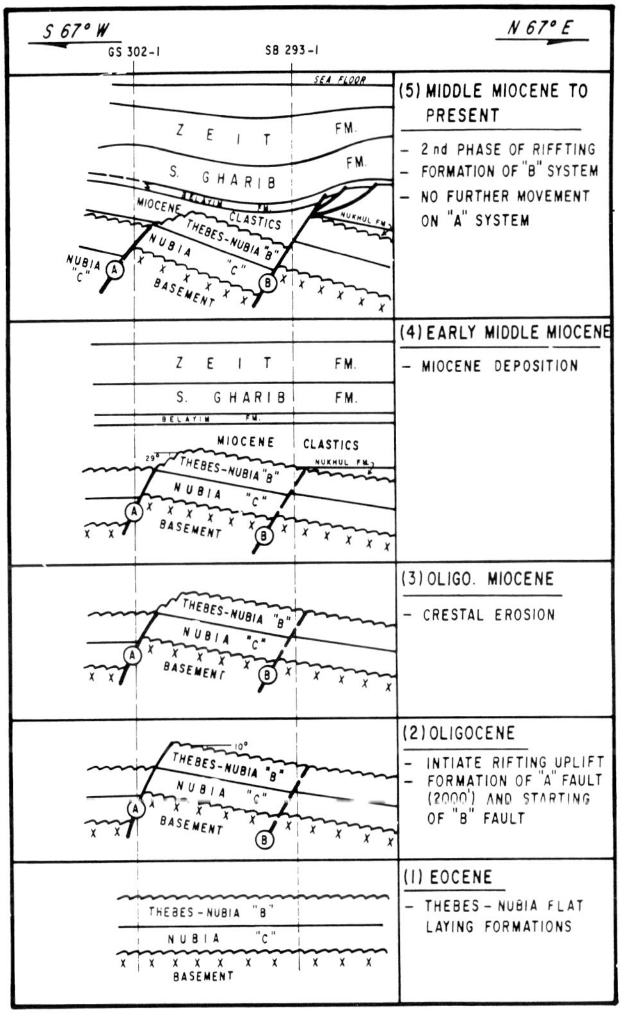

Figure 8. The geologic evolution of the Northern Zone of Ramadan field. No post-Oligocene movement is associated with fault A, so there has been no deformation of the overlying Belayim Formation in this zone.

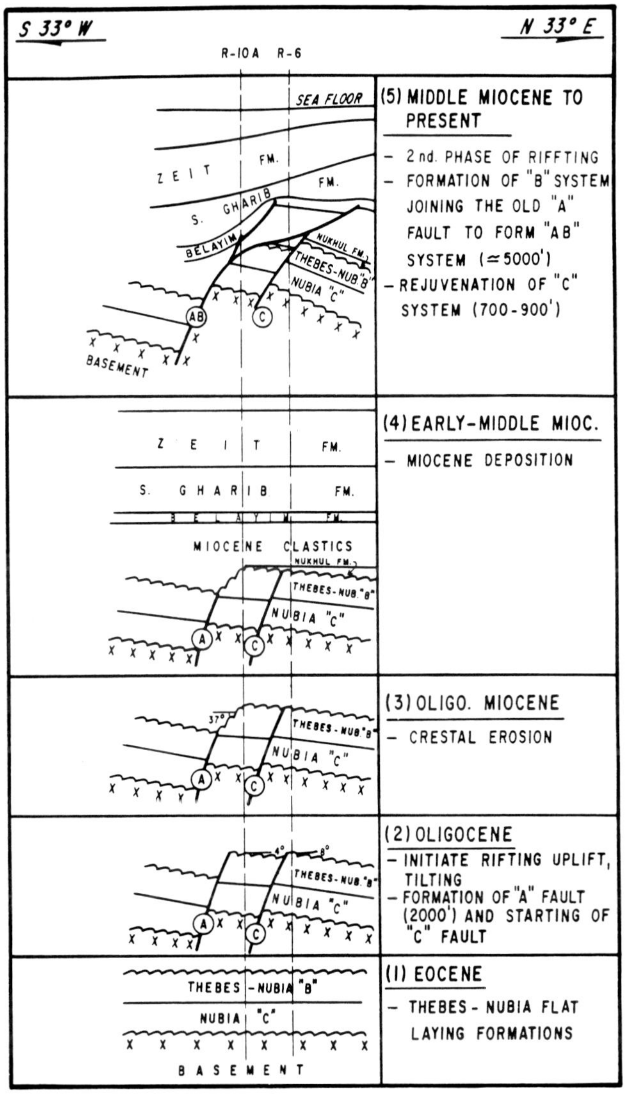

Figure 9. The geologic evolution of the Southern Zone of Ramadan field. Here the rejuvenation of fault A has resulted in a significant structural displacement at the Belayim surface.

to develop, over which the overlying Miocene sediments could have moved, thus resulting in the present-day fault relationship across that surface.

Present-Day Structural Setting

The following describes the present-day structural setting in each of the three recognized zones: Northern Zone, Transfer Zone, and Southern Zone.

Northern Zone Structure

The structure of the Northern Zone is dominated by the presence of two separate fault systems, "A" and "B" (Figure 7). The "B" system forms the main bounding fault of Ramadan field, while the "A" system bounds a more westerly block. The historical relationship of these two systems can best be discussed through a detailed examination of the schematic geoseismic cross section (Figure 10). The line of section is constructed along seismic line W85-D65 (Figure 11), tying the wells GS 302-1 and SB 293-1 and intersecting both faults "A" and "B."

Fault "A" is a two fault system associated with at least 649 m (2130 ft) of missing section. The uppermost fault displays 67 m (220 ft) of missing section between the Matulla and Wata formations. The lower fault separated the Nubia "A" and Nubia "C" sandstones with a minimum of 584 m (1916 ft) of missing section.

The "B" fault system, encountered in the well SB 293-1, is associated with a minimum of 869 m (2850 ft) of missing section between the Miocene Rudeis Formation and the sands of the upthrown Nubia "C" sandstone.

The block between the "A" and "B" fault systems, as mapped on the pre-Miocene unmigrated time structure map (Figure 11), is considered to be continu-

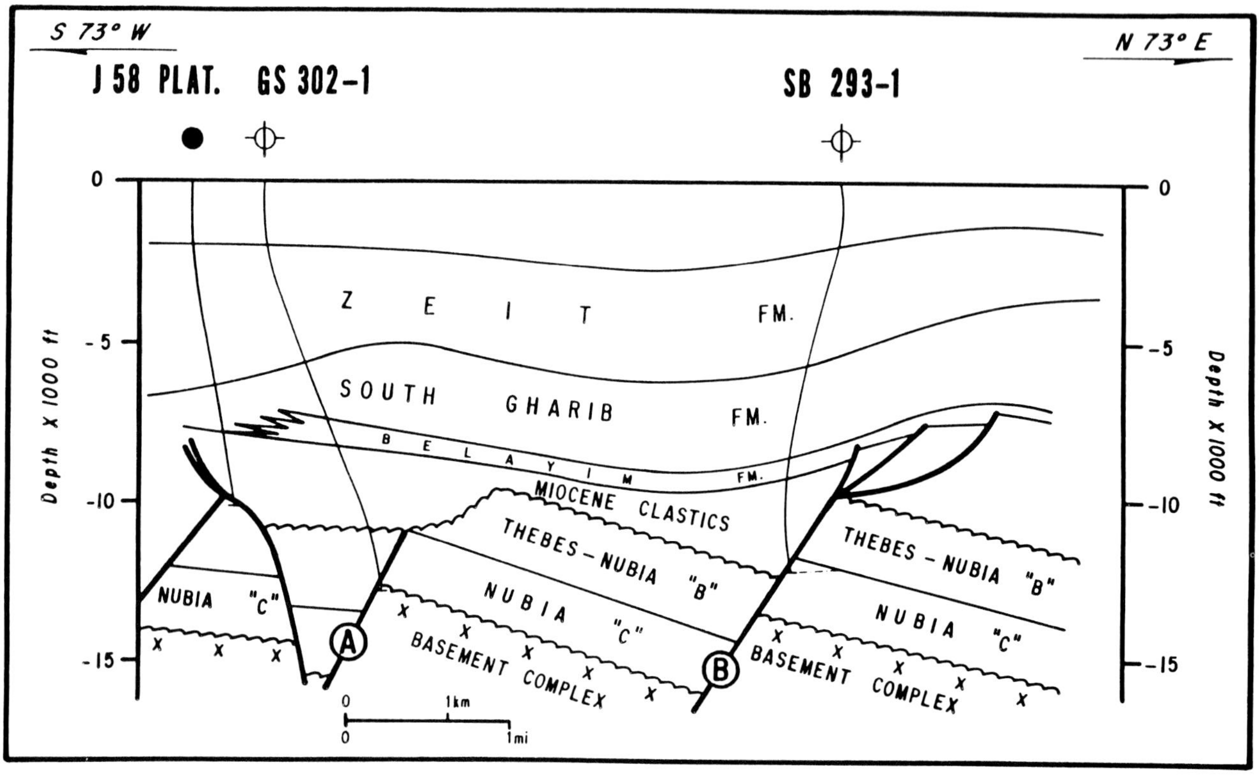

Figure 10. The geo-seismic cross section along seismic line W88-D65. Movement on the B fault system bounding northern Ramadan field has caused a significant structural flexure of the overlying Belayim Formation. No such event has occurred over fault A. This, together with the onlap of the Miocene clastics, indicates that the block to the east of fault A was a pre-existing "high" later buried during Miocene sedimentation. Movement on fault A probably pre-dated Miocene sedimentation and certainly pre-dated the major displacement along the fault B system.

ous. The dip on this block, shown as 18° toward the northeast, is controlled geologically by the conformable basement contact and the "B" fault throw, estimated from the SB 293-1 wellbore data. Modeling of this dip fits well with the seismic data.

An angular unconformity is evident in the seismic data along the top of this pre-Miocene block (Figure 12). This is consistent with geologic data from the SB 293-1 wellbore where an intra-Miocene isopach is some 107 m (350 ft) thinner than the comparable isopach upthrown to fault "B," indicating onlap of the Miocene sediments to the pre-existing pre-Miocene structure. Further support for onlap is present on the Nukhul Formation isopach map (Figure 13), which shows the Nukhul Formation to thicken to the east.

Evidence to support the relative ages of the two fault systems is present in the relationship of the position of the present-day Belayim surface to the underlying pre-Miocene structural configuration. On the seismic data, a very reliable Belayim event is present between the two well control points, providing excellent control on the Belayim structure. Although both the underlying "A" and "B" fault systems display an excess of 610 m (2000 ft) of offset, the Belayim above the two faults behaves quite differently. Over fault "B," the Belayim horizon drapes some 610 m (2000 ft) from the crest of the Ramadan feature to the synclinal axis between July and Ramadan fields, whereas there is no expression whatsoever at the Belayim horizon of fault "A." Therefore, movement along fault "A" must have pre-dated deposition of the Belayim Formation, whereas major motion along fault "B" had to occur after Belayim deposition. The Miocene onlap, discussed above, points to the existence of fault "A" prior to any Miocene sedimentation. Furthermore, any attempt to reconstruct the stratigraphic section above the basement requires the Belayim horizon to be some 610 m (2000 ft) higher in the section than observed over fault "A."

Transfer Zone Structure

This portion of the field is characterized by the presence of numerous cross-elements, through which a portion of the throw associated with the "B" fault system is transferred to the "A" system, forming the "AB" bounding fault system of southern Ramadan field. The remaining throw on the "B" system is accounted for by movement along the "C" fault system that trends into the "B" system of northern Ramadan and separates the eastern and western productive blocks of the field. A corner is formed in this

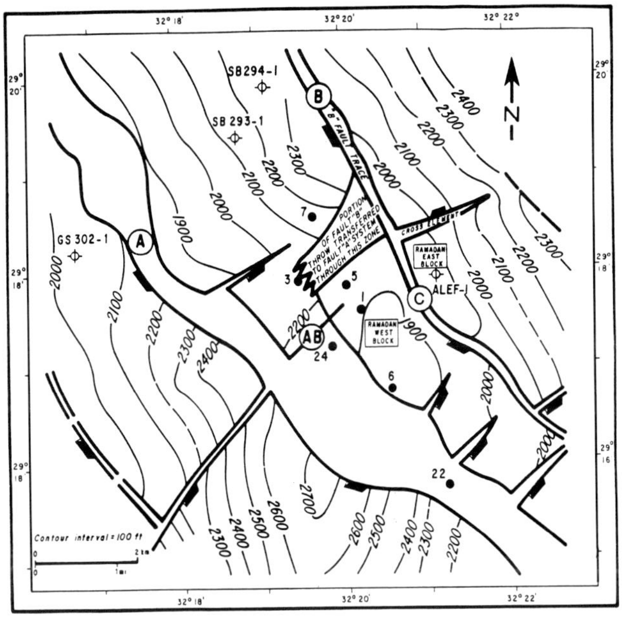

Figure 11. Pre-Miocene unmigrated time structure map showing that in the Northern Zone the block between the "A" and "B" fault systems is undisturbed, while in the Transfer Zone it is characterized by the presence of several cross-elements through which a portion of the throw of the "B" fault system is transferred to the "A" system; in the Southern Zone the "A" and "B" fault systems have come together to form the "AB" system. The remaining flow of the "B" system is accounted for by movement along the "C" fault.

zone where the "B" fault system intersects a series of cross-elements in which three downthrown blocks are trapped and where northwest dip is developed as a consequence.

This zone is typified by cross section MP-70 (Figure 14), named for the coincident seismic line (Figure 4) oriented along an azimuth of 58°. It traverses the central portion of the field, joining wells R-19A and Alef-1. In this section, the "A," "B," and "C" fault systems are cut by the wells R-19A, R24-27A, and R1-29. Fault "A" is cut by the R-19A wellbore within the Miocene clastics, where 442 m (1450 ft) of section is missing. In well R24-27A, 976 m (3200 ft) of section is missing at the pre-Miocene contact, which represents a combination of components from both the "A" and "B" fault systems. The "C" fault system was encountered in the R1-29 wellbore where a total of 210 m (700 ft) of section are missing in the pre-Miocene section at the intersections with two faults. The "C" fault is terminated by a low-angle Miocene normal fault, which indicates that the "C" system was there prior to that Miocene structural failure. The small block defined between these two faults in the R1-29 wellbore displays 18° of westerly dip, indicating block rotation within the fault zone. Between the "A" and "B" fault systems, well R-19A encountered a block displaying 20° westerly dip. This block has been caught up between the "A" and "B" fault systems and rotated toward the west.

The Belayim Formation is shown to drape, not fault, over the underlying structure. This conclusion was based on seismic mapping of the Belayim horizon where no evidence of a fault is found. Geologic evidence for drape can be recognized in many crestal wells where a measured thickness of 845 m (2772 ft) of unrepeated Belayim section was penetrated, which display formation dips from 35° to 45° toward the west. When this information is combined with the well R-19A data, which show 18° to 20° of dip toward the southwest off the western flank of the structure, it becomes clear that no fault is required structurally at the Belayim level. There is a slight thickening of the Hammam Faraun and Feiran members on the downthrown side of the system that could be associated with growth during the later stages of Belayim deposition. However, it appears that the major movement affecting the Miocene section occurred after deposition of Belayim Formation.

The 61 m (200 ft) displacement fault associated with the "B" fault in the R24-27A and R24-28 wells is consistent with the displacement of the fault zone resulting from the proximity of the "A" and "B" fault systems. As these systems converge, the number of faults associated with the bounding fault is seen to increase. In well R36-39 to the south, five faults were associated with this zone.

Southern Zone Structure

It is within the Southern Zone that the "A" and "B" fault systems are merged to form the "AB" bounding fault (Figure 15). The resulting throw within the system is in excess of 1523 m (5000 ft), as seen in well R6-39.

The pre-Miocene subcrop map (Figure 16) over Ramadan field shows a distinct difference between that of the Southern Zone, bounded by the "AB" system, and that of the remainder of the field. Over the Southern Zone, the pre-Miocene subcrop has a much wider expression in plan, indicating a lower angle with respect to the horizontal on the pre-Miocene surface. This is consistent with the assumption that the "A" system pre-dated Miocene sedimentation and that a significant erosional scarp was formed at its upthrown edge (Figure 9-5). From well control over this zone, the scarp is known to dip between 24° and 28° and to strike near 125°. The bounding fault has been measured at 54° along an azimuth of 127°. The real challenge over the Southern Zone of the field is to predict the location of the intersection of the bounding fault with the pre-Miocene erosional scarp, since this intersection defines the apex of the trap containing the Nubia "C" reserves.

GENERAL STRATIGRAPHY

The stratigraphic succession of Ramadan fields is summarized in Figure 5. As noted above, the sandstones of the Nezzazat Group (Cenomanian to lower Senonian), Nubia "A" (Lower Cretaceous), and Nubia "C" (Lower Carboniferous) act as reservoirs while, as

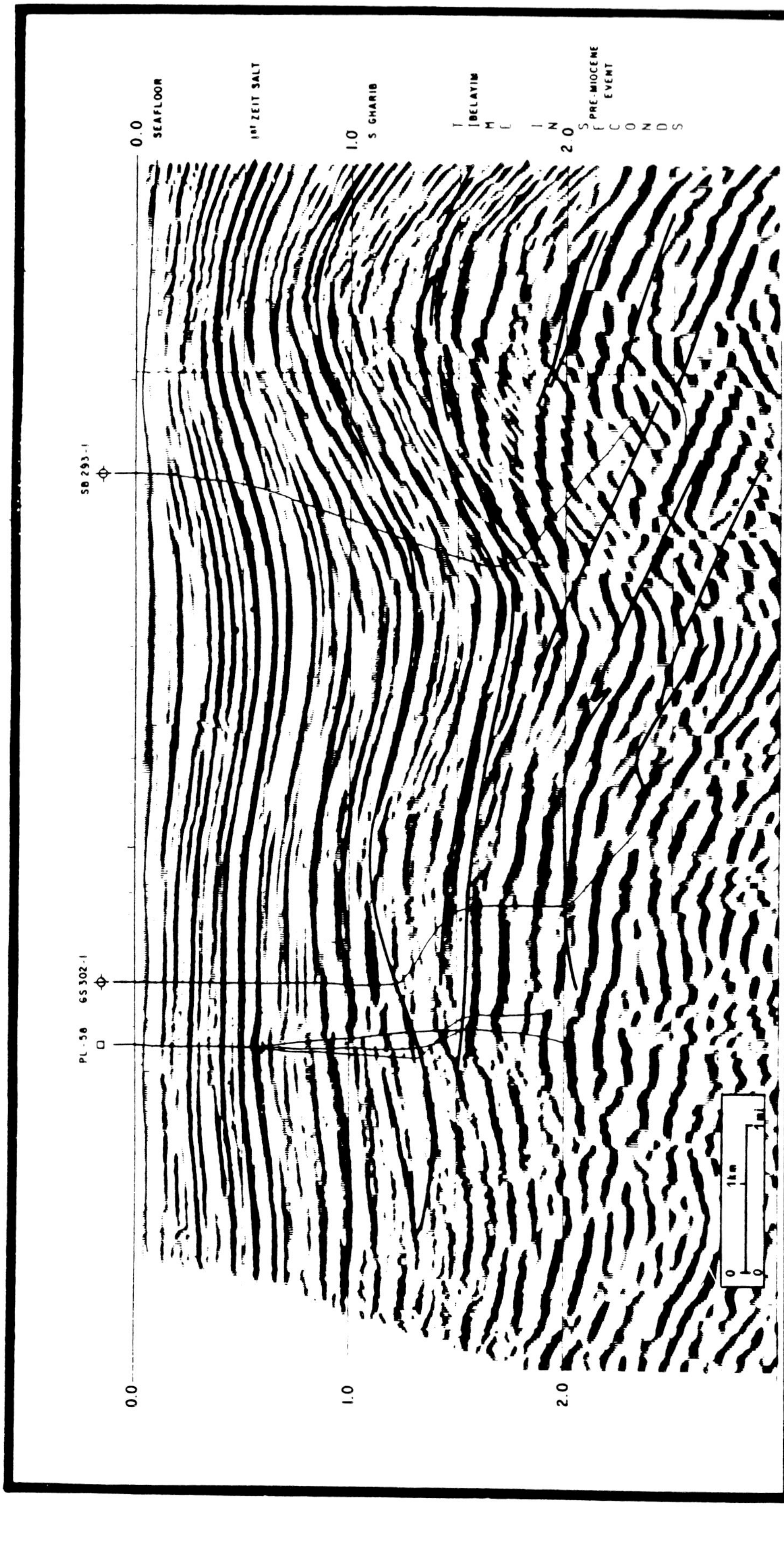

Figure 12. Interpreted unmigrated seismic line W88-D65 showing over 400 milliseconds of structural drape at the SB293-1 location in the Northern Zone of the Ramadan field and ramp dip up to the GS302-1 location over the fault A system.

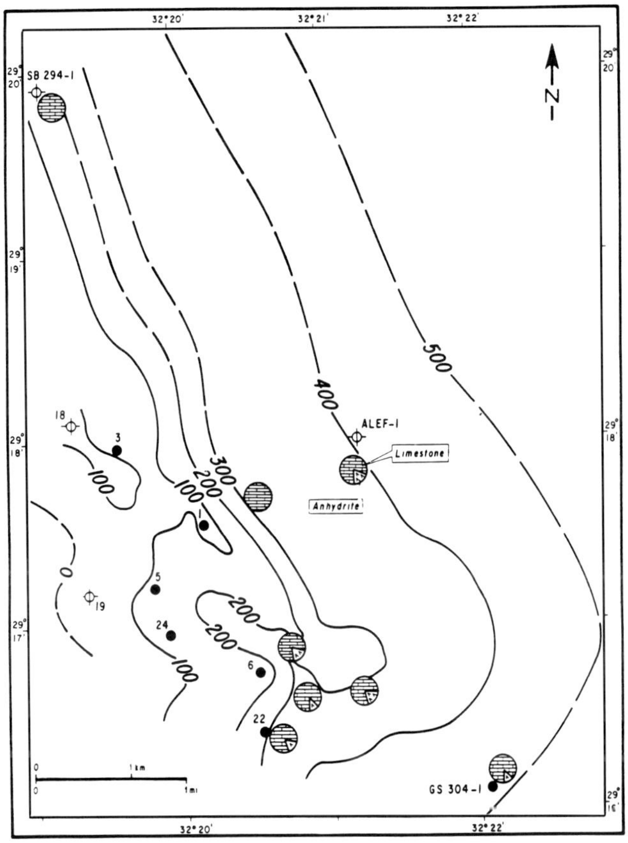

Figure 13. The basal Miocene Nukhul Formation isopach map. The Nukhul Formation is shown to thicken to the east, supporting an onlap on a northeasterly dipping pre-Miocene block. Contour interval is 100 ft (30.5 m).

it will be shown in the geochemistry section below, the limestones and shales of the Thebes (Eocene), Esna (Paleocene), and Sudr (Campanian) formations and the Nubia "B" (Lower Carboniferous) shales display good source potential. In this field, the Miocene section from the Nukhul through Zeit formations is considered the ultimate seal for the underlying pre-Miocene reservoirs.

TRAP

Ramadan field is a complex of tilted fault blocks with reservoirs truncated and sealed by overlying impermeable beds. The field is contained in a northeasterly dipping regime (Figures 11 and 12) bounded to the west by a Clysmic trending fault system with between 854 and 1677 m (2800–5500 ft) of throw. The southern and northern limits of the field are also structurally controlled by northeasterly trending oblique-slip faults. The undip seals for all reservoirs are impermeable shales, limestones, and evaporites of Miocene age. The Nubia "C" reservoir is sealed vertically by the Nubia "B" massive shales, while the vertical seal for the Nubia "A" and Nezzazat reservoirs is probably massive limestone of the Sudr Formation.

Reservoir Characteristics and Facies

In Ramadan field, additional production has been established from the sandstones of the Nubia "A" and Nezzazat Group (Matulla, Wata, and Raha, respectively), but primary production is still mainly from the Nubia "C" sandstone.

Work is just beginning toward a better understanding of the Nezzazat and Nubia "A" reservoirs. Neither reservoir has yet been mapped nor adequately examined from a petrographical or petrophysical standpoint to provide enough detail to support a drilling process. The Nubia "C" reservoir, on the other hand, has received more attention, thereby allowing a relatively comprehensive reservoir description.

The factors affecting permeability within the Nubia "C" reservoir have been studied by GUPCO. These factors include the presence of authigenic clay minerals (Figure 17), asphaltic matter (Figure 18), and silica overgrowths (Figure 19) within the pore space, as well as the effects of compaction. In the central field, the presence of authigenic clay minerals has greater effect on porosity and permeability than does the presence of asphaltic matter. This is due to the overall decrease in the amount of asphaltic matter from the north to the south of the field. In the north, asphaltic matter can represent as much as 30% of the total rock volume.

The Nubia "C" reservoir section has been subdivided into three zones based on permeability, porosity, and lithology (Figure 20). The uppermost zone (Zone 1) is characterized by the presence of uncorrelatable shale units within a massive sandstone section. It ranges in thickness from 30 to 76 m (100–250 ft). The permeability of this zone varies as a function of the amount of asphaltic matter present. This asphaltic matter affects mainly the northern portion of the field. The intermediate zone (Zone 2), composed of massive sandstone with very minor discontinuous shale interbeds, displays good permeability and porosity (Figure 20) over a thickness of 76 to 210 m (250–690 ft) of section. The lowermost interval (Zone 3), although made up of massive sandstone, displays an overall increase in clay and asphalt content resulting in an overall decrease in permeability and porosity.

Through petrophysical analysis it has been shown that the new cut-off within the reservoir occurs when (1) porosity drops below 11%, (2) permeability decreases to less than 10 md, or (3) the percentage of clay in the rock matrix becomes greater than 9%.

Through correlation of lithology and core permeability, 10 md of permeability correlates to a combination of clay and asphaltic matter in the rock matrix of greater than 14% of the total (Figures 21 and 22). This is probably closer to reality than the petrophysical clay estimate since the gamma-ray tool is not sensitive to koalinite. Petrographic work (Figure 23) also has shown that 10 md of core permeability is highly correlatable with a bulk density of 2.45g/cm^3 throughout the reservoir.

The Nubia sandstones are characterized by (1) fining-upward, large-scale, high-angle cross-bedding; (2)

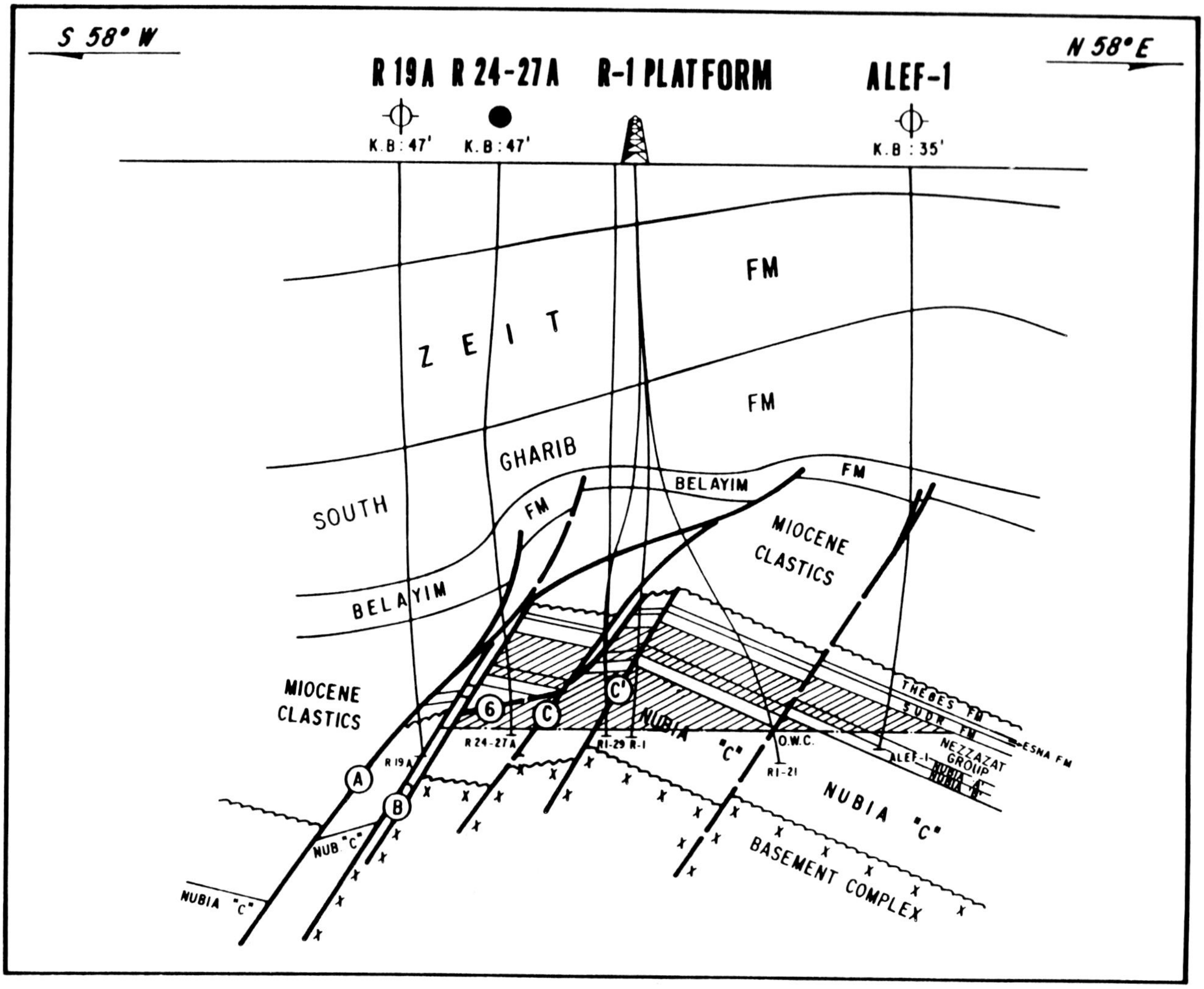

Figure 14. Schematic cross section on seismic line MP-70 (Figure 4), through the Transfer Zone in Ramadan field, shows the complexity of the structure and the drape of Belayim Formation over the underlying structure. Oil-bearing intervals are hachured.

the presence of carbonaceous fragments; and (3) the absence of marine fauna. The Nubia sands are believed to have been deposited in braided streams that were widespread and that intersected to produce sheet sand with lenses of clayey and silty impermeable sandstones characteristic of overbank and levee environments. The frequently observed black and reddish shale interbeds may be attributed to flood plain origin (Hassouba, personal communication).

Source, Maturation, and Migration

Ramadan field geochemistry is based on GUPCO's evaluation of potential source rocks in the field and the maturation modeling of the hydrocarbon-generating basin in the vicinity of the field.

Potential Source Rocks

Total organic carbon determinations in weight percent (TOC wt.%) of ditch samples (Figure 24) indicate intervals of variable good to excellent organic content within the Thebes, Esna, and Sudr formations and in the Nezzazat Group. Results of "Rock Eval" pyrolysis indicate that these intervals show variable capabilities for generating oil and minor gas (mainly from type II kerogen). Intervals of fair organic content within the lower Miocene clastics were tested and found to possess poor to fair generating potential from gas-prone type III kerogen.

The Eocene Thebes Formation displays an average TOC of 1.8 wt.% (maximum 1.95 wt.% TOC), type II kerogen, a vitrinite reflectance (R_o %) of 0.68, and an average thickness of 122 m (400 ft). The Paleocene Esna Formation, a shale of 15 m (50 ft) average thickness, has a TOC of 0.85 wt.%, type III kerogen, and R_o of 0.69%. The Campanian Sudr Formation has an average TOC of 1.57 wt.% (maximum 3.03 wt.%), contains both type I and II kerogen, has an R_o of 0.71%, and averages 91 m (300 ft) in source potential thick-

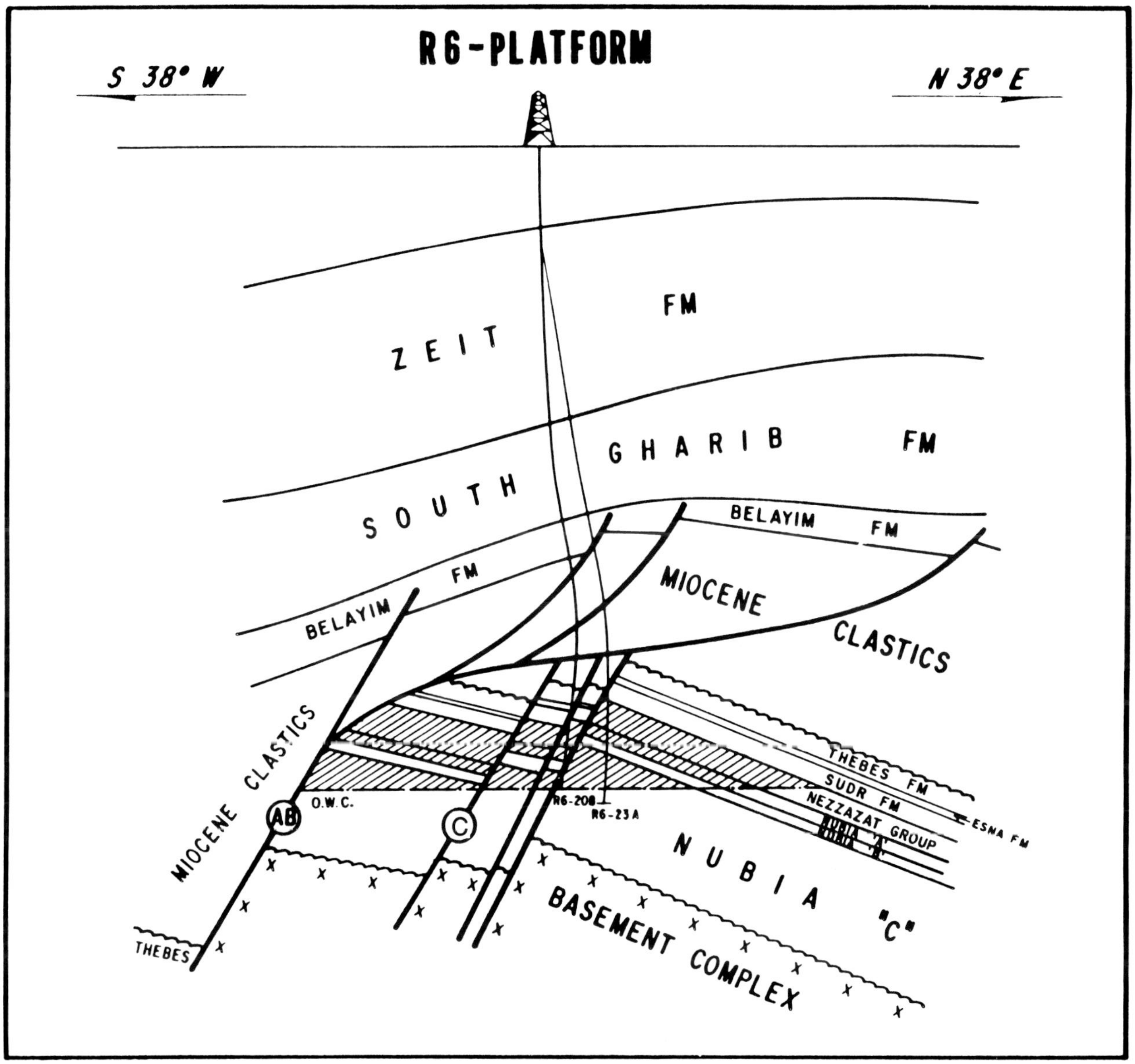

Figure 15. Schematic cross section on seismic line W86-82 through the Southern Zone (Figure 7) of the Ramadan field. Here the "A" and "B" fault systems have come together to form "AB" bounding systems with noticeably less drape of Belayim Formation over the underlying structure. Oil-bearing intervals are hachured.

ness. The Matulla Formation has an average TOC of 0.97 wt.% (maximum 1.08 wt.%), contains types II and III kerogen with an R_o of 0.77%, and an average thickness of 30 m (100 ft) of source potential.

Maturation and Migration

An explorationist generally is interested more in when and where the oil was expelled and migrated than in the beginning of generation. Therefore, in this work, the oil window is defined as the depth interval between the depth of peak generation (not at the onset of generation) and the depth of the oil floor (the depth below which liquid oil may not generate). A thermal burial history model was generated for the South Belayim trough based on the Lopatin (1971)–Waples (1980, 1985) method. The South Belayim trough (Figure 25) is located downdip to the north and east from Ramadan field and considered to be the source of Ramadan oil. The R_o % data were used to adjust the magnitudes of local tectonic events and erosions in the model. The geothermal gradient employed is based on bottom-hole temperatures corrected for loss of heat during mud circulation (Figure 25).

In the field, the geothermal gradient is about 1.6°F/100 ft (28.9°C/km). In the basinal model (Figure 26), a lower geothermal gradient of 1.3°F/100

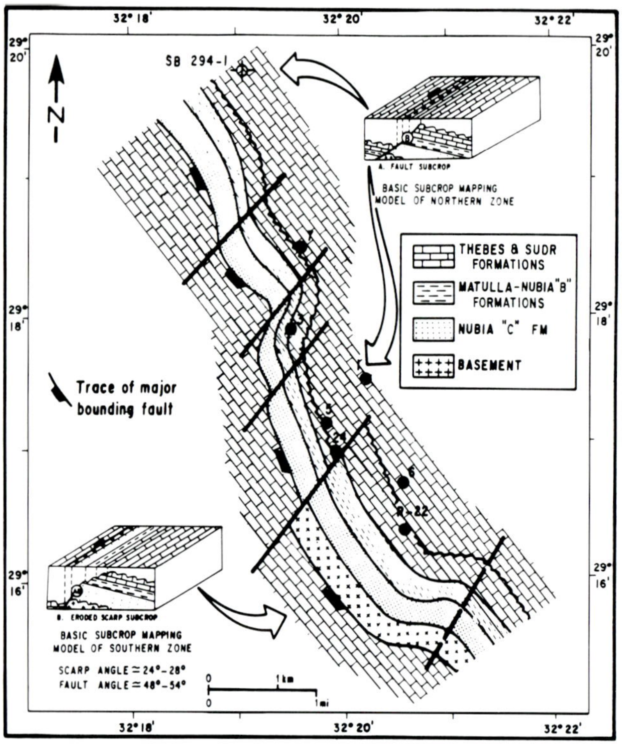

Figure 16. Pre-Miocene subcrop map over Ramadan field shows that the Southern Zone bounded by "AB" system has a much wider expression in plan, indicating a lower-angle pre-Miocene surface than in the remainder of the field.

Figure 17. Thin-section showing fair quality reservoir. The rock is composed mainly of quartzose sandstone, "Q," with kaolinite, "K," partially filling the pore spaces, "P." × 110.

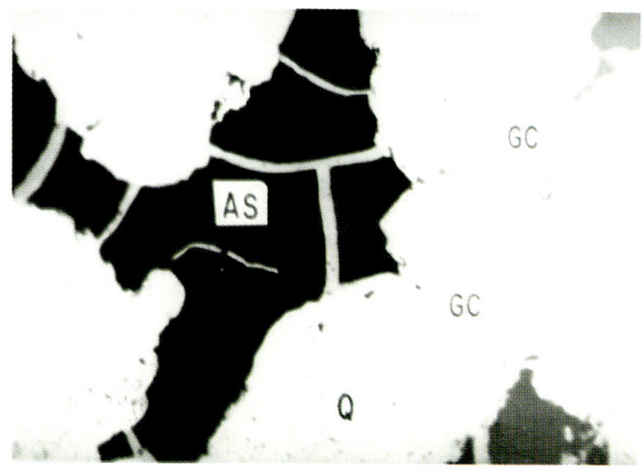

Figure 18. Thin-section showing cracked asphaltic matter "AS" produced by sudden cooling of asphaltine when core reached the surface. "GC" indicates grain-to-grain contact. × 65.

ft (23.7°C/km), as derived from the geothermal gradient map (Figure 25), was employed. The uplift and erosion are replaced by a hiatus (or a continuous period of low sedimentation rate). The model suggests that the pre-Miocene formations of the Nezzazat Group reached the stage of peak hydrocarbon generation and began to expel oil some 8 Ma, very close to the average timing of migration in the Gulf of Suez (Shahin, 1988).

Such timing post-dates the deposition of thousands of feet of Miocene evaporites that must have accumulated as the basin subsided. These evaporites, as well as providing the burial depth necessary for kerogen maturation and hydrocarbon generation, provided the ultimate seal in the Gulf of Suez. Presently, the pre-Miocene source intervals are shown to have passed through the peak generation stage (at R_o = 0.8%). The Nubia "B" shales are presently overmature as they are shown in Figure 26 below the oil floor.

The present-day depth of peak oil generation is 4938 m (16,200 ft), while the end of oil generation (oil floor) is shown at 5822 m (19,100 ft).

The term *spent source rock* may be applied here in describing the Nubia "B" shales at present, whereas the Matulla, Wata, Sudr, Esna, and Thebes formations remain "active" within the hydrocarbon generation limits. Similar conclusions pertaining to the timing of oil expulsion and migration and the pre-Miocene sediments as the main source in the northern and central Gulf of Suez have been reached by Shahin and Shehab (1984), Chenet et al. (1984), Ungerer et al. (1986), RRI (1984), and Shahin (1988).

The northeasterly dip and the northern bounding cross-element, which brings into juxtaposition the Nubia "C" sandstone of Ramadan field and the pre-Miocene sediments to the northeast, provides the most likely oil migration pathway into the structure. The southwestern corner of the mapped structure shows a potential spill point at 3659 m (12,000 ft) where the Nubia "C" sand comes in contact with the Nezzazat Group on the downthrown side. The original OWC observed in the field was 3674 m (12,050 ft).

Field Characteristics

Oil produced in Ramadan field is paraffinic-naphthenic (Tissot and Welte, 1984) and non-biodegraded (Figure 27). It is 30.8° APT, 0.88 wt.% sulfur, and has

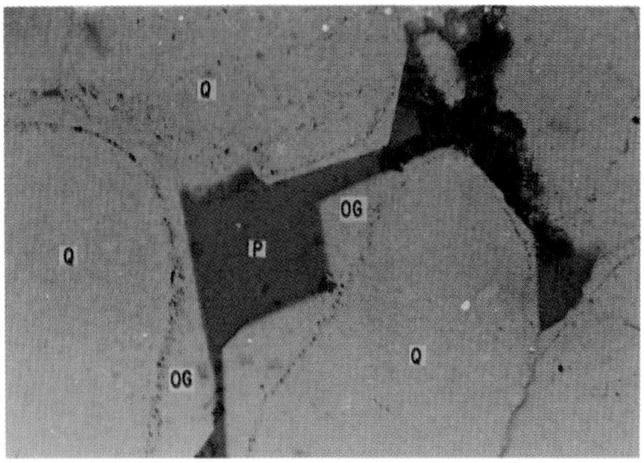

Figure 19. Thin-section showing rectangular porosity, "P," produced by development of silica overgrowth, "OG," in optical continuity with quartz grains, "Q." Note at the upper right corner the development of authigenic kaolinite clays that absorb asphaltic matter. × 80.

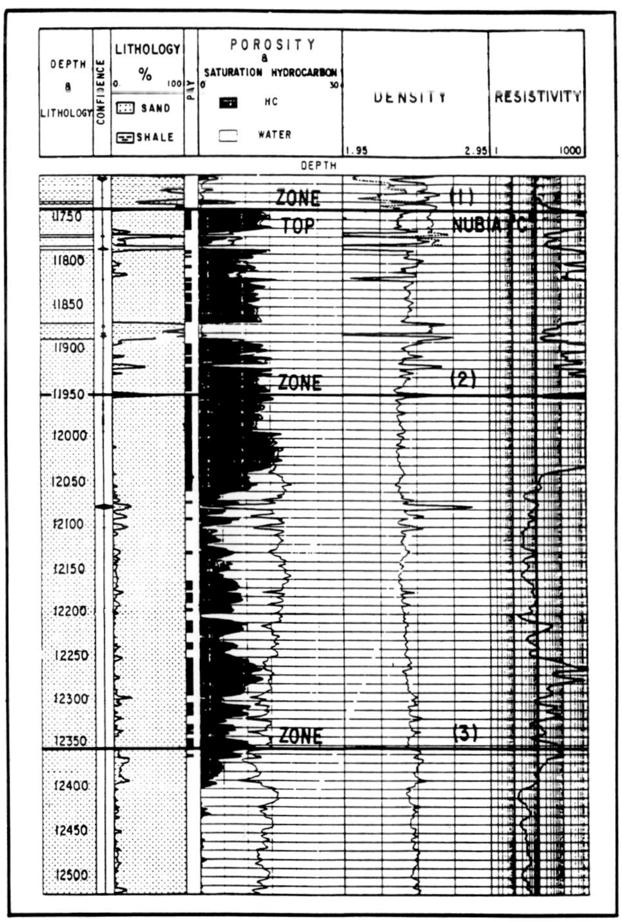

Figure 20. Electric log response of the three zones comprising the Nubia "C" reservoir in well R5-30. Subdivision into three zones is based on permeability, porosity, and lithology.

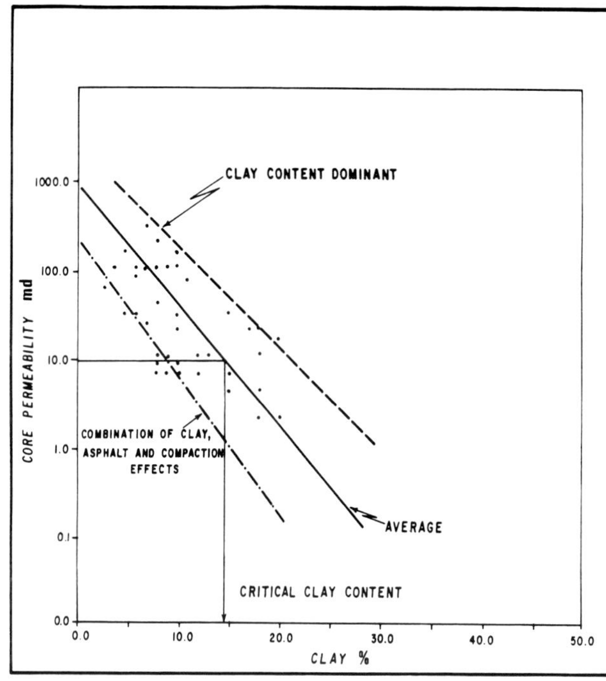

Figure 21. Crossplot of permeability vs. clay content of examined samples from the cored well R1-34, Ramadan field.

an estimated oil viscosity of 0.5 cp. Oil analyses have shown that trace amounts of what may be pyrobitumen are consistently present.

High temperature melting-point wax is also present in all analyzed samples in sufficient quantity to form a sludge in the bottom of the sample containers. These waxes, although insoluble at room temperature, may be soluble at reservoir temperatures, 295°F (146°C). Based on carbon isotopes vs. optical rotation plots, all Ramadan oils are classified as being of one type. A similar conclusion was reached by Rohrback (1982).

Production began December 1974 at a rate of 25,000 BOPD, no water, and a solution GOR of 500 ft^3/bbl. The initial pressure within the field was 5869 psi. As the field was further developed, the production rate increased (Figure 28) to a peak rate of 130,000 BOPD in January 1979 from 13 wells. The average oil column thickness within the Nubia "C" sandstones was 305 m (1000 ft) with a connate water content of 7.8%. Oil production remained fairly constant at rates near 100,000 BOPD until 1981 when production began a steady decline. Drilling of additional wells along with some successful water shutoff workovers performed in 1982 and 1983 arrested this decline for a short period of time. However, in 1984 the field again began a decline in production and reached a low of 30,000 BOPD in November 1986.

Three new wells have been recently completed in the west block of the field and the oil production rate of the field has been increased close to 50,000 BOPD. It has also been recognized that the majority of the oil

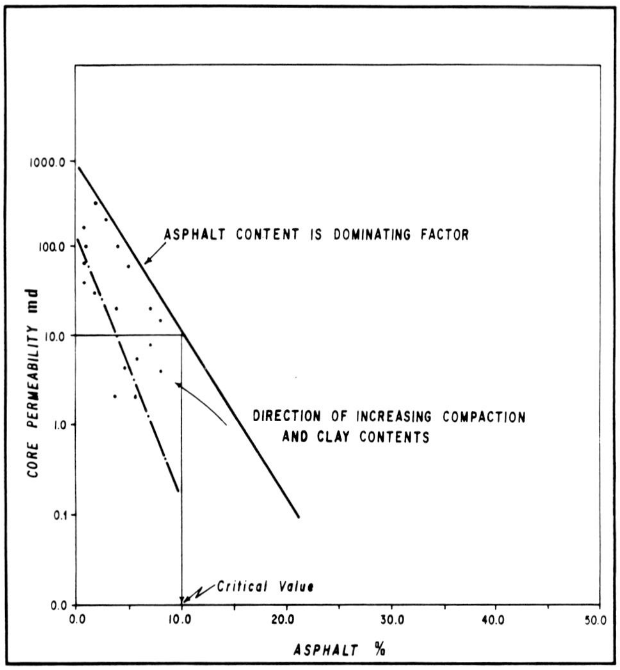

Figure 22. Crossplot of permeability vs. asphaltic content of examined core samples from well R1-34, Ramadan field.

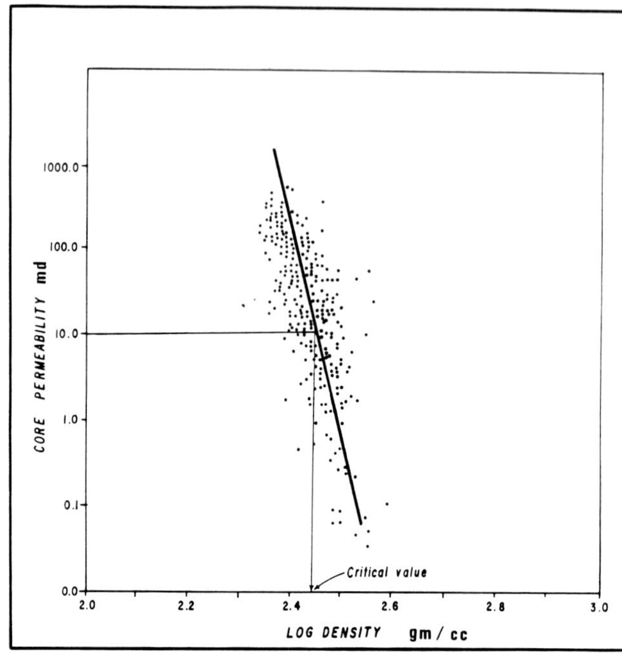

Figure 23. Relation between permeability and density of cored interval from well R1-34. Ramadan field core permeability of 10 md is highly correlatable with a bulk density of 2.45 g/cm^3 throughout the reservoir.

produced from the Nubia "C" reservoir has come from Zone 2. Presently its pressure is 4250 psi with a gradient of 0.3 psi/ft at a temperature of 146°C (295°F). Production tests have verified that when Zones 2 and 3 are co-mingled, Zone 2 is responsible for over 90% of the observed production. This appears to be due to the overall lower permeability of Zone 3. It has been recognized that within Zone 3, however, there are intervals with reservoir parameters equivalent to those of Zone 2 and that, through a selective perforation program, production from Zone 3 may be realized at a commercial rate.

EXPLORATION AND DEVELOPMENT CONCEPTS

Ramadan structure is a scarp resulting from crestal erosion of the upthrown edge of tilted pre-Miocene fault blocks during the Oligocene. Tilting of the uplifted pre-Miocene fault blocks was revealed by a three-arm dipmeter survey obtained from Alef-1 and by the seismic interpretation of a deep pre-salt structure. Reserves are defined by the intersection of the main bounding fault with the pre-Miocene erosional scarp. Adopting the concept of eroded scrap has allowed a westward repositioning of the interpreted updip limit of the Nubia "C" reservoir, resulting in an increase in the estimate of recoverable reserves in excess of 50 million barrels of oil.

Ramadan field has been produced under a strong water-driven, primary recovery mechanism. Significant water production first began in mid-1979 and has continued to increase as various intervals are flooded out. However, a continuing workover program to plug off watered-out intervals while perforating additional oil pay intervals has been successful. Future performance will be characterized by increasing water-cuts as intervals are watered out. Although the potential still exists for water shutoff workovers, this practice will decrease water production and increase oil production only temporarily.

Drilling will continue to the south and southwest until all remaining reserves within the Nubia "C" reservoir are identified. Infill drilling will continue in order to optimize areal sweep. As water production from the Nubia "C" increases, many wells will be recompleted in the secondary and tertiary reservoirs of the Nubia "A" and Nezzazat sections. The field may be a candidate for an enhanced oil recovery (EOR) program.

ACKNOWLEDGMENTS

The Egyptian General Petroleum Corporation (EGPC) and AMOCO permitted the publication of this paper.

The study was prepared under supervision of A. Shawki Abdine (Chairman of the Board), Wafik Meshref (Exploration General Manager), and Safi Wasfi (Chief Geologist), Gulf of Suez Petroleum Company (GUPCO).

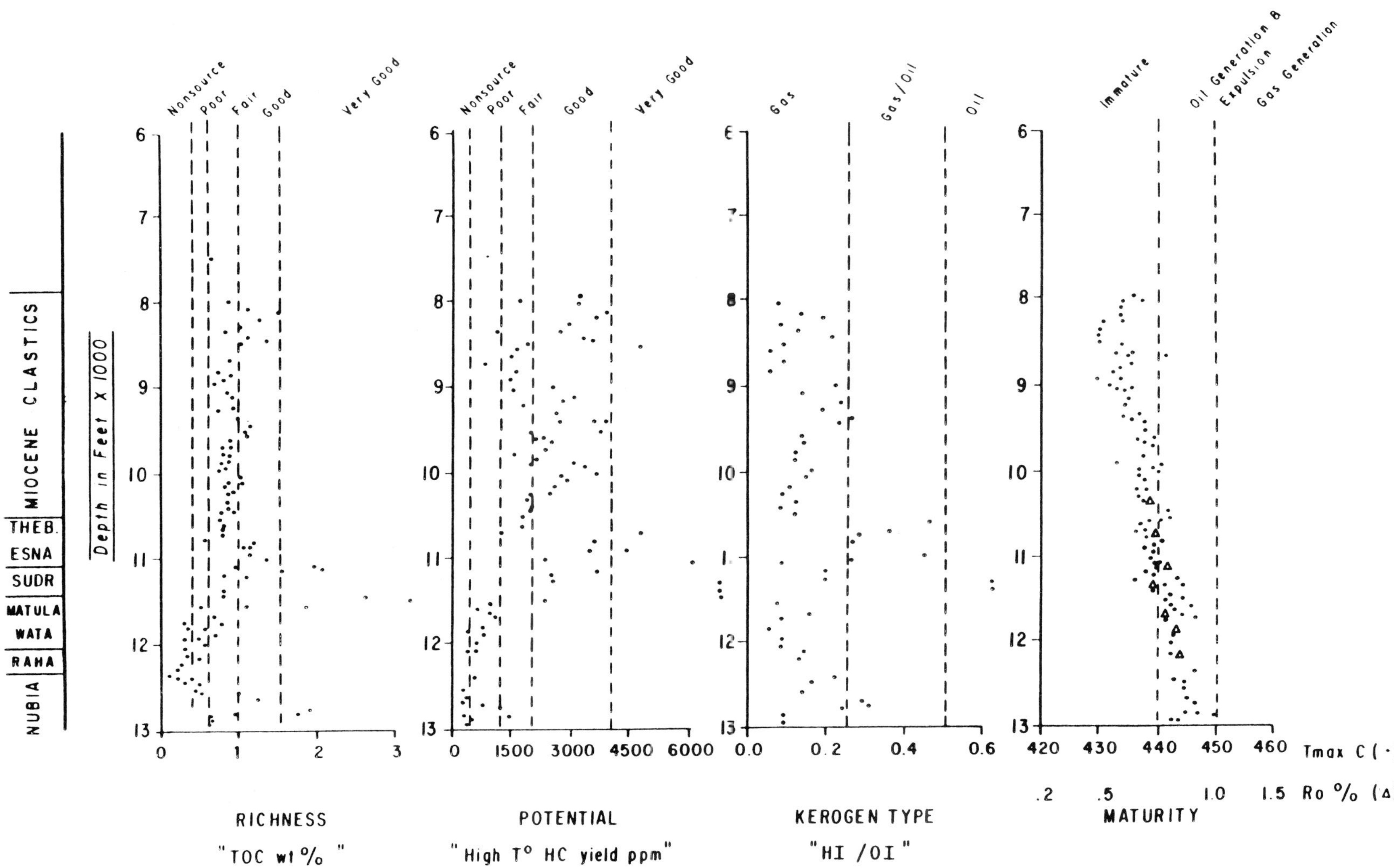

Figure 24. The Campanian lower Sudr Formation and intervals within the Thebes, Esna, Matulla, Wata, and Nubia formations are potential source rocks in the vicinity of Ramadan field as revealed by TOC wt.% and the pyrolysis-derived genetic potential, Hydrogen Index/Oxygen Index, T_{max}, as well as R_o measurements.

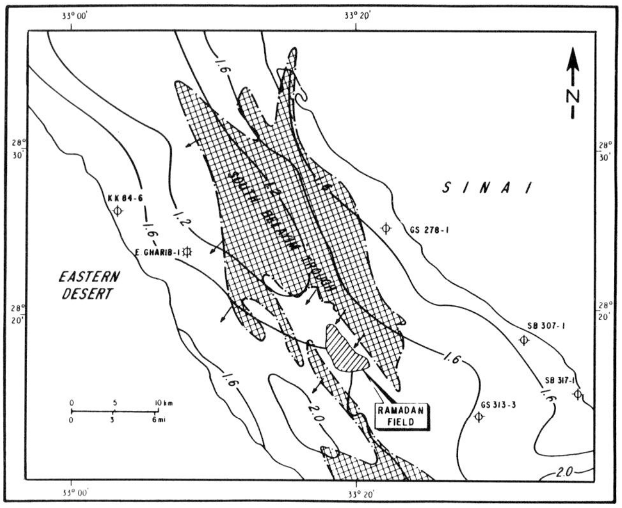

Figure 25. Geothermal gradient map suggesting the South Belayim trough, located downdip to the north and east from Ramadan field, as the likely generating basin for the Ramadan structure. Arrows represent potential migration pathways, generally updip from the generating basin. Contour interval is 0.4°F/100 ft.

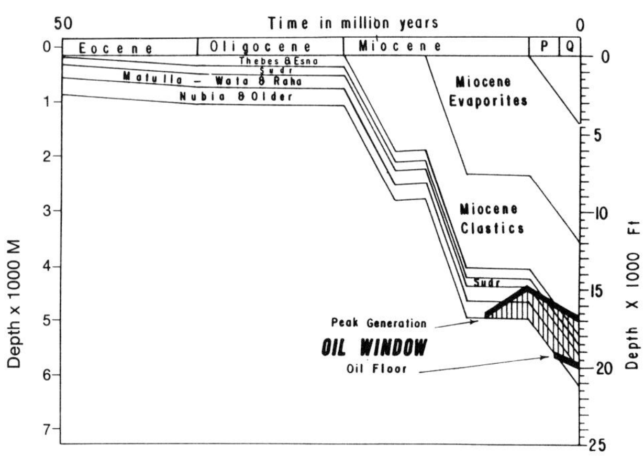

Figure 26. Thermal burial history model of the South Belayim trough (Figure 25), which is considered to be the source of Ramadan oil. The pre-Miocene source rocks reached the stage of peak generation after deposition of the Miocene evaporites, the ultimate seal in the Gulf of Suez.

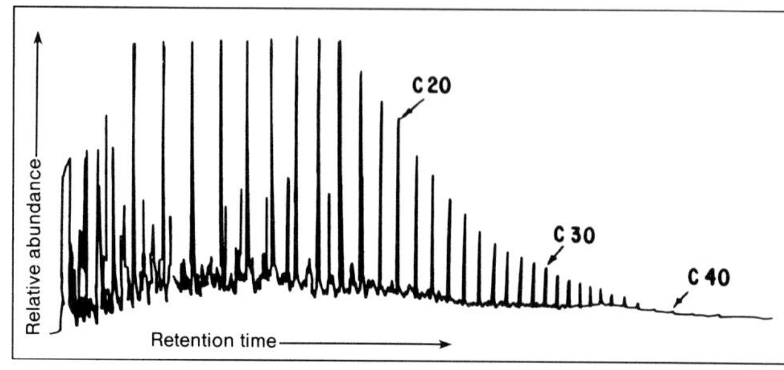

Figure 27. A representative gas chromatogram of Ramadan field oil, revealing its paraffinic-naphthenic non-biodegraded, and mature nature.

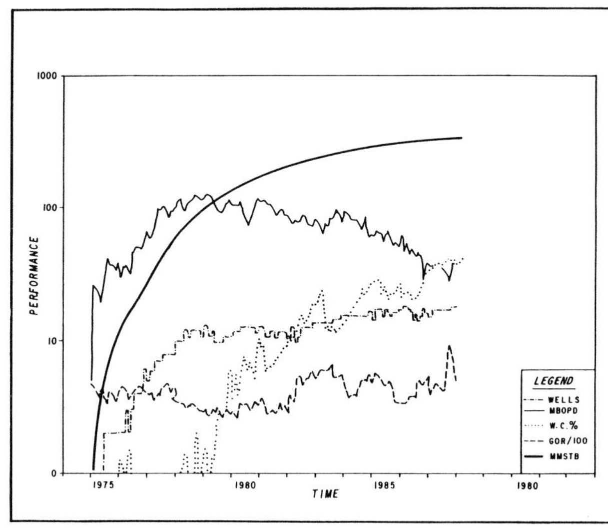

Figure 28. Nubia "C" reservoir production performance, Ramadan field. MBOPD, thousand barrels of oil per day; W.C.%, water cut %; GOR, gas-oil ratio in SCFG/100 per STB (stock tank bbl of oil); MMSTB, million stock tank barrels of oil, cumulative.

The study was edited by A. Nabil Shahin and compiled by Paul Garossino and Salah Shazly.

In addition to cited literature, several unpublished GUPCO reports were the basis for the Ramadan field study. These reports are authored by A. Abdine, A. Badawi, M. Bahr, R. Brown, M. Fathy, P. Garossino, M. Hamawy, H. Hammouda, A. Hassouba, H. Hataba, H. Mansour, A. Mokhtar, W. Meshref, M. Sarawy, A. Shahin, B. Torkelson, and S. Shazly.

The manuscript was reviewed by H. Dalton, A. Gad, A. Hassouba (who also provided reservoir petrology and environment), H. McDowell, R. Nelson, J. Jenner, T. Russell, and T. Thompson. Thanks are extended to AAPG reviewers E. Beaumont, J. Lowell, and W. Brownfield for their critical review and valuable comments. We are, however, responsible for any negative qualities that remain.

Technical assistance was provided by A. Salem, M. Morsi, M. Aman, A. Badawi, M. Laboudy, H. Mostafa, and G. Moustafa.

REFERENCES CITED

Abdine, A. S., 1981, Egypt's petroleum geology: good grounds for optimism: World Oil, v. 92, p. 99-112.

Brown, R. N., 1978, Ramadan oil field: Egyptian General Petroleum Corporation, 5th International Exploration Conference.

Chenet, P., J. Letouzey, and E. Zaghloul, 1984, Some observations on the rift tectonics in the eastern part of the Suez Rift: Egyptian General Petroleum Corporation, 7th International Exploration Conference, Cairo.

Lopatin, N. V., 1971, Temperature and geologic time as factors in coalification: USGS translation from Akad. Nauk. SSR. IZV. Geol. Series no. 3, p. 95-106.

Rohrback, B. G., 1982, Crude oil geochemistry of the Gulf of Suez: Egyptian General Petroleum Corporation, 6th International Exploration Conference, Cairo.

RRI (Robertson Research International), 1984, The Gulf of Suez area, stratigraphy, petroleum geochemistry, and petroleum geology, Consultant Report: Egyptian General Petroleum Corporation.

Shahin, A. N., and M. M. Shehab, 1984, Petroleum generation, migration and occurrence in the Gulf of Suez, offshore South Sinai: Egyptian General Petroleum Corporation, 7th International Exploration Conference, Cairo.

Tissot, B. P., and D. H. Welte, 1984, Petroleum formation and occurrence: New York, Springer-Verlag, 537 p.

Ungerer, P., P. I. Chenet, I. Moretti, A. Chiarelli, and J. L. Ondin, 1986, Modeling of oil genesis and migration in the southern part of the Suez Rift. Egypt: Egyptian General Petroleum Corporation, 8th International Exploration Conference, Cairo.

Waples, D. W., 1980, Time and temperature in petroleum formation, application of Lopatin's method to petroleum exploration: American Association of Petroleum Geologists Bulletin, v. 64, n. 6, p. 916-926.

Waples, D. W., 1985, Geochemistry in petroleum exploration: Boston, International Human Resources and Development Corporation, 232 p.

APPENDIX 1. FIELD DESCRIPTION

Field name Ramadan field

Ultimate recoverable reserves NA

Field location:
- **Country** Egypt
- **Basin/Province** Gulf of Suez

Field discovery:
- **Year first pay discovered** Nubia "C" sandstone, discovered and on production 1974
- **Year second pay discovered** Nezzazat Group, discovered 1974, first production 1983
- **Third pay** Nubia "A" sandstone, discovered 1974, first production 1986

Discovery well name and general location:
- **First pay** GS 303-1, lat. 28°17′31″N, long. 33°18′52″E; 174 km (108 mi) to the southeast of Suez City; 31 km (19 mi) east of Ras Shukheir base
- **Second pay** Same as discovery well
- **Third pay** Same as discovery well

Discovery well operator GUPCO
- **Second pay** GUPCO
- **Third pay** GUPCO

IP in barrels per day and/or cubic feet or cubic meters per day:
- **First pay** 25,000 BOPD (well GS 303-1, December 1974)
- **Second pay** 2420 BOPD (well R4-12, April 1983)
- **Third pay** 3200 BOPD(well R3-25, November 1986)

Geologic concept leading to discovery and method or methods used to delineate prospect

Prior to the discovery well GS 303-1 (R-1), Alef-1 was drilled in October 1985 on the crest of the Miocene evaporite sequence (Brown, 1980; Abdine, 1981). This well was abandoned in Nubia "A" Formation without any hydrocarbon indications. Additional seismic data obtained in 1971–1972 allowed the interpretation of the deeper pre-salt structure. In addition, the three-arm dipmeter obtained from Alef-1 well, although poor in quality, indicated a northeasterly dip in the pre-Miocene strata. All above led to the subsequent decision to drill well GS 303-1, which is updip from Alef-1. Well GS 303-1 was later called Ramadan-1.

Structure:

Province/basin type Bally III, Klemme IIIA; Gulf of Suez
Tectonic history
After the deposition of the pre-Miocene sediments, the entire area was subjected to the first phase of rifting during the Oligocene, accompanied by uplifting and tilting and followed by an Oligocene-Miocene crestal erosion. This was followed by a Miocene transgressive phase that resulted in the deposition of the Miocene clastics and evaporites. A second phase of rifting started during the late Miocene, resulting in the rejuvenation of the older faults.

Regional structure
Ramadan oil field is located in the central province of the Gulf of Suez. This province is characterized by northeast dip regime bounded to the north and south by southwesterly dipping provinces.

Local structure
Northeast dipping pre-Miocene faulted tilted blocks bounded from the west by northwest-trending fault of 152 to 610 m (500–2000 ft) average throw.

Trap:
Trap type(s) Structural, faulted tilted blocks

Basin stratigraphy (major stratigraphic intervals from surface to deepest penetration in field):

Chronostratigraphy	Formation	Depth to Top in ft (m)
Middle Miocene	Zeit	1400 (427)
Middle Miocene	South Gharib	4800 (1463)
Middle Miocene	Belayim	6650 (2027)
Middle Miocene	Kareem	7100 (2165)
Lower Miocene	Rudeis	7600 (2317)
Lower Miocene	Nukhul	9100 (2774)
Eocene	Thebes	9170 (2796)
Paleocene	Esna Shale	9540 (2909)
Upper Cretaceous	Sudr	9570 (2918)
Upper Cretaceous	Nezzazat	9930 (3027)
Lower Cretaceous	Nubia "A"	10,600 (3232)
Paleozoic	Nubia "B"	10,700 (3262)
Paleozoic	Nubia "C"	10,841 (3305)

Reservoir characteristics:
Number of reservoirs 3
Formations Nubia "C," Nubia "A," and Nezzazat Group (Matulla, Wata, and Raha formations sandstones)
Ages Paleozoic, Early and Late Cretaceous, respectively
Depths to tops of reservoirs Nubia "C," 3475 m (11,400 ft); Nubia "A," 3400 m (11,160 ft); Nezzazat Group, 3195 m (10,480 ft)
Gross thickness (top to bottom of producing interval) 1000 ft (304.8 m)
Net thickness—total thickness of producing zones
Average 750 ft (228.6 m)
Maximum 900 ft (274.3 m)
Lithology
Nubia "C": fine- to medium-grained, occasionally coarse-grained sandstone with kaolinite matrix
Nubia "A": same as Nubia "C"
Nezzazat Group: very fine to fine sandstone, moderately sorted, with glauconite and some pyrite crystals
Porosity type Primary intergranular
Average porosity 15%
Average permeability 100 md

Seals:
Upper
Formation, fault, or other feature Sudr limestone, Nubia "B" shale
Lithology Massive limestones and shales
Lateral
Formation, fault, or other feature Fault
Lithology NA

Source:

- **Formation and age** Thebes (Eocene), Sudr (Campanian), Matulla (early Senonian–Turonian)
- **Lithology** Thebes (limestone), Sudr (brown limestone), Matulla (shales and marls)
- **Average total organic carbon (TOC)** Thebes, 1.8%; Esna, 0.9%, Sudr, 1.6%; Matulla, 1.0%
- **Maximum TOC** Thebes, 1.95%; Esna 0.85%; Sudr, 3.03%; Matulla, 1.08%
- **Kerogen type (I, II, or III)** Thebes, II; Esna, III; Sudr, II, I; Matulla, II, III
- **Vitrinite reflectance (maturation)** R_o = 1.02 at 12,800 ft (3900 m)
- **Time of hydrocarbon expulsion** 10 Ma (middle Miocene)
- **Present depth to top of source** 9000 ft (2740 m) avg.
- **Thickness** Thebes, 400 ft (122 m); Esna 50 ft (15 m); Sudr, 300 ft (91 m); Matulla, 100 ft (31 m)
- **Potential yield** NA

APPENDIX 2. PRODUCTION DATA

Field name Ramadan field

Field size (data given for Nubia "C"; not available for Nubia "A" or Nezzazat Group):

- **Proved acres** 2851
- **Number of wells all years** NA
- **Current number of wells** 40
- **Well spacing** 40
- **Ultimate recoverable** NA
- **Cumulative production** 360 million bbl
- **Annual production** 13.6 million bbl
- **Present decline rate** 25%
  - **Initial decline rate** Nil
  - **Overall decline rate** 25%
- **Annual water production** (1987) 8.8 million bbl
- **In place, total reserves** NA
- **In place, per acre-foot** NA
- **Primary recovery** NA
- **Cumulative water production** 50 million bbl (as of February 1988)

Drilling and casing practices:

- **Amount of surface casing set** 395 ft
- **Casing program** 24 in. to 350 ft; 13⅜ in. to 3500 ft; 9⅝ in. to 10,150; 7 in. to TD
- **Drilling mud** Sea water salt saturated, oil-based mud
- **Bit program** Varies
- **High pressure zones** Miocene clastics

Formation evaluation:

- **Logging suites** BHC, GR, FDC, CNL, and DLL
- **Testing practices** Typically production tested
- **Mud logging techniques** Basic online logging for development wells and pressure detection inline logging for exploratory wells

Oil characteristics:

- **Type** Paraffinic-naphthenic
- **API gravity** 30.8°
- **Base** NA
- **Initial GOR** 500
- **Sulfur, wt.%** 0.88
- **Viscosity, SUS** 0.5 cp
- **Pour point** NA
- **Gas-oil distillate** NA

Field characteristics:

Average elevation 11,500 ft
Initial pressure 5870 psi
Present pressure 4250 psi
Pressure gradient 0.3
Temperature 295°F (146°C) (surface 80°F; 27°C)
Geothermal gradient 1.6°F/100 ft (29.2°C/km)
Drive Natural water influx
Oil column thickness 1000 ft
Oil-water contact 12,050 ft subsea
Connate water 7.8%
Water salinity, TDS NA
Resistivity of water NA
Bulk volume water (%) NA

Transportation method and market for oil and gas:
Submarine pipelines from offshore platforms to Ras Shukheir facility (land tanks) then pumping and shipping through the port by shipping facilities.

Examples of Other Rift Basins

Chapter 4

Structure, Stratigraphy, and Petroleum Geology of the Pripyat and Dnieper-Donets Basins, Byelarus and Ukraine

G.F. Ulmishek
U.S. Geological Survey
Denver, Colorado, U.S.A.

V.A. Bogino
Byelarussian Geologic Exploration Organization ("Byelarusgeologia")
Minsk, Byelarus

M.B. Keller
Committee of Geology and Use of Mineral Resources
of the Russian Federation
Moscow, Russia

Z.L. Poznyakevich
Belorussian Geological Prospecting Research Institute
Minsk, Byelarus

ABSTRACT

The Dnieper-Donets and Pripyat basins are underlain by a Devonian rift system that extends southeastward into the deformed Donbas fold belt. The two basins are probably among the geologically oldest oil- and gas-productive rift basins of the world. Gas was first discovered in the Dnieper-Donets basin in 1950; in the Pripyat basin the first oil discovery was made only in 1964. At present, both basins are in a declining stage of production; however, the exploration potential has not been exhausted.

The basement dips southeastward along the rift strike from a maximum of about 5 km (3.1 mi) in the Pripyat basin to possibly as much as 16–18 km (9.9–11.1 mi) in the southeast part of the Dnieper-Donets basin. The basement and rift fill are cut by a large number of faults into many tilted fault blocks. Salt tectonics further complicates the structure. The basins are separated by a synrift uplift composed largely of Devonian volcanics.

The sedimentary cover consists of prerift platform (Middle Devonian–lower Frasnian), synrift (upper Frasnian–Famennian), postrift sag (Carboniferous–Lower Permian), and postrift platform (Triassic–Cenozoic) sequences separated by major unconformities. The sequences are composed of a variety of clastic and carbonate rocks, local Devonian volcanics, and three thick salt formations—two in the synrift Devonian sequence and one in the Lower Permian. A Riphean–Vendian (Proterozoic) rift fill occurs at base of the sedimentary cover in the western Pripyat basin. Another Riphean rift filled with unknown rocks possibly underlies the Devonian in the southeast part of the Dnieper-Donets basin.

The Late Devonian rift was formed due to clockwise rotation of the Ukrainian shield relative to the rest of the Russian craton. Hercynian collision along the southern edge of the craton resulted in strong deformation in the Donbas fold belt. Northwestward, the intensity of deformation decreases rapidly, and no signs of compression are observed in the Pripyat basin.

The Pripyat basin produces oil from carbonate rocks of the prerift platform and synrift sequences. Most of reserves are in faulted structural traps along crests of tilted fault blocks. Source rocks have not been geochemically identified, but they are probably present in both productive sequences. The source rocks are mature in the north zone of the basin where all fields are concentrated. In the rest of the basin, source rocks are largely immature except possibly in the deepest depressions. Time of maturation is uncertain, but may be late Paleozoic or late Mesozoic–Cenozoic.

Almost all hydrocarbon reserves of the Dnieper-Donets basin have been found in the postrift sag sequence, largely in clastic reservoirs. Salt-assisted structural traps dominate, although structural traps related to basement tectonics and stratigraphic traps are present. The Lower Permian salt seal controls the major part of the reserves; however, new discoveries come chiefly from the deeper, Lower Carboniferous section. Source rocks have not been geochemically identified; probably the most important source rocks occur in the lower part of the postrift sequence in the central and southeastern areas of the basin. The synrift Devonian sequence also contains as yet unidentified source rocks. Both Devonian and lowermost Carboniferous source rocks are overmature in much of the basin, resulting in the dominance of gas in hydrocarbon reserves. In addition, gas could have been generated by coal and coaly shales in the Carboniferous section. In most areas, the maximum maturation had been achieved by pre-Late Permian time. Both lateral and long-range vertical migration were involved in formation of fields.

INTRODUCTION

The Donbas-Pripyat rift is one of the best-studied Paleozoic structures of its type. The rift cuts into the southern part of the Russian (East European) craton, and was a tectonotype of aulacogens, as originally proposed by Shatsky (1964). The rift is more than 1300 km (810 mi) long and 70–130 km (44–81 mi) wide and consists of three major parts: the Pripyat basin on the northwest, the Dnieper-Donets basin in the central part, and the Donbas fold belt on the southeast (Figure 1). (Actually, "Donbas" is an abbreviation that stands for "Donets basin," implying coal basin; however, we use this abbreviation to avoid confusion with the Dnieper-Donets petroleum basin.) The Donbas part of the rift was structurally inverted, folded, and thrust northward at the end of the Early Permian. Carboniferous Donbas coals have been extensively mined for decades.

The Voronezh and Belorussian regional highs are north of the Dnieper-Donets and Pripyat basins, respectively (Figure 1). Paleozoic rocks of the highs gradually thicken and dip toward the basins, but the abrupt increase in the thickness is related to the boundary faults of the rift. On the Ukrainian shield south of the basins, the crystalline basement is exposed on the surface or overlain by thin Mesozoic sediments. Paleozoic rocks are also thin on the Poles structural saddle, which separates the Pripyat basin from the Brest (Podlasie-Brest) basin to the west. The latter basin straddles the Byelarus–Poland boundary.

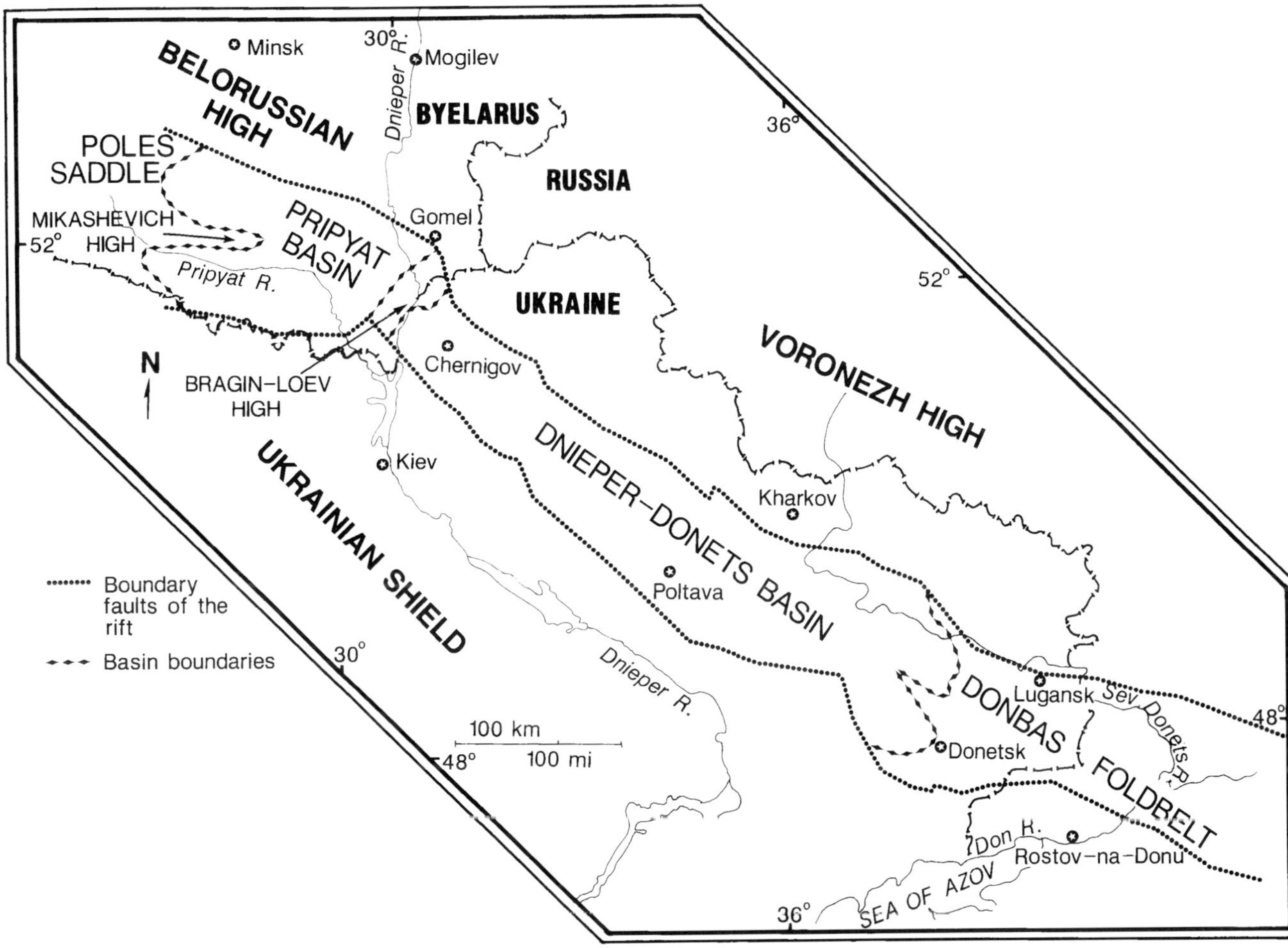

Figure 1. Location map of the Pripyat and Dnieper-Donets basins.

About 200 oil and gas fields have been discovered in the Dnieper-Donets and Pripyat basins after deep drilling was initiated in the early 1950s. About one-quarter of the fields are located in the Pripyat basin, and all but one of the fields are oil fields. In the Dnieper-Donets basin, oil fields are concentrated in its shallowest, northwestern part, whereas gas and gas condensate fields dominate in the central and southeastern areas of the basin.

Both basins are rather significantly explored to depths of 3.5–4 km (2.2–2.5 mi). More than 1200 exploratory and delineation wells totaling 4.3 million m (14.1 million ft) have been drilled in the Pripyat basin. The amount of exploratory drilling in the Dnieper-Donets basin (about 3000 wells) exceeds 7 million m (23 million ft). The remaining exploration objectives in the latter basin are deep; about one-half of the wells drilled during the last 10–15 yr are deeper than 4.5 km (2.8 mi).

REGIONAL SETTING

The Dnieper-Donets basin is located in the eastern part of Ukraine; the Pripyat basin is almost completely in Byelarus (former Belorussian Republic of the USSR). The boundary between the two states runs along the Bragin-Loev structural saddle that separates the basins (Figure 1).

The northern and southern boundaries of the Pripyat basin are drawn along boundary faults of the Late Devonian rift. Across these faults, thickness of the Paleozoic sedimentary rocks abruptly decreases or pinches out toward the Belorussian high and Ukrainian shield. The rocks within the basin thin westward along the rift strike onto the Mikashevich uplift, which is the easternmost projection of the Poles saddle. The Bragin-Loev high east of the basin is a Late Devonian uplift and volcanic center. Sedimentary and volcanic rocks are thin on the southern part of the uplift and dip to its northern part where a structural strait is located that connects the Pripyat and Dnieper-Donets basins.

The Devonian rift structure occupies most of the Dnieper-Donets basin. In Russian literature the rift is known as the Dnieper graben. The basin also includes the northern shoulder of the rift where thick Visean and younger rocks overlie the crystalline basement and gradually pinch out to the north. The southeastern boundary of the basin with the Donbas fold belt is uncertain. Anticlines of the Donbas gradually dip under younger sediments of the basin losing their structural expression. This boundary is commonly drawn along the outcrop of Tournaisian–lower Visean rocks (Gavrish, 1989).

The basement of the Pripyat and Dnieper-Donets basin is formed by Archean–lower Proterozoic crystalline rocks similar to the basement of the rest of the Russian craton. South and north of the basins, these rocks are well studied in exposures of the Ukrainian shield and in shallow wells and some quarries on the top part of the Voronezh regional high. Deeply metamorphosed Archean rocks constitute several large blocks separated by relatively narrow early Proterozoic fold belts. These structures are cut by Archean and Proterozoic granites. The prominent part of the lower Proterozoic section is formed by the Kursk series of ferruginous quartzites 2–5 km (1.2–3.1 mi) thick that produce a strong magnetic anomaly. The dominant structural grain of basement rocks is north to south on the Ukrainian shield and northwest to southeast on the Voronezh high (Milanovsky, 1987). This structural grain is strongly discordant relative to the strike of the Donbas-Pripyat rift.

In the western Pripyat basin, the crystalline basement is covered by upper Proterozoic (Riphean and Vendian) rocks that underlie the Devonian and younger sedimentary section of the basin. Red continental clastics of the Riphean Poles series, up to 400 m (1312 ft) thick, occur at the base. They are overlain by Vendian tillites of up to 200 m (656 ft) thick (Vilchan series) and by upper Vendian volcanics and clastics up to 120 m (394 ft) thick (Volyn and Valday series). These rocks were deposited in a trough of the northeastern strike; the origin of this trough, the Volyn-Orsha aulacogen, is related to middle–late Proterozoic rifting of the Russian craton (Garetsky, 1979) (Figure 2).

Proterozoic sedimentary rocks also may be present in the southeastern part of the Dnieper-Donets basin. Deep seismic sounding data indicate that a thick (4–4.5 km or 2.5–2.8 mi) section of rocks with velocities of 5.8–6.2 km/s occurs there at depths of 10–12 km (6.2–7.5 mi) between the basement and the Paleozoic basinal sedimentary cover. These rocks fill a narrow graben in the axial zone of the Devonian rift. Most geologists interpret the rocks as a sedimentary fill of a late Proterozoic rift that possibly served as a zone of weakness during the following Late Paleozoic rifting event (Dolenko et al., 1981; Chirvinskaya and Sollogub, 1980). The Proterozoic graben fill extends into the Donbas fold belt where its thickness reaches 10 km (6.2 mi) (Gavrish, 1987).

Devonian and younger rocks of the Pripyat and Dnieper-Donets basins unconformably overlie the Proterozoic or crystalline basement. Thickness of these rocks increases from 2–3 km (1.2–1.9 mi) in the western areas of the Pripyat basin to as much as 12–15 km (7.5–9.3 mi) in the southeastern part of the Dnieper-Donets basin. The Dnieper-Donets basin dips to the southeast an average of 30 m/km, or about 2° (Gavrish, 1985). The principal increase of the thickness is found in Visean–Lower Permian rocks. The Devonian and younger sedimentary section of the basins may be divided into the prerift platform, synrift, and postrift sag sequences, and the overlying platform sequence of post-Early Permian age. The stratigraphic positions of boundaries of the sequences somewhat vary, which is related to the observed propagation of rifting in the northwestern direction (Garetsky et al., 1990a; Konishchev, 1984).

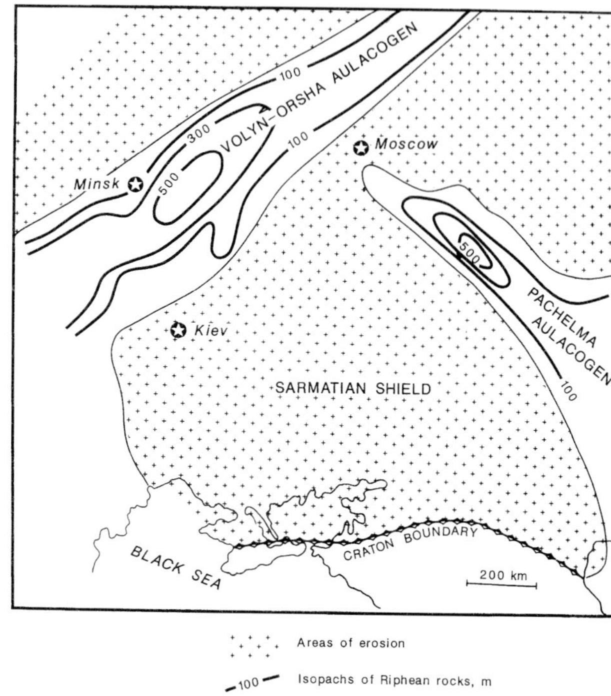

Figure 2. Structure of the southwestern Russian craton at end of the Proterozoic (modified from Dolenko et al., 1981).

STRATIGRAPHY

Prerift Sequence

The prerift sequence was deposited under platform conditions; where preserved from later erosion, these rocks extend outside the rift without noticeable change in thickness and facies. The stratigraphy of this sequence has much in common with that of the Volga-Ural province. The stratigraphic range of the prerift platform sequence in the Pripyat basin is from the Middle Devonian to the Rechitsa horizon of the middle Frasnian (Figure 3). In the Dnieper-Donets basin, the upper boundary of the sequence descends into the lower part of the Frasnian stage (Sargay horizon).

Prerift Devonian rocks have been studied from drilling data mainly in the Pripyat basin; in the Dnieper-Donets basin, they have been penetrated only in shallower marginal areas, primarily in the northwestern part of the basin. In the Pripyat basin, Middle Devonian and basal Frasnian (Lan horizon) rocks are clastics, red and continental in the middle part and gray and marine in the lower and upper parts. Thin beds of dolomite and anhydrite are present in the

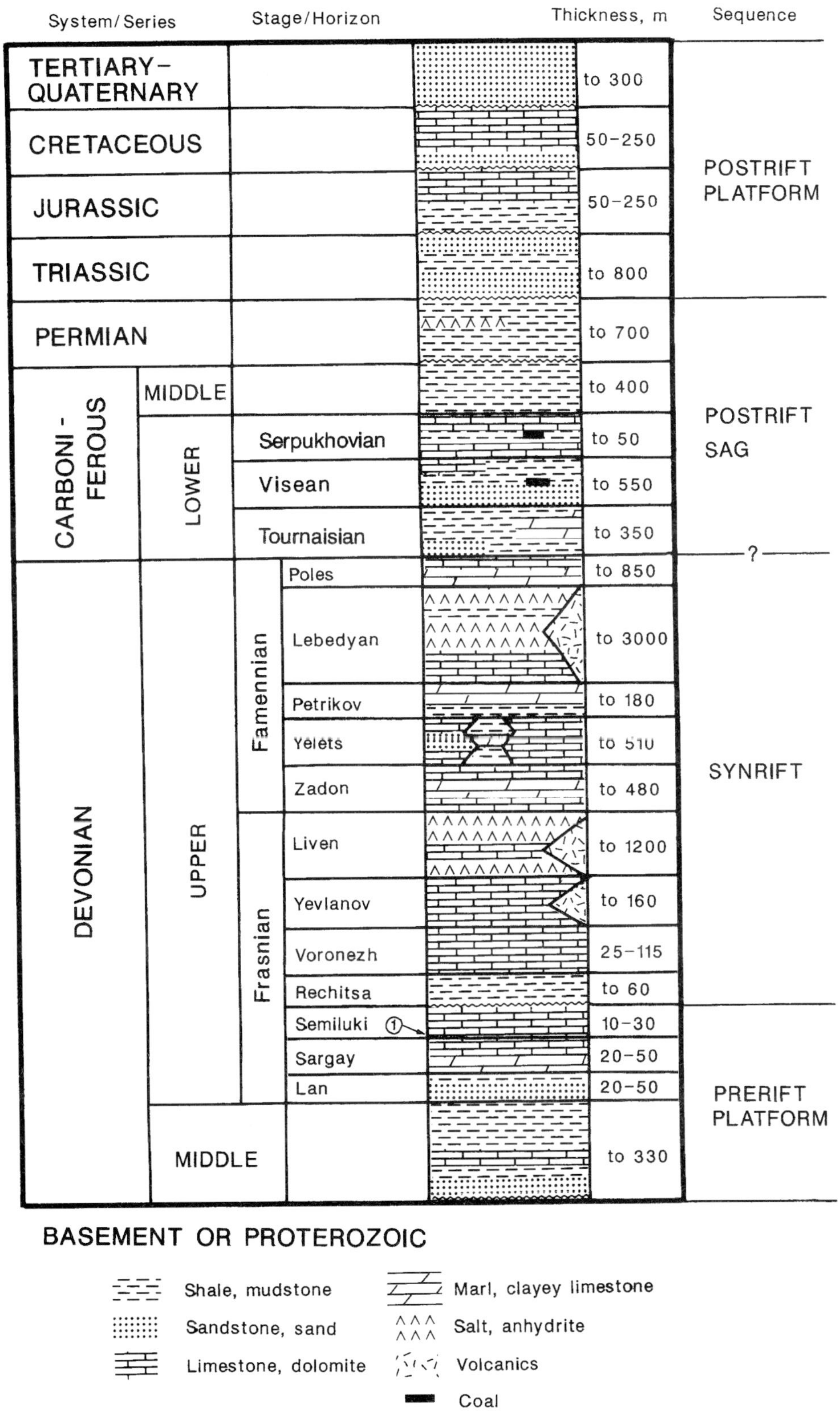

Figure 3. Columnar stratigraphic section of the Pripyat basin. Number 1 in circle shows the Moiseev bed.

lower half of the Middle Devonian. In exploration practice, these rocks are known as the subsalt clastic section. Thickness of the section decreases from 280 m (919 ft) northwest of the basin to 100 m (328 ft) in the southeast. The paleostructural grain that controlled thickness changes was probably oriented in an approximately north-south direction roughly perpendicular to the trend of the younger rift.

Lower Frasnian carbonate rocks of the Sargay and Semiluki horizons occur at the top of the prerift platform sequence of the Pripyat basin. Marl and dolomite, containing thin beds and inclusions of anhydrite, compose the lower part of the section. The marl and dolomite are overlain by massive, commonly reefal limestones and dolomites, with numerous fossils. A 5–10-m- (16–33-ft-) thick organic-rich shale bed (Moiseev bed) is present at the base of the Semiluki horizon. The carbonates are vuggy because of leaching, especially in the upper part of the section underlying the pre-Famennian unconformity. Thickness of the section ranges from 40 to 80 m (131 to 262 ft).

The composition and thicknesses of the prerift platform sequence in drilled areas of the Dnieper-Donets basin are rather similar to those in the Pripyat basin. However, volcanic material is abundant in the upper part of carbonate rocks (Sargay and Semiluki horizons) in the northwestern part of the basin, especially along the northern boundary fault. The presence of this volcanic material suggests the earlier commencement of rifting in the Dnieper-Donets basin compared to the Pripyat basin. In the Donbas fold belt, volcanic rocks are known in strata as old as the Middle Devonian; therefore, the rifting could have begun there at that time (Maydanovich and Radzivill, 1984; Garetsky et al., 1990a; Shpak, 1990). The prerift sequence is locally absent in the Dnieper-Donets basin. On some drilled synsedimentary basement uplifts in the northwestern part of the basin, the basement is overlain by rocks ranging from the Lan to the Voronezh horizons.

Synrift Sequence

In the Dnieper-Donets basin, judging from seismic data, thickness of the synrift sequence may reach 5–6 km (3.1–3.7 mi) (Kabyshev, 1987). The upper boundary of the sequence is commonly defined by overlap of upper Visean rocks onto the rift shoulders across the boundary faults (Dolenko et al., 1981); the distribution of older Carboniferous and Upper Devonian rocks is mainly limited by these faults. However, the major unconformity at the base of Carboniferous rocks suggests that rifting was terminated by the end of the Devonian. The absence of Tournaisian–lower Visean strata on the rift shoulders may be partly related to erosion beneath the pre-late Visean unconformity. In the Pripyat basin, the synrift sequence of up to 4 km (2.5 mi) thick constitutes the principal part of the sedimentary fill. The upper stratigraphic boundary of the sequence is poorly defined because of regional thinning of Carboniferous and younger rocks and deformation of Devonian salt. The transition from the synrift to the sag sequence there is gradual. Probably, the sequence boundary occurs at about the same stratigraphic level as in the Dnieper-Donets basin, that is, at top of Devonian rocks.

The synrift sequence of the Pripyat basin contains two salt formations of late Frasnian (Liven horizon) and late Famennian (Lebedyan horizon) age. The salt formations are underlain and separated by mainly carbonate rocks and are overlain by uppermost Famennian shales and carbonates (Figure 3).

In the central and eastern areas of the basin, the sequence begins with variegated shales with layers of tuff of the Rechitsa horizon that unconformably cover prerift rocks. Westward, the shales pinch out under younger carbonates of the Voronezh and Yevlanov horizons. The carbonates contain variable amounts of clay material, beds of anhydrite, and thin layers of salt in the upper part of the section. Organic-rich black shales are present in latitudinal synsedimentary depressions, but their distribution is poorly known (Makhnach et al., 1984). Voronezh and Yevlanov carbonate rocks, together with underlying carbonates of the prerift sequence, are designated as the subsalt carbonate section in the petroleum geology literature.

Facies changes in Voronezh-Yevlanov rocks indicate the beginning of formation of the Pripyat basin. Marls and clayey carbonates with anhydrite beds dominate in the section in the central areas. To the northern and western margins, these marls and carbonates pass into purer carbonate rocks having less anhydrite. Layers of sandstones are present along the southern margin, indicating the appearance of the clastic provenance on the Ukrainian shield. Volcanics of the basaltic composition are present on the Bragin-Loev uplift east of the basin.

The lower salt formation of the Liven horizon in the central areas of the basin is composed of rock salt with carbonate and shale beds and anhydrite. On the northwestern and southwestern basin margins, salt pinches out and the section consists of carbonate and clastic rocks. Sandstones are abundant on the southern margin. On the Bragin-Loev uplift, the section is composed of massive volcanics. Thickness of the formation increases from 300 m (984 ft) on the north to more than 900 m (2953 ft) in the southern part of the basin. The salt is moderately deformed into domes and pillows; the intensity of deformation increases to the south.

The intersalt carbonate formation of the Pripyat basin includes rocks of the Zadon, Yelets, and Petrikov horizons of the lower Famennian Stage. The rocks were deposited during active rifting and formation of tilted fault blocks. The distribution of facies generally reflects the paleostructural plan. At various stratigraphic levels of the formation, shallow-water carbonate facies, including reefal deposits, occupy the upper portions of the tilted blocks, whereas off-reef and basinal facies were deposited in troughs. The formation is commonly absent on upper edges of the blocks due to nondeposition and erosion (if the tectonic tilt resulted in subaerial exposure) or later sliding of the carbonate plates down the blocks between underlying and overlying salts because of continuing tilting and deformation of the lower salt.

The sea that covered the basin during deposition of intersalt carbonates progressively deepened with time. Organic-rich, thin-bedded black shales and carbonates accumulated in deep water. The black shale deposition

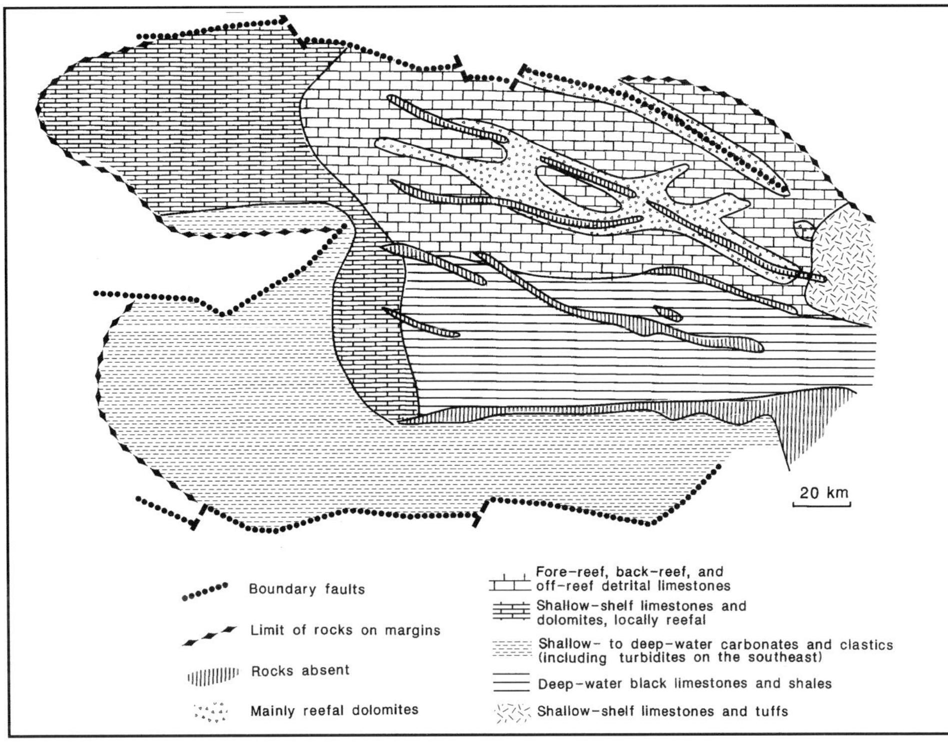

Figure 4. Facies of the Upper Devonian Zadon horizon in the Pripyat basin (modified from Makhnach et al., 1984).

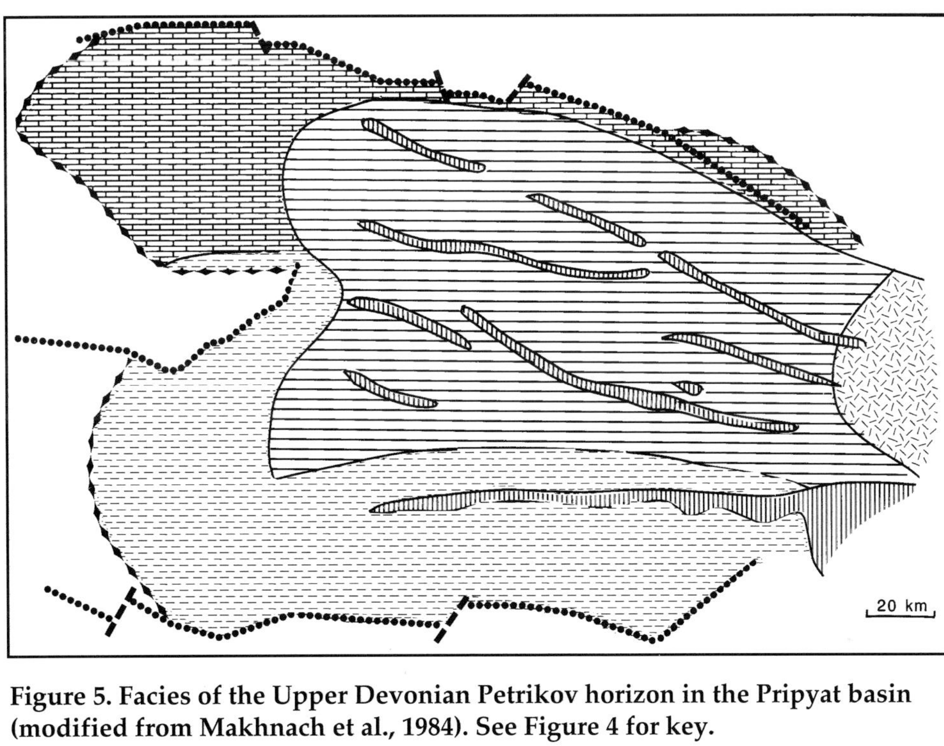

Figure 5. Facies of the Upper Devonian Petrikov horizon in the Pripyat basin (modified from Makhnach et al., 1984). See Figure 4 for key.

(Zadon horizon) was originally limited to a rather narrow axial zone of the basin (Figure 4). Later black shale facies gradually expanded northward and overlay shallow-water carbonates. During Petrikov deposition, most of the basin was covered by the black shale facies (Figure 5). Shallow-water carbonate sedimentation was preserved only on the northwest and in a narrow zone along the northern basin boundary. Thick clastics, mainly sandstones, derived from the Ukrainian shield, alternate with carbonate rocks in the southern zone of the basin. Volcanics are interbedded with carbonates on the extreme east of the basin.

Thickness of the intersalt carbonate formation varies substantially. The thickness commonly ranges from 200 to 600 m (656 to 1968 ft) in the northern areas, from 200 to 350 m (656 to 1148 ft) in the basin center, and reaches 950 m (3117 ft) on the south. Lesser thickness in the central zone of the basin is related to the lack of material to compensate for tectonic subsidence (Makhnach et al., 1981).

The upper salt formation that overlies intersalt carbonates is very thick in the Pripyat basin, and consists of two subformations. The lower subformation is mainly composed of rock salt and includes beds of algal limestones in its lower part in northern areas of the basin. The upper subformation consists of abrupt alternation of salt, anhydrite, shale, and carbonate rocks; in the western basin, this subformation contains very large commercial deposits of potassium salts. The lower subformation is significantly deformed; thickness of the salt in domes reaches 3 km (1.9 mi). The upper subformation occurs primarily in depressions between the domes. Thick sandstone beds are present in the salt in the south, along the Ukrainian shield. On the Bragin-Loev uplift, the salt formation passes into thick volcanics.

Suprasalt Devonian (Poles horizon) rocks are carbonates and shales with anhydrite beds in the lower part and with highly organic-rich oil shale beds in the upper part. Upward in the section, the rocks pass into lowermost Carboniferous continental variegated clastics. The suprasalt rocks occur mainly in depressions between salt domes; their thickness increases eastward, toward the Dnieper-Donets basin, from about 100 to more than 700 m (328 to 2297 ft). As mentioned, the upper boundary of the synrift sequence in the Pripyat basin is poorly defined. Synrift faults do not cut through the upper salt formation and thicknesses of suprasalt Devonian and Lower Carboniferous rocks primarily reflect salt deformation.

In the Dnieper-Donets basin, synrift Devonian rocks have been drilled only in relatively shallow marginal and northwestern areas. Limited data suggest that thickness of the synrift sequence is generally similar to that of the Pripyat basin. The subsalt part of the sequence in the northwest Dnieper-Donets basin contains abundant volcanics alternating with clastics and carbonates. Farther southeast, these rocks pass into carbonates and salt (Figure 6). Salt beds appear in rocks as old as the Voronezh horizon. The lower salt formation is far thicker (to 3 km or 1.9 mi) than in the Pripyat basin, whereas the upper salt is thinner and is known mainly in narrow grabens along the boundary faults of the rift. Whether this salt is present in the deeply buried axial zone of the basin is uncertain. Salt tectonics of the Dnieper-Donets basin is dominantly related to the lower salt formation. Both salt formations probably pass into other lithologies in the Donbas fold belt. Salt domes appear just northwest of its boundary with the Dnieper-Donets basin.

The intersalt formation of the Dnieper-Donets basin comprises clastic, carbonate, and volcanic rocks; maximum drilled thickness exceeds 1.5 km (0.9 mi) (Garetsky et al., 1990a). Very thick clastics derived from the Ukrainian shield form the section along the southwestern basin margin. On the northeastern margin, the lower part of the section consists of carbonates containing reefs and beds of shales. The upper part of the section is composed of poorly sorted flysch-like clastics. The clastics are limited mainly to a narrow graben along the boundary fault; they thin and possibly pinch out basinward (Shpak, 1989a). Thick (to 500 m or 1640 ft) volcanic beds are widespread in the northwestern part of the basin. The section contains layers of organic-rich bituminous black shales and carbonates in many drilled wells (Khomenko, 1977), but their areal and stratigraphic distributions are poorly understood. Deep-water black-shale facies may occupy the central zone of the basin, whereas thick clastics along the northeastern and southwestern boundaries are turbidites that constitute the marginal clastic fill of the deep-water sea.

The suprasalt part of the synrift sequence of the Dnieper-Donets basin is more well known than underlying rocks. The stratigraphic development was affected by formation of a deep-water depression in the central and southeastern part of the basin. Organic-rich shales and marls and clastic turbidites of latest Devonian, Tournaisian, and early Visean age, up to 2500 m (8202 ft) and possibly 3500 m (11,483 ft) thick, were deposited in the depression. Shelf carbonates that include reefs and shales bound the depression on the southwest and northeast. Northwestward, along the strike of the basin, the deep-water facies pass into shallow-marine clastics and carbonates, and then into continental red beds and volcanics. The pre-Carboniferous and pre-late Visean unconformities are pronounced in the marginal zones and on the northwest and resulted in the local absence of uppermost Devonian and Tournaisian rocks.

Postrift Sag Sequence

The postrift sag sequence is relatively thin (although stratigraphically fairly complete) in the Pripyat basin. The sequence thickens dramatically southeastward into the Dnieper-Donets basin. Farther southeast, in exposures of the Donbas fold belt, the Carboniferous part of the sequence is more than 11 km (6.8 mi) thick. The lower boundary of the sequence is defined in the Dnieper-Donets basin by a major unconformity at the base of Carboniferous rocks, an abrupt decrease in sedimentation rate, and the disappearance of volcanics. The top of the sequence is marked by a regional pre-Triassic unconformity. Overlying Mesozoic and Tertiary sediments form a gentle depression (Ukrainian syneclise of Russian geologists), which extends far beyond the outlines of the Paleozoic basins.

The largest thicknesses of the sequence in the Pripyat basin are limited to its southern part and do not exceed 1.5 km (0.9 mi). The sequence is present in depressions between salt domes and thins or pinches out on the tops of the domes. Lowermost Carboniferous continental sandstones, shales, and marls occur at the base of the sequence. Upper Visean–Serpukhovian, dark gray, coaly shales with beds of sandstone, limestone, and coal about 200 m (670 ft) thick overlie older rocks. This section is absent in the northern area of the basin. Unconformably overlying Middle Carboniferous rocks up to 430 m (1430 ft) thick are variegated shales and sandstones containing limestone layers and up to 20 coal beds, each of which is 1–2 m (3.3–6.6 ft) thick. Upper Carboniferous rocks have not been identified in the Pripyat basin. Lower Permian rocks are red and variegated clastics with a few limestone beds. In a limited area in the central part of the basin, the section contains rock salt with beds of potassium salts.

In the Dnieper-Donets basin, Tournaisian–lower Visean strata are characterized by complex facies relationships. In the northwest, the section is primarily composed of continental clastics with coal beds and bauxite. Southeastward, along the basin strike, the continental clastics pass into marine clastic and carbonate rocks. A deep-water depression inherited from the latest Devonian occupied the axial central and southeastern zone of the basin. Flysch-like fine clastics, hundreds of meters thick, were deposited in the depression. On the margins of the depression, the section is composed of thinner shallow-marine shales and carbonates that include reefs.

An overlying, very thick (possibly to 8–10 km or 5–6.2 mi in the southeast) section in the stratigraphic interval from the upper Visean through the lower part of the Upper Carboniferous is formed by a cyclic paralic series (Figure 6). The series is dominantly composed of clastic rocks ranging from marine to continental, and includes numerous, but thin, beds of limestone and coal. Lower Bashkirian transgressive marine carbonates and shales devoid of coals unconformably overlie Lower Carboniferous rocks and divide the paralic series into two parts. Alluvial and deltaic clastics dominate in cyclothems in the upper part of the series, whereas marine clastics are more abundant in its lower part. The series is stratified into a number of local suites based mainly on the distribution of sandstones.

Rocks deposited during the Late Carboniferous and Early Permian reflect a progressive change of climate from humid to arid. The upper part of the Upper Carboniferous consists of red and variegated sandstones with thin and infrequent limestone beds. They pass upward into similar lowermost Permian clastics containing beds of dolomite and anhydrite and characterized by stratiform copper mineralization. Overlying Lower Permian rocks consist of cyclic alternation of carbonate rocks, anhydrite, and salt. Salt bed thickness increases upward. Organic buildups are known in pinch-out zones of salt (Lukin and Galitsky, 1974). The upper part of the section is composed mostly of salt (including potassium salts) and red clastics. The salt-bearing formation is up to 1200 m (4000 ft) thick in depressions of

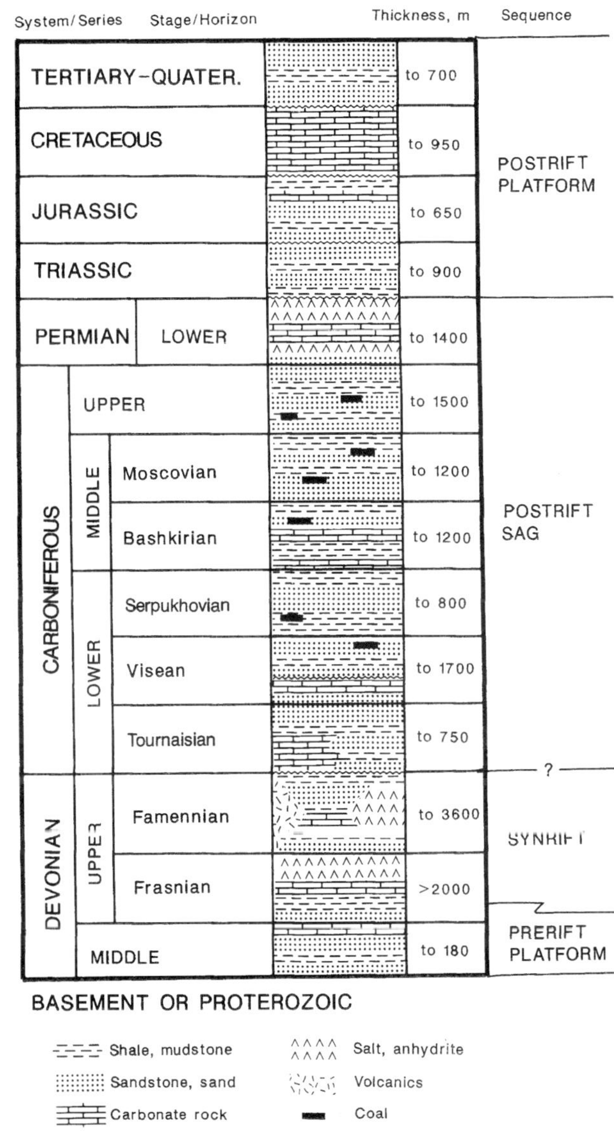

Figure 6. Columnar stratigraphic section of the Dnieper-Donets basin.

the southeastern Dnieper-Donets basin; to the northwest and on the basin margins, salt beds pinch out of the section and the formation thins to several tens of meters.

The areal extent of the postrift sag sequence is controlled by pre-Triassic truncation. Triassic and younger sediments overlie various rocks ranging from the Lower Permian in the axial zone of the rift, to the Middle and Lower Carboniferous on the basin margins, to the crystalline basement outside the basins.

Postrift Platform Sequence

Continental red and variegated clastics, with intercalations and inclusions of gypsum in their lower part and beds of freshwater carbonates in their upper part, occur at the bottom of the postrift platform sequence. Most of the clastics are of Triassic age, but the basal beds do not contain fossils and their age is uncertain—perhaps Late Permian (Beznosov, 1987) or Early Triassic (Gavrish, 1989).

Rhaetian-Liassic continental clastics are found only in the southeastern Dnieper-Donets basin. The overlying Toarcian–Middle Jurassic section is stratigraphically more complete in the same area; there, Toarcian–Aalenian marine shales and carbonates unconformably cover the continental red beds. Younger Middle Jurassic rocks onlap northwestward along the basin strike and are composed of Bajocian–lower Bathonian marine clastics and upper Bathonian continental tuffaceous shales.

The Upper Jurassic is a transgressive marine series composed of clastics and limestones (in the Oxfordian). Lower Cretaceous and Cenomanian marine clastics unconformably overlie the Jurassic; the section contains an upper Aptian formation of continental kaolinic shale. The Upper Cretaceous mostly is composed of marl and chalk. The Paleogene section unconformably overlies the Upper Cretaceous and older rocks and consists of marine glauconitic sand, mudstone, and limestone. Neogene continental and marine clastics are locally present.

The postrift platform sequence covers the Pripyat and Dnieper-Donets basins, the western part of the Voronezh regional high, and the northern slope of the Ukrainian shield. To the northwest, the sequence extends into the Polish–Lithuanian depression of the southern Baltic basin (Ulmishek, 1990). To the east, the sequence is connected with the North Caspian basin through a foredeep north of the Donbas fold belt. Thickness of the platform sequence in the basins is significantly controlled by salt tectonics; the sequence is thin or absent on the tops of salt domes and plugs and may be more than 2 km (1.2 mi) thick in depressions in the southeastern Dnieper-Donets basin. The present-day limits of the sequence are erosional; the original depositional area was much wider.

STRUCTURE

Pripyat Basin

The typical rift structure of the Pripyat basin is well exhibited in rocks underlying the upper salt formation of late Famennian age. Rift faults (except for boundary faults) do not penetrate upward through this formation; therefore, the suprasalt structure is dominantly controlled by deformation of the salt.

The presalt structure of the basin consists of a series of tilted fault blocks extending roughly parallel to the strike of the rift (Figure 7). Six main near-latitudinal blocks separated by major faults are pres-ent. Three of the northern fault blocks dip to the north and two of the southern blocks dip to the south (Figure 8). The axial Petrikov-Khoynin zone is relatively uplifted and both northern and southern dips are present in separate smaller blocks. The zone consists of two uplifts, the Petrikov uplift on the west and the Khoynin uplift on the east, which dip toward each other along the strike of the zone. The northern and southern marginal steps are narrow structures transitional between the rift shoulders and deep depressions of the adjacent blocks. The

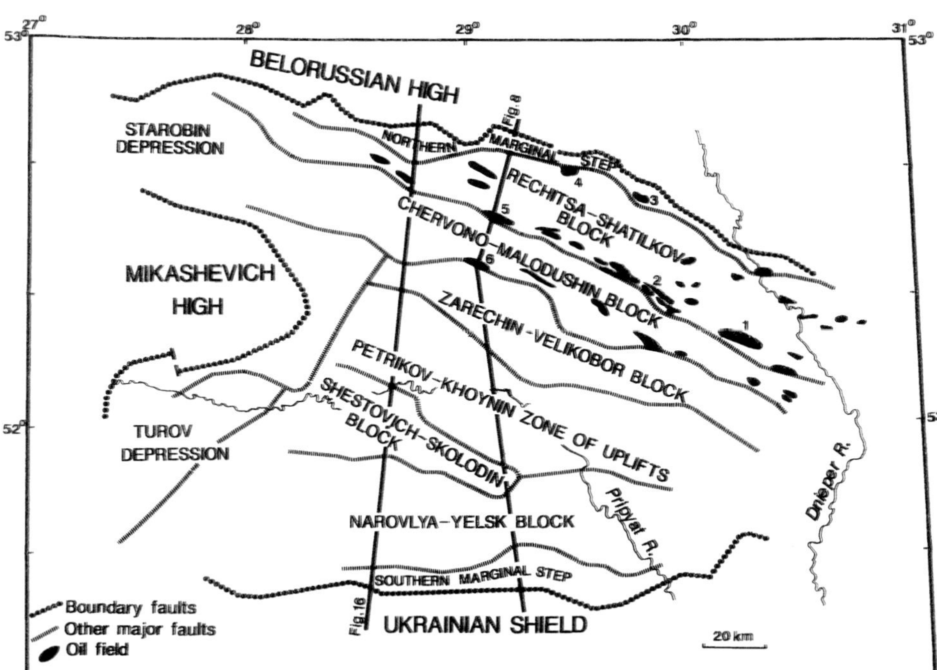

Figure 7. Main structural units and oil fields of the Pripyat basin. Numbers indicate oil fields: 1 = Rechitsa, 2 = Ostashkovich, 3 = Berezin, 4 = Sudovits, 5 = Vishan, 6 = Oktyabr.

structures are faulted into a number of small blocks that dip in various directions.

A structural map of the top of the crystalline basement is shown in Figure 9. Major features of this structure also are characteristic of structure on the top of the prerift sequence, although the regional tilt of the latter is slightly different owing to the appearance and gradual westward thickening of upper Proterozoic sedimentary rocks. In the deepest depressions adjoining the northern and southern marginal steps, the surface of the basement is buried to more than 5 km (3.1 mi) below sea level. This surface occurs at only 2–3 km (1.2–1.9 mi) on the Petrikov and Khoynin uplifts in the axial zone of the basin. The map also demonstrates complex faulting of the basement and subsalt rocks. Major fault blocks shown in Figure 7 are cut by variously oriented subordinate faults into a large number of small blocks. The faulting is far more complex on the south side of the basin, especially in the axial zone, than in the northern areas. Some geologists believe that at least part of the subordinate faults that are transverse to the main fault blocks inherit the north–northeastern structural grain of the Archean–early Proterozoic basement (Garetsky, 1976, 1979). However, these strike directions are not evident on the map except for the northeasterly trending Bragin-Loev high east of the Pripyat basin.

The structure on the top of the intersalt carbonate formation (Figure 10) is similar, in its basic features, to that of the basement (Figure 9). Some differences are seen mainly in the south where the lower salt is more strongly deformed than in the central and northern areas. Many small fault blocks shown in Figure 9 are not identified in the structure of the intersalt carbonate formation partly because of much uncertainty in correlating reflectors recorded from different parts of the formation in various areas.

No significant faults (except for boundary faults) have been identified in suprasalt rocks. The structure of uppermost Devonian and younger Paleozoic rocks is completely controlled by deformation of the upper salt formation. In some depressions between salt domes, the base of suprasalt rocks is subsided deeper than 3 km (1.9 mi), however the base occurs at depths of 400–1200 m (1312–3937 ft) over most of the salt domes. On the tops of a number of salt domes, Paleozoic rocks are absent and the salt (or cap rock) is directly overlain by Mesozoic sediments.

Dnieper-Donets Basin

The basement and subsalt rocks occur very deep in the Dnieper-Donets basin and are less well known than those of the Pripyat basin. Most information on the deep structure was obtained from refraction seismic surveys; only in recent years were limited areas studied by deep reflection seismic (Gavrish, 1989).

The northeastern and southwestern boundary faults of the Dnieper-Donets basin separate the rift from the rift shoulders (Figure 11). Depths to the basement on the shoulders reach 3–4 km (1.9–2.5 mi) on the southwest and 4–5 km (2.5–3.1 mi) on the northeast. The

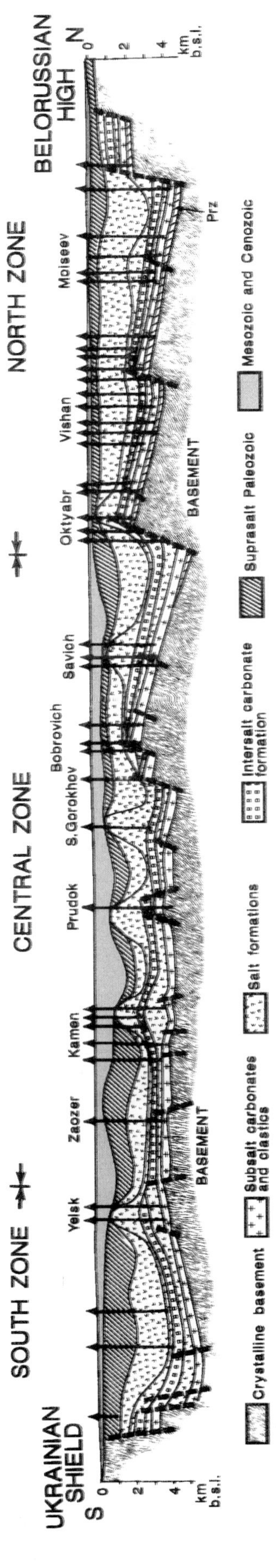

Figure 8. Cross section through the Pripyat basin (adapted from Konishchev, 1984). A stratigraphic boundary inside the upper salt formation separates the lower and upper subformations. Location of the cross section is shown in Figure 7. b.s.l. = below sea level.

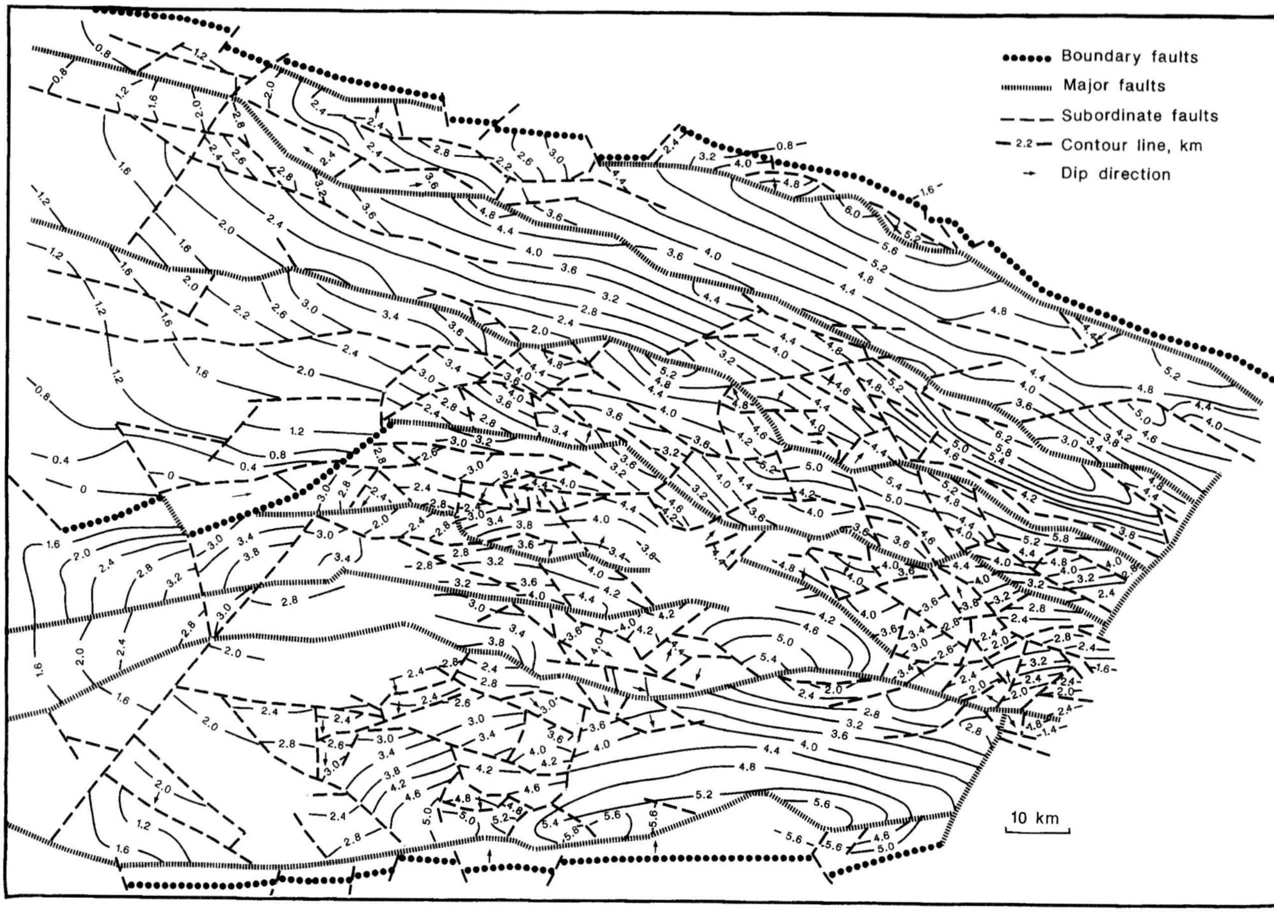

Figure 9. Contour map on top of the crystalline basement of the Pripyat basin (modified from Garetsky, 1979). Structural contours in meters below mean sea level. Contour interval is 0.4 km.

downthrow of the basement surface across the boundary faults ranges from several hundred meters to 2–3 km (1.2–1.9 mi) and more (Figure 12).

Judging from refraction seismic data, the basement of the rift is faulted into a large number of uplifted and downfaulted blocks. The main structural grain consists of a series of axial depressions and two subordinate zones of narrow depressions along the northeastern and southwestern boundary faults (Figure 13). The depressions of the northeastern margin are better developed; on the southwestern margin, the zone of depressions is interrupted by structural terraces descending to the basin axis. The zones of marginal and axial depressions are separated by chains of uplifted blocks. From the northwest to the southeast, along the rift strike, the basement surface dips from 4–5 km (2.5–3.1 mi) to more than 18 km (11.1 mi) (Figure 11).

Interpretation of refraction seismic data by some Ukrainian geologists, especially in recent publications, suggests that the lower several kilometers of the sedimentary cover below a depth of 8–11 km (5–6.8 mi) in the southeastern one-half of the basin is made of upper Proterozoic rocks filling an older and more narrow rift (Gavrish, 1989; Shpak, 1989a; Sollogub, 1986). This interpretation is based on the apparent splitting of a refractor correlated with the basement surface observed on some profiles of deep seismic sounding. The intervening sequence of rocks is characterized by boundary velocities of 5.8–6.2 km/s (Dolenko et al., 1981). This sequence extends southeastward from the Dnieper-Donets basin into the Donbas fold belt; however, the sequence has not been identified in the eastern half of the fold belt (Sollogub, 1986). The interpretation of the sequence as a Proterozoic graben fill remains uncertain, and more data, especially deep seismic reflection surveys, are needed. A longitudinal cross section along the axis of the rift (Figure 14) does not indicate the presence of the Proterozoic graben.

Uplifts and depressions in structure of the basement surface are separated by faults with throws ranging from several hundred meters to 1–2 km (0.6–1.2 mi). Longitudinal faults roughly parallel to the rift strike are the longest and dominate the structure. Basement uplifts separating the axial zone of depressions from the marginal zones are probably the synrift horsts, at least in the northwestern one-half of the basin. This was proven by drilling on the larger Koshelev uplift on which postrift upper Visean rocks directly overlie the crystalline basement (Dolenko et al., 1981).

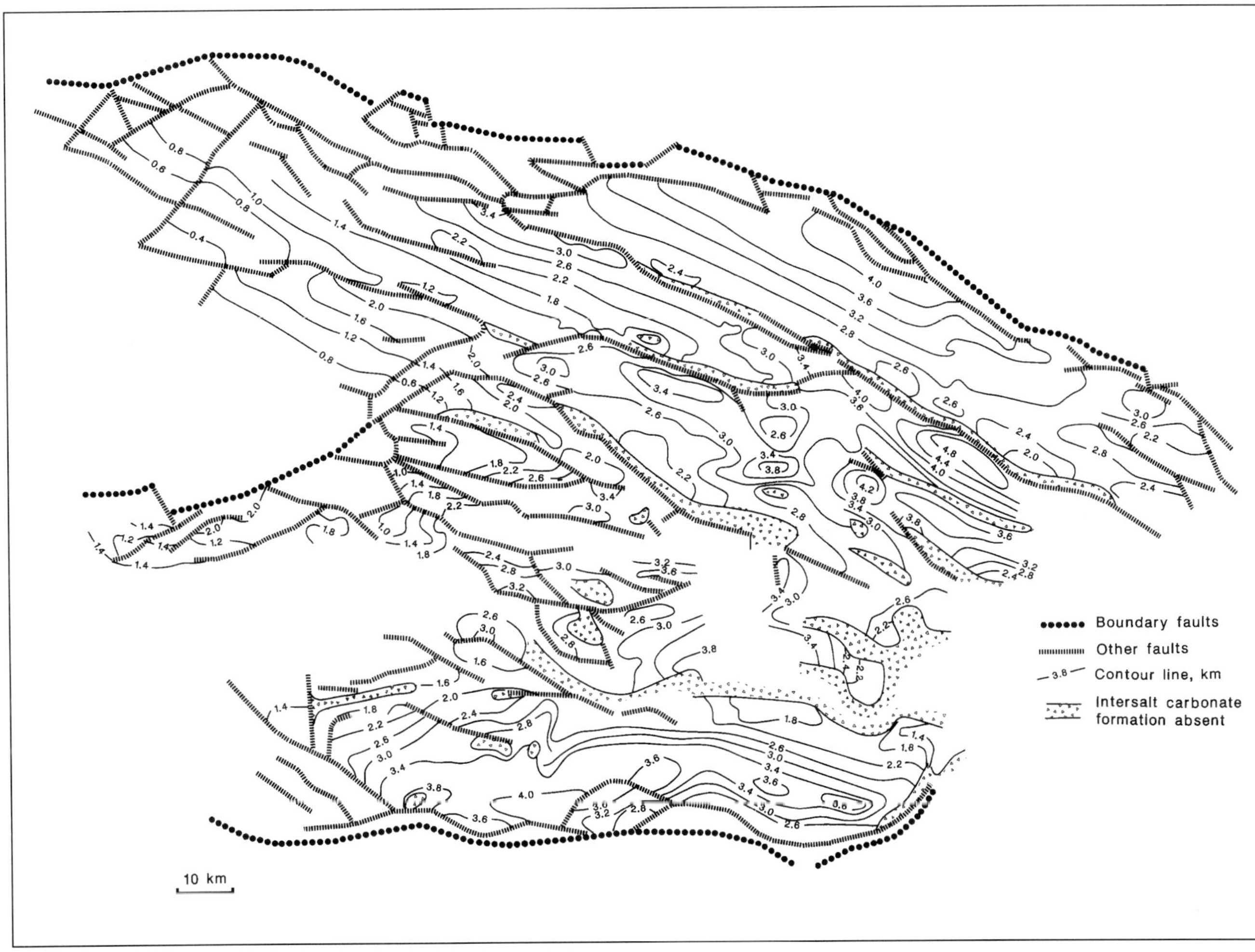

Figure 10. Contour map on top of the Devonian intersalt carbonate formation of the Pripyat basin. Structural contours in meters below mean sea level. Contour interval 0.4 km.

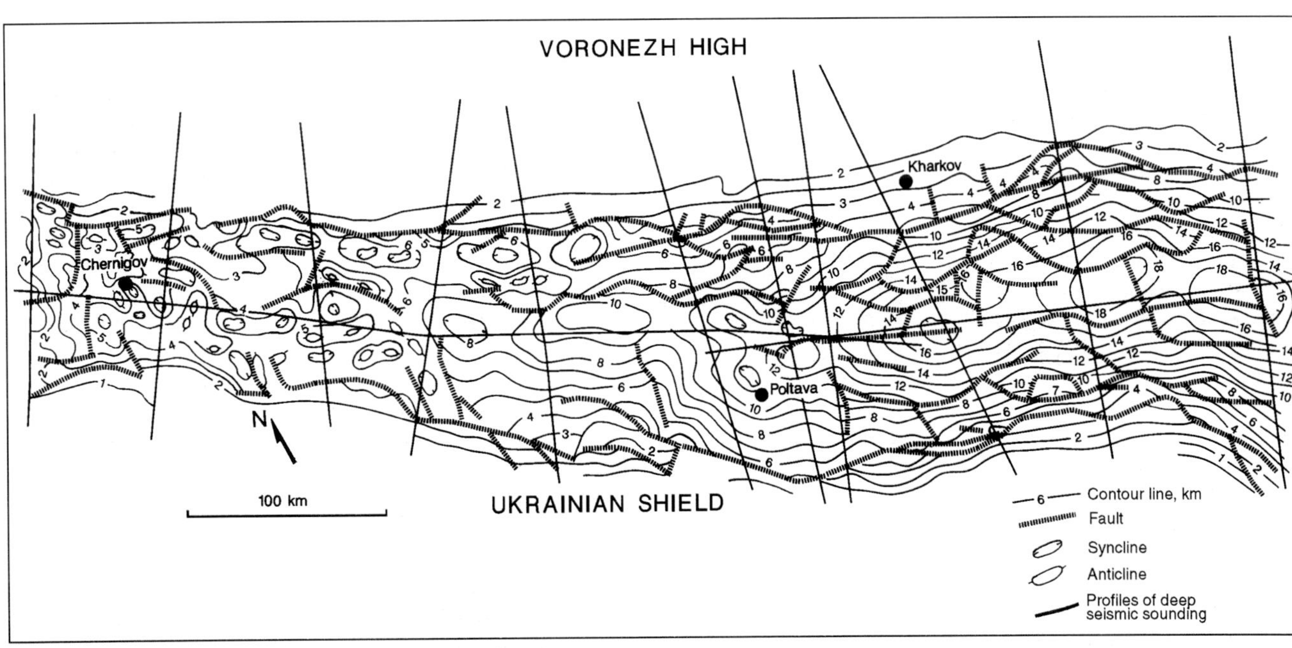

Figure 11. Contour map of the top of the basement of the Dnieper-Donets basin (adapted from Shpak, 1989a). Structural contours in meters below mean sea level. Contour interval 1 km.

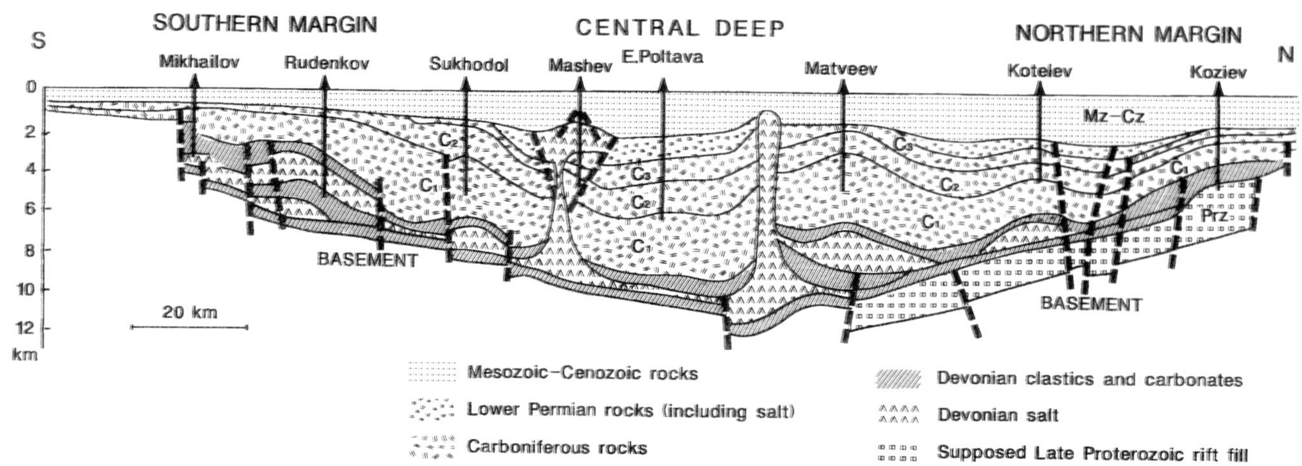

Figure 12. Transversal cross section of the Dnieper-Donets basin (adapted from Gabrielyants, 1991). C1, C2, and C3 are the Lower, Middle, and Upper Carboniferous, respectively. Location of the cross section is shown in Figure 19.

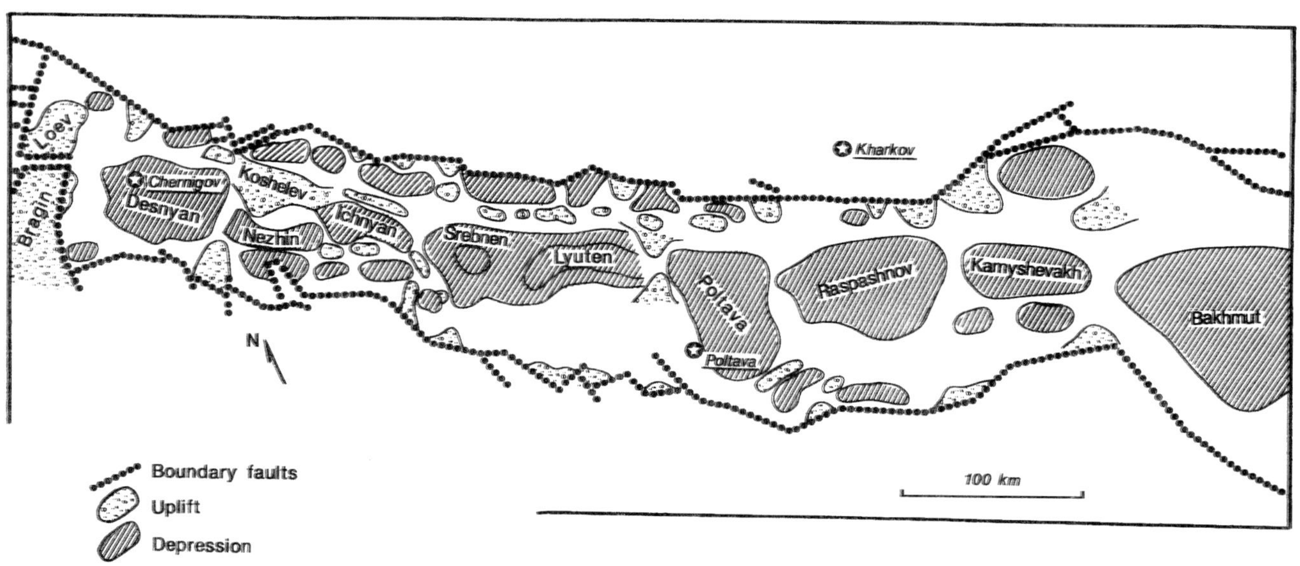

Figure 13. Structural map of the Dnieper-Donets basin showing major depressions and uplifts of the basement surface (modified from Konishchev, 1984).

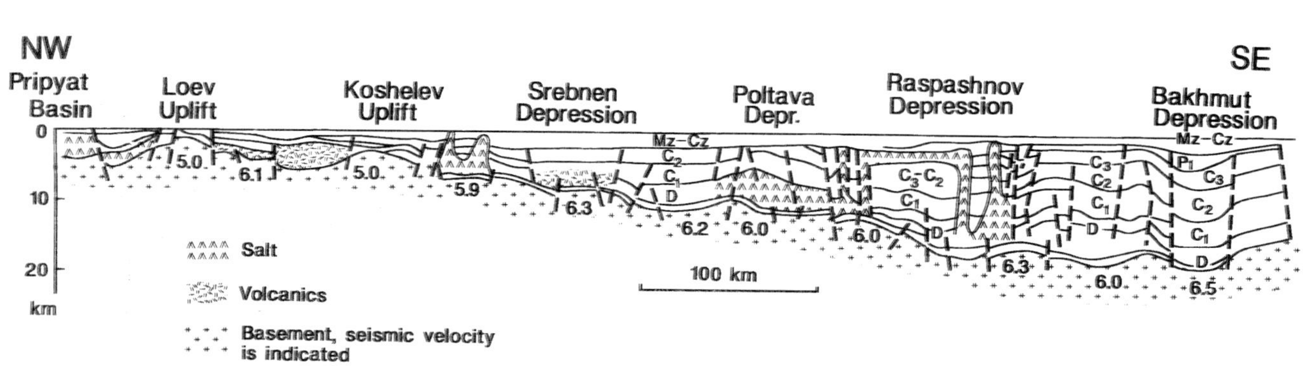

Figure 14. Longitudinal cross section of the Dnieper-Donets basin (adapted from Shpak, 1989a). The cross section is interpreted primarily from deep seismic sounding data. C1, C2, and C3 are the Lower, Middle, and Upper Carboniferous, respectively. P1 is the Lower Permian. Approximate location of the cross section is shown in Figure 19. Numbers, e.g., 5.0, indicate seismic velocity (km/s); Mz–Cz = Mesozoic–Cenozoic.; D = Devonian.

Analysis of refraction seismic data led many Ukrainian geologists to interpret axial and marginal zones of depressions as structures of different ages. The marginal zones are thought to be older and to have formed as synrift grabens at a time when the axial zone of the rift was largely uplifted (Shpak, 1989a). The axial depressions in this interpretation were formed mainly during the Carboniferous due to intense postrift sagging. However, recent detailed CDP (common-depth-point) seismic surveys in the Srebnin depression (Figure 13) showed gross incompatibility of structural pictures derived from reflection and refraction seismic surveys (Gavrish, 1989). The key refractor, interpreted as the basement surface, is nearly horizontal and crosses reflectors that dip to the center of the depression. Earlier interpretation suggested that the depression was formed over a synrift basement uplift indicated by a positive gravity anomaly. The reflection seismic survey identified a Devonian section more than 4 km (2.5 mi) thick in the center of the depression. The depth to the basement appeared to be as much as 13 km (8 mi) and may be 15 km (9.3 mi), which is far more than the depth of 8–9 km (5–5.6 mi) shown in Figure 11. These results question the previous mapping, including the identification of the late Proterozoic rift, which was based dominantly on deep seismic sounding data. The results also seem to indicate that the axial graben was formed on the synrift stage of development.

The cause of divergence of the "basement" refractor from the basement surface is unclear. Possibly, the divergence is related to the appearance of Devonian volcanics in the axial graben of the rift. The comparison of geophysical and drilling data in the northwestern part of the basin indicates that seismic characteristics of Devonian volcanics are similar to those of the basement (Shpak, 1990).

Structure of subsalt Devonian rocks is poorly known; only locally is the resolution of seismic data sufficient for mapping. Where mapped, these rocks mostly drape over basement highs and generally repeat the basement structure. Tilted fault blocks that dip mostly toward the rift boundaries have been mapped along the basin margins (Gavrish, 1989).

The structure of suprasalt Devonian to Lower Permian rocks is quite different from the basement structure and essentially reflects the salt tectonics. Faulting is more related to salt tectonics than to the basement faults. Regionally, all rocks dip southeastward along the basin strike (Figure 15). For example, the base of upper Visean rocks deepens from 2–3 km (1.2–1.9 mi) in the northwest Dnieper-Donets basin to 8–10 km (5–6.2 mi) in the southeast part of the basin. Linear anticlines, mostly parallel to the basin strike, isometric local uplifts, and gentle depressions dominate the structural pattern. Many of the folds have Devonian salt in their cores. Closures of the folds gradually decrease upward in the section and only gentle uplifts and depressions are seen in the structure of Lower Permian rocks.

Mesozoic rocks are deformed generally conformably with underlying Carboniferous and Permian strata despite significant unconformity between these two sequences. The inheritance of the late Paleozoic structure in Mesozoic rocks is characteristic of the axial zone of the basin, whereas generally monoclinal Mesozoic rocks are only slightly deformed into structural bays and noses and some gentle closed anticlines on the northeastern and southwestern rift margins. Mesozoic deformation is chiefly related to continuous movement of Devonian salt and possibly to compaction of underlying sediments. Tertiary sediments are almost undeformed. They are slightly tilted to the basin axis and gradually dip southeastward along the strike of the basin. Boundary faults of the rift are reflected in Cenozoic strata as gentle flexures. Closures of mapped local structures generally do not exceed 10–15 m (33–49 ft), but commonly are much less. Only a limited number of salt domes and plugs continued to develop during the Cenozoic and locally produced structural relief of up to 40 m (131 ft) in Tertiary strata.

BASIN DEVELOPMENT

The Donbas-Pripyat rift was formed in the Late Devonian in the crestal area of a large regional arch, which remained mostly uplifted and was subjected to erosion during the late Proterozoic and early Paleozoic. The arch occupied much of the southwestern part of the Russian craton and included the present-day Ukrainian shield and Voronezh regional high (Figure 2); this arch is known in Russian literature as the Sarmatian shield (Sarmatians are ancient, presumably Iranian-language people that populated plains north of the Black Sea). In the northwest and northeast, the arch was bounded by the upper Proterozoic Volyn-Orsha and Pachelma aulacogens. However, if a presumed graben filled with upper Proterozoic rocks exists at the base of the Dnieper-Donets basin, the Sarmatian shield would have been broken by rifting long before the Paleozoic. A narrow margin of the Russian craton covered by upper Proterozoic and Paleozoic rocks was located southwest and south of the arch.

Formation of the Donbas-Pripyat rift is believed to have been caused by a mantle plume (Kabyshev, 1987; Konishchev, 1984). However, a model of passive rift, in which stretching of the crust was a leading cause of rifting and the mantle arose passively to fill the space in the crust, seems to better fit the observed regional geology, especially the rotational component of horizontal plate movements. Recent CDP seismic surveys, in combination with deep seismic sounding along two north-south profiles crossing the Pripyat basin, showed a typical picture of tensional listric faulting penetrating into the mantle (Garetsky and Klushin, 1987, 1989; Shishkin and Shishkina, 1989). Three mantle faults are identified in the northern part of the rift. One of the faults is expressed as a boundary fault in the structure of the basement surface; the two other faults separate northern fault blocks from each other and from the rest of the rift (Figure 16). Only one listric fault penetrates into the mantle on the south, resulting in asymmetry of the rift possibly

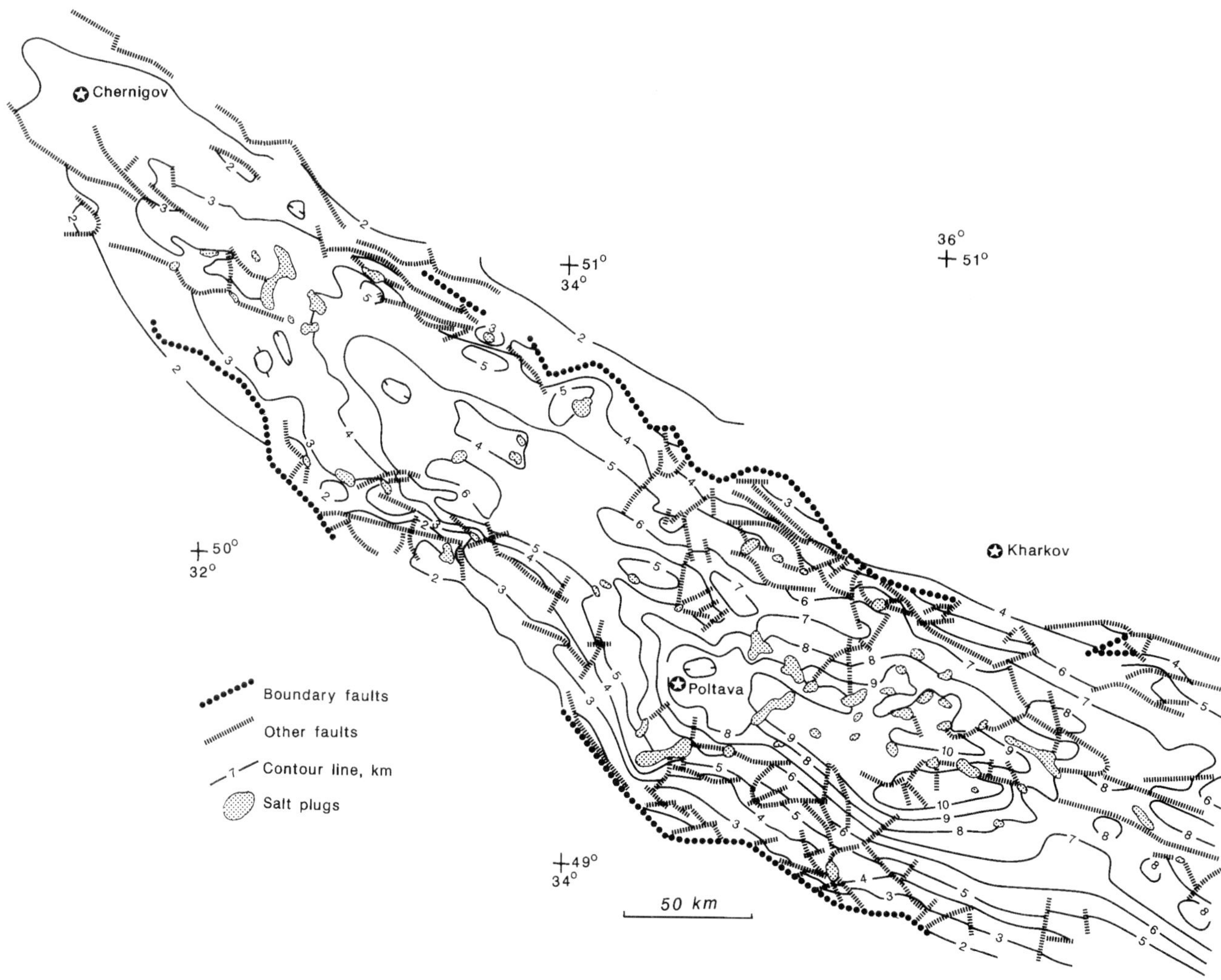

Figure 15. Contour map on the base of upper Visean rocks of the Dnieper-Donets basin (adapted from Andreeva et al., 1984). Structural contours in kilometers below mean sea level. Contour interval is 1 km.

related to plate motion to the north (Klushin et al., 1989). Satellite listric faults merge with main faults or die out in the crust; they also are present on the rift shoulders. In the Pripyat rift, spreading evaluated at 9–14 km (5.6–8.7 mi) (Garetsky et al., 1988) was not large enough to provide for opening to oceanic crust, but did result in thinning of the continental crust to 34–36 km (21.1–22.3 mi) above the uplift of the Moho surface in the basin center, as compared with 45–50 km (27.9–31.1 mi) on the Ukrainian shield and Belorussian high.

Resolution of deep seismic sounding data in the Dnieper-Donets basin is insufficient to identify listric faulting; however, this part of the rift probably is similar to the Pripyat part. The amount of tension and spreading increased southeastward, as evidenced by thickness of the crust beneath the sedimentary cover decreasing from 30 to about 15 km (18.6 to 9.3 mi) from the northwest to the southeast of the basin (Gavrish, 1989). Farther southeast, into the Donbas fold belt, thickness of the crust again increases to 40 km (24.8 mi), reaching up to 50 km (31.1 mi), but the granite layer is absent there.

The composition and structure of the crust in the Donbas-Pripyat rift indicate that breaking of the Sarmatian shield was associated with clockwise rotation of the Ukrainian shield relative to the rest of the Russian craton. The commonly suggested southeast continuation of the Donbas-Pripyat rift into the Karpinsky ridge, which is covered by Mesozoic sediments, is unlikely; very thick (to 20 km or 12.4 mi) Carboniferous–Lower Permian, marine, fine clastics composing the ridge may be interpreted as sediments deposited in a passive-margin regime (probably behind oceanic volcanic arcs) on the continental slope of Laurasia. This scenario implies that the Donets-Pripyat rift is not a failed arm of a triple junction, as was supposed by Zonenshain et al. (1990), but rather a rift of the Red Sea type (but with far less opening). Hercynian collision of microcontinents with the southern margin of Laurasia resulted in a change from the tensional to the compressional regime with folding in the Karpinsky ridge, Donbas, and the extreme southeastern part of the Dnieper-Donets basin. The rest of the Dnieper-Donets basin and the Pripyat basin, however, experienced very little or no compression.

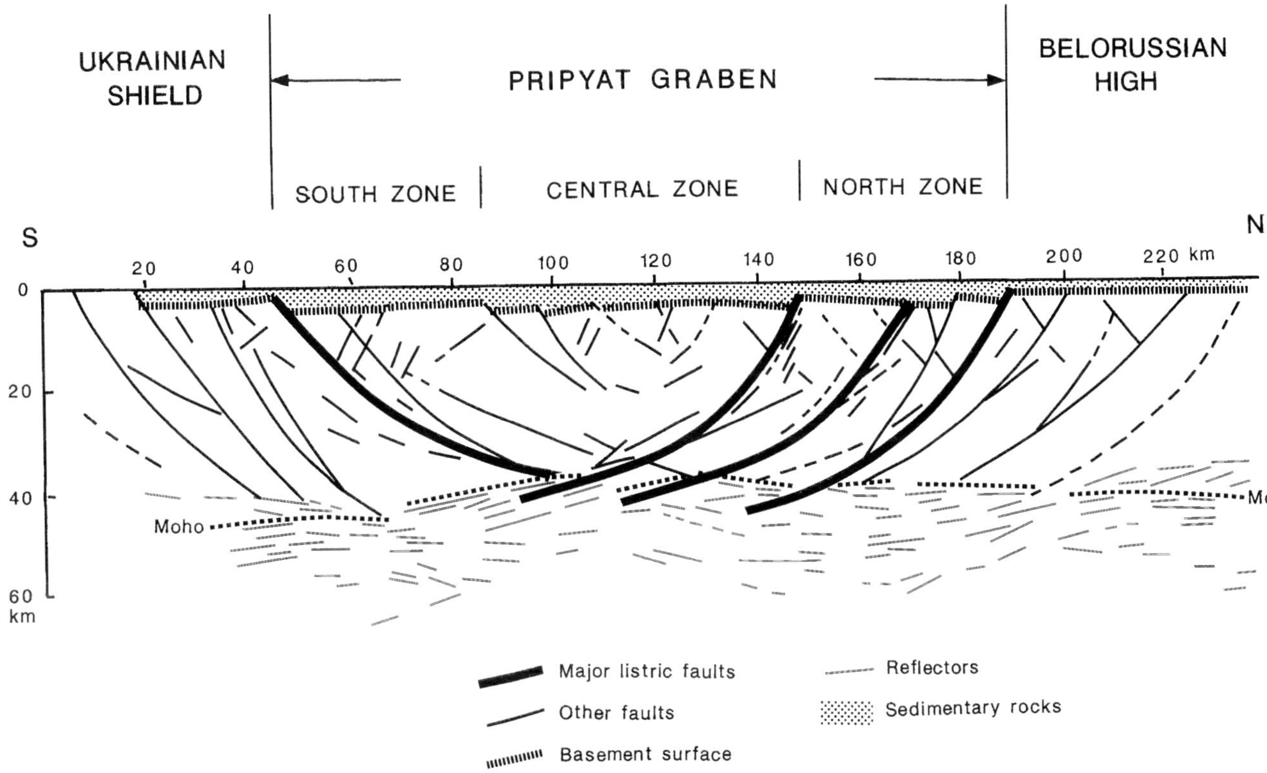

Figure 16. Cross section of the earth's crust of the Pripyat basin based on interpretation of deep common-depth-point seismic survey (adapted from Ayzberg et al., 1988). Location of the cross section is shown in Figure 7.

The Donbas-Pripyat rift cuts across the northern to northeastern structural grain of the Archean–lower Proterozoic basement. The role of transverse faults of the basement in the development of the rift (these faults separate Archean massifs from lower Proterozoic folded zones) is widely debated in Russian and Ukrainian literature (Gavrish, 1985, 1989; Dolenko et al., 1981; Kabyshev, 1987; Maydanovich and Radzivill, 1984). Although interpretation of different authors varies, all geologists agree that transverse faults were active during rift development. In the Pripyat basin, a fault of the northeastern strike separates the plunge of the Mikashevich high and the Turov depression from the rest of the basin (Figure 7), but the structural role of this fault does not seem significant, and its genetic relation to rifting is doubtful. Clearly expressed transverse faults separate the Bragin-Loev uplift with its thick Devonian volcanics from the Pripyat and Dnieper-Donets basins, but their relation to the internal structure of the basement is unclear. Probably, the uplift represents a rift gap with an associated volcanic center as classified by Nelson et al. (1992) on data from the East African rift system. Another well-expressed transverse fault with a significant strike-slip component crosses the northwestern plunge of Donbas (Smishko, 1985). We suppose that this fault bounds a yet unidentified Devonian paleouplift similar to the Loev-Bragin uplift between the Dnieper-Donets and Pripyat basins. Devonian salt formations pinch out against this uplift and do not extend into the Donbas, where an open-marine basin was located at that time (Shumlyansky, 1985). Several other north-trending faults were mapped in the basement structure in exposures of the Ukrainian shield or were interpreted from magnetic and gravimetric data from its northern slope. Supposedly, these faults affected distribution of thicknesses and facies of the Dnieper-Donets basin during rift development; however, the data are very uncertain because of the deep occurrence of synrift rocks. A number of northeast-trending and northwest-trending faults of the Pripyat basin separate subordinate tectonic blocks and have a strike-slip component reaching several kilometers (Ayzberg et al., 1991). Some of these diagonal faults slightly offset the main longitudinal faults. Formation of transverse faults in both Pripyat and Dnieper-Donets basins probably had a largely passive character and was related to the heterogeneity of the basement and to changes of the tensional stress along the strike of the rift.

The rift began to form in the middle Frasnian after a period of platform deposition in the Middle Devonian and early Frasnian. The appearance of volcanics in the Dnieper-Donets basin and Donbas earlier than in the Pripyat basin may suggest propagation of rifting northwestward, but this remains to be proven by more extensive stratigraphic and structural data. Active rift development took place during the late Frasnian and Famennian, and was accompanied by extrusions of alkalic and alkalic-basalt magmas similar in composition to those of the East African rift system.

The transition from rift to postrift sag stage of development is not well dated in both Pripyat and Dnieper-Donets basins. Tournaisian and lower Visean rocks do not extend outside the boundary faults of the rift, except for the easternmost part of the southeastern margin; yet, within the rift, the rate of sedimentation decreased by an order of magnitude and volcanics are absent (Kabyshev, 1987). In addition, a significant pre-Carboniferous unconformity is recorded in many areas. These rocks can be interpreted as a transitional series between the rift stage and a typical sag development that began after pre-late Visean uplift and erosion.

Formation of the sag most probably was related to the thermal contraction and isostatic subsidence. Thickness of the sag sequence increases greatly southeastward along the rift, i.e., in the direction of thinning of the crust, and especially its granite layer. Probably, at the end of the Devonian, the structure of the Dnieper-Donets basin was similar to that of the Pripyat basin and dips of the basement blocks were directed to the rift boundaries. Some of these dips are still preserved in marginal depressions of the Dnieper-Donets basin (Gavrish, 1989). However, the depocenter of the sag sequence was located along the basin axis, resulting in the present-day structural dips being reversed from the basin boundaries to the axial area.

Deformation in the Donbas fold belt could have begun as early as the Late Carboniferous (Dolenko et al., 1981), but the principal compression and folding there and in the southeastern Dnieper-Donets basin occurred in the Late Permian. Since that time, the Donbas fold belt remained uplifted; Lower Carboniferous through lowermost Permian rocks are exposed on the surface (Figure 17). The Dnieper-Donets and Pripyat basins also were uplifted and eroded; the depth of pre-Triassic erosion on uplifted structures of the Dnieper-Donets basin exceeded 1000 m (Goncharenko et al., 1984).

Deposition of Mesozoic–Cenozoic rocks took place in a large regional depression that extends far beyond the Paleozoic basin boundaries. The relationship of this depression with the Pripyat and Dnieper-Donets basins is uncertain. Maximum thicknesses are found in the Dnieper-Donets basin, but they may be related to formation of salt domes and filling of compensational depressions between the domes.

Salt tectonics plays an important role in structure of the Pripyat and Dnieper-Donets basins. In the latter, anticlines having a salt core control most oil and gas fields. In the Pripyat basin, Devonian salt forms domes; in the Dnieper-Donets basin, both domes and piercement plugs are present. Lower Permian salt of the Dnieper-Donets basin is undeformed and constitutes a seal for major gas accumulations. Deformation of Devonian salt in both basins began as early as the end of the Devonian and continued until the present (Dolenko et al., 1981; Konishchev, 1984). In the Pripyat basin, major deformation of the upper salt began as early as the time of deposition of its upper

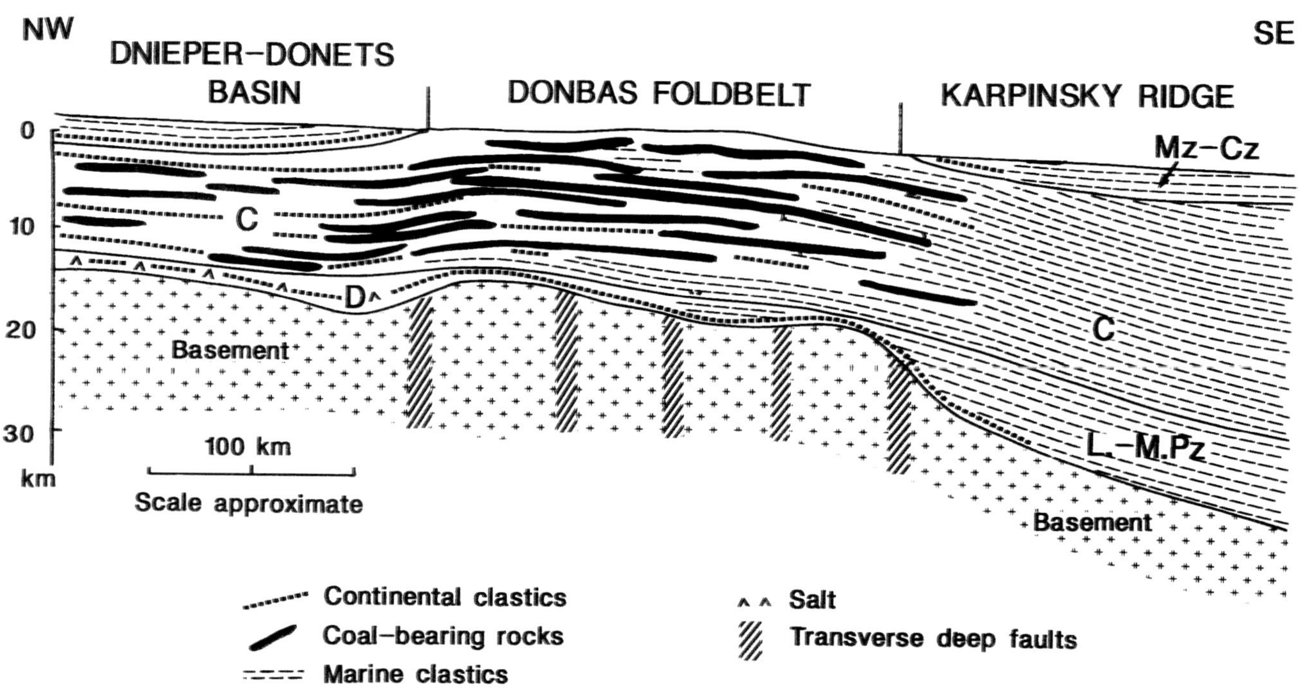

Figure 17. Cross section along the strike of the Dnieper-Donets basin, Donbas fold belt, and Karpinsky ridge showing relationships in thickness of the sedimentary cover and in amounts of coal in the section (adapted from Maydanovich and Radzivill, 1984). The cross section is based on interpretation of deep seismic sounding data. Note the increase of thickness of the Carboniferous section southeastward and pinch out of Devonian salt in the Donbas fold belt. The precise location of the cross section is not available.

subformation, which consists of alternating rock salt with anhydrite, shale, and carbonate beds. This subformation primarily occurs between salt domes formed by purer salt of the lower subformation (Figure 8). The salt domes continued to grow during deposition of suprasalt rocks.

In the Dnieper-Donets basin, salt domes are mainly located along the basin margins, whereas plugs are found in the axial zone. Tops of the majority of salt structures in the basin are located at three main unconformities in the stratigraphic succession (pre-Carboniferous, pre-Triassic, and pre-Tertiary) and at the present-day surface (Kurilyuk et al., 1991). This placement is the result of the abrupt increase in the rate of the salt dome growth during regional uplifts and related gaps in sedimentation, probably due to stress from the basement block tectonics. However, the synsedimentary growth of salt structures also has been documented in a number of areas (Kabyshev, 1979). The earliest termination of salt growth occurred in the deepest southeastern part of the basin, apparently because of depletion of salt resources between the domes (Figure 18). Updip, along the basin axis, the salt dome growth continued into the Mesozoic and, farther northwestward, into the Cenozoic (Khnykin, 1992). The salt tectonics is still in its incipient stage in the shallowest areas of the basin.

PETROLEUM GEOLOGY

Exploration Results

Oil shows in Dnieper-Donets basin were first observed in a cap rock of the Romen salt dome (Figure 18) in 1936 in the course of exploration for gypsum. Shallow drilling resulted in the discovery of a small oil pool. However, the first significant commercial discoveries were not made until 1950 when the giant Shebelinka gas field and large Radchenkov oil and gas field were found in Carboniferous and Lower Permian rocks (Figure 19). By 1989, about 150 oil, gas, and gas-condensate fields had been discovered in the basin. A characteristic feature of exploration was the constant increase of drilling depths through time as a number of prospects at shallower depths diminished. Forty fields contain pays at depths exceeding 4500 m (14,764 ft), and 25 of them produce from depths of more than 5 km (3.1 mi) (Kabyshev and Shpak, 1988). The deepest oil flow was obtained from Serpukhovian rocks in the Kairakozov field (Figure 19) at 4981–5018 m (16,342–16,436 ft); the deepest gas discovery was in lower Visean rocks at depths of 5848–6057 m (19,186–19,872 ft) (Shpak, 1989b).

Despite significant exploration efforts, most of the fields discovered at great depths are small. Three-quarters of original recoverable oil reserves and 55% of gas reserves of the basin occur at depths of less than 3 km (1.9 mi) (Dolenko et al., 1981). Gas strongly predominates over oil and constitutes about 80% of hydrocarbon reserves (in barrels of oil equivalent). The majority of discovered reserves is heavily concentrated in three gas fields (Shebelinka, West Krestishchen, and Yefremov) and three oil fields (Lelyakov, Gnedintsev, and Pogarshchin) (Figure 19). The original recoverable reserves of the largest field of the basin, the presently significantly depleted Shebelinka field, exceed 18 tcf of gas (Zhabrev, 1983).

In the Pripyat basin, exploratory drilling into suprasalt rocks began soon after World War II, but was unsuccessful for a long time. Only in 1964, after the drilling was targeted at deeper rocks below the upper salt formation, was the first commercial discovery made in the Rechitsa field (Figure 7). To date, 46 fields have been discovered. Unlike in the Dnieper-Donets basin, all but one of the fields are oil fields. Oil pays occur at depths of 1650 to 4500 m (5413 to 14,764 ft); however, the dominant amount of original reserves is found at depths of 2000 to 3500 m (6562 to 11,483 ft). About 40% of oil reserves are concentrated in two large fields, the Ostashkovich and Rechitsa (Figure 7). The majority of other fields are small.

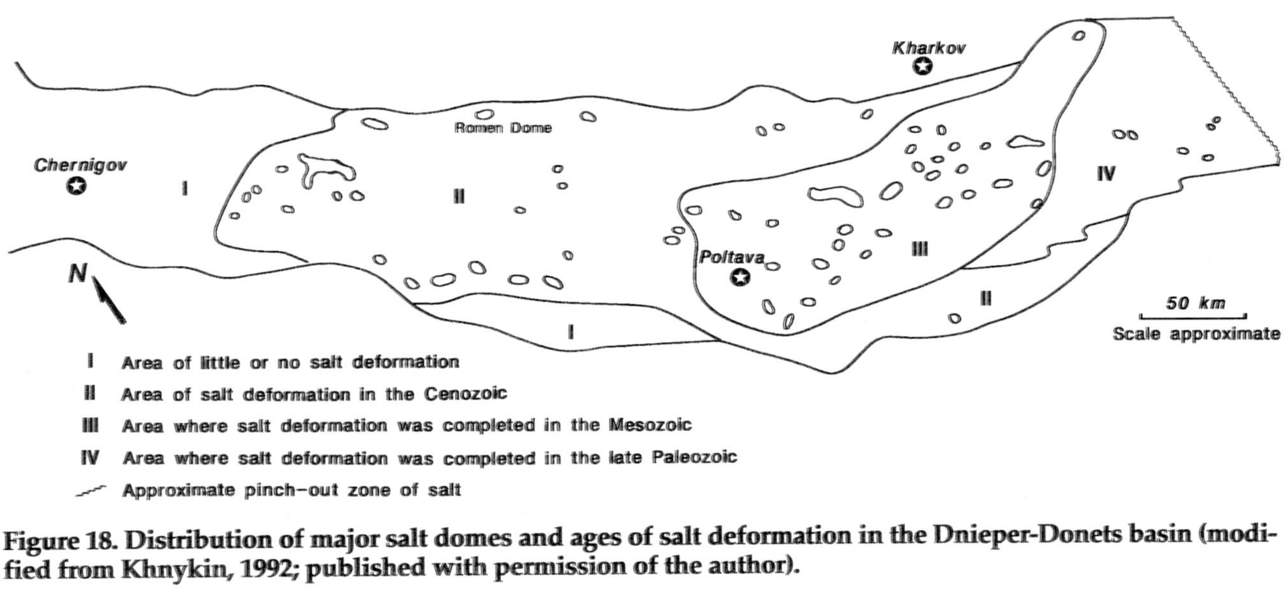

Figure 18. Distribution of major salt domes and ages of salt deformation in the Dnieper-Donets basin (modified from Khnykin, 1992; published with permission of the author).

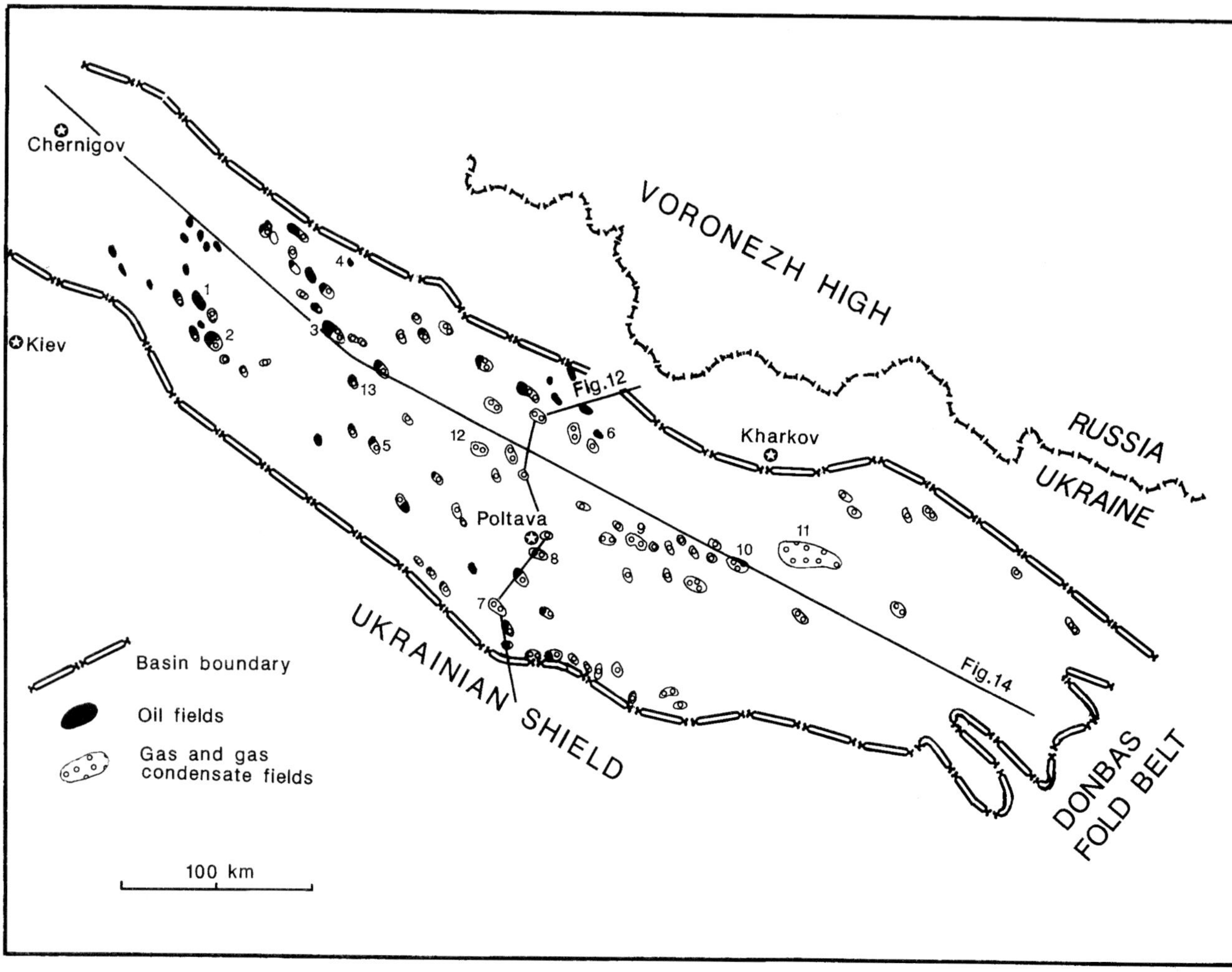

Figure 19. Oil and gas fields of the Dnieper-Donets basin (modified from Maksimov, 1987). Oil and gas fields indicated by numbers: 1 = Lelyakov, 2 = Gnedintsev, 3 = Pogarshchin (Glinsk-Rozbyshev), 4 = Romen, 5 = Radchenkov, 6 = Kairakozov, 7 = Rudenkov, 8 = Mashev, 9 = West Krestishchen, 10 = Yefremov, 11 = Shebelinka, 12 = Solokhov, 13 = Yablunov.

Reservoir Rocks and Stratigraphic Distribution of Oil and Gas

The main oil reserves of the Pripyat basin are found in carbonate reservoir rocks and are about equally divided between the Upper Devonian intersalt and subsalt carbonate formations. All commercial fields are located in the northern part of the basin. In the prerift and early synrift subsalt carbonate formation, best reservoir properties are characteristic of vuggy dolomites of the Semiluki horizon beneath the pre-late Frasnian unconformity. Vugs were formed by leaching of corals and other fossils (Makhnach et al., 1984). Supposedly, the leaching took place not only during postsedimentary subaerial exposure, but also due to later synrift faulting and penetration of high-salinity water through the faults (Pakholchuk et al., 1991). Both dolomitization and tectonic stress produced widespread fracturing. Porosity of the dolomites ranges from 4 to 20% or more, and averages 7 to 10%. Fracture permeability is high. The effective thickness commonly ranges from 10 to 20 m (33 to 66 ft). Younger rocks of the subsalt carbonate formation have poorer reservoir properties. These younger reservoir rocks are areally limited and related chiefly to dolomitization of algal and other organic limestones.

In three fields of the Pripyat basin, the lower pay is found in basal Frasnian sandstone of the Lan horizon. Porosity of the sandstone ranges from 7 to 14%, and permeability is from a few to a few tens of millidarcys. The oil reserves are small. The lower salt formation constitutes a regional seal for both carbonate and clastic reservoirs in subsalt rocks.

In the synrift intersalt carbonate formation, the best reservoir rocks are commonly associated with organic carbonate facies developed along the crests of the tilted fault blocks (Figure 4). Algal oncolitic and stromatolitic dolomites are intensively fractured and possess high (15–20%) porosity and permeability. Reservoir rocks outside the carbonate banks and reefs are chiefly detrital limestones with porosity not exceeding 10%. Permeability varies widely and depends primarily on tectonic fracturing. In a few fields, oil flows also were obtained from interbeds of algal carbonates in the upper salt formation.

The stratigraphic range of productivity in the Dnieper-Donets basin is much greater than in the Pripyat basin. Oil and gas pools are known from Devonian through Jurassic rocks. Two fields also have been discovered in the weathered top of the crystalline basement just north of the northeastern boundary fault of the rift (Kabyshev, 1991). Oil fields are located mainly in the northwestern area of the basin, whereas gas fields dominate in the southeast (Figure 19). Most of the fields are multistrata; the maximum number of pays in a single field reaches 27. However, the concentration of hydrocarbon reserves in the sedimentary section is uneven. Very few hydrocarbon pools have been discovered in Devonian, Triassic, and Jurassic rocks. Middle Carboniferous rocks contain only 5.5% of original reserves (in barrels of oil equivalent), whereas Lower Carboniferous rocks are much richer and contain 37% of original discovered reserves in about 120 productive fields (Shpak, 1989a). The largest amounts of both oil and gas reserves are found in Upper Carboniferous–Lower Permian rocks, although they are productive in only 24 fields.

This high concentration of reserves at the top of Paleozoic rocks is obviously related to the presence of an excellent regional seal formed by Lower Permian salt and/or Lower Triassic shale. Other widespread seals composed of relatively thick, although variable, shale beds occur at the top of the lower Visean and at the top of the Lower Carboniferous. These seals separate the sedimentary succession into productive sections; however, many oil and, especially, gas pools are sealed by rather thin (from a few to a few tens of meters) shale beds inside the productive sections. Devonian salt formations and thick shales at the top of the Devonian also can be important regional seals, but the Devonian section is almost unexplored because of great depths of occurrence over most of the basin.

Reservoir rocks of the Dnieper-Donets basin are diverse. In the poorly known Devonian section, the main exploration potential is related to reefal facies that are present in the lower part (Zadon horizon) of the intersalt carbonate formation and in the suprasalt Upper Devonian to lowermost Carboniferous section. Three fields and a number of shows have been discovered in this facies. Nonreefal carbonates and sandstones in the Devonian section have poor reservoir properties. The distribution of reefs is unknown; however, the analogy with the Pripyat basin suggests that the facies trends should be related to the synrift tectonics and that reefs are probably developed along the crests of the tilted fault blocks and boundary faults. Porosity of the fractured reefal dolomites as measured in drilled wells ranges from 8 to 15%.

Carbonate reservoirs dominate in the lower–middle Tournaisian section. The distribution of facies in this, and possibly also in underlying uppermost Devonian beds, is controlled by a deep-water marine depression formed in the southeastern and central axial zone of the basin. Basinal siliceous shales were deposited in the depression, whereas shallow-water and reefal carbonates were formed on surrounding shelves.

The paleogeographic situation changed in the late Tournaisian–early Visean with the appearance of a large river in the northwest part of the basin. Alluvial to nearshore marine facies were formed in that area. Deep-water marine sedimentation continued in the more southeastern axial zone of the basin and carbonate shelves were preserved north and south of this zone. Most of the following Carboniferous time was characterized by cyclic sedimentation. Thin marine carbonate and shale beds were deposited during transgressive parts of the cycles. A complex set of alluvial, deltaic, swamp, and nearshore marine clastic facies was formed during regressive parts of the cycles. Clastic material was brought by the major river from the northwest and from the Ukrainian shield by numerous rivers from the south. Carbonate shelf sediments were periodically deposited only in the eastern area of the northeastern margin. A marine bay that accumulated mainly prodeltaic shales and turbidites permanently existed in the axial central and southeastern zone (Sokiran et al., 1978). Reservoir rocks of the Carboniferous section are dominantly sandstones. Seventy-four sandstone pays separated by beds of shale, carbonate, and coal are present in the section and are assigned local nomenclature (Dolenko et al., 1981). Thickness of reservoir beds ranges up to 65 m (213 ft). Reservoir properties of the sandstones vary significantly and depend on the type of facies and diagenetic changes related to the depth of burial. Better prime intergranular porosity is characteristic of more sorted, less cemented, significantly quartzose bar, delta-front, and channel sandstones (Samarskaya and Polyak, 1990). During burial, these sandstones preserved good porosity (10–17%) and permeability (up to 300 md) through the oil window to depths of 4–5 km (2.5–3.1 mi) (Shpak, 1989b). Partly, this preservation is due to secondary dissolution of carbonate cement; reservoir rocks with higher original porosity are more susceptible to these diagenetic changes. Deeper, in the wet-gas window, secondary porosity and fracturing control the reservoir properties; this type of reservoir is present to depths of 6–7 km (3.7–4.3 mi). Other kinds of alluvial and turbidite sandstones lose their porosity at depths of only 3–4 km (1.9–2.5 mi).

Uppermost Carboniferous and lowermost Permian section is composed of mainly alluvial clastics. High porosity (15–18%) and permeability (150–300 md) are characteristic of the northwestern areas, but porosity and permeability decrease southeastward where intergranular porosity commonly does not exceed 12% and fracture permeability dominates. The clastic section is overlain by fractured carbonates and anhydrites that, together with the clastics, contain massive hydrocarbon pools in the principal fields of the basin. Lower Permian reefs are present on the margins of salt-bearing depressions. Vuggy dolomites of the reefs have excellent reservoir properties (porosity to 28%, permeability more than 700 md) and contain oil in

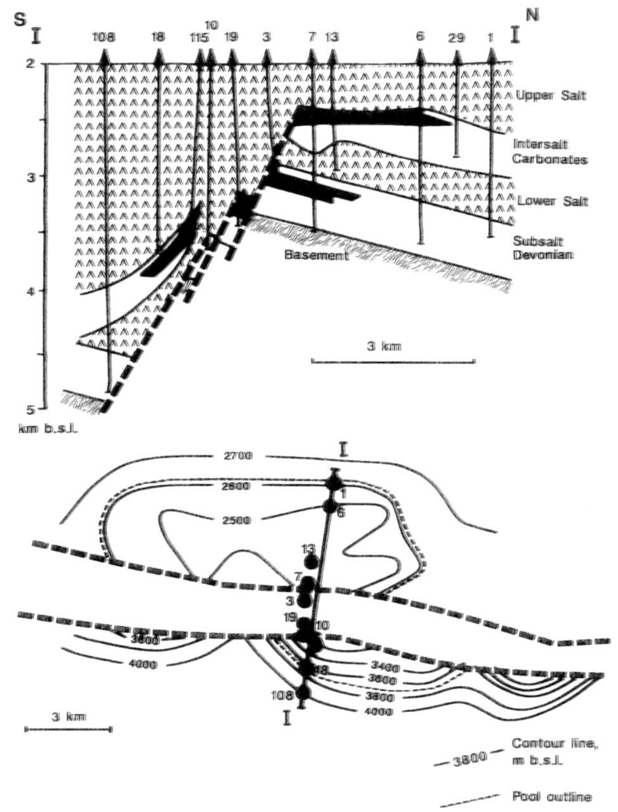

Figure 20. Cross section and contour map of the Ostashkovich oil field in the Pripyat basin (adapted from Mikhaylov, 1986). Contours are on top of the lower Famennian Yelets horizon. Location of the field is shown in Figure 7. b.s.l. = below sea level.

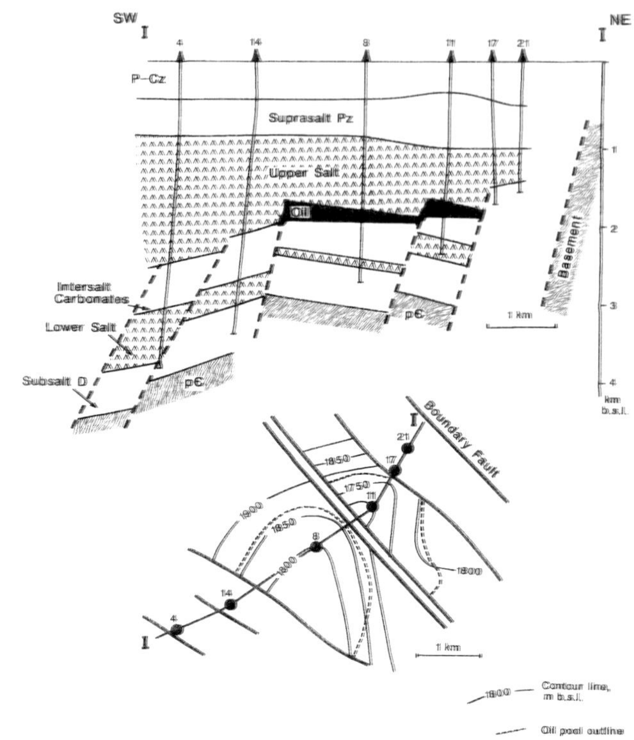

Figure 21. Cross section and contour map of the Berezin oil field in the Pripyat basin (contour map is modified from Makhnach et al., 1984). Contours are on top of the intersalt carbonate formation. Location of the field is shown in Figure 7. b.s.l. = below sea level.

several fields, including the large Lelyakov field (Figure 19) (Lukin and Galitsky, 1974). A few small gas pools found in the Triassic and Jurassic are in poorly consolidated sands with very high porosity.

Traps

The dominant amount of oil in the Pripyat basin is found in traps at the crests of tilted fault blocks. In the subsalt section, oil pools are in upthrown blocks and are sealed updip by salt against the fault plane as, for example, in the largest Ostashkovich field (Figure 20). Smaller oil accumulations are found in some fields in narrow intermediate blocks between the major pools and the footwalls. Stratigraphic control on the pool limits is minor; only in a narrow zone along the fault planes is the pore space of some reservoir beds filled by anhydrite and salt.

Traps in the intersalt carbonate section are more complex. In the majority of fields the carbonate section is absent in a zone extending along the crest of the tilted blocks. The width of the zone is larger than the expected horizontal throw. To some extent, this absence may be related to nondeposition and/or erosion of the carbonates due to synsedimentary tilting; however, the main cause of the absence of carbonate rocks is probably deformation of the lower salt expressed as injection of the salt from the downthrown block upward along the fault plane. Gravitational sliding of the carbonate bed between the lower and upper salts also is conceivable. Oil pools sealed updip by the pinch-out zone of the carbonate formation are found both in the hanging wall and the footwall. The large Ostashkovich field shown in Figure 20 presents a typical example of these relationships.

Unlike the subsalt carbonate formation, oil pools in the intersalt carbonates commonly have a significant stratigraphic control owing to changes in reservoir properties. Primarily, these changes are related to transition of reefal secondary dolomites into nonpermeable off-reef facies (Makhnach et al., 1984). The stratigraphic control on reservoir distribution significantly decreases the drilling success rate. Purely stratigraphic pools are uncommon, although small accumulations of this type have been found in the uppermost part of the intersalt carbonates and in carbonate beds in the upper salt formation (Mikhaylov, 1986).

Another type of a structural trap has been found in the Berezin field (Figure 21). The field is located in the vicinity of the northern boundary fault (Figure 7) and is cut by subordinate synthetic faults. The oil accumulation is in the intersalt carbonate formation and is controlled by a rollover structure. A trap in the

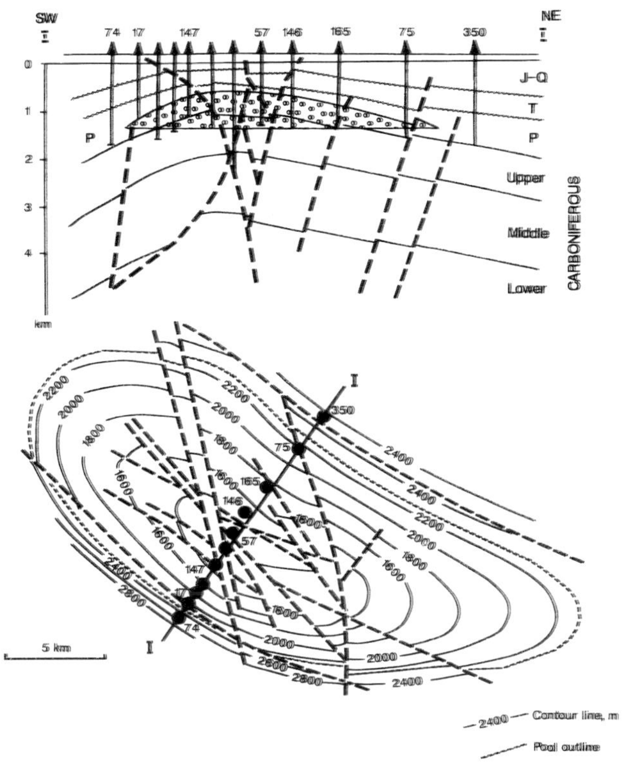

Figure 22. Cross section and contour map of the Shebelinka gas field in the Dnieper-Donets basin (adapted from Maksimov et al., 1990). Contours are on a pay near the top of the Lower Permian productive section. Location of the field is shown in Figure 19.

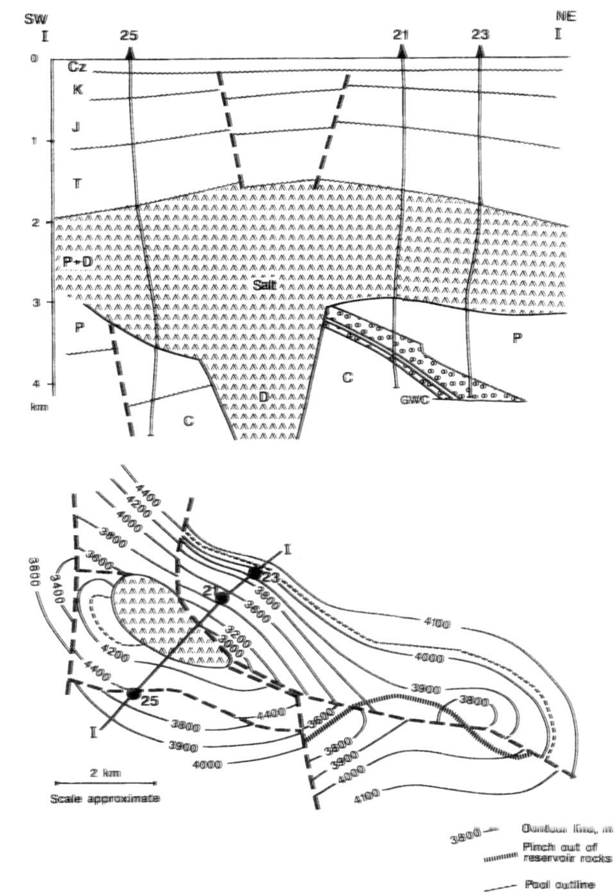

Figure 23. Cross section and contour map of the Mashev gas field in the Dnieper-Donets basin (adapted from Shpak, 1989a). Contours are on a pay near the top of the productive section. Location of the field is shown in Figure 19.

Sudovits field (Figure 7), which produces from both subsalt and intersalt carbonates, probably also is of this type.

In the Dnieper-Donets basin, main oil and gas reserves are found in salt-assisted structural traps. The morphology of a trap depends on the size and depth of the underlying salt body and on the time of salt flowage. The Shebelinka structure (Figure 22), which contains the largest gas field in the basin, is an anticline above a salt dome buried to a depth of 6–7 km (3.7–4.3 mi). Salt flowage took place late in the geologic history and the anticline is present in Paleozoic through Mesozoic strata. The productive section includes 14 pays in faulted Upper Carboniferous and Lower Permian rocks. All the pays have a single gas-water contact.

The medium-size Mashev gas field (Figure 23) is controlled by an anticline related to a Devonian salt plug. The plug is terminated upward in Lower Permian salt that forms a regional seal. Laterally, the gas pool is sealed by the salt plug, faults, and pinch-out zone of reservoir rocks. Most of the other oil and gas fields of the Dnieper-Donets basin are of types similar to the Shebelinka and Mashev fields, although the position of the salt core may vary substantially. In addition to the salt-assisted structures, structural traps related to the basement tectonics are present in the relatively shallow, peripheral parts of the basin where Devonian salt is thin or absent. However, these traps contain smaller hydrocarbon reserves.

Although the majority of oil and gas fields in the Dnieper-Donets basin are found in faulted structural uplifts, lateral stratigraphic sealing of pools is very common. Complicated facies relationships in suprasalt Paleozoic rocks, consisting of a wide variety of facies ranging from continental to deep marine, resulted in widespread lateral discontinuity of reservoir rocks. Actually, a large number of pools is controlled by combination rather than structural traps. Purely stratigraphic traps in sandstones pinching out laterally and updip also are known, although very little exploration has been specifically targeted at these traps. They are primarily characteristic of marginal monoclines and slopes of large depressions. The medium-size Rudenkov field contains gas pools in several sandstone beds in the Visean section (Figure 24). The sandstones pinch out laterally and southward onto the southeastern marginal monocline of the basin. The sandstones probably were deposited as a clastic fan derived from the Ukrainian shield.

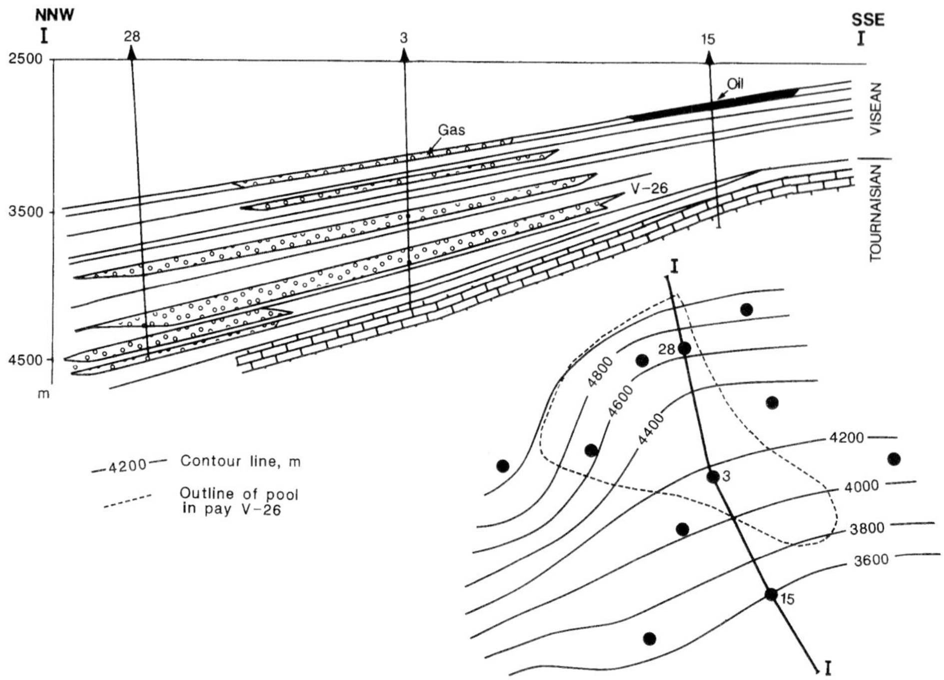

Figure 24. Cross section and contour map of the Rudenkov gas field in the Dnieper-Donets basin (adapted from Mikhaylov, 1986). Contours are on a reflector in lower Visean rocks. Two-thirds of gas reserves are in pay V-26. The upper pay contains a small oil pool. Location of the field is shown in Figure 19.

Analogous pools in the Visean have been found on slopes of the Srebnin depression (Lazarchuk et al., 1984). The amount of discovered hydrocarbons in purely stratigraphic traps is small, but this type of trap has a significant exploration potential.

Source Rocks

Petroleum source rocks are poorly identified in both Pripyat and Dnieper-Donets basins. Existing data are mostly limited to the fractional composition of extractable bitumen and to the TOC (total organic carbon) and hydrocarbon contents in rocks. Somewhat more data are available on composition of oils; however, only two oils of the Pripyat basin have been studied using gas chromatography/mass spectrometry techniques (Petrov, 1988). No attempts of oil-to-source rock correlation are found in the literature. A geochemical study of oils and potential source rocks of the Pripyat basin is now under way in the U.S. Geological Survey. After completion, results of this study will be published elsewhere. At present, identification of source rocks in both basins is speculative.

The most probable source rocks in the Pripyat basin are dark-colored shales and carbonates of the Famennian intersalt carbonate formation, especially in the upper part of the formation. This deep-water facies deposited off the reefal buildups (which are the prime exploration targets) is sparsely cored. Maximum TOC contents from Rock-Eval analyses reach 3–4% and hydrocarbon indices reach 400–500 mg/g. Kerogen of the organic-rich samples is of type II. In structural depressions where very few wells have been drilled, this type of organic-rich facies probably constitutes the bulk of the intersalt carbonate formation (Figures 4, 5). These rocks are the likely source of oils in shallow-water intersalt carbonates deposited along the crests of tilted fault blocks.

However, structural considerations suggest that, at least in some fields, subsalt pays are unlikely to receive oil from a source in the intersalt carbonate formation. In these fields, even in uplifted blocks adjacent to faults, the subsalt pays are structurally lower than the potential source rocks of the intersalt formation in downthrown blocks and are separated from these rocks by the lower salt. In addition, preliminary geochemical data indicate the presence of two genetically different types of oils characterized by dissimilar sets of biomarkers (J. Clayton, personal communication, 1993). A source for subsalt (and possibly some intersalt) oils is most probably located below the lower salt formation. Only a few samples in an analyzed collection of subsalt carbonates of the Yevlanov horizon show TOC values in excess of 1%, and hydrocarbon indices are low. The carbonates were deposited in a shallow-water sea and basinal facies are absent.

The most promising candidate for source rock in the subsalt section is the Moiseev bed at the bottom of the Frasnian Semiluki horizon that has not been cored, but is exposed on basin margins. This 5–10-m-(16–33-ft-) thick bed, characterized by typically high x-ray readings on well logs, is composed of alternation of dark carbonates and black shales, apparently very rich in organic matter. The rocks are visually very similar to the contemporaneous Domanik Formation, which sourced the great majority of oil reserves in the Volga-Ural and Timan-Pechora provinces on the eastern Russian craton (Ulmishek, 1988). Thickness of the Domanik Formation also is small, commonly not exceeding 20–25 m (66–82 ft); however, this formation has generated very large amounts of oil (more than 40 billion bbl have been extracted from the Volga-Ural province alone). Samples from an exposure of the Moiseev bed on the northern basin margin are now being analyzed by the U.S. Geological Survey.

Organically very rich oil shales deposited in apparently brackish water conditions (Makhnach et al., 1984) are present in the Devonian suprasalt section of the Pripyat basin, but these rocks are thermally immature.

Distribution of source rocks in the Dnieper-Donets basin may be inferred only on the basis of the facies composition of the Devonian–Carboniferous sequence because no data on oil/source rock correlation exist. In the poorly known Devonian section, organic-rich (TOC to 3%) clayey carbonates and shales have been encountered in the subsalt carbonate formation (Shpak, 1989a). Beds of these rocks increase in number toward the top of the formation. Judging from the great thickness of the lower salt, far exceeding that in the Pripyat basin, deposition of subsalt carbonates and shales in the Dnieper-Donets basin probably occurred in a semienclosed deep-water sea during the early stage of rifting. These potential source rocks may be present throughout the basin.

Facies differentiation in the intersalt carbonate formation apparently was significant. In parts of the basin where the upper salt is present, the paleogeographic situation was similar to that in the Pripyat basin. Deep-water organic-rich shales and carbonates with TOC reaching 5.8% were drilled in marginal grabens of the Dnieper-Donets basin (Shpak, 1989a). On edges of the grabens, these rocks pass into reefal carbonates. In areas where the upper salt is absent, the presence of source rocks is less probable because the grabens there were rapidly filled with flysch-like clastics and volcanics.

The paleogeographic situation during the latest Devonian, much of the Early Carboniferous, and the early Bashkirian was characterized by terrestrial, deltaic, and nearshore environments in the northwestern part of the basin and on its margins, whereas deep-water basinal facies occupied the central and southeastern axial zone. In general, the boundary between nearshore and basinal facies shifted southeastward with time. The northwestern association of facies is characterized by the dominance of coaly organic matter and the presence of coal beds. Basinal facies consists mostly of shales that include organic-rich (measured TOC to 3.5%) beds with primarily sapropelic organic matter. Thickness of the basinal facies is great and these deeply buried, organic-rich rocks are a likely source for much of the oil and gas in the basin.

The Carboniferous system of the Dnieper-Donets basin contains thick sections of rocks significantly enriched by coaly organic matter and containing coal beds. These rocks were deposited in paralic, and to a lesser extent, in terrestrial conditions. In the Early Carboniferous, these facies occupied northwestern areas of the basin and parts of its margins; later, they dominated in the axial zone of the basin. Coaly shales contain up to 10–12% organic matter. Hundreds of coal beds, some of which are more than 1 m (3 ft) thick, are present in the section, although the amount of coal is variable over the area (Shpak, 1990; Lapchinsky and Dzhamalova, 1987). Much coal in the Lower Carboniferous contains a significant amount of liptinite material, mostly presented by spores; higher in the section the liptinite component is small. The coaly organic matter may be a source of some gas in the basin, but its importance is difficult to evaluate. Generation of oil by this organic matter is of low probability, as indicated by geochemical data (Belokon et al., 1991), although an opposite point of view was expressed by some geologists (Ilinskaya and Kulaeva, 1979).

MATURATION AND FORMATION OF FIELDS

Uncertainties in identifying source rocks, probable large changes in heat flow during geologic history, and temperature variations related to postdepositional deformation of salt make maturation modeling very difficult and uncertain in both Dnieper-Donets and Pripyat basins.

All oil fields discovered in the Pripyat basin are heavily concentrated in its North zone (Figure 7). Drilling on more than 70 prospects in the Central and South zones has not resulted in commercial discoveries. The cause of this distribution of fields is apparently related to the differences in thermal maturity of potential source rocks between these zones.

Analyzed oils from noncommercial flows in the Central and South zones of the basin are strongly immature judging from distribution of normal alkanes and extremely high concentrations and characteristics of biomarkers (Petrov, 1988). Vitrinite reflectances measured in rocks associated with these flows are 0.41–0.43%. Several samples from the productive North zone analyzed by the U.S. Geological Survey demonstrate the presence of both mature and immature oils (J. Clayton, personal communication, 1993). Measurements of vitrinite reflectance in rocks at the top of the intersalt carbonate formation indicate a significant difference in values between samples from the North zone and the rest of the basin (Ammosov et al., 1980; Grechishnikov, 1985). In the North zone, maximum R_a values reach 7.8–8.5% (R_o = 0.7–0.94) in an elongated zone just north of and parallel to the chain of the main oil fields (Rechitsa, Ostashkovich, etc.). In the rest of the basin, R_a values at approxi-

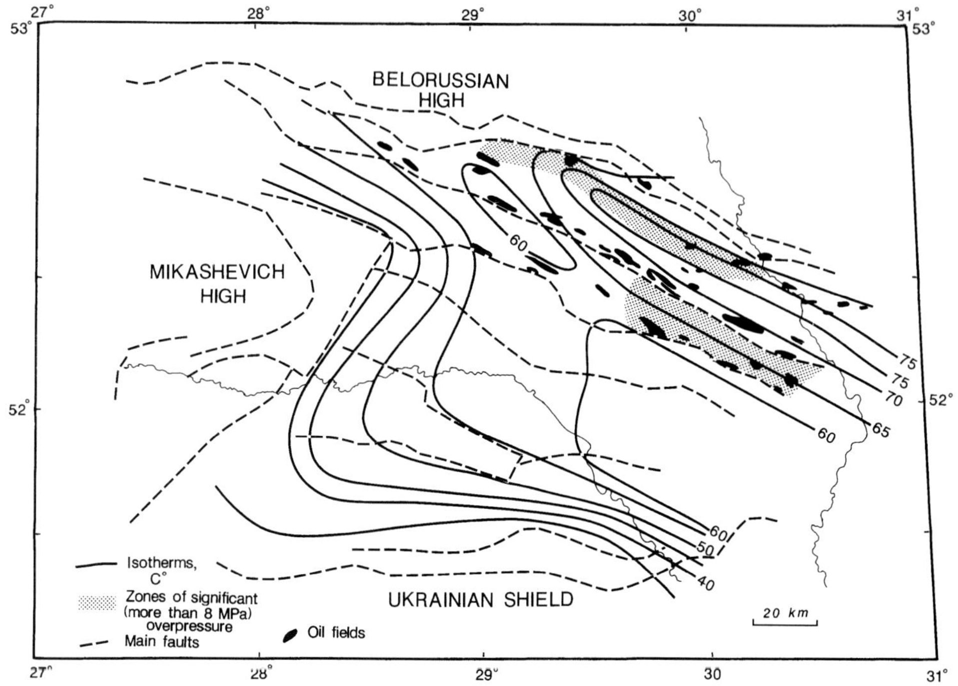

Figure 25. Temperature in rocks at a depth of 3 km below seal level in the Pripyat basin (modified from Garetsky et al., 1990b). Zones of significant overpressure are also shown.

mately the same depths are in the range of 6.0 to 7.0% (R_o = 0.29–0.5) and lower.

The distribution of present-day temperatures at depth of 3 km (1.9 mi) b.s.l. (below sea level) is shown in Figure 25. The map clearly indicates that temperatures in the North zone far exceed those in the Central and South zones. The probable cause of the higher temperatures on the north is believed to be a magmatic body in the underlying crust (Garetsky et al., 1990b). Deep seismic sounding indicates the presence of two low-velocity zones beneath the northern part of the basin. The lower zone, at a depth of about 60 km (37.3 mi), is interpreted as a zone of melting in the crust.

Described data suggest two possible maturation models in the Pripyat basin. The first model supported by most geologists (e.g., Kleschev, 1981) infers that maximum paleotemperatures and maturation levels were reached soon after the termination of rifting (in the latest Devonian–earliest Carboniferous) when the heat flow related to the rifting was at its peak. Later, subsidence occurred contemporaneously with the regional decrease of heat flow. However, some geologists suggest that maximum temperatures could have been achieved by the time of pre-Late Permian uplift and erosion (Ammosov et al., 1980). The uncertainty is related to the fact that neither rate of cooling nor thickness of suprasalt Paleozoic rocks removed by pre-Late Permian erosion are known. Higher maturation in the North zone is explained by the presence of three major listric faults (Figure 16) that produced high heat flow due to convective transmission of heat (Garetsky et al., 1990b). The concentration of Devonian volcanism just east of the North zone and its decrease southward may be an indirect indication of unevenness of the thermal regime during the rifting stage. Later emplacement of the magmatic body in the crust may be related to the same zone of weakness, but it did not result in significant additional maturation. If the model is correct, the heat flow on the north of the basin had to be extremely high. Despite significant depths to potential source rocks by the end of the Devonian (up to 3.5–4 km or 2.2–2.5 mi), about three-quarters of the overlying strata are composed of salt characterized by low geothermal gradients because of very high thermal conductivity.

The second maturation model suggests that petroleum source rocks in most areas of the basin did not subside into the oil window during the rifting stage and subsequent Paleozoic deposition. Thermal maturity in the North zone by the end of the Paleozoic was similar to that observed at present in the rest of the basin (Ammosov et al., 1980). The present-day maturation in the North zone was achieved only in the late Mesozoic or Cenozoic and is related not to tectonic subsidence, but to formation of the magmatic body and associated increase of heat flow. The model is attractive because it explains why the largest fields of the basin are narrowly located in the zone of presently high temperatures. Overpressures measured in a number of reservoirs (Figure 25) support the model. The model does not exclude that, in some of the deepest depressions, maturation could have been achieved in the Paleozoic. In addition, the Pripyat basin contains sizable (in tens of million barrels) commercial

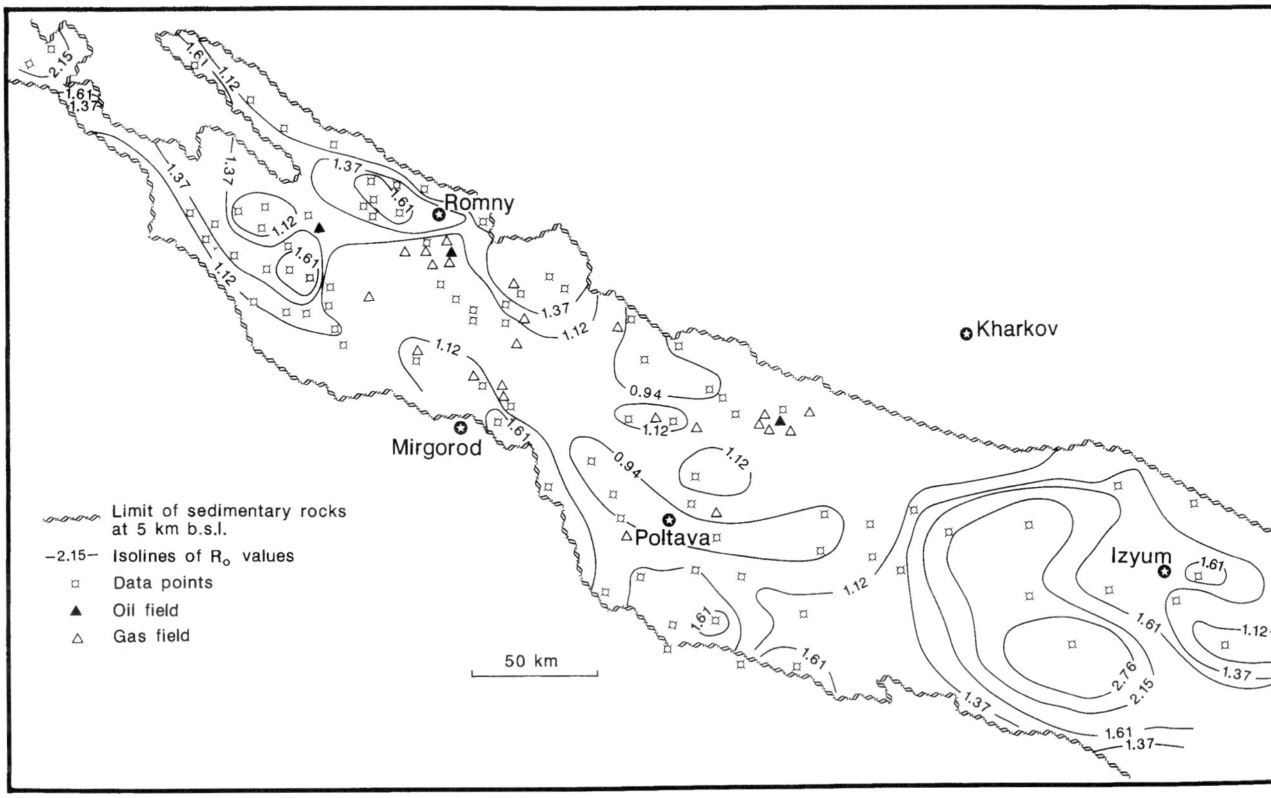

Figure 26. Vitrinite reflectances in rocks at a depth of 5 km below sea level in the Dnieper-Donets basin (modified from Shpak, 1989a). Vitrinite reflectances in air are recalculated to those in oil (after Ammosov et al., 1977).

accumulations of immature oils that apparently were generated at temperatures corresponding to R_o not exceeding 0.5%.

The distribution of hydrocarbon accumulations in the Dnieper-Donets basin is characterized by concentration of oil fields in its northwestern areas and dominance of gas fields in the deep southeastern areas (Figure 19). At depths below 4–4.5 km (2.5–2.8 mi) significant oil accumulations are absent and only gas and gas-condensate fields are known (Shpak and Lukin, 1986). In general, this distribution is obviously related to the increase of depth and maturity of source rocks along their dip to the southeast.

Present-day temperatures at 3.5 km (2.2 mi) b.s.l. in the Dnieper-Donets basin range from 70 to 120°C (Shpak, 1989a). Minimum temperatures are measured in the central and northwestern axial zones of the basin where they range from 70 to 90°C. The temperatures increase toward the margins, especially toward the southern margin, to 90–110°C. They also are high in the extreme southeast, where they reach 100–120°C. There, the high temperatures probably are related to the pre-Paleogene tectonic activity in Donbas and adjacent areas of the basin. Paleocene rocks on uplifted structures of that area unconformably overlie various horizons of the Carboniferous. Local anomalies in the temperature distribution are believed to be related to salt plugs and convective flow of hot fluids from deep overpressured zones upward (Shpak, 1989a). Temperatures at the basement surface in the deepest areas may reach 500–600°C.

The distribution of vitrinite reflectances at 5 km (3.1 mi) b.s.l. is rather similar to that of the present-day temperatures (Figure 26). Rocks from the Devonian (in marginal areas on the northwest) to the Middle Carboniferous (on the southeast) are found at this depth. The lowest values ranging from R_a = 8.0–8.5% (R_o = 0.76–0.94) are found in the central area of the basin. Toward the margins, vitrinite reflectance increases significantly and rocks at this depth occur in the gas window. Relatively few measurements in the southeastern part of the basin indicate high overmaturity reaching R_a = 12–14% (R_o = 2.76–4.26). The high reflectance values in the southeastern region indicate deep subsidence of the rocks before pre-Paleogene uplift and erosion. As much as 2.5 km (1.6 mi) of rocks could gave been removed by this erosion (Ivanova, 1991).

The distribution of vitrinite reflectances and thicknesses of stratigraphic units indicate that Devonian potential source rocks are generally overmature. They occur in the upper part of the gas window in the northwest and probably below the gas window on the southeast (Shpak and Lukin, 1986). Only in the shallowest areas of the basin margins might the upper part of the Devonian section occur in the oil window. Devonian source rocks are believed to have reached the oil window as early as the Early Carboniferous in most of the basin. By the Middle Carboniferous, the rocks had subsided into the gas window zone (Shpak and Lukin, 1986).

The thermal transformation of Lower Carboniferous rocks varies from immaturity on the margins to high overmaturity in the southeast. In the axial part of the basin where oil-source basinal facies are present, the rocks may be in the oil window only in the northwestern area. To the southeast, they occur in the gas window and probably below the gas window in the area adjacent to the Donbas fold belt; the zone of nitrogen and carbon dioxide gases has been identified in coal mines of Donbas (Lapchinsky and Dzhamalova, 1987). The maturity of Lower and Middle Carboniferous coals also ranges from immaturity to high overmaturity in the extreme southeast. The present-day distribution of oil and gas fields in the basin seemingly follows the maturity trend in Lower Carboniferous rocks, suggesting that these rocks were the main source for oil and probably also for gas. The input of gas from the coal source could have been significant in the southeastern part of the basin.

Most Lower Carboniferous source rocks in the central and southern parts of the basin reached the oil window soon after deposition owing to rapid Carboniferous subsidence, and remained in the oil window until the later Carboniferous or Early Permian. Figure 27 shows subsidence and associated heating curves for two wells in the Pogarshchin and Solokhov fields located in the Lyuten and Poltava depressions, respectively (Figures 13, 19). In both wells, Tournaisian source rocks reached the oil window in the Early Carboniferous owing to the great thickness of Visean–Serpukhovian clastics in these depressions. The source rocks remained in the oil window until the middle of the Early Permian in the Pogarshchin field and only until the Middle Carboniferous in the Solokhov field situated in the deeper Poltava depression.

Maximum temperatures in the Paleozoic sedimentary cover of the basin were probably reached by the end of the Early Permian (Ammosov et al., 1980; Grigorchuk, 1989). Following uplift and erosion resulted in significant cooling of rocks. Renewed subsidence during the Triassic through the Tertiary did not result in additional heating and maturation over most of the basin except maybe for local areas (Ivanova, 1991).

The Lower Permian salt seal played a very important role in preserving hydrocarbon accumulations. Oil and gas generated before deposition of the seal have been preserved mainly in Lower Carboniferous reservoirs. Accumulations in the Lower Carboniferous display statistical relationships with the age of trap formation—almost all discovered reserves are in traps that already existed in the Early Permian (Kabyshev, 1987; Shpak, 1989a). On average, only about 40% of the trap volume is filled with hydrocarbons. Evidently, oil and gas that migrated vertically upward before deposition of the Permian salt seal were largely lost. Heavy and extra-heavy oil accumulations in Carboniferous reservoirs known in several

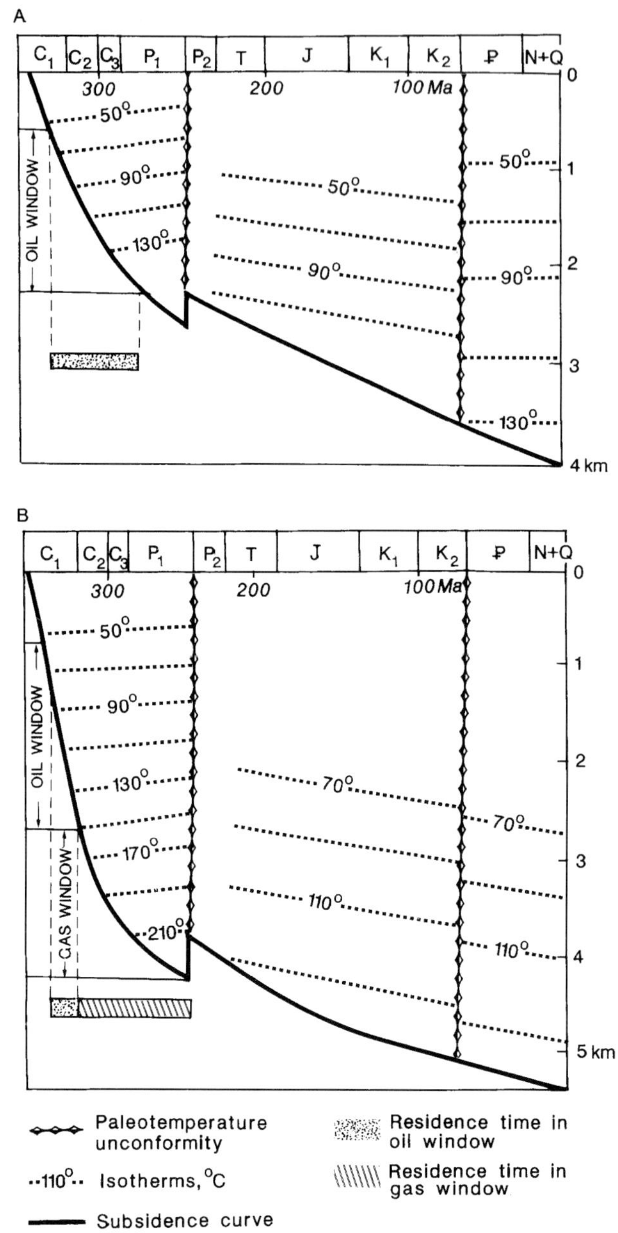

Figure 27. Temperature history and maturation of Tournaisian rocks in (A) well 475 of the Pogarshchin field and (B) well 49 of the Solokhov field of the Dnieper-Donets basin (adapted from Grigorchuk, 1989).

fields (e.g., Yablunov field, Figure 19) probably result from the early, pre-Permian, stage of migration (Shpak and Lukin, 1986).

Hydrocarbon accumulations in the Upper Carboniferous–Lower Permian section that contain the majority of discovered reserves in the basin obviously were formed after deposition of the Lower Permian salt seal. Rocks of the section are regionally immature; only in the deepest areas do they occur in the upper part of oil window (Ivanova and Yatselenko, 1988). Many of the traps (including the largest Shebelinka field) are of Mesozoic–Cenozoic age and are filled almost to the spill point. Evidently, active growth of salt domes during the pre-Triassic, pre-Cretaceous, and pre-Paleogene uplift and erosion and associated faulting and fracturing aided vertical migration of hydrocarbons. Distribution of oil and gas fields in this section clearly reflects the maturity of Lower Carboniferous source rocks. Only gas and gas condensate accumulations are found in the deep central zone of the basin. Oil fields are present mainly in the northwest and on the margins where oil could have migrated updip from adjacent source rocks that occurred in the oil window. Middle Carboniferous accumulations are known only in those fields in which the Lower Carboniferous is also productive. Probably, they also were formed in the post-Early Permian owing to vertical migration of hydrocarbons. Long-range vertical migration from the Lower Carboniferous source into Middle Carboniferous and Lower Permian reservoirs was aided by the absence of significant regional seals between the top of the Lower Carboniferous and the Permian salt. In a few localities where the salt is thin or absent, hydrocarbons migrated through Lower Permian rocks and basal Triassic shales and formed small gas accumulations in the Mesozoic.

FUTURE POTENTIAL

Both the Pripyat and Dnieper-Donets basins are presently in the mature stage of exploration. New discoveries are relatively small and do not replace produced reserves, resulting in a steady decline of production. However, the exploration potential is far from exhausted.

In the North zone, crests of the tilted fault blocks, which contain almost all the discovered oil in the Pripyat basin, have been thoroughly explored and only a limited number of small new discoveries may be expected in this play. Future potential of this zone where mature source rocks are certainly present is mainly related to a search for stratigraphic and structural traps on slopes of the principal fault blocks. Structural traps may be related to subordinate faults and to roll-over structures on the downthrown sides of larger listric faults. Prospects in both subsalt and intersalt carbonates are probable. Sliding of intersalt carbonate formation downdip from the crests of the tilted blocks could have possibly resulted in deformation of the sliding plates. Whether this type of deformation is sufficient to produce structural traps remains to be seen. Stratigraphic traps are more probable in the synrift intersalt carbonate formation and may be related to debris aprons of shallow-water carbonates eroded from exposed crests and upper slopes of the tilted blocks and from northern shelf and transported into depressions. Analogous debris aprons were described, for example, downslope of the Devonian reefs of Alberta, Canada (Hopkins, 1977). Better resolution of seismic data in depressions is critical for exploration.

Numerous attempts to explore crests of the tilted fault blocks in the Central and South zones of the Pripyat basin have been unsuccessful. More than 70 prospects have been drilled, but no commercial discoveries have been made. Flows of heavy immature oil suggest that the problem likely is insufficient maturity of source rocks. However, the presence of movable (although noncommercial) oil in several shallower structures indicates that a higher degree of maturity could have been reached in almost undrilled, deep depressions between the crests. Modern geochemical studies directed to precise identification of potential source rocks, especially in the deeper subsalt section, and modeling of their maturation are crucial for further exploration in these zones. If pockets of mature source rocks are identified in the depressions, the exploration should be targeted at adjacent traps along the probable migration paths. In addition to structural traps, sandstone turbidites derived from the Ukrainian shield and transported into southernmost deep depressions may be an attractive prospect similar to the Permian Spraberry Sandstone play in the Midland basin of Texas.

The Dnieper-Donets basin has been thoroughly explored to depths of about 3.5 km (2.2 mi). This depth interval almost completely includes Upper Carboniferous–Lower Permian rocks that presently contain the largest portion of discovered reserves, as well as older Carboniferous rocks on the northwest and on the basin margins. Exploration potential of these rocks seems to be low; it is mainly related to subtle traps adjacent to Devonian salt plugs and beneath salt overhangs. The potential of the Middle Carboniferous section, even at larger depths, is moderate at best. This section is characterized by the absence of significant regional seals and local seals control only infrequent small pools. Presently, this section contains only a few percent of discovered reserves.

During recent years, the majority of new discoveries (32 fields were discovered in the basin in 1986–1990) were made in the Lower Carboniferous section at depths of 3.5–5.5 km (2.2–3.4 mi). This section is believed to contain at least two-thirds of undiscovered resources of the basin (Kabyshev et al., 1992). Best exploration targets at these depths are concentrated in the central part of the basin, particularly in the Srebnen and Poltava depressions (Figure 13). Although structural traps also are present, probably more important are stratigraphic traps related to synsedimentary pinch out of sandstones updip on the slopes of the depressions (Figure 24).

Exploration potential of other types of stratigraphic traps in the Lower Carboniferous and the lower part of the Middle Carboniferous sections also is significant. The sections are characterized by a transition from alluvial facies on the northwest to deep-water basinal facies in the central area of the basin. A variety of stratigraphic traps related to river channels, deltas, sandstone bars, etc., may be expected. This play is virtually unexplored. In addition, a reef play in a transitional zone from carbonate shelves to deep-water sediments is almost certainly present, especially in the Tournaisian and lower Bashkirian sections.

The potential of Devonian rocks, despite the presence of indigenous source rocks and reef reservoirs, is limited by great burial depths in most of the basin. Devonian plays are largely concentrated on the northeastern and southwestern margins outside of areas where the Devonian section is largely composed of volcanics.

In conclusion, the most targets for future exploration in the Dnieper-Donets basin are located at great depths. In assessment of Ukrainian geologists, almost 70% of undiscovered hydrocarbon resources are in rocks buried to depths of more than 4 km (2.5 mi) and 40% of undiscovered hydrocarbons are below 5 km (3.1 mi) (Kabyshev et al., 1992). All plays at these depths are strongly gas prone and the economics of deep exploration should be carefully evaluated.

REFERENCES CITED

Ammosov, I.I., et al., 1977, Paleogeothermal criteria of distribution of oil fields (Paleogeotermicheskiye kriterii razmeshcheniya neftyanykh zalezhey): Moscow, Nedra, 158 p.

Ammosov, I.I., V.I. Gorshkov, and N.P. Grechishnikov, 1980, Paleotemperatures of transformation of oil- and gas-bearing rocks (Paleotemper-atury preobrazovaniya neftegazonosnykh otlozheniy): Moscow, Nauka, 111 p.

Andreeva, R.I., et al., 1984, Atlas of the geology and petroleum productivity of the Dnieper-Donets depression (Atlas geologicheskogo stroyeniya i neftegazonosnosti Dneprovo-Donetskoy vpadiny): Ministry of Geology of the Ukrainian SSR, Kiev, 86 p.

Ayzberg, R.E., et al., 1988, Tectonics of oil-productive complexes of the Pripyat paleorift and its relation to the deep structure of the earth crust: Sovetskaya Geologiya, no. 12, p. 3–14.

Ayzberg, R.E., et al., 1991, Paleogeodynamic reconstructions of cratonic basins (Paleogeodinamicheskie rekonstruktsii platformennykh basseynov): Minsk, Byelarus, BelNIGRI, 182 p.

Belokon, T.V., et al., 1991, Geologic-geochemical characteristics of the section drilled by Dnieper-Donets superdeep well: Sovetskaya Geologiya, no. 8, p. 95–102.

Beznosov, N.V., ed., 1987, Reference book on the stratigraphy of petroleum provinces of the USSR (Spravochnik po stratigrafii neftegazonosnykh provintsiy SSSR): Moscow, Nedra, 336 p.

Chirvinskaya, M.V., and V.B. Sollogub, 1980, Deep structure of the Dnieper-Donets aulacogen on geophysical data (Glubinnaya struktura Dneprovsko-Donetskogo avlakogena po geofizicheskim dannym): Kiev, Ukraine, Naukova Dumka, 178 p.

Dolenko, G.N., et al., 1981, Tectonics and petroleum productivity of the Dnieper-Donets depression (Tektonika i neftegazonosnost Dneprovsko-Donetskoy vpadiny): Kiev, Ukraine, Naukova Dumka, 228 p.

Gabrielyants, G.A., ed., 1991, Regional geology of petroleum regions of the USSR (Regionalnaya geologiya neftegazonosnykh territoriy SSSR): Moscow, Nedra, 284 p.

Garetsky, R.G., ed., 1976, Tectonics of Belorussia (Tektonika Byelorussii): Minsk, Byelarus, Nauka i Tekhnika, 199 p.

Garetsky, R.G., ed., 1979, Tectonics of the Pripyat depression (Tektonika Pripyatskogo progiba): Minsk, Byelarus, Nauka i Tekhnika, 175 p.

Garetsky, R.G., and S.V. Klushin, 1987, Structure of the Pripyat paleorift at depth: Doklady Akademii Nauk SSSR, v. 297, no. 6, p. 1438–1442.

Garetsky, R.G., and S.V. Klushin, 1989, Listric faults in the Pripyat paleorift: Geotektonika, no. 1, p. 48–60.

Garetsky, R.G., V. Ya. Kozhenov, and V.G. Nikolaev, 1988, Comparative analysis of history of tectonic development of the Pripyat, central North Sea, and Baikal grabens: Byulleten Moskovskogo Obshchestva Ispytateley Prirody, Otdel Geologicheskiy, v. 63, no. 1, p. 3–12.

Garetsky, R.G., et al., 1990a, Uncompensated depressions of the East European platform (Nekompensirovannye progiby Vostochno-Yevropeyskoy platformy): Minsk, Byelarus, Nauka i Tekhnika, 103 p.

Garetsky, R.G., V.V. Pechernikov et al., 1990b, Reflection of deep structure of the Pripyat paleorift in thermal conditions of the sedimentary cover: Sovetskaya Geologiya, no. 3, p. 93–99.

Gavrish, V.K., ed., 1985, Petroleum provinces of Ukraine (Neftegazonosnye provintsii Ukrainy): Kiev, Ukraine, Naukova Dumka, 172 p.

Gavrish, V.K., ed., 1987, Geology and petroleum productivity of the Dnieper-Donets depression—methods of study of the deep structure and petroleum potential (Geologiya i neftegazonosnost Dneprovo-Donetskoy vpadiny—Metodika izucheniya glubinnogo stroeniya i neftegazonosnosti): Kiev, Ukraine, Naukova Dumka, 147 p.

Gavrish, V.K., ed., 1989, Geology and petroleum productivity of the Dnieper-Donets depression—deep structure and geotectonic development (Geologiya i neftegazonosnost Dneprovo-Donetskoy vpadiny—Glubinnoye stroeniye i geotektonicheskoye razvitiye): Kiev, Ukraine, Naukova Dumka, 203 p.

Goncharenko, B.D., M.B. Keller, and A. Yu. Makharinsky, 1984, Regional unconformities and petroleum productivity: Geologiya Nefti i Gaza, no. 4, p. 11–17.

Grechishnikov, N.P., 1985, Causes of the uneven distribution of oil pools in the Pripyat depression from the point of view of vitrinite reflectance measurements: Geologiya Nefti i Gaza, no. 2, p. 11–14.

Grigorchuk, K.G., 1989, Evolution of reservoir properties of rocks during lithogenesis: Geologiya Nefti i Gaza, no. 7, p. 20–25.

Hopkins, J.C., 1977, Production of foreslope breccia by differential submarine cementation and downslope displacement of carbonate sands, Miette and Ancient Wall buildups, Devonian, Canada, *in* H.E. Cook and P. Enos, eds., Deep-water carbonate environments: Society of Economic Paleontologists and Mineralogists Special Publication, no. 25, p. 155–170.

Ilinskaya, V.V., and T.P. Kulaeva, 1979, Composition of organic matter in Paleozoic rocks of the Dnieper-Donets depression and the character of oil and gas generation, *in* N.B. Vassoevich and P.P. Timofeev, eds., Oil-source formations and principles of their identification (Neftematerinskiye svity i printsipy ikh diagnostiki): Moscow, Nauka, p. 155–160.

Ivanova, A.V., 1991, Use of vitrinite thermometry in geology of oil and gas: Sovetskaya Geologiya, no. 3, p. 11–15.

Ivanova, A.V., and V.S. Yatselenko, 1988, Location of oil window in the sedimentary cover of the Dnieper-Donets depression, *in* V.V. Semenovich, S.G. Neruchev, and O.K. Bazhenova, eds., Conditions of oil and gas generation at great depths (Usloviya neftegazoobrazovaniya na bolshikh glubinakh): Moscow, Nauka, p. 116–120.

Kabyshev, B.P., 1979, Rate of growth of salt structures in the Dnieper-Donets depression: Geotektonika, no. 3, p. 25–31.

Kabyshev, B.P., 1987, Paleotectonic studies and petroleum productivity of aulacogens (Paleotektonicheskiye issledovaniya i neftegazonosnost v avlakogennykh oblastyakh): Leningrad, Nedra, 192 p.

Kabyshev, B.P., 1991, Quantitative assessment of oil and gas potential of crystalline basement rocks: Geologiya Nefti i Gaza, no. 3, p. 2–5.

Kabyshev, B.P., and P.F. Shpak, 1988, Conditions of oil and gas generation and accumulation in deeply buried Paleozoic rocks of the Dnieper-Donets depression, *in* V.V. Semenovich, S.G. Neruchev, and O.K. Bazhenova, eds., Conditions of oil and gas generation at great depths (Usloviya neftegazoobrazovaniya na bolshikh glubinakh): Moscow, Nauka, p. 108–112.

Kabyshev, B.P., et al., 1992, Directions of petroleum exploration in the Dnieper-Donets depression: Geologiya Nefti i Gaza, no. 5, p. 14–18.

Khnykin, V.I., 1992, Evolution of salt basins and salt thicknesses on platforms and in foredeeps. Poster presentation at the International Symposium on Geodynamic Evolution of Sedimentary Basins, Moscow, May 18–23, 1992. Published in Abstracts and Program of the Symposium, Moscow, p. 64.

Khomenko, V.A., 1977, Lithology of Devonian rocks of the Dnieper-Donets depression (Litologiya devonskikh otlozheniy Dneprovsko-Donetskoy vpadiny): Kiev, Ukraine, Naukova Dumka, 148 p.

Kleschev, K.A., 1981, Geologic regularities in distribution of zones of oil accumulation in the Pripyat depression, *in* Formation of oil and gas pools at various stages of development of sedimentary basins (Formirovaniye zalezhey nefti i gaza na razlichnykh etapakh razvitiya sedimentatsionnykh basseynov), Trudy VNIGRI, v. 230, p. 150–157.

Klushin, S.V., Z.L. Poznyakevich, and A.M. Sinichka, 1989, Asymmetry of the Pripyat paleorift—specifics of geodynamics and structural styles: Izvestiya Akademii Nauk SSSR, Seriya Geologicheskaya, no. 12, p. 100–114.

Konishchev, V.S., 1984, Comparative tectonics of salt-tectonic regions of ancient platforms (Sravnitelnaya tektonika oblastey galokineza drevnikh platform): Minsk, Byelarus, Nauka i Tekhnika, 189 p.

Kurilyuk, L.V., et al., 1991, Paleozoic salt formations of the Dnieper-Donets depression: Sovetskaya Geologiya, no. 4, p. 15–21.

Lapchinsky, Y.G., and Kh. F. Dzhamalova, 1987, Gas productivity and geochemical peculiarities of gases in coal-bearing regions of eastern Ukraine, *in* A.A. Trofimuk, N.I. Nesterov, and I.P. Zhabrev, eds., Regularities in distribution of hydrocarbon gases and associated components (Zakonomernosti razmeshcheniya uglevodorodnykh gazov i soputstvuyushchikh im komponentov): Moscow, Nauka, p. 98–104.

Lazarchuk, Y.G., N.V. Doganchuk, and V.A. Baladogly, 1984, New exploration target in Visean rocks of the Dnieper-Donets depression: Geologiya Nefti i Gaza, no. 4, p. 7–11.

Lukin, A.E., and I.V. Galitsky, 1974, Bioherms in the Lower Permian of the Dnieper-Donets depression: Doklady Akademii Nauk SSSR, v. 215, no. 1, p. 170–173.

Makhnach, A.S., I.I. Uryev, and S.A. Kruchek, 1981, Devonian intersalt formation of the Pripyat depression—regional regularities of the framework and composition (Devonskaya mezhsolevaya tolshcha Pripyatskoy vpadiny—regionalnye zakonomernosti stroyeniya i sostava): Minsk, Byelarus, Nauka i Technika, 217 p.

Makhnach, A.S., et al., 1984, Organic buildups in the Devonian of Belorussia (Organogennye postroyki devona Belorussii): Minsk, Byelarus, Nauka i Technika, 236 p.

Maksimov, S.P., ed., 1987, Oil and gas fields of the USSR (Neftyanye i gazovye mestorozhdeniya SSSR), v. 1: Moscow, Nedra, 357 p.

Maksimov, S.P., et al., 1990, Geology of oil and gas of the East European craton (Geologiya nefti i gaza Vostochno-Yevropeyskoy platformy): Moscow, Nedra, 275 p.

Maydanovich, I.A., and A. Ya. Radzivill, 1984, Tectonics of coal basins of Ukraine (Osobennosti tektoniki ugolnykh basseynov Ukrainy): Kiev, Ukraine, Naukova Dumka, 120 p.

Mikhaylov, I.M., ed., 1986, Hydrocarbon accumulations of the non-anticlinal type in the Pripyat-Donets aulacogen (Zalezhi uglevodorodov neanticlinalnogo tipa Pripyatsko-Donetskogo avlakogena): Moscow, Nauka, 79 p.

Milanovsky, E.E., 1987, Geology of the USSR, part 1—ancient platforms and metaplatformal regions (Geologiya SSSR, chast 1—drevniye platformy i metaplatformennye oblasti): Moscow, Moscow University, 415 p.

Nelson, R.A., T.L. Patton, and C.K. Morley, 1992, Rift-segment interaction and its relation to hydrocarbon exploration in continental rift systems: AAPG Bulletin, v. 76, no. 8, p. 1153–1169.

Pakholchuk, A.A., I.I. Uryev, and P.B. Tsalko, 1991, Mechanism of formation of carbonate reservoir rocks in the Semiluki horizon of the Pripyat depression: Geologiya Nefti i Gaza, no. 8, p. 6–11.

Pechernikov, V.V., and I.M. Shakhnovsky, 1983, Causes of uneven distribution of oil fields in the Pripyat depression: Geologiya Nefti i Gaza, no. 9, p. 31–35.

Petrov, A.A., 1988, Oils of early stages of generation: Geologiya Nefti i Gaza, no. 10, p. 50–53.

Samarskaya, E.V., and R.Y. Polyak, 1990, Distribution of reservoir rocks in upper Visean sandstone V-18 in the Dnieper-Donets depression: Geologiya Nefti i Gaza, no. 9, p. 23–25.

Shatsky, N.S., 1964, Troughs of the Donets type, *in* N.S. Shatsky, ed., Selected works (Izbrannye Trudy), v. 2: Moscow, Nauka, p. 22–34.

Shishkin, E.I., and T. Yu. Shishkina, 1989, Possible mechanism of formation of the Pripyat paleorift: Geotektonika, no. 6, p. 72–80.

Shpak, P.F., ed., 1989a, Geology and petroleum productivity of the Dnieper-Donets depression—petroleum productivity (Geologiya i neftegazonosnost Dneprovsko-Donetskoy vpadinyneftegazonosnost): Kiev, Ukraine, Naukova Dumka, 204 p.

Shpak, P.F., 1989b, A problem of oil and gas presence within deep-seated sedimentary formations: UN International Workshop on Petroleum Field Geology and Geochemistry, Preprint, Kiev, Ukraine, IGN ANU, 46 p.

Shpak, P.F., ed., 1990, Coal-bearing formations and the composition of coals of the Dnieper-Donets depression (Uglenosnye formatsii i veshchestvennyi sostav ugley Dneprovsko-Donetskoy vpadiny): Kiev, Ukraine, Naukova Dumka, 218 p.

Shpak, P.F., and A.E. Lukin, 1986, Conditions of oil and gas generation and formation of hydrocarbon pools in Paleozoic rocks of the Dnieper-Donets depression, *in* N.A. Yeremenko et al., eds., Evolution of oil and gas generation in history of the Earth (Evotutsiya neftegazoobrazovaniya v istorii Zemly): Moscow, Nauka, p. 119–123.

Shumlyansky, V.A., 1985, Boundary between the Dnieper-Donets depression and the Donbas fold system: Geotektonika, no. 6, p. 109–114.

Smishko, R.M., 1985, Main faults of the Donets basin and related strike-slip movements: Geotektonika, no. 2, p. 120–125.

Sokiran, N.P., V.A. Krivosheya, and N.V. Tkhorovskaya, 1978, Turbidites in upper Serpukhovian rocks of the Dnieper-Donets depression: Geologicheskiy Zhurnal, no. 1, p. 132–138.

Sollogub, V.B., 1986, Lithosphere of the Ukraine (Litosfera Ukrainy): Kiev, Naukova Dumka, 184 p.

Ulmishek, G.F., 1988, Upper Devonian–Tournaisian facies and oil resources of the Russian craton's eastern margin, *in* N.J. McMillan, A.F. Embry, and E.J. Glass, eds., Devonian of the world, volume 1: regional syntheses: Canadian Society of Petroleum Geologists Memoir 14, p. 527–549.

Ulmishek, G.F., 1990, Geologic evolution and petroleum resources of the Baltic basin, *in* M.W. Leighton et al., eds., Interior cratonic basins: AAPG Memoir 51, p. 603–632.

Zhabrev, I.P., ed., 1983, Gas and gas condensate fields (Gazovye i gazokondensatnye mestorozhdeniya): Moscow, Nedra, 375 p.

Zonenshain, L.P., M.I. Kuzmin, and L.M. Natapov, 1990, Tectonics of lithospheric plates of the USSR, Vol. 1, 327 p. (Tektonika litosfernykh plit territorii SSSR, tom 1): Moscow, Nedra.

Chapter 5

Recôncavo Basin, Brazil: A Prolific Intracontinental Rift Basin

A.M.F. de Figueiredo, Editor
Exploration Department, DEPEX
Rio de Janeiro, Brazil

This chapter is the result of the contributions of several geologists and geophysicists of the PETROBRÁS Exploration Department who have been working with the Recôncavo basin. Specific contributions were given by

J.A.E. Braga, J.C. Zabalaga, J.J. Oliveira, G.A. Aguiar, O.B. Silva, L.F. Mato, L.M.F. Daniel, L.P. Magnavita
Exploration District, DEXBA
Salvador, Bahia, Brazil

C.H.L. Bruhn
Exploration Department, DEPEX
Rio de Janeiro, Brazil

REGIONAL SETTING AND GENERAL OVERVIEW

The Recôncavo basin is located in northeastern Brazil, and occupies an area of 11,500 km^2. It corresponds to the southern portion of an elongated intracontinental aulacogen that extends to the north, encompassing the Tucano and Jatobá basins (Figure 1). The basin is limited to the east by the Salvador fault, with a maximum dip slip of around 5000 m, and to the west by the Maragogipe fault, with an average vertical displacement of 200 m. To the north, the basin is separated from the Tucano basin by the Aporá high, while to the south its exposed portion is delimited by the Itacaré high (Figure 2). This asymmetric graben is basically filled with Upper Jurassic and Lower Cretaceous sediments, which characterize two major sedimentary sequences: the first, prerift, is composed of an arid alluvial fan system and associated facies; the second, synrift, is composed of alluvial, fluvial, and deltaic lacustrine sediments representing the complete record of infilling of a tectonically active lake basin. A thin package of Upper Cretaceous and Tertiary postrift coarse clastics partially covers the present-day basin. As a whole, the sedimentary section reaches a maximum thickness of about 6000 m, deposited over Archean granulites and, locally, over slightly metamorphosed Proterozoic rocks and Permian–Carboniferous sediments.

Petroleum exploration began in the Recôncavo basin in 1937, when a few shallow wells were drilled in Lobato, in the neighborhood of Salvador, close to a number of oil seeps discovered during the construction of a railroad. Two years later, in 1939, wildcat 1-L-3-BA was completed and became the first oil producer in Brazil.

Following the success of Lobato, a broad program of geologic surface mapping oriented exploration activities for nearly 15 yr; during this phase, 36 wildcats and 27 stratigraphic wells were drilled, which led to the discovery of nine oil fields, including Candeias (1941), Dom João (1947), and Água Grande (1951). These three fields hold 316.0 million m^3 (1988 million bbl; 42%) of present-day proven in-place oil volume in the basin.

In the mid-1950s, with the creation of PETROBRÁS, a program for the systematic exploration of the basin was adopted, based on seismic reflection and refraction surveys, gravimetry, and electroresistivity, in addition to photogeologic studies and structural drilling. The use of these tools led to the development of an exploratory drilling program, whose pace increased progressively until 1967. During this period, 439 exploratory wells were drilled, including 182 wildcats and, as a result, 35 oil and gas fields were discovered, almost entirely in structural traps mapped by seismic data. The proven oil volume allocated during this period reached 348.5 million m^3 (2192 million bbl), equivalent to 47% of the total oil volume in the basin; 84% of this volume was concentrated in only four fields: Taquipe, discovered in 1958; Buracica, in 1959; and Araçás and Miranga, in 1965.

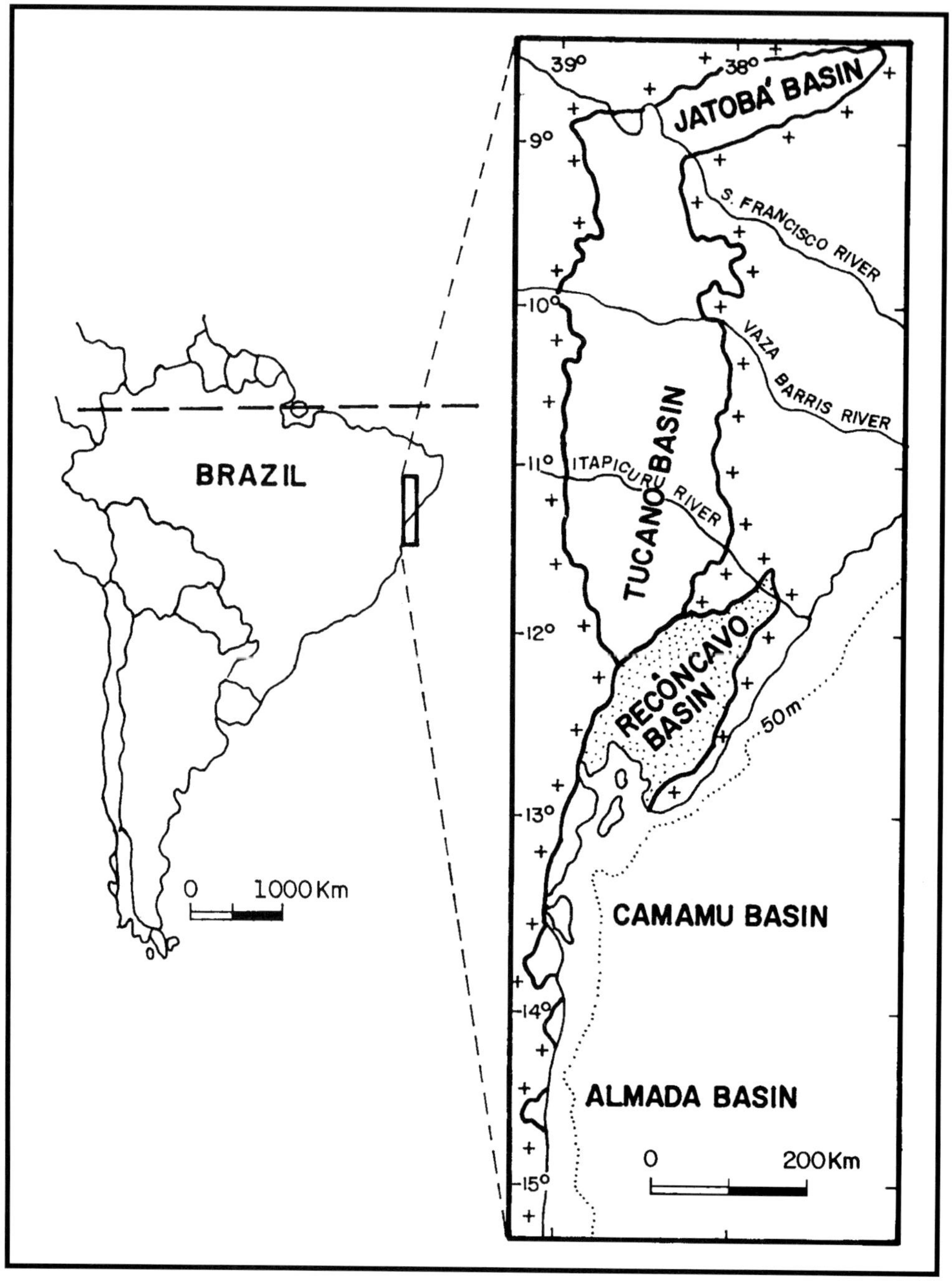

Figure 1. Location map of Recôncavo basin, Brazil.

Following this period of major discoveries, the basin entered a phase of exploratory decline due to the exhaustion of traditional prospects (structural highs mapped by seismic methods), which became less and less conspicuous. Thus, over the next 13 yr of activities, only 24.0 million m^3 (151 million bbl) of oil were added, less than 3% of proven oil volume.

Starting in 1980, exploration was redirected toward areas previously considered as less attractive and that were located at the edges of the basin and in regional lows. Prospectivity of these areas was made feasible due to enhanced seismic resolution. Results have been highly encouraging and the drilling of 80 wildcats led to the incorporation of nearly 61.2 million m^3 (385 million bbl) of proven oil volume, corresponding to approximately 8% of the basin total.

Exploration efforts and increases in proven volume over time are illustrated in Figure 3. By late 1990, 22,000 km of seismic profiles had been acquired and 5050 wells had been drilled, of which 978 are exploratory and 4082 are production wells. The total in-place hydrocarbon volume discovered in the basin amounts to 761 million m^3 (4,786.7 million bbl) of oil and 82 billion m^3 (2.9 tcf) of natural gas, with a cumulative production of about 195 million m^3 (1,226.6 million bbl) of oil and 37 billion m^3 (1.2 tcf) of gas. The remaining proven reserves are about 37 million m^3 (232.7 million bbl) of oil and 25 billion m^3 (882.75 bcf) of gas.

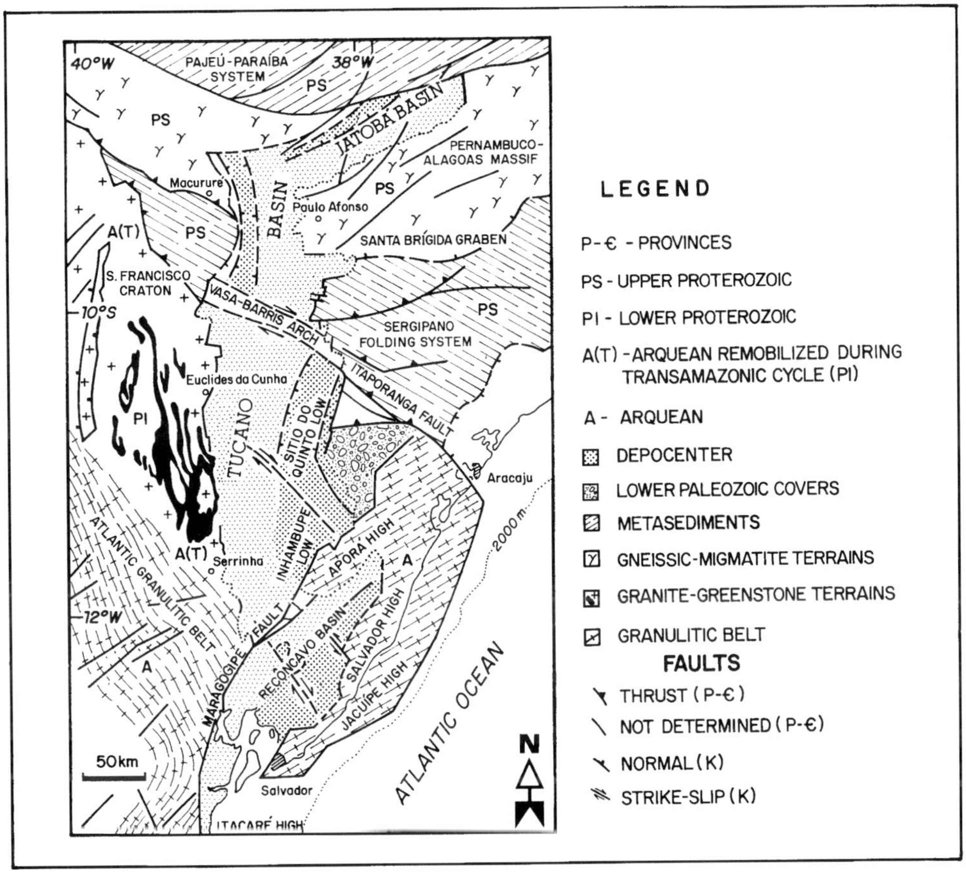

Figure 2. Regional tectonic setting of Recôncavo basin (Milani, 1984).

TECTONICS

Plate Setting

The Recôncavo graben constitutes the southern portion of the Recôncavo-Tucano-Jatobá rift, a Late Jurassic to Early Cretaceous system of half-grabens whose tectonic and sedimentary history reveals a close relationship to the early evolutionary stages of the South Atlantic, from which it was aborted as a secondary branch during the Aptian.

The development of this intracontinental rift, a unique feature on the eastern Brazilian margin, occurred as a response to the same extensional stress field that broke up Gondwana and definitively separated the African and South American continents.

Under the Jurassic–Cretaceous extensional stress field and molded by preexisting crustal weakness zones, a small microplate known as the East Brazilian microplate (Szatmari et al., 1985) became individualized. Its counterclockwise movement in relation to the South American continent, referred to a pole located near the eastern border of the Jatobá basin, permitted the development of an intracontinental rift in that area (Figure 4).

In the Aptian, the main rifting axis was located in the region of the present continental margin; thus, microplate movement ceased and tectonic activity consequently came to a close without the development of any expressive sedimentary record of post-Aptian age in the Recôncavo-Tucano-Jatobá rift.

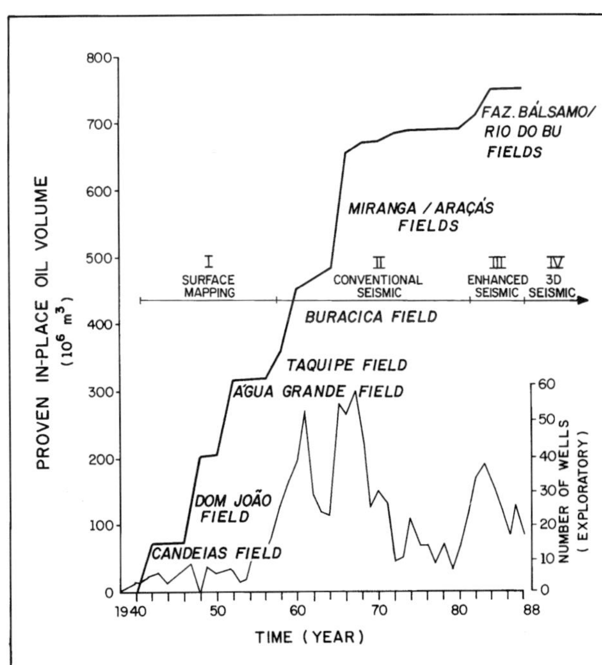

Figure 3. Evolution of proven oil reserves (in millions of cubic meters of oil) and number of exploratory wells drilled per year in Recôncavo basin.

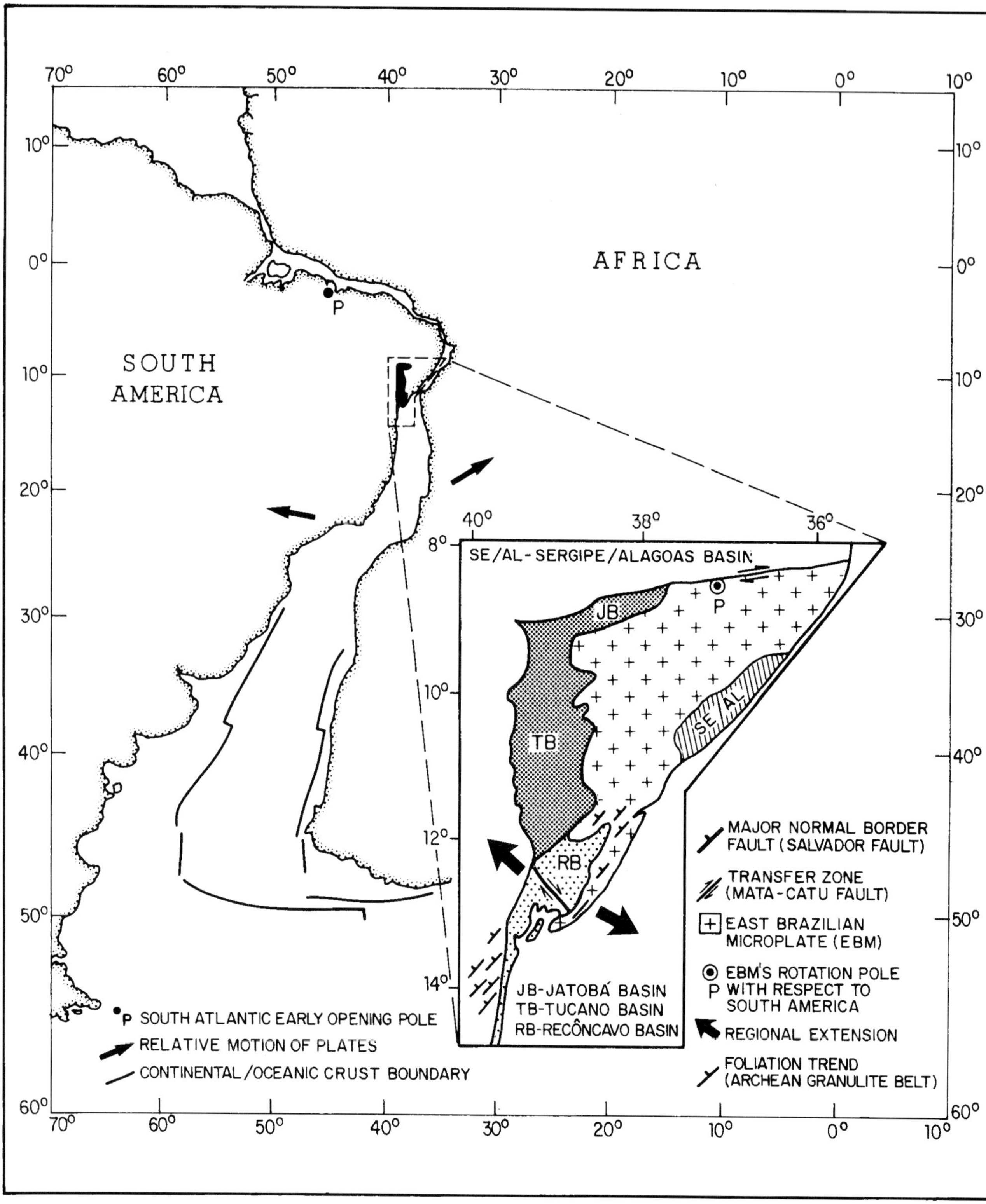

Figure 4. Paleoreconstruction of continental configuration in the late Aptian. Inset shows tensile stress field in Recôncavo basin as deduced from structural analysis of rift and coincident with interpreted motion of the east Brazilian microplate (Milani et al., 1987).

Structural Configuration

In the region of the Recôncavo basin, the crystalline basement consists of Archean granulitic gneiss with vertical foliation trending N30°E (Figure 2). As a tectonic heritage of the intensive planar anisotropy of the metamorphic complex, the structural framework of the Recôncavo is defined by a set of normal faults par-

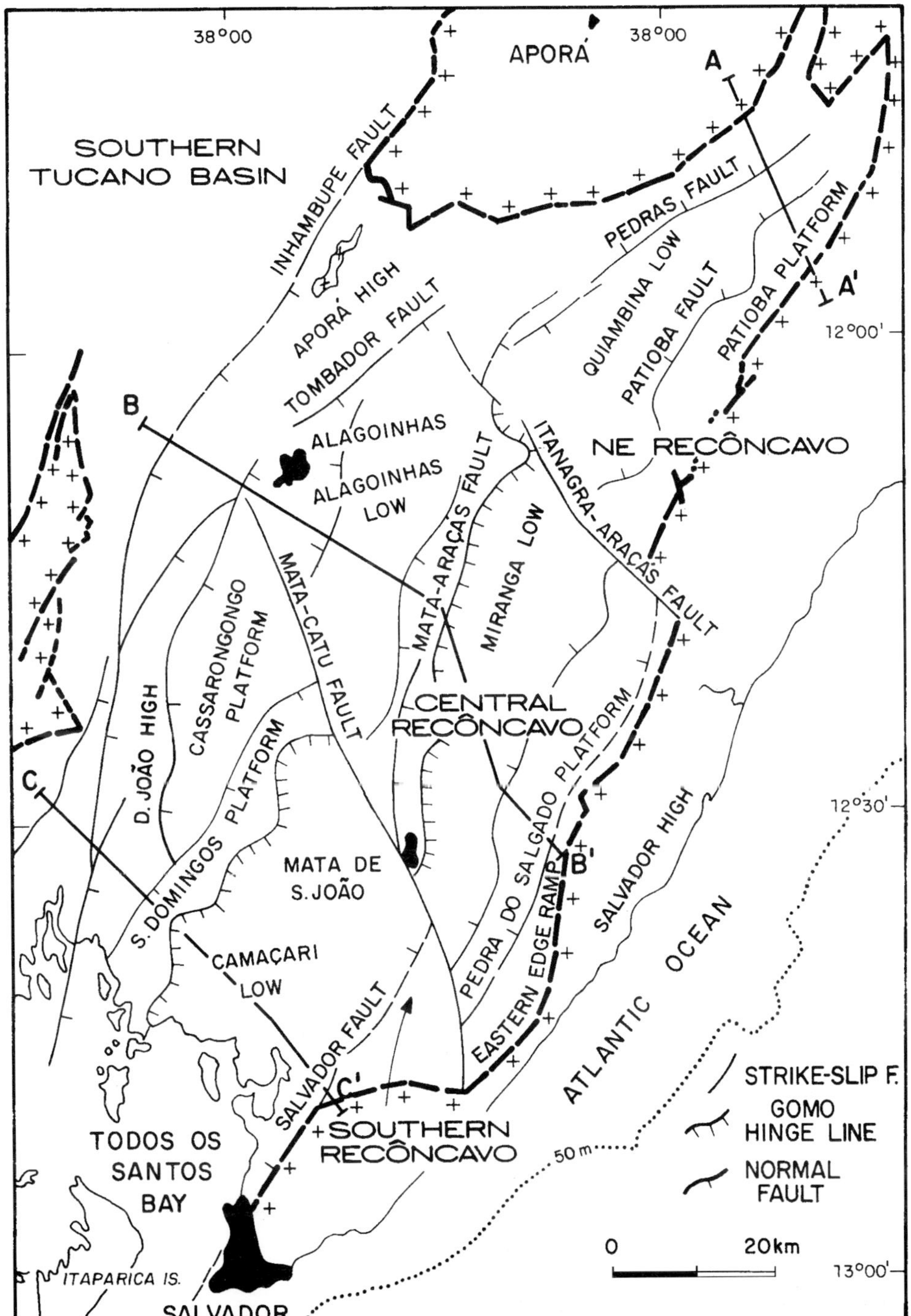

Figure 5. Structural configuration of Recôncavo basin compartments. A–A′, B–B′, and C–C′ represent seismic sections (seismic sections shown in Figures 7A, B, C, respectively).

allel to the foliation of the basement; the major fault—the Salvador—limits the graben to the southeast and controls the onset of its depocenter. Due to differing rates of crustal stretching, the basin was cut across by northwest-trending transfer faults, the major one being the Mata-Catu fault (Milani and Davison, 1988). The principal motion of this fault is believed to be vertical, but evidence of horizontal displacement and transpressional movement may be observed along the strike of the Mata-Catu fault (Netto et al., 1984). The northwest-southeast–trending faults, particularly the Mata-Catu and Itanagra-Araçás faults, allow the basin to be divided into three compartments—the northeastern, central, and southern—each displaying its own structural and stratigraphic peculiarities (Figure 5). The Bouguer gravity map shown in Figure 6 provides a

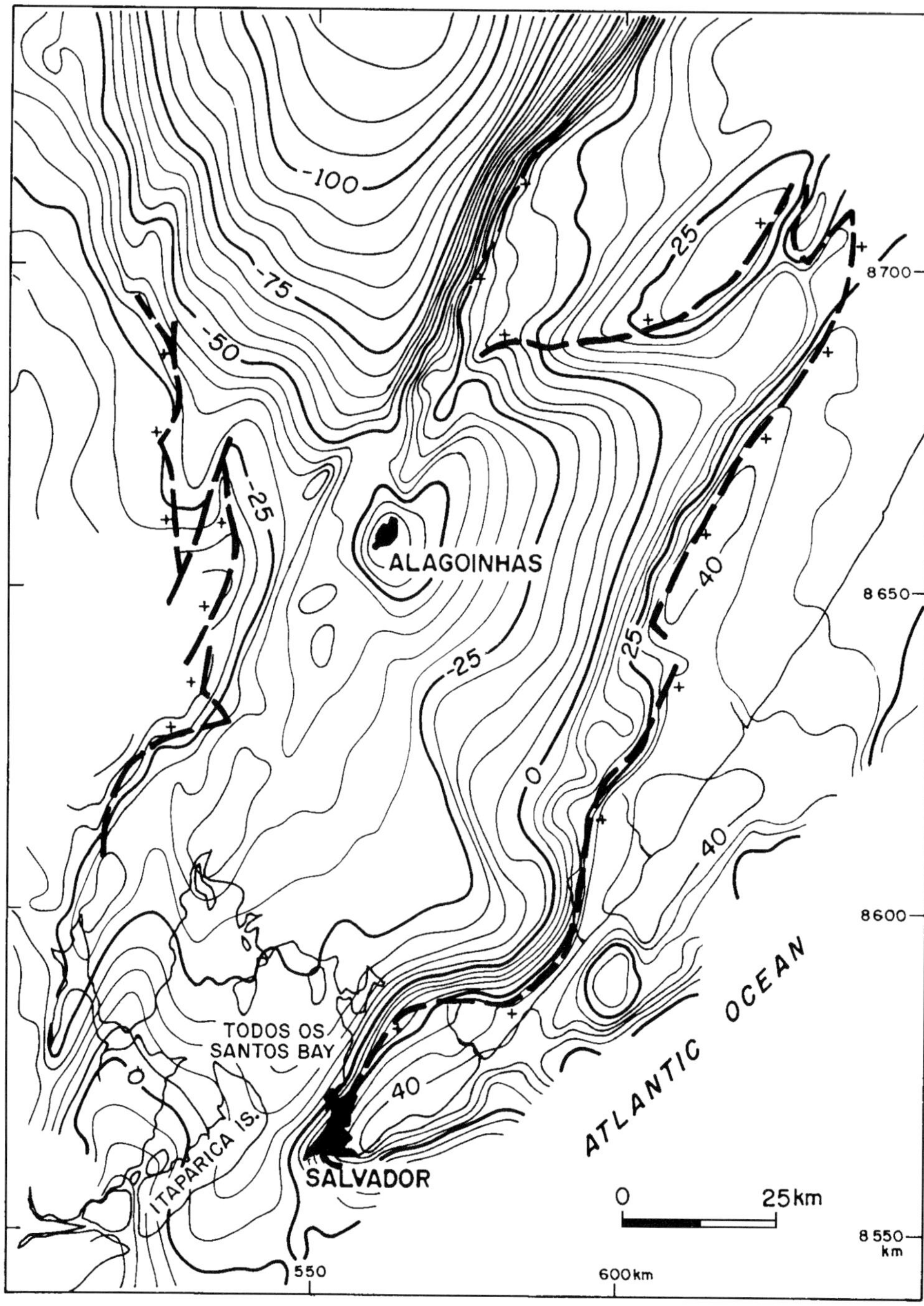

Figure 6. Bouguer gravity map of Recôncavo basin.

fine portrait of the main structural features of the basin, described in greater detail in the following sections.

Northeastern Recôncavo

The northeastern subbasin occupies an area of approximately 1600 km^2. It is limited to the east-southeast by the Salvador fault, to the north-northeast by the Aporá high, and to the southeast by the Itanagra-Araçás fault. Antithetic faults, possibly of late Neocomian age, are predominant, with the downthrown blocks dipping southeast. The major structural features, shown in seismic section A–A′ (see Figures 5, 7A) are the Patioba platform and the Quiambina low.

Patioba Platform

The Patioba platform occupies an extensive strip and is limited to the west by a fault of the same name and to the east by the Salvador fault, the eastern limit of the basin. Several oil fields, producing from Upper Jurassic sediments, occur along the Patioba fault, which reaches a maximum dip slip of nearly 1500 m. Smaller grabens within this feature result in the occurrence of combined traps, as in the case of the Riacho da Barra field. Domal structures,

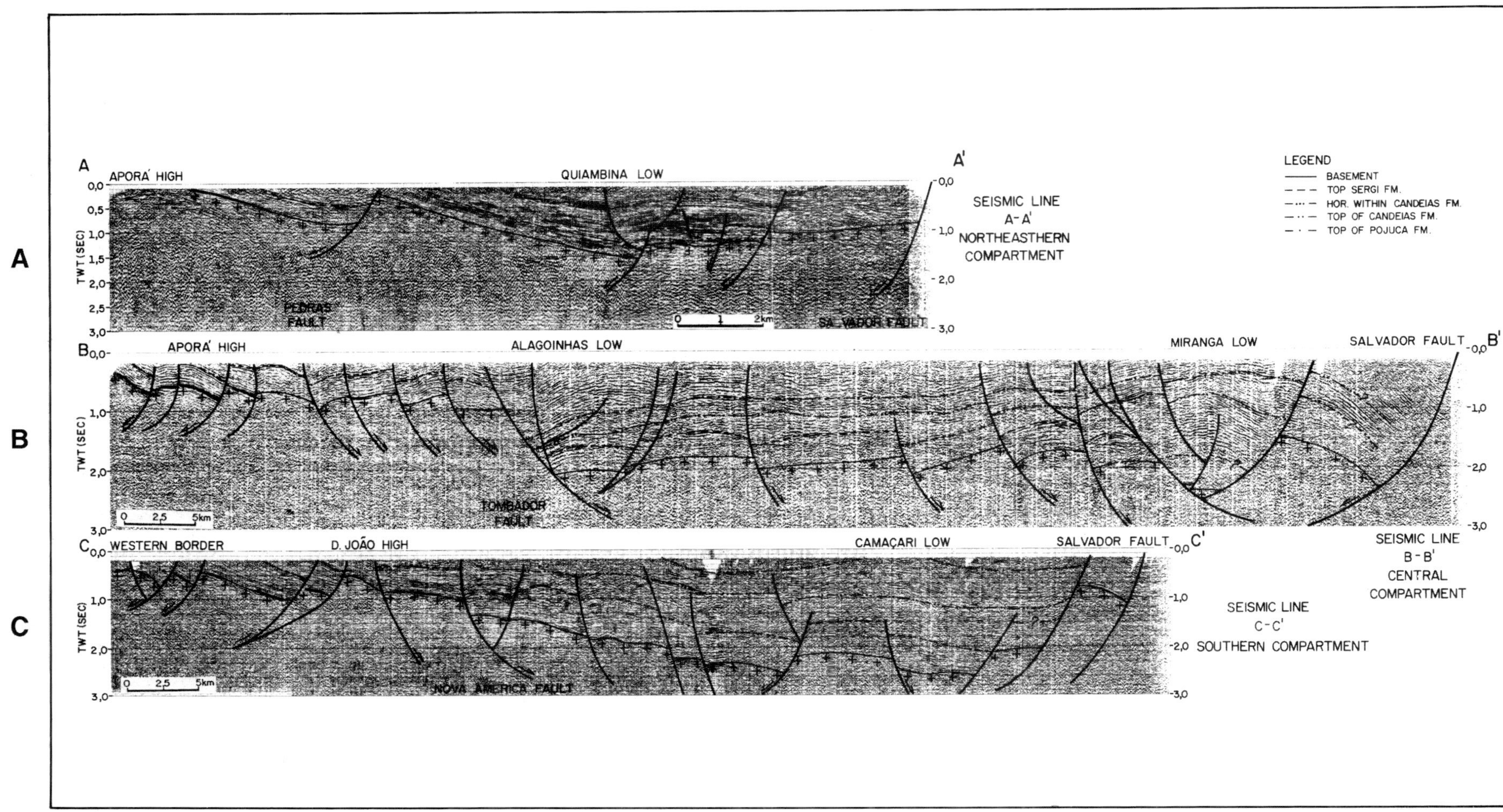

Figure 7. (A) Compressed seismic section A–A´ showing major structural elements of Northeastern Recôncavo subbasin. (B) Compressed seismic section B–B´ showing major structural elements of Central Recôncavo subbasin. (C) Compressed seismic section C–C´ showing major structural elements of Southern Recôncavo subbasin. (See Figure 5 for locations.)

possibly of compressive origin, are also mapped along this feature.

Quiambina Low

The Quiambina low occurs between the Patioba platform and the Pedras fault and is limited to the south by the Itanagra-Araçás fault. Secondary faults with downthrown blocks to the west result in the occurrence of important structural alignments, as, for example, the upthrown block in the Pedras fault. Gravitational faults in the rift section are associated with incipient shale movement (Figure 7A).

Central Recôncavo

The central compartment is limited to the northeast and southwest by transfer faults and occupies an area of 3500 km^2. Northwest-tilted blocks are important features and form traps for several oil fields, such as Água Grande, Boa Esperança, and Araçás. During the late Neocomian/Barremian, block-tilting produced a small depocenter in the western portion of this subbasin, constituting a feature known as the Alagoinhas low (Figures 5, 7B) and prompting an inversion in the direction of normal basin tilt. Domal and anticlinal structures, shale diapirs, and other features are also present, suggesting the occurrence of compressive stress, as displayed in seismic section B–B´ (Figure 7B). Among major features are the Aporá high, Alagoinhas low, Mata-Araçás platform, Miranga low, Pedra do Salgado platform, and the ramp of the eastern edge.

Aporá High

The Aporá high feature separates the Recôncavo and Tucano Sul basins, close to parallel 12S, north of the city of Alagoinhas (Figure 5). The high is formed by a series of small horst and grabens and is not well known due to the lack of seismic and well data.

Alagoinhas Low

The Alagoinhas low is a depression of late Neocomian/Barremian age, directly related to the Mata-Catu transfer fault. The poor quality of seismic records makes mapping of the early Neocomian fault very difficult. The Tombador fault separates this feature from the Aporá high.

Mata-Araçás Platform

This feature's western limit is not well defined, and there is a gradual smooth transition to the Alagoinhas low. To the east, a substantial fault separates this platform from the Miranga low. At its extremities, near the Água Grande and Araçás oil fields, the sedimentary section dips northwest, making the area extremely attractive for oil exploration.

Miranga Low

The Miranga low is the prolongation of the Quiambina low in the northeastern Recôncavo. The sedimentary sequence becomes thicker in this depression, generating strong diapiric movement and associated growth faults. The narrowing of this feature in the southern portion and the presence of a large number of shale diapirs and anticlinal structures could indicate compressive stress in the area (Figures 5, 7B).

Pedra do Salgado Platform

This feature represents the southern extension of the Patioba platform, displaced by the Itanagra fault. The platform is limited to the west by the Pedra do Salgado fault and to the east by the Salvador fault. During the late Neocomian/Barremian, this feature was subjected to strong eastward tilting. Within the limits between this platform and the Miranga low, the layers dip west, generating excellent potential hydrocarbon traps against the flanks of diapirs or even against fault planes.

Ramp of the Eastern Edge

This is a shallow platform dipping smoothly west, in the direction of the Salvador fault. The Barremian sediments are thought to rest upon the crystalline basement, which reaches maximum depths of 1500 m.

Southern Recôncavo

This subbasin encompasses the entire Recôncavo basin south of the Mata-Catu fault with approximately 5000 km^2 (Figure 5). The Dom João high separates two compartments, with an antithetic fault to the west and a synthetic fault to the east, as shown in seismic section C–C´ (see Figures 5, 7C). The major features are the Dom João high, São Domingos platform, Camaçari low, and Taquipe canyon.

Dom João High

This major feature is composed of small horsts, such as the Orobó and Dom João. The faults that limit this feature seem to be listric and of early Neocomian age.

São Domingos Platform

This platform consists of an extensive strip located between the Dom João high and the Camaçarí low, with the layers exhibiting a smooth bending toward the northeast-southwest. In the northern portion, close to the town of Buracica and near the Mata-Catu fault, seismic data indicate the presence of complex faulted features, suggesting the occurrence of local transpressional stress.

Camaçari Low

This feature is similar to the Miranga low, where a thick sedimentary section is disturbed by shale diapirs and growth faults. However, the difference between the two is remarkable, the Camaçari low being characterized by structures that display lower relief and less intense diapirism.

Taquipe Canyon

The southern Recôncavo compartment is the only one displaying sizable erosional features. Known as the Taquipe Paleocanyon, this set of features was eroded in the late Neocomian and filled with an essentially clay-rich section.

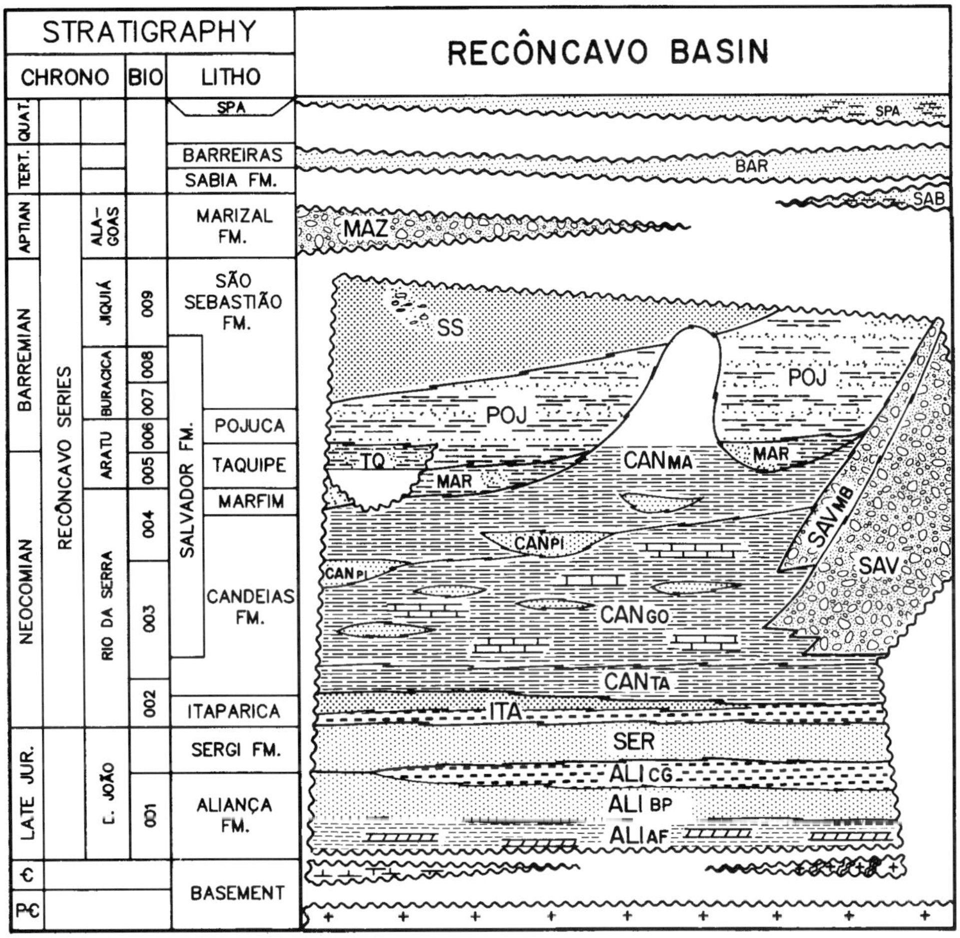

Figure 8. Stratigraphic column of Recôncavo basin.

STRATIGRAPHY

In standard chronostratigraphy, Upper Jurassic–Lower Cretaceous sediments of Recôncavo basin (Figure 8) may be tentatively correlated to the Portland/Volgian and Lower Cretaceous sediments of England and Germany, based on their abundant flora and nonmarine ostracodes. The paleontologic record of these sediments, however, does not correlate well with the international stratigraphic column. This limitation led PETROBRÁS to adopt regional stages, designated as the Dom João, Rio da Serra, Aratu, Buracica, Jiquiá, and Alagoas, valid not only for the Recôncavo, but also for other basins in northeastern Brazil. Ostracode biozones, associated with lithologic markers of regional distribution, were used for the purpose of delimiting these stages (Viana et al., 1971). As a rough approximation, in this chapter the Dom João is correlated to the Volgian, the Rio da Serra to the early Neocomian, the Aratu and Buracica to the late Neocomian, the Jiquiá to the Barremian, and the Alagoas to the Aptian.

Lithostratigraphy and Depositional Systems

The sedimentary section is represented by the following formations: Aliança, Sergi, Itaparica, Candeias, Marfim, Taquipe, Pojuca, São Sebastião, Salvador, Marizal, Sabiá, and Barreiras (Figure 8). The first three formations, plus the Tauá Member of the Candeias Formation, belong to the prerift phase. The others, with the exception of the Marizal, Sabiá, and Barreiras formations, are representative of the synrift phase.

Aliança Formation

The Aliança Formation is characterized by the presence of continental red shales and sandstones, with the subordinate occurrence of evaporitic beds (anhydrite and halite). The formation is subdivided into three members: Afligidos, Boipeba, and Capianga.

The Afligidos Member is composed of sandstones and evaporitic beds in the lower part and red shales in the upper part. These sediments were deposited in a playa-lake environment (Figures 9A, B), grading upward to a lacustrine system. The Boipeba Member is characterized by a clastic wedge composed of sandstones, conglomerates, and minor red shales and constitutes the middle portion of the formation. It represents the record of an alluvial fan system that prograded into the Afligidos lake from the west (Figure 9C, D). A thick section of red shales of lacustrine origin, named the Capianga Member, is found in

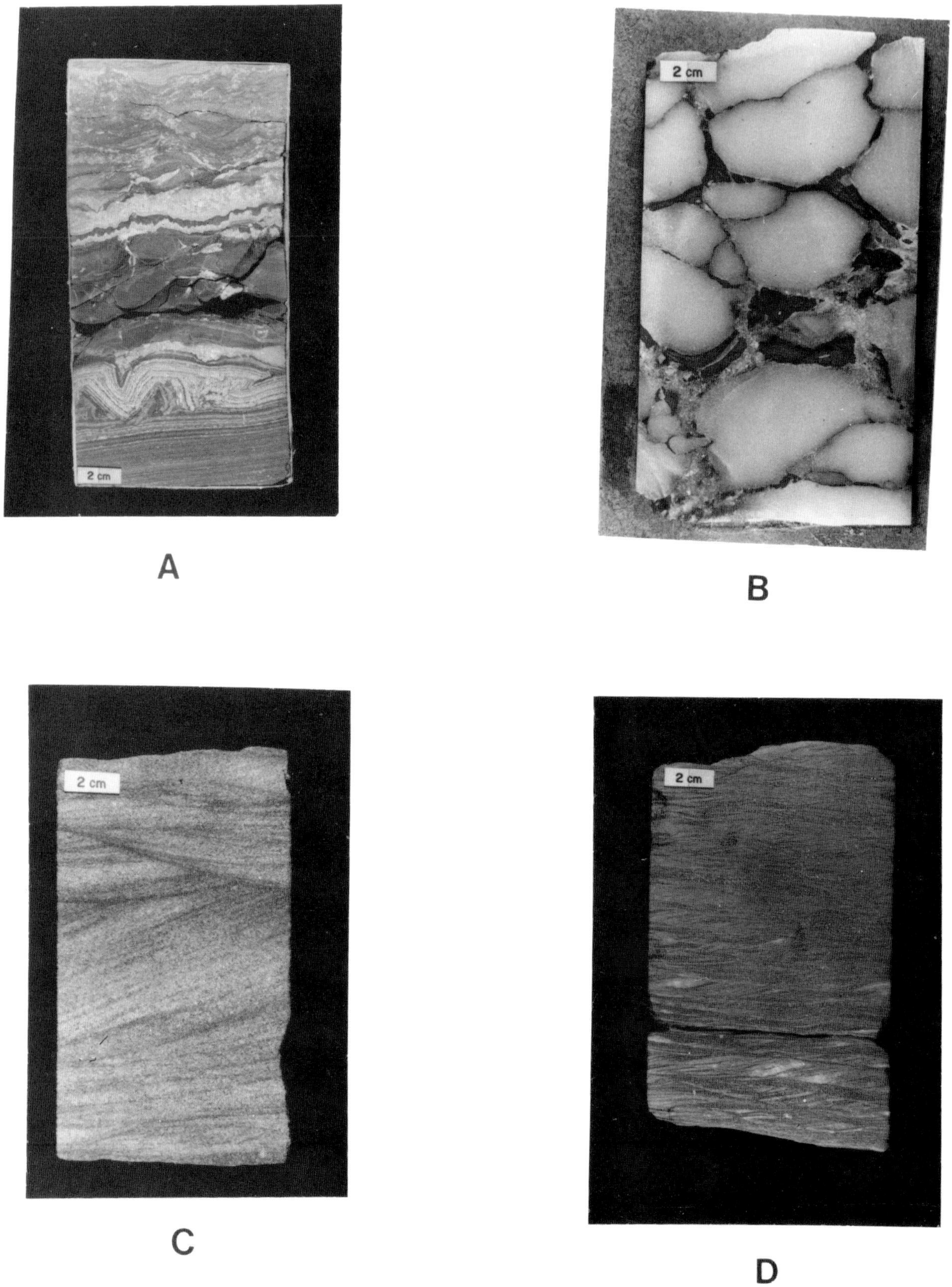

Figure 9. Aliança Formation core slabs. A = sabkha to subtidal laminated anhydrite and dolomite marl deformed by microfolds and enterolithic structures; B = nodular anhydrite in red shale sabkha; C = cross-stratified coarse-grained fluvial sandstone; D = ripple-drift cross-laminated very fine sandstone (upper point-bar facies).

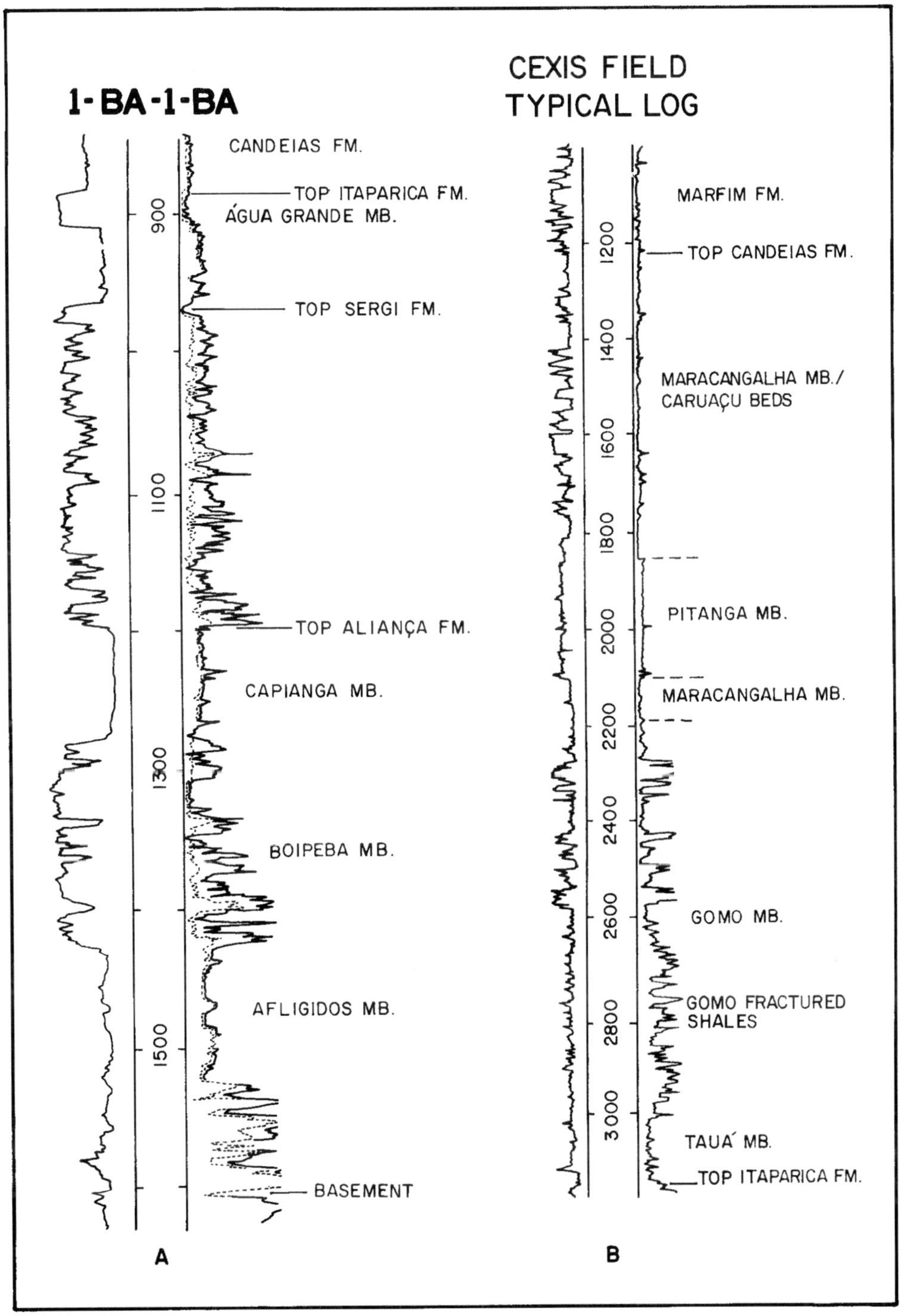

Figure 10. Typical electric logs of Aliança, Sergi, Candeias, and Marfim formations. A = Aliança and Sergi formations; B = Candeias and Marfim formations.

the upper part still associated with the alluvial fan complex of the Boipeba Member (Figure 10A).

Sergi Formation

The Sergi Formation is basically composed of alluvial to fluvial, fine-grained to conglomeratic, light-colored sandstones with sporadic intercalations of red shales at its base. This package is the record of a broad system of amalgamated alluvial fans wedging upward toward the east that probably extended beyond the current limits of the basin.

Sedimentary processes within the present-day basin have been ascribed to braided rivers in proximal areas grading to a coarse-grained meander belt type in more distal facies (Figures 10A, 11, 12A). Associated eolian facies display regional distribution and occur particularly at the top of the Sergi Formation, where they reach thicknesses of up to 20 m (Figure

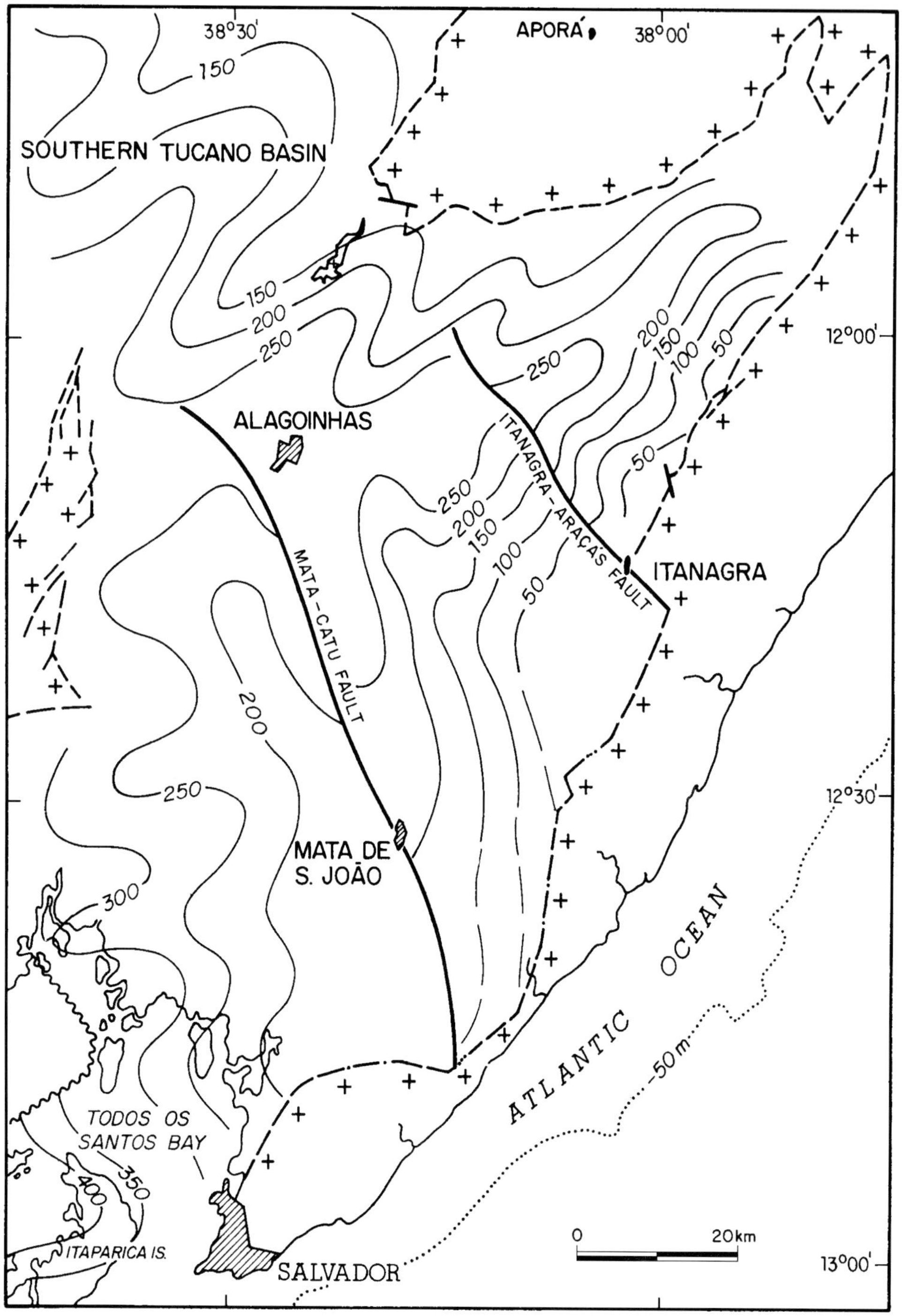

Figure 11. Isopach map of Sergi Formation, Recôncavo basin. Contour interval in meters (adapted from Netto et al., 1984).

12B). Figure 11 shows the total thickness map for this unit.

Itaparica Formation

The Itaparica Formation is characterized by cyclic sedimentation of sandstones, siltstones, and shales, constituting a fining- and thinning-upward sequence. This trend is suddenly interrupted by the presence of a very persistent fine- to coarse-grained sandstone of regional distribution, designated the Água Grande Member.

This unit represents the transitional phase from a typically continental to a lacustrine environment. In the upper section, a progradational coarse-grained fluvio-deltaic system with strong eolian reworking is recorded (Figures 12C, D).

Candeias Formation

The Candeias Formation is composed chiefly of shales, with lenses of sandstone, siltstone, and some lenticular limestone at its base. The Candeias

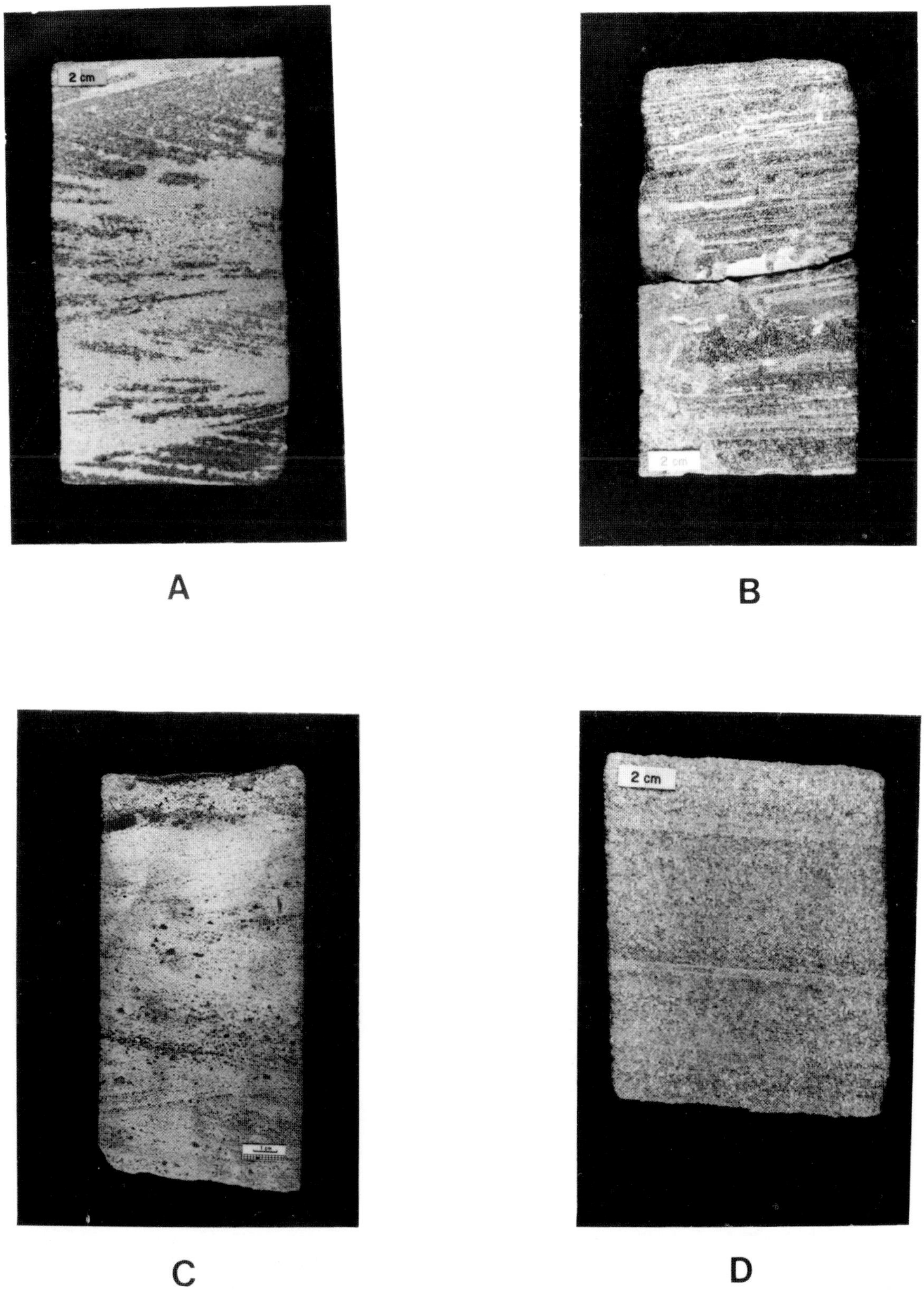

Figure 12. Sergi Formation and Água Grande Member core slabs. A = Sergi Formation: conglomeratic to coarse-grained sandstone with trough cross-stratification representing fluvial deposits; B = Sergi Formation: bimodal medium- to fine-grained eolian sandstone deposit; C = Água Grande Member, Itaparica Formation: coarse-grained, poorly sorted sandstone with medium-scale cross-bedding related to fluvial deposits; D = Água Grande Member, Itaparica Formation: laminated fine- and medium-grained sandstone, well-sorted on each lamina, related to an eolian environment.

Formation is subdivided into the Tauá, Gomo, Pitanga, and Maracangalha members (Figures 8, 10B).

The Tauá Member is the basal unit, consisting of dark-gray micaceous, splintery, hard, and fossiliferous shales with parallel lamination. Its external geometry shows little variation and the thickness oscillates around 80 m. The lower contact of Tauá Member shales with Água Grande Member sandstones or with the brown shales of the Itaparica Formation is usually abrupt and displays good contrast in SP (spontaneous potential) and electric-log resistivity curves; the same occurs with its upper contact when passing from the low-resistivity to high-resistivity shales of the Gomo Member. The dark clays and muds of the Tauá Member were deposited over marshy, lacustrine plain during a hot, humid period.

The Gomo Member is the clay-rich part of the middle portion of the Candeias Formation, and is characterized by dark-gray to greenish gray, very calcareous and, locally dolomitic, shales, with frequent concretions and lenses of micritic limestones. In electric logs, they correspond to arcuate-shaped, oscillating high-resistivity curves, informally designated the "Candeias belly." The lower contact with the Tauá Member is abrupt, whereas the upper contact is gradational with the Maracangalha Member, but sometimes abrupt with the Pitanga Member. Its thickness varies from a few hundred meters in platform areas to more than 2000 m in the deeper parts of the basin (Figure 13). The shales and carbonates of the Gomo Member are considered lacustrine deposits, with some associated siliciclastic turbidites (Figures 14A, B).

Figure 13. Isopach map of Gomo Member, Candeias Formation, Recôncavo basin (adapted from Netto et al., 1984).

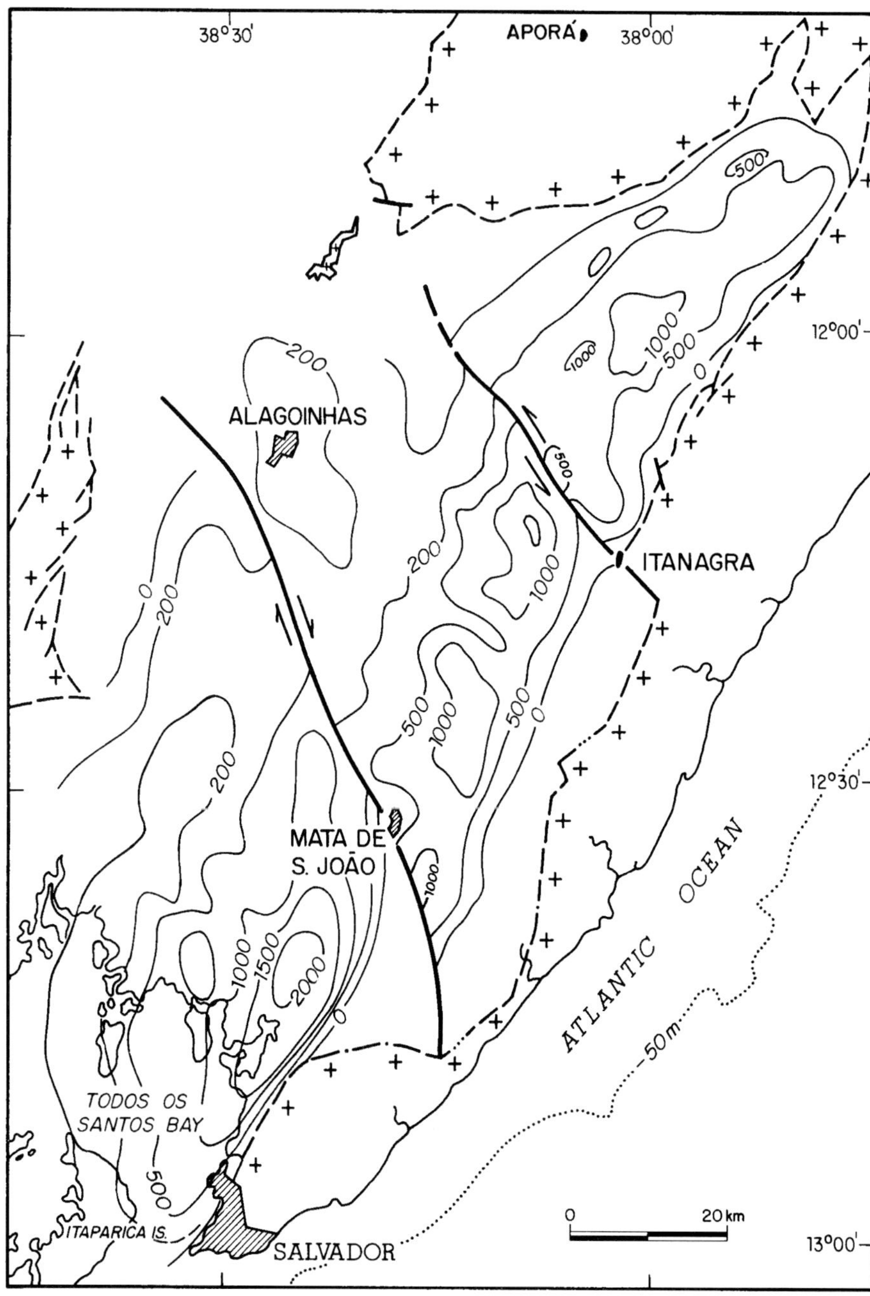

Figure 14. Candeias Formation core slabs. A = Gomo Member: fine- to medium-grained massive sandstone; B = Gomo Member: fine-grained sandstone displaying Tbc Bouma sequence; C = Pitanga Member: fine-grained muddy and massive sandstone; D = Pitanga Member: very fine grained muddy massive sandstone, presenting shale clasts.

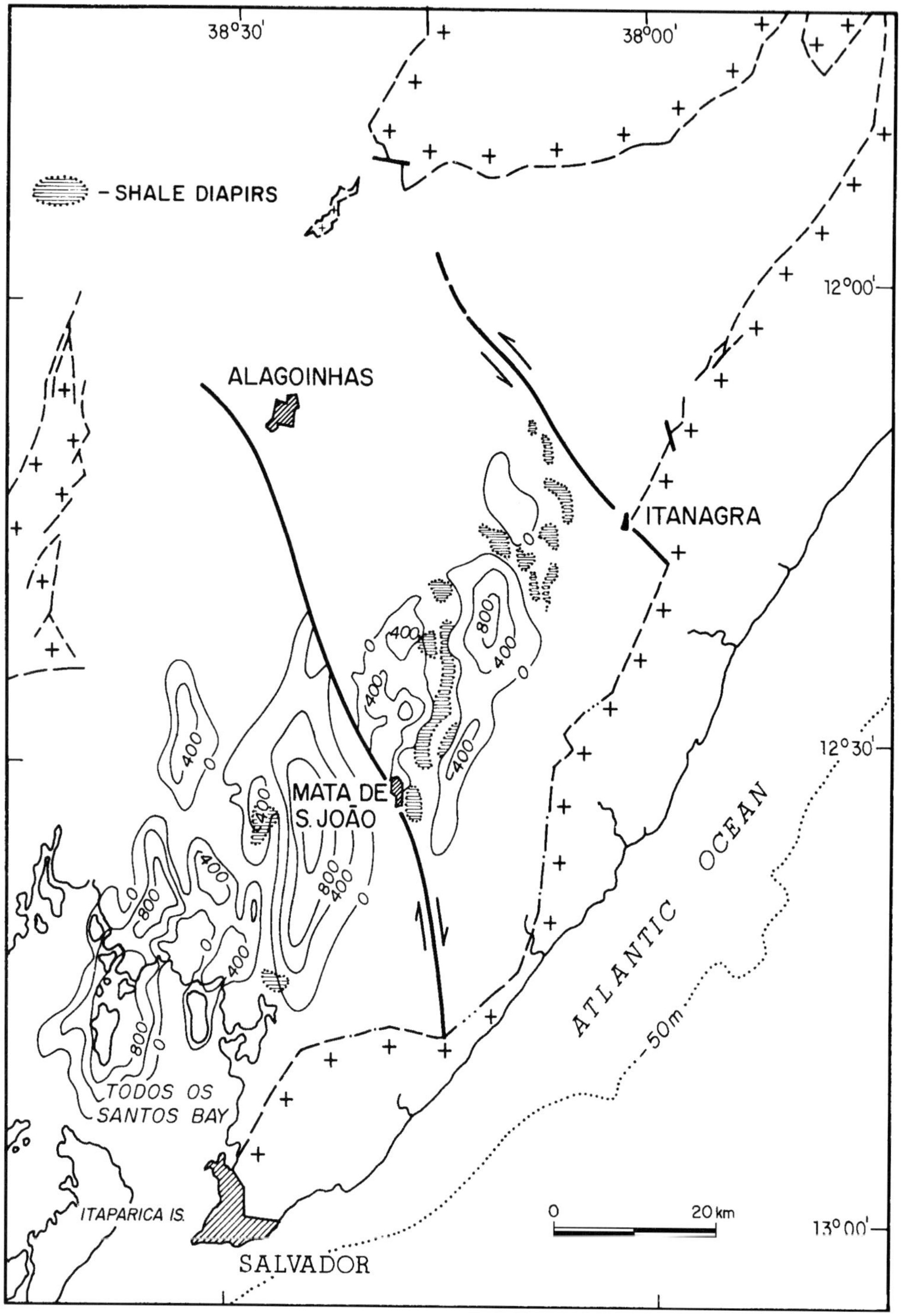

Figure 15. Isopach map of Pitanga Member, Candeias Formation, Recôncavo basin (Netto et al., 1984).

The Pitanga Member is a set of yellowish and greenish, very fine grained, highly argillaceous massive sandstones (Figures 14C, D). In outcrops, these sandstones present frequent slump and water-escape structures. In electric logs, they correspond to uniform, very distinctly identified, high-resistivity curves. They occur intercalated within the shales of the Maracangalha Member, exhibiting irregular shapes and sharp variations in thickness, reaching up to 1200 m (Figure 15). The lateral and vertical contacts are abrupt.

The Maracangalha Member represents the clay-rich part of the upper portion of the Candeias Formation. It consists of light- to dark-greenish shales, exhibiting low resistivities in electric logs. These shales are believed to be the source of basin diapirs. The Maracangalha Member's upper and lateral contacts

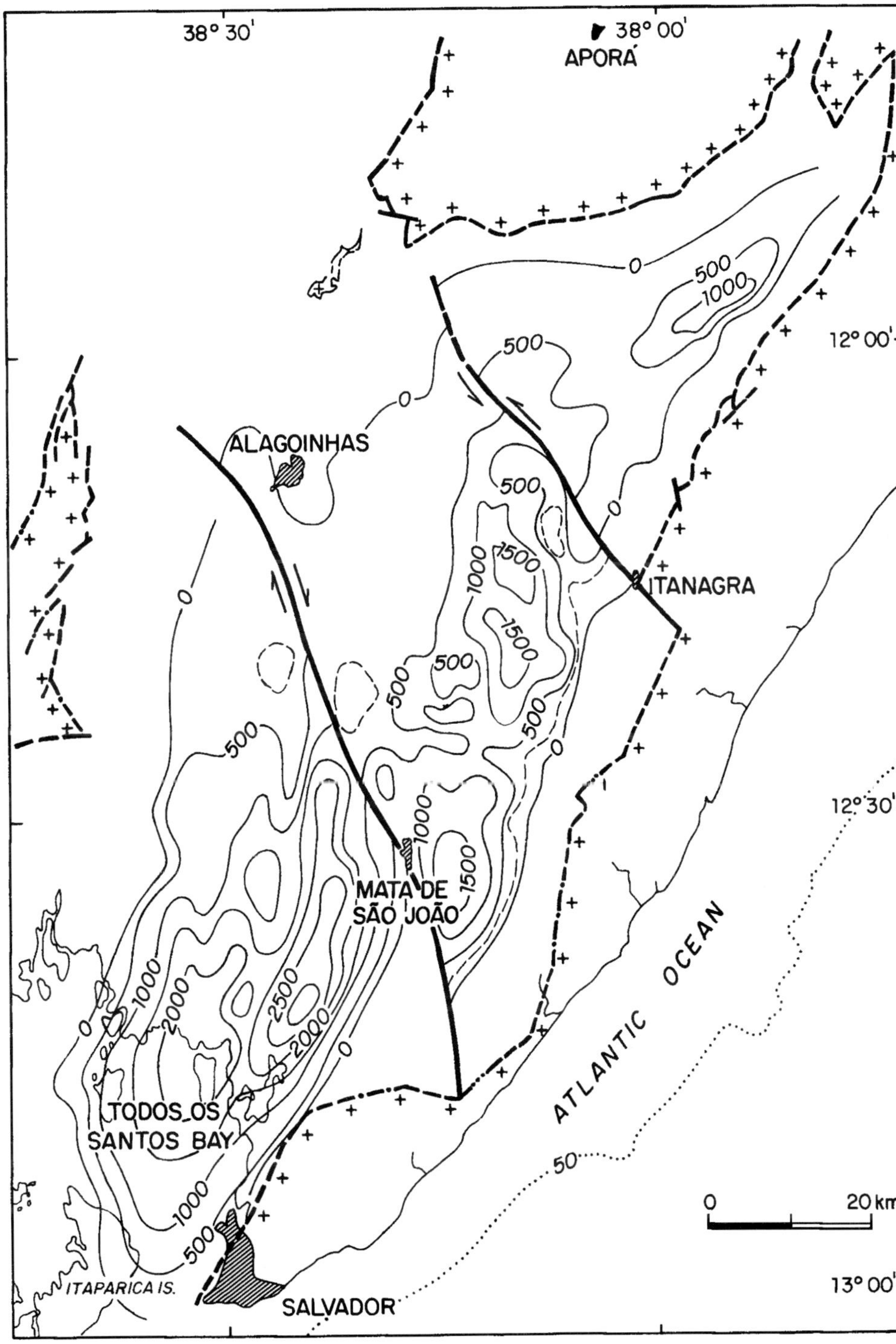

Figure 16. Isopach map of Maracangalha Member, Candeias Formation, Recôncavo basin (Netto et al., 1984).

with the Marfim Formation are conformable and sometimes display interdigitation with the sandstone of the Caruaçu beds of the Marfim Formation. The shale geometry is irregular and difficult to restore due to argillokinetic deformation. The member's thickness varies, reaching more than 3000 m in the most subsided areas (Figure 16). The muds of the Maracangalha Member, deposited in the deepest parts of the lake, suffered plastic deformation due to several factors, including basin tilting and episodic earthquake activity. As earthquakes occurred, the massive sandstones of the Pitanga Member were deposited between argillaceous masses of the Maracangalha Member. Being less dense than the massive sandstones, the shales moved upward and created diapirism in the basin.

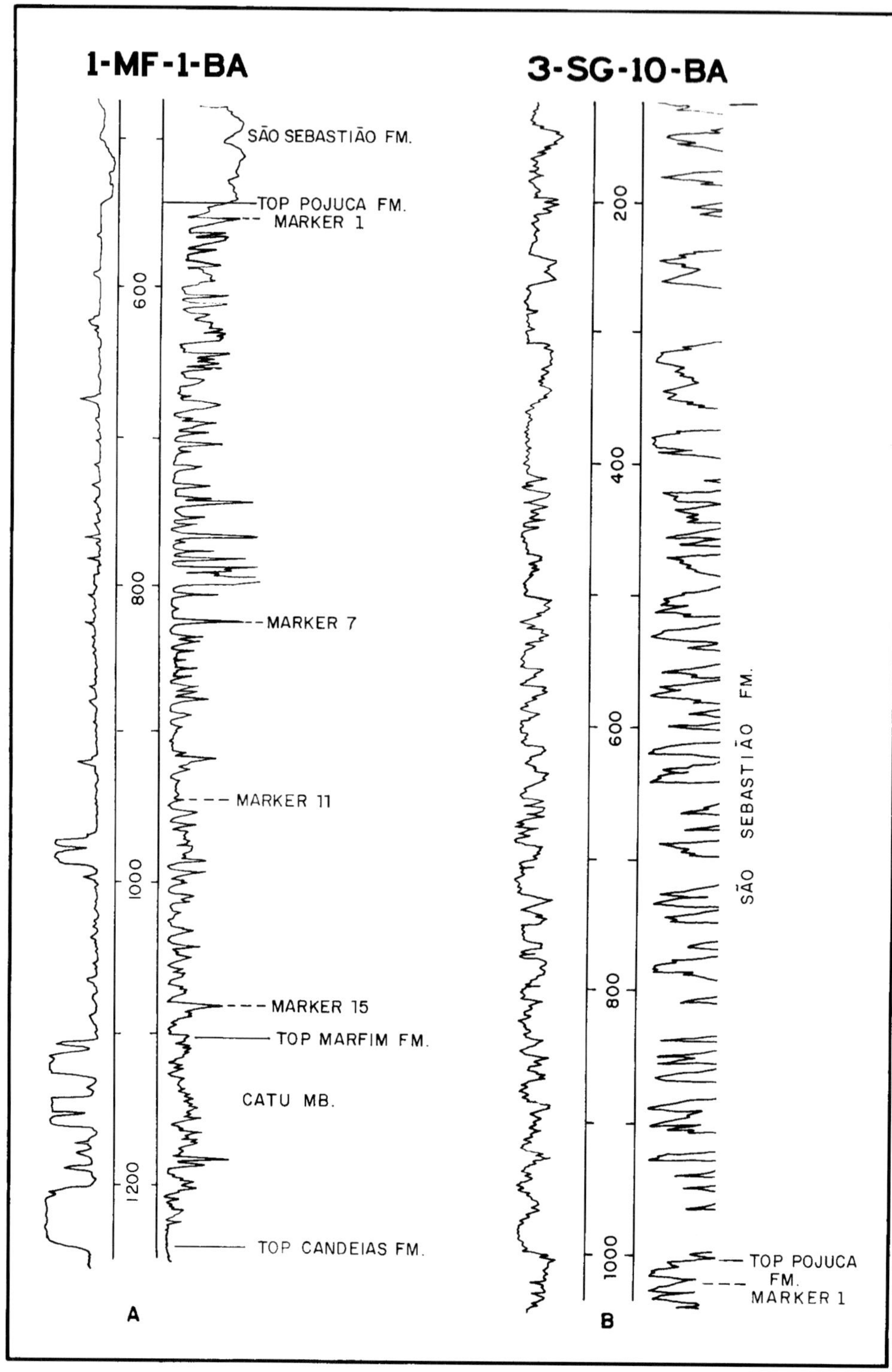

Figure 17. Typical electric logs of Marfim, Pojuca, and São Sebastião formations. A = Marfim (Catu Member) and Pojuca formations, with electric log markers; B = São Sebastião Formation.

The predominant source of Candeias sediments is located to the west-northwest in the southern and central compartments of the Recôncavo, whereas in the northeastern compartment an additional significant contribution is verified from the northeast and east.

Interdigitation of these sediments occurs in the clastic wedge of fanglomerates at the eastern edge of the basin. These fanglomerates belong to the Salvador Formation and will be described in a following section.

Marfim Formation

The Marfim Formation consists of light-gray and greenish, medium- to fine-grained sandstones, exhibiting current ripples and slump structures. Intercalations of greenish siltstones and lenses of greenish gray and brownish, micaceous, not very calcareous shales occur in its middle and basal parts. The Marfim is subdivided into the Caruaçu beds and the Catu Member (Figures 17A, 18).

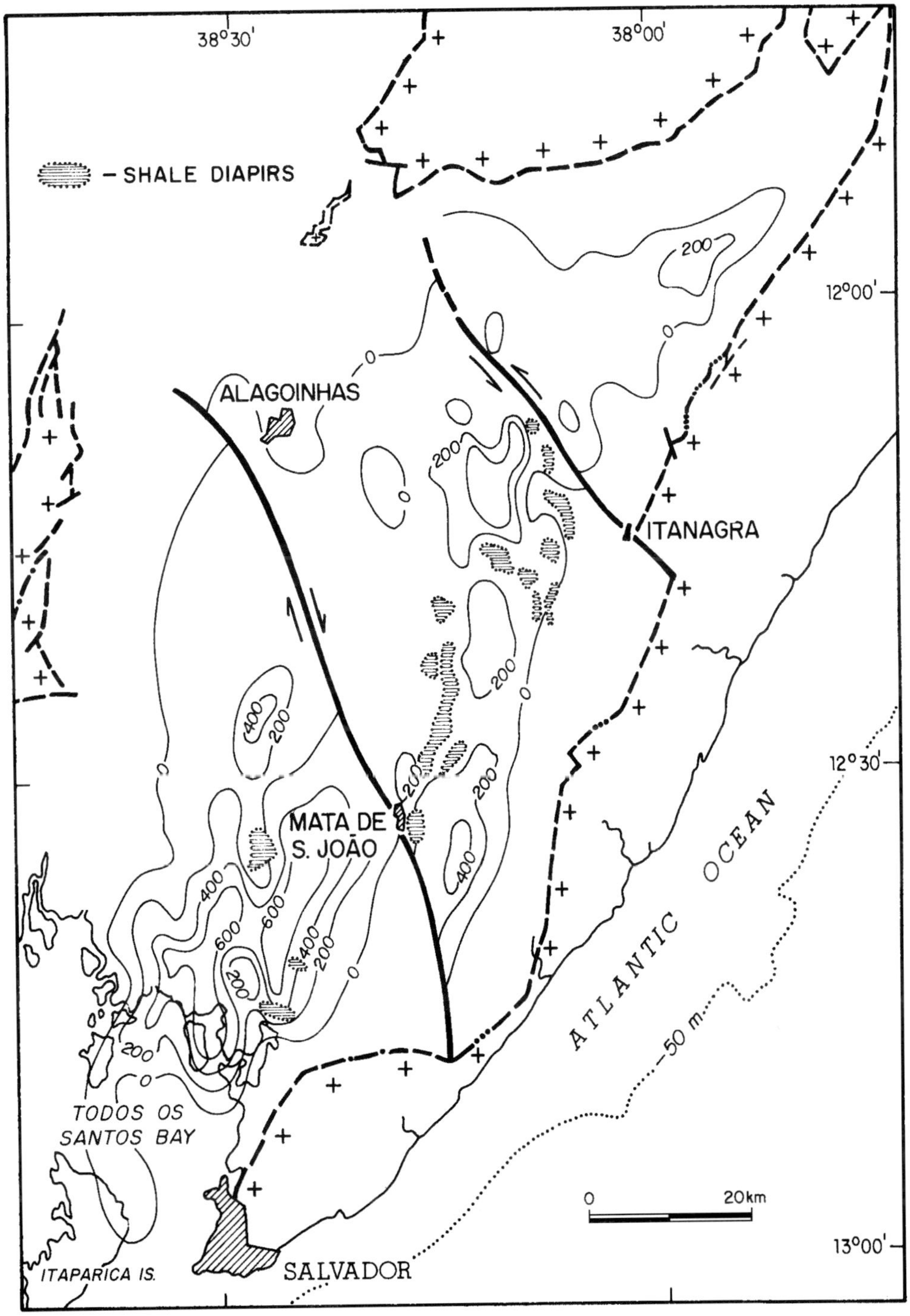

Figure 18. Isopach map of Caruaçu Beds, Marfim Formation, Recôncavo basin (Netto et al., 1984).

The Caruaçu beds consist of lenticular sandstones intercalated with the Maracangalha shales of the Candeias Formation. They underlie the Catu Member, from which they differ by their characteristic bell-shaped, often serrated SP curves. The isolith values of the Caruaçu sandstones reach more than 600 m in the most subsided areas of the basin (Figure 18). These sandstones were deposited in the transitional zone between the Marfim and Candeias formations and their outcrops to the south of the basin exhibit centimetric to decimetric layers with complete and incomplete Bouma sequences (where the most preserved facies is the Tbe) and load structures at the base of each cycle. Flute casts, convoluted beds, sandstones dikes, and, in some cases, antidunes are also frequent (Figures 19A, B, C).

The Catu Member occurs in the upper part of the Marfim Formation and is composed of light-gray, fine- to medium-grained, arkosic, well-sorted sandstones (Figure 19D). Its thickness is around 100 m,

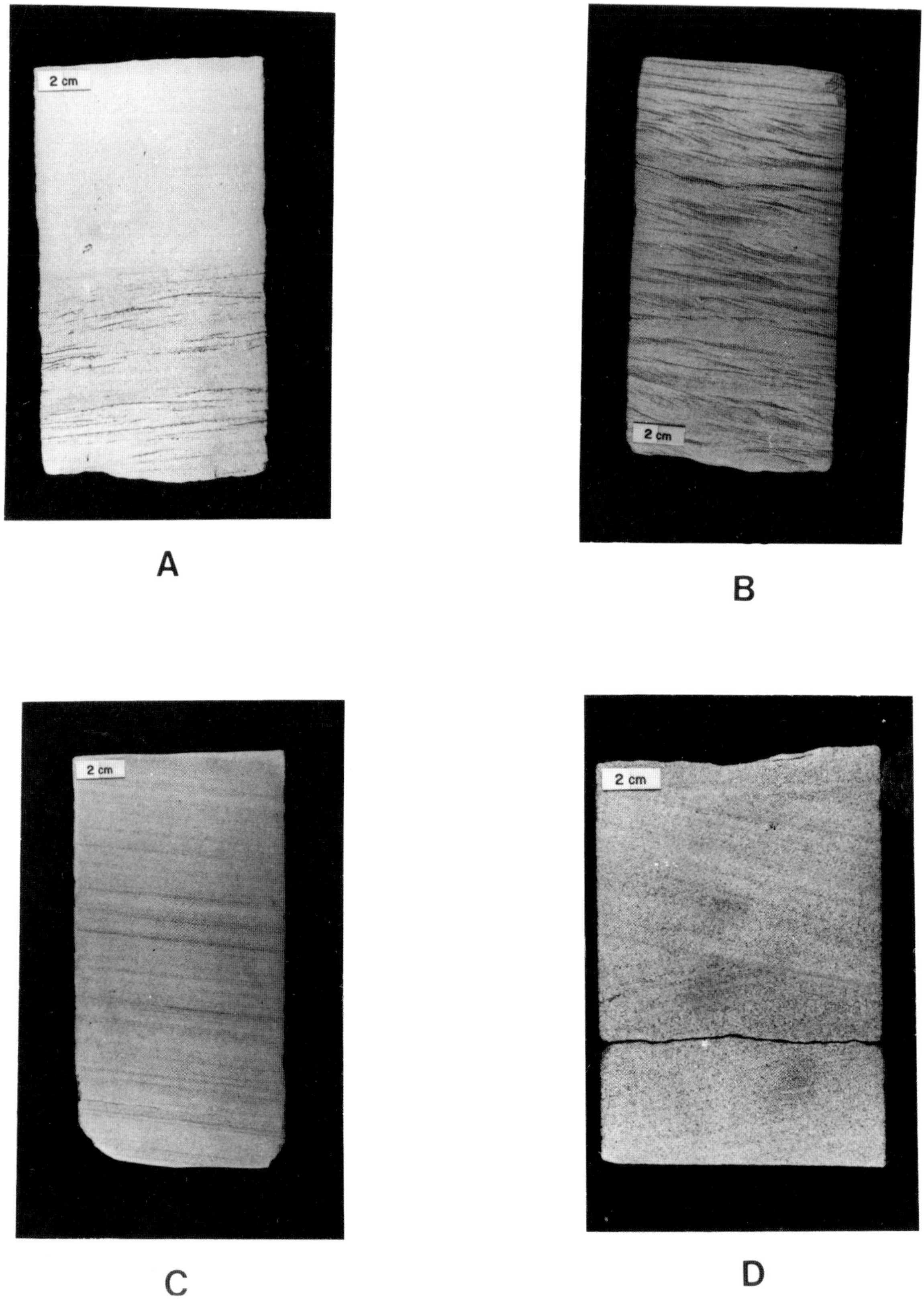

Figure 19. Marfim Formation core slabs. A = Very fine grained sandstone with climbing ripple stratification at base and parallel lamination at top; B = very fine grained sandstone with small-scale cross-stratification and climbing-ripple stratification; C = fine-grained sandstone with parallel stratification; D = Catu Member: Medium-grained sandstone with medium-scale trough cross-stratification.

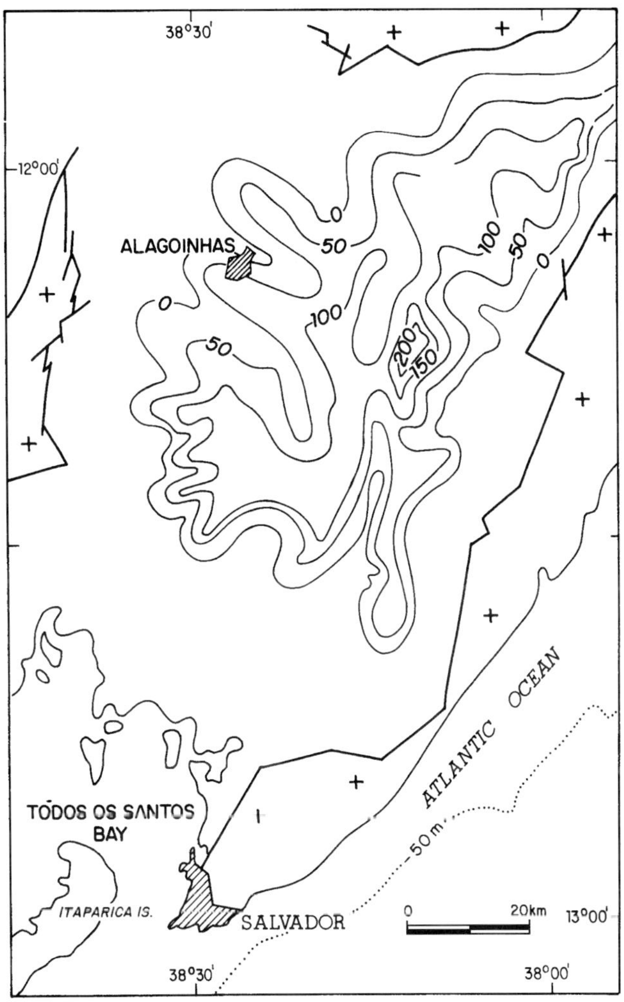

Figure 20. Isopach map of Catu Member, Marfim Formation, Recôncavo basin (adapted from Netto et al., 1984).

exceeding 200 m in areas of the central compartment that have experienced the most subsidence in the Recôncavo basin (Figure 20). The upper contact with the Pojuca Formation is conformable, whereas the lower contact with the remainder of the Marfim Formation and with the Candeias Formation is gradational. The Catu Member, with its geometry and box-like SP pattern, constitutes a record of the inception of a lacustrine delta system deposited within the basin. The deltaic lobes coalesced and formed a sheet covering two-thirds of the basin area (Figure 20).

Paleocurrent studies in outcrops of the Caruaçu beds in the southern Recôncavo indicate a source from the north. In the case of the Catu Member, a deltaic progradation toward the south is interpreted, with some lateral influence of the Salvador fault scarp fanglomerates (Figures 18, 20).

Pojuca Formation

The Pojuca Formation formed as a result of lacustrine deltaic cyclic sedimentation and is composed of sandstones, shales, siltstones, and limestones. The sandstones are whitish to greenish gray, fine- to medium-grained, well-sorted, and calcareous. The shales are gray to green, sometimes black, calciferous and fossiliferous, grading to calcilutites. The limestone beds are formed of ostracodal calcarenites and calcilutites (Figure 17A).

The lower contact of the Pojuca Formation is conformable with the Marfim Formation and, in some areas, with the Candeias Formation; laterally, at the eastern edge of the basin, the Pojuca undergoes interdigitation with the fanglomerates of the Salvador Formation and its distal sandstones (Sesmaria Member). In the southwest area of the basin, the Pojuca is partially truncated by the São Sebastião Formation.

The Pojuca Formation exhibits a uniform geometry in platform areas, reaching a thickness of about 500 m; in basin areas it exceeds a thickness of 1000 m. Deltaic sedimentation progressively filled in the basin, but was periodically interrupted by phases of inundation. During these phases, shales and thin beds of ostracodal limestones were deposited; these presently correspond to the correlation markers of the Pojuca Formation, numbered from 1 to 15 (Figure 17A). In outcrops and cores are records of root marks and mud cracks in the clay-rich part and small-scale cross-bedding passing to climbing ripples in the sandy part, evidence of the predominance of a shallow-lacustrine environment. Some outcrops in the basin exhibit progradational structures (sigmoidal lobes) typical of homopycnal inflow, which characterize a delta-front setting (Figures 21A, B).

During deposition of the Pojuca Formation, filling of the basin proceeded from the northeast and north. The construction of a large polycyclic deltaic system, intercalated with lacustrine interludes, occurred in the northeastern and central compartments of the Recôncavo basin.

The prodeltaic environment was well developed in the southern compartment. At the eastern edge of the basin, the Salvador fault system showed continued rejuvenation, allowing the conglomerates of the Salvador Formation to pile up and interfinger with the deltaic sedimentation.

Taquipe Formation

The Taquipe Formation includes sediments that filled in a paleocanyon incised in the southern compartment of the Recôncavo. The Taquipe is elongated, displaying a length of 60 km and a variable width that increases toward the south (Netto and Oliveira, 1985). Close to the Todos os Santos Bay, it reaches a thickness of 600 m (Figure 22).

The formation consists of laminated gray shales, containing intercalations of marls. Lenses of very fine to medium-grained sandstones, locally rich in ostracod shells, also are common. The basal portion contains shale breccia and lenses of conglomerate composed of oncolite pebbles and shale clasts. The most common sedimentary structures are cross-stratification, plane-parallel bedding, fluidization, and bioturbation.

The basal contact of the unit is discordant with the Candeias, Marfim, and Pojuca formations, depending upon the greater or lesser excavation of the paleo-

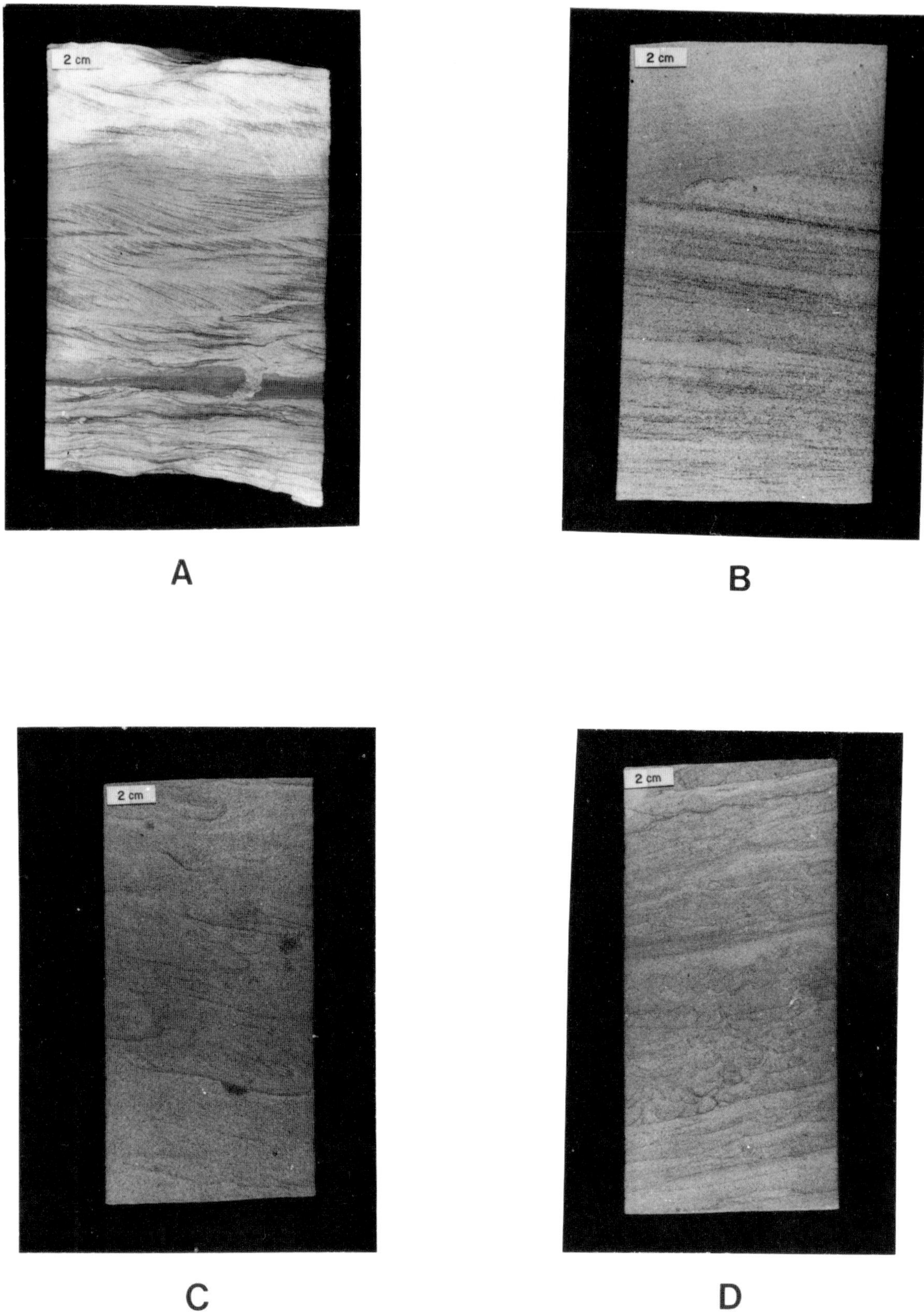

Figure 21. Pojuca and Taquipe Formation core slabs. A = Pojuca Formation: very fine grained sandstone with climbing-ripple stratification; B = Pojuca Formation: fine-grained sandstone exhibiting tabular cross-stratification (sigmoidal feature); C and D = Taquipe Formation: fine-grained sandstone disturbed by water-escape structures.

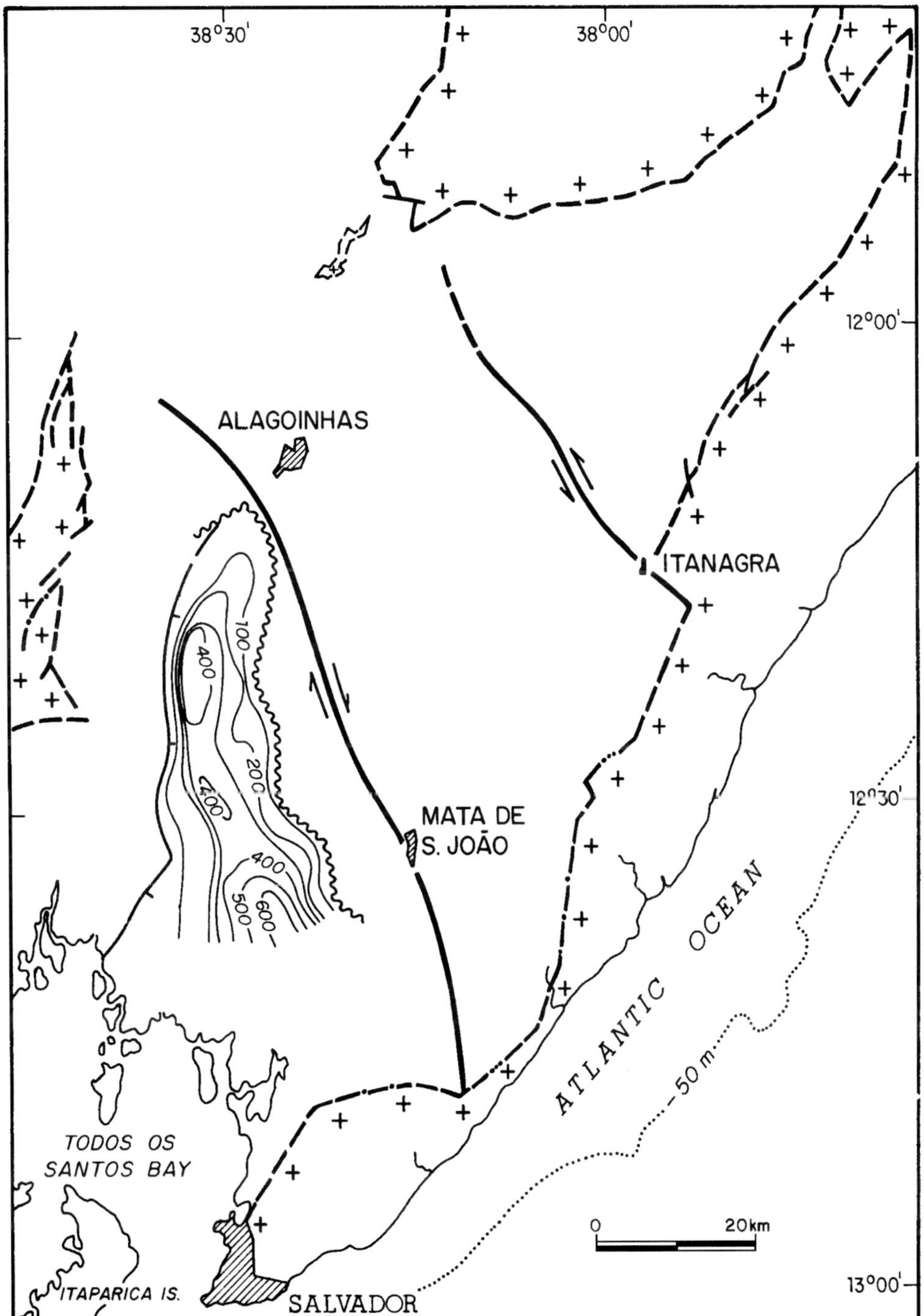

Figure 22. Isopach map of Taquipe Formation, Recôncavo basin (adapted from Netto et al., 1984).

canyon. Its upper contact is transitional with the upper one-third of the Pojuca Formation. Its major source is the head of the canyon to the north, but the walls of the canyon itself also make a substantial lateral contribution.

São Sebastião Formation

The São Sebastião Formation consists of medium- to coarse-grained, yellowish and reddish, friable, quartzose, and feldspathic sandstones, with intercalations of brown, purple, and red silty clay. The sandy bodies exhibit good lateral continuity and constitute an excellent freshwater source. The shales in the basal portion are greenish gray and dark-gray to black, passing to cyclic intercalations of reddish and brownish clays in the upper two-thirds of the formation.

The basal contact of the São Sebastião with the Pojuca Formation is gradational (Figure 17B), whereas its top displays an unconformity that was later covered by the Aptian Marizal Formation and by the

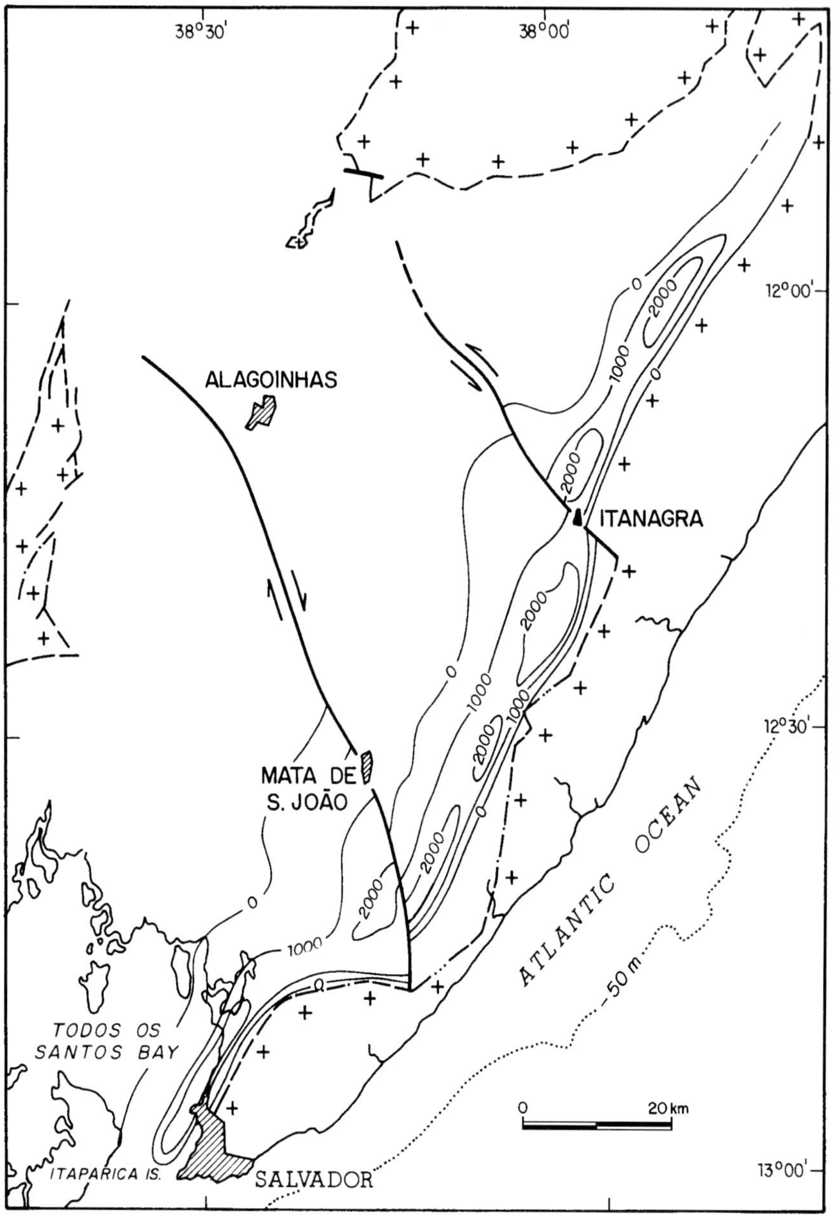

Figure 23. Isopach map of Salvador Formation, Recôncavo basin (adapted from Netto et al., 1984).

Pliocene Barreiras Formation. The presence of good electric log markers, well tied to ostracodal biozones, helps correlation within the São Sebastião Formation. Its maximum thickness is 1900 m, which is reached in the Alagoinhas low.

During deposition of the São Sebastião Formation the deltaic plain evolved into a fluvial plain. As in the case of the Pojuca Formation, the fluvial cycles of the São Sebastião were interrupted by periods of inundation, which were responsible for the deposition of clay-rich beds providing good regional correlation. Braided longitudinal bars, as well as point bars of the fluvial environment, were preserved. Paleocurrent studies in basin outcrops indicate a sedimentary source from the north and northwest.

Salvador Formation

The Salvador Formation consists of polymictic conglomerates, composed of granules, pebbles, and boulders of granulitic rocks, limestones, quartzites, sandstones, and shales, disseminated in a silty-argillaceous matrix. The proportion of gravel decreases from east to west, grading into the distal sandstones designated as the Sesmaria Member.

The Salvador Formation occurs in a 10-km-wide strip and interfingers with sediments of all rift-phase units. The conglomeratic alluvial fan complex, stacked along the entire eastern edge of the basin, reaches a thickness of about 2000 m wedging to 0 toward the west (Figure 23).

Marizal Formation

The Marizal Formation consists of conglomerate with medium- to coarse-grained matrix, with phenoclasts of gneiss in a sandy matrix. Its thickness ranges from 20 to 30 m both in the Camaçari (southern compartment) and Alagoinhas (central compartment) lows, whereas in other portions of the Recôncavo, its thickness is somewhat reduced. These sediments filled paleovalleys in the São Sebastião Formation, which were sculptured by Aptian erosion.

Sabiá Formation

The Sabiá Formation is composed of marine shales, sandstones, and limestones of early Miocene age. It occurs only in the area of the Miranga low and represents a marine transgression preserved in this depression.

Barreiras Formation

The Barreiras Formation is a basal stratified conglomerate, composed of rounded granules of milky quartz, altered gneiss, and abundant pebbles of clays and sandstones, as well as intercalations of poorly sorted coarse-grained sandstones of various colors, red predominating over yellow. White, purple, and yellow to reddish and purplish clays are also present.

In the Recôncavo basin, sediments of the Barreiras Formation were deposited discordantly over all units of the Bahia Supergroup and of the Marizal and Sabiá formations.

Pollen analysis and the study of 38 species of Dicotyledonease confirm the assignment of a late Tertiary age (Pliocene) to this typical continental section.

Biostratigraphy

The biostratigraphy of the Recôncavo is based on nonmarine ostracodes and palynomorphs. Other fossils in the sedimentary section—such as small Gastropoda, Conchostraca, fish remains, etc.—do not favor detailed work, but may be used as ecologic indicators.

Ostracoda (Phyllum Arthropoda: Class Crustacea; Subclass Ostracoda)

The ostracodes of the Recôncavo basin are benthonic and belong mainly to the suborder Podocopina (Sars, 1866) and to the families Cyprididae (Baird, 1845) and Cytheridae (Baird, 1850). In addition, the local genera *Petrobrasia, Reconcavona, Salvadoriella*, and *Ilhasina* were described by K. Krömmelbein in 1958.

Figure 24 shows the nine zones and 26 subzones that constitute the biostratigraphy of the Recôncavo-Tucano basin (NRT). Elaboration of these units was based on the stratigraphic behavior of 112 species, 35 of which are diagnostic.

Palynomorphs

A palynomorph superzone is recognized (PC-108), divided into five zones. Figure 24 also shows the relationship between palynological and ostracodal zonation.

Chronostratigraphy

The first paleontologic dating of sediments in the Recôncavo basin was based on fish, shellfish, and Crocodylidae(?) remains, which were considered as belonging to the Neocomian (Lower Cretaceous), Wealden stage.

The ostracodal fauna, whose stratigraphic value was discovered much later, also belongs to the Neocomian (Wealden stage) and Purbekian (Upper Jurassic), exhibiting identical taxon forms in Germany, Gabon, Congo, and Zaire.

The chronostratigraphic units as a whole constitute the Recôncavo Series and are based on biozones (Oppel's zones) defined by ostracodes, many of them coincident with lithologic markers. The chronostratigraphic units are divided into five stages, designated by local names. From the oldest to the youngest, these are Dom João, Rio da Serra, Aratu, Buracica, and Jiquiá (see Figures 8, 24). The limits of the Dom João stage coincide with those of the Brotas Group, that is, the base of the Aliança Formation and the top of Sergi Formation. Biozone NRT 001 is restricted exclusively to the shales at the top of the Aliança Formation (Capianga Member).

The Rio da Serra stage has its lower limit at the top of the Sergi Formation, which is also the base of biozone NRT 002, and its upper limit corresponds to the excellent marker 15, also the upper limit of biozone NRT 004. The Rio da Serra includes the Itaparica, Candeias, and Marfim formations. In some cases, the latter is absent and the Candeias Formation can extend to the upper stages. This stage may also contain the base of the Pojuca Formation, and its maximum known thickness exceeds 2900 m, reaching more than 3500 m in depositional lows.

The Aratu stage encompasses sediments deposited between marker 15 and the upper level, known as marker 1, which includes the Pojuca and Taquipe formations and, sometimes, in the southern compartment, the base of the São Sebastião Formation. The Aratu's thickness reaches more than 1000 m in depositional lows, and its strata include two biozones, designated NRT 005 and NRT 006.

The Buracica stage may be characterized by two biozones, NRT 007 and NRT 008. It usually includes sediments of the São Sebastião Formation and sometimes those of the top of the Pojuca Formation. The lower and upper limits are indicated by electric log markers.

The Jiquiá stage contains the upper portion of the São Sebastião Formation and coincides with biozone NRT 009. In the Recôncavo basin, the Jiquiá is restricted to the Camaçari and Alagoinhas lows.

BASIN EVOLUTION

The evolution of the Recôncavo basin is closely related to the initial stages of the rupture of the Gondwana paleocontinent and the formation of the South Atlantic Ocean (Ponte, 1971; Asmus, 1975).

The area presently occupied by the basin initially was located over the axis of an extensive geanticline,

Period	Series	Local Stage	Ostracode Code		Ostracode Zone	Ostracode Subzone	Palynomorph Code	Palynomorph Superzone	Palynomorph Zone	Palynomorph Code
EARLY CRETACEOUS	RECÔNCAVO	JIQUIÁ	NRT 009	3	Petrobrasia diversicostata	C. (Pseudocypridina) faveolata	P – 100	CEDRITITES ? SP.	Stellatopollis S.P. 1	P-190
				2		C.(S.) devexa				
				1		B. postangularis				
		BURACICA	NRT 008	3	Cypridea (Sebastianites) fida minor	C.(S?) miro			Dicheiropollis etruscus	P-180
				2		Petrobrasia capivarensis			Caytonipollenites Pallidus	P – 160
				1		C.(S?) sostensis sostensis				
			NRT 007	4	Coriacina coriacea	Paracypridea quadrirugosa Weberi				
				3		"Metacypris" sp. 6				
				2		Petrobrasia marfinensis				
				1		Cypridea eminens				
		ARATU	NRT 006	2	Cypridea (Morinina?) bibullata bibullata	Cypridea (Morinina?) bibullata tribullata				
				1		Reconcavona triebelli				
			NRT 005	5	Paracypridea obovata obovata	Reconc. gastracantha				
				4		Reconc. uniacantha				
				3		Reconcavona uncinata				
				2		Paracypridea elegans elegans				
				1		Ilhasina remanei cuneiformis				
		RIO DA SERRA	NRT 004	5	Paracypridea brasiliensis	Cypridea ventronodata			Leptolepidites major	P–140
				4		Paracypridea maacki				
				3		Paracypridea bicallosa				
				2		Reconcavona? polita				
				1		Cyp. salv. nodifer				
			NRT 003	2	Cypridea (Morininoides) candeiensis	Cypridea (Morininoides) hadronodosa			Allisporites ? S.P.	P–120
				1		Cypridea sellata				
			NRT 002	2	Theriosynoecum varietuberatum	Cypridea cf. C primaria				
				1		Cypridea kegeli				
LATE	JURASSIC	D.JOÃO	NRT 001		Bisulcocypris pricei					

Figure 24. Correlation chart of ostracode and palynomorph zones, Recôncavo basin (adapted from Regali and Viana, 1989).

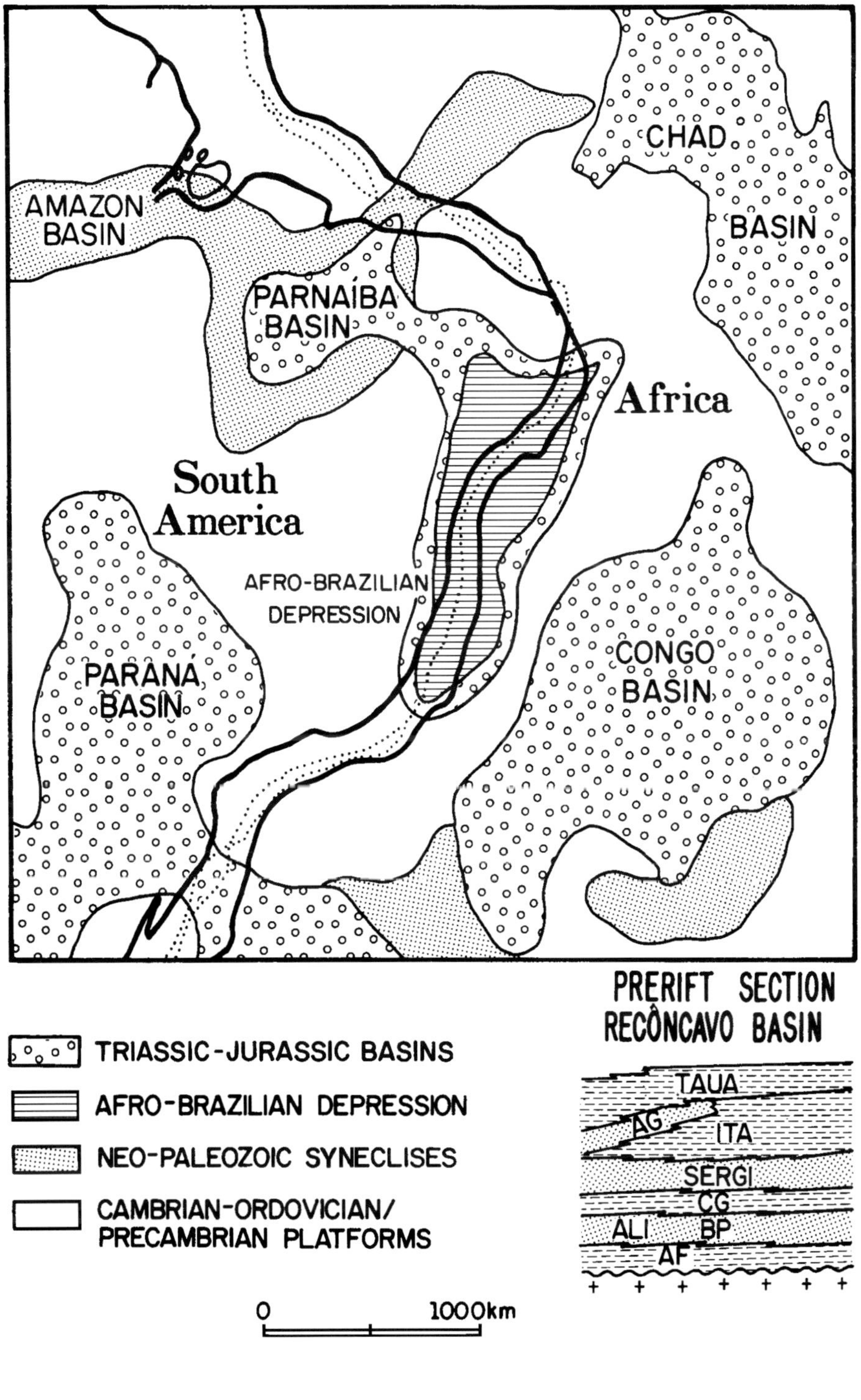

Figure 25. Paleogeologic reconstruction of Afro-Brazilian Depression in the Late Jurassic (adapted from Ponte, 1971).

developed probably during the end of Paleozoic, and extended for nearly 1500 km. During this uplift, almost all Paleozoic rocks were eroded, with only a few sparse remnants of these sediments having been preserved (Estrella, 1972).

At the end of the Jurassic, a shallow, narrow north-south–trending depression, designated the Afro-Brazilian depression, was formed along the crest of this geanticline (Ponte, 1971). This depression (Figure 25) received most of the sedimentary material eroded off the uppermost portions of the geanticline. These sediments were deposited by a system of coalescing, arid alluvial fans, which are represented in the Recôncavo basin by the red shales, evaporites, and sandstones of the Brotas Group (Aliança and Sergi formations), believed to have been deposited over present rift limits (Ghignone, 1979). Sedimentation in the Afro-Brazilian depression continued until the beginning of the Rio da Serra stage (Santo Amaro Group) and produced the fluvial and lacustrine sedimentation of the Itaparica Formation and the Tauá Member of the Candeias Formation (Figure 26A).

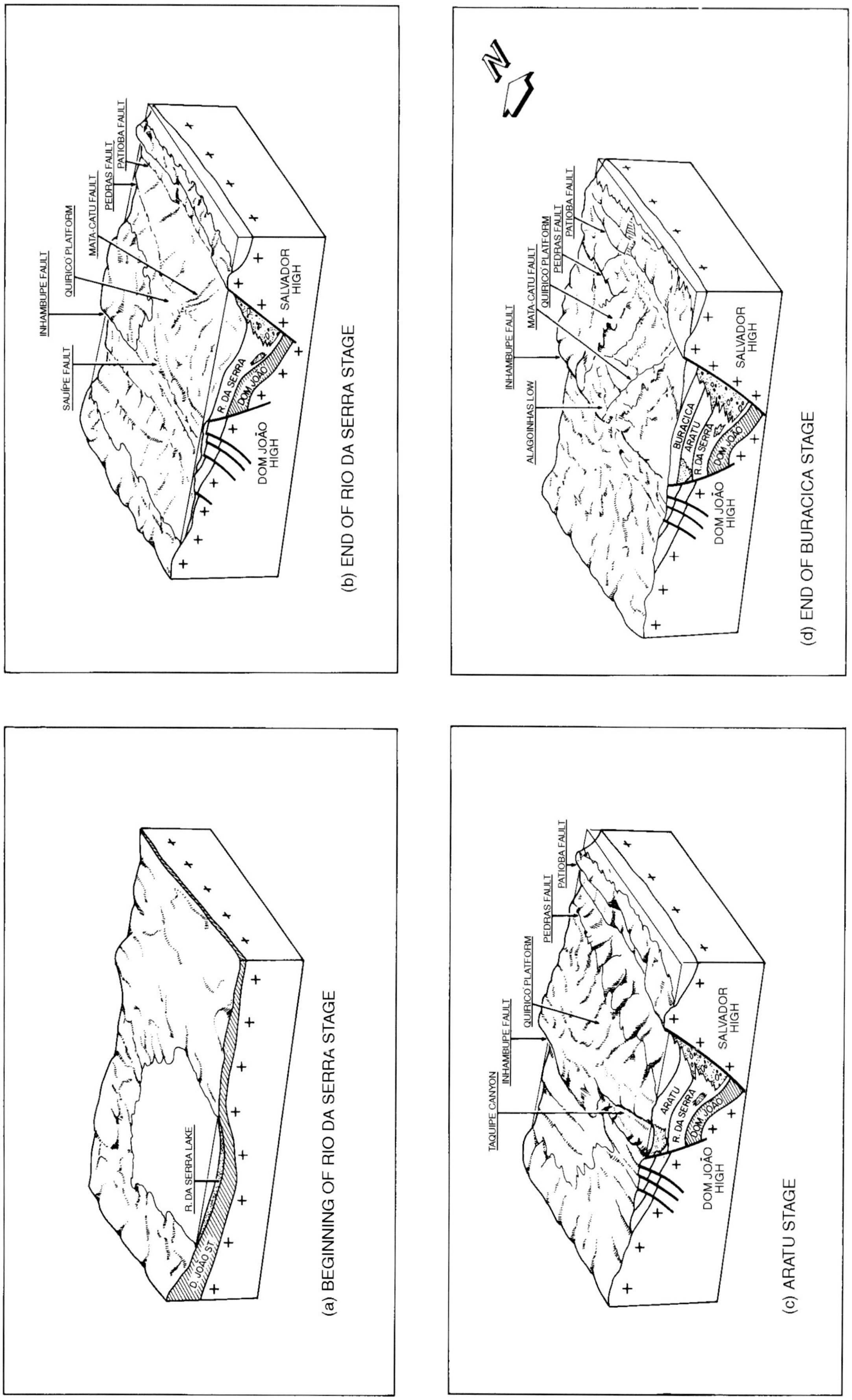

Figure 26. Schematic geologic evolution of Recôncavo basin. A = Beginning of Rio da Serra stage; B = end of Rio da Serra stage; C = Aratu stage; D = end of Buracica stage.

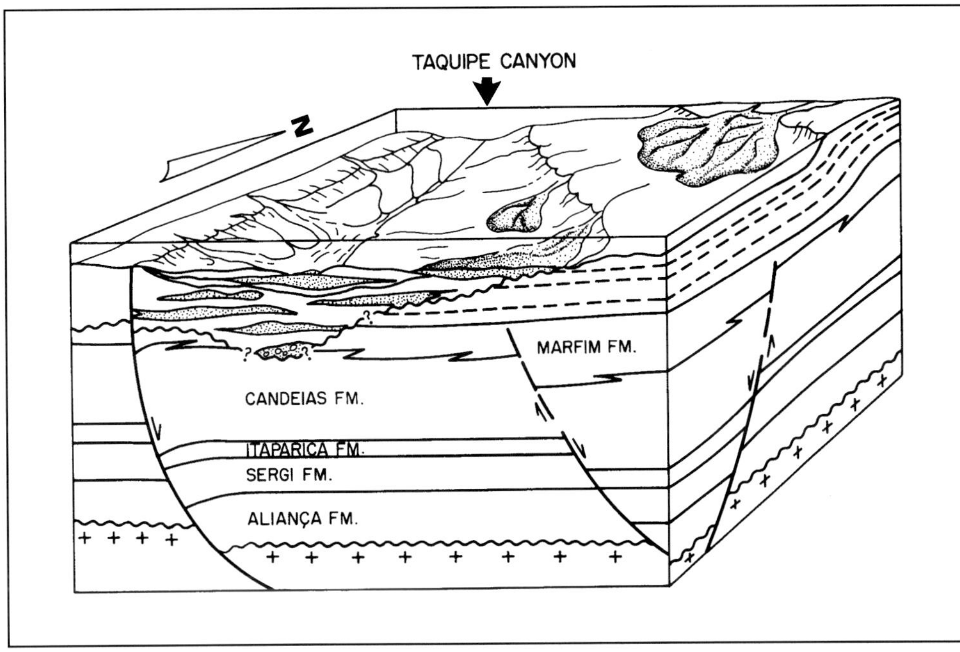

Figure 27. Tectonic and depositional model of Taquipe Canyon, Recôncavo basin.

The fluvial plain, where sandstones of the Água Grande Member of the Itaparica Formation were deposited, remained exposed long enough for an eolian plain to develop at the beginning of the Early Cretaceous. Subsequently, this arid surface sank slowly and was inundated following the establishment of a more humid climate. Dark clays and muds of the Tauá Member of the Candeias Formation were deposited in that shallow, marshy, limnic environment.

During the middle Rio da Serra stage, the rift underwent its true evolution, with the development of the East Brazilian microplate, as described by Milani (1984). Because the sedimentation rate of the Gomo Member was much lower than the subsidence rate of the basin, the Recôncavo lake grew deeper and deeper. At that time, the physiography already displayed the Quiricó and Patioba shallow platforms, as well as the Camaçari, Miranga, and Quiambina lows and intermediate platforms.

At the end of the Middle Rio da Serra stage (Figure 26B), the previously mentioned platforms and lows were affected by tectonism. Episodic seismic activity prompted the remobilization of thick masses of platform silty sandstones (Pitanga Member) of the Candeias Formation and simultaneously triggered the argillokinesis of the basin shales of the Maracangalha Member. The diapirism of these shales occurred primarily in the most subsided areas of the central and southern compartments of the basin, and the close relationship between these diapirs and the occurrence of thick bodies of massive sandstones is very well documented.

During the Middle Rio da Serra stage, conglomerates began to be stacked along the Salvador fault at the eastern edge of the basin, resulting in a clastic wedge designated the Salvador Formation (Figure 26D).

In the Late Rio Serra stage, when the lake was still moderately deep, high-density turbidity currents deposited the sandstones of the Caruaçu beds (Marfim Formation), intercalated with basin shales of the Candeias Formation. The shallowing of the basin permitted progradation of the deltaic system of the Catu Member, marking the beginning of a new depositional style, which prevailed throughout the entire Aratu and Buracica stages. The sandstone bodies of the Marfim and Pojuca formations were deposited as a set of essentially sand-rich deltaic systems, with tabular geometry and reasonable lateral continuity, interrupted by lacustrine transgressive cycles. Each of these cycles was marked by the deposition of argillaceous or calcareous highly fossiliferous horizons, which are widely distributed across the basin and constitute Pojuca Formation marker beds.

In the subsiding lows of the Aratu stage, the cyclic sedimentation of the Pojuca Formation and the Catu Member of the Marfim Formation exhibit thickness variations that were controlled by growth faults. The best examples are located in the Miranga low (see Figure 7B).

During the Aratu stage, the greatest interruption of sedimentation occurred prior to the deposition of marker 7 (Pojuca Formation). Because of a sudden variation of the base level and/or a major tectonic event along the eastern border of the Dom João high, favorable conditions for the development of subaqueous erosional events were created in the southern compartment of the basin (Figure 26C). This excavation formed a canyon that was soon filled by sediments during the upper Aratu stage (NRT 005.4); these sediments were designated the Taquipe Formation (Figure 27). In the remaining areas of the basin not affected by erosion, the deltaic sedimentation of the

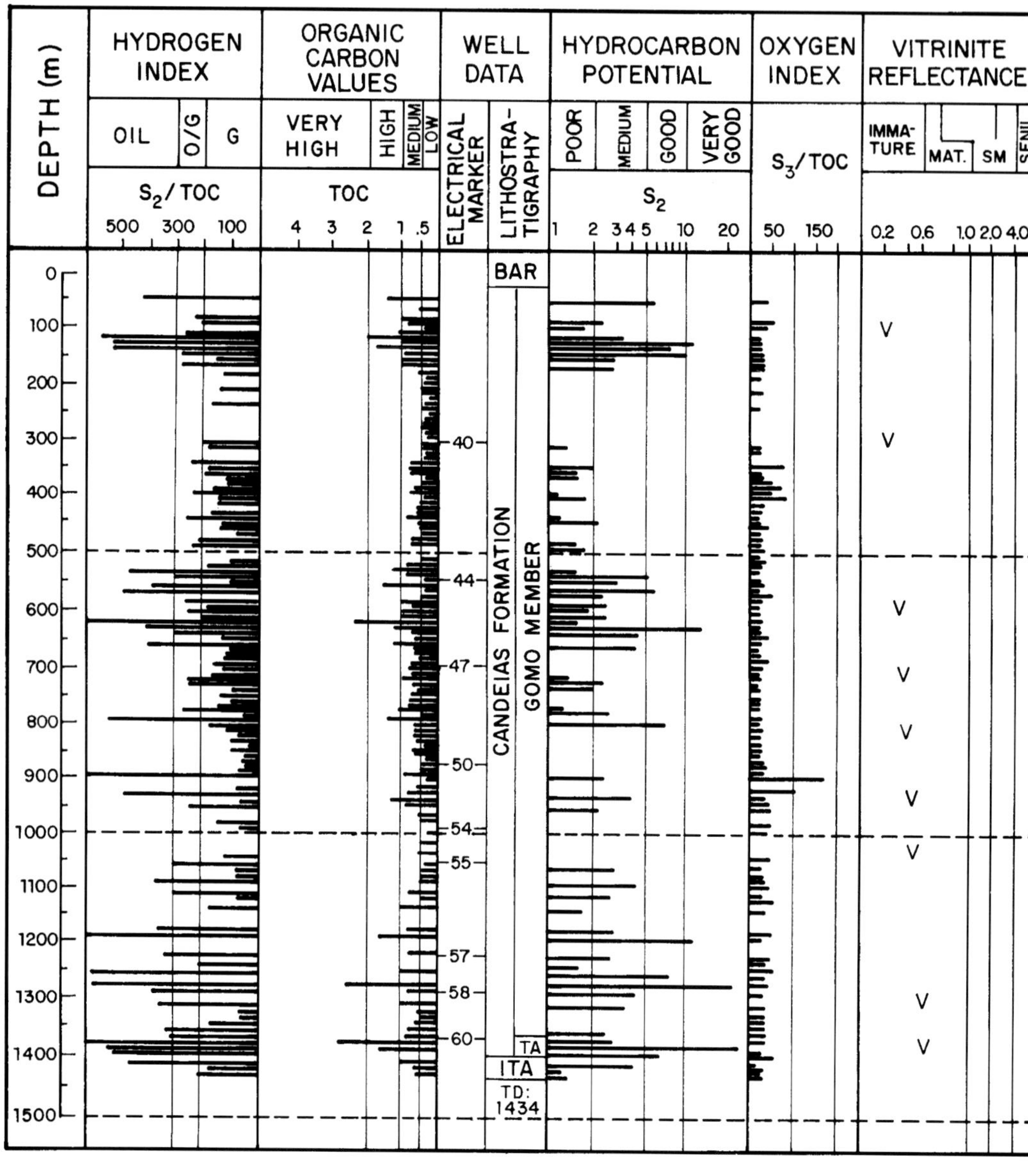

Figure 28. Typical geochemical log of Recôncavo basin.

Pojuca Formation proceeded normally and the Alagoinhas low in the central compartment was created at the end of the Aratu stage.

At the beginning of the Buracica stage, the shallow basin was dominated by fluvial sedimentation, which was periodically interrupted by floodings, producing key layers characterized by easily correlated electric log markers. The sediments of the São Sebastião Formation are very sandy, and beds are thicker and coarser than those of the Pojuca Formation. In addition, they exhibit a higher sandstone/shale ratio in the northeastern and central compartments of the basin, becoming more argillaceous to the south. The regions that had begun subsiding during the Rio da Serra stage were still active during the Buracica.

During and after deposition of Buracica stage sediments, a new phase of tectonic instability reshaped the structural framework of the basin. Fault reactivation accentuated basin subdivision along transfer fault zones and allowed sedimentation to occur in the Jiquiá stage. These deposits are still preserved in the Camaçari low.

OIL AND GAS SYSTEMS

Source Rock

Geochemical evaluation of the Recôncavo basin identified the shale section of the Candeias Formation (Gomo and Tauá members) as the main hydrocarbon source rocks of the basin. For the argillaceous sediments of the Gomo Member, average total organic carbon (TOC) content oscillates at around 1% in regional lows, whereas residual generating potential reaches an average of 5 kg HC/ton of rock in the deepest portions of the basin and, locally, more than 10 kg HC/ton of rock (Figure 28). Despite the present high TOC and

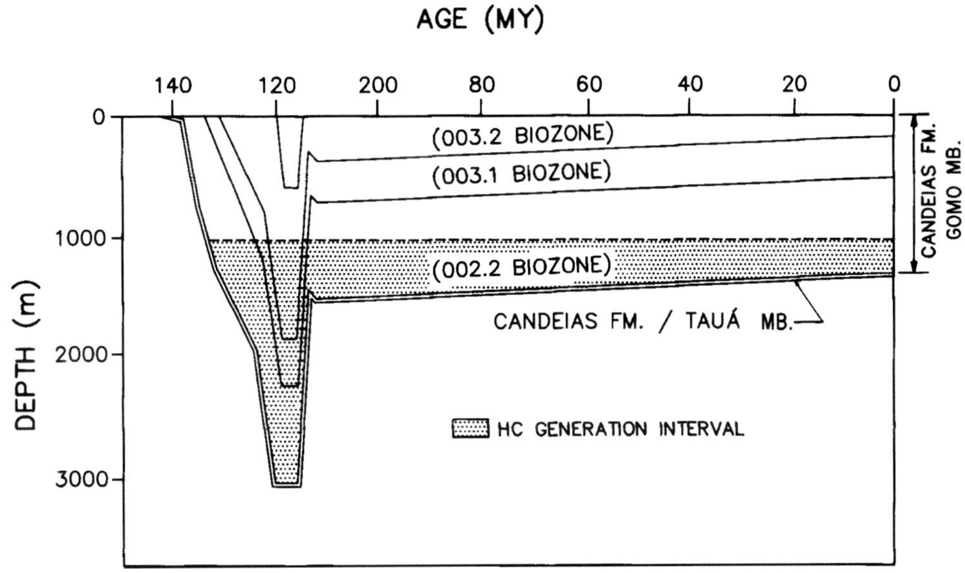

Figure 29. Typical burial diagram of source-rock section in northeastern subbasin, Recôncavo basin.

generating potential values, the Pojuca Formation shales are immature throughout the entire basin.

The low organic carbon content and relatively poor values obtained for generating potential are interpreted to be related to the high rate of convertibility of organic matter into hydrocarbons due to the advanced maturation stage of Candeias shales. The top of the mature zone is found at shallower levels at the edges of the basin than in depocenters. Close to the eastern edge, the top is even shallower due to higher thermal flow, probably caused by crustal thinning at the edge or by lateral heat loss from narrow rifts. In deeper depositional lows, source rocks may have reached senility.

Vitrinite reflectance data and thermal subsidence studies indicate that hydrocarbon generation in the Candeias Formation started in the early Aptian (Figure 29).

Migration and Accumulation Models

Three basic models of hydrocarbon migration and accumulation are proposed for the Recôncavo basin (Figure 30).

(1) Structural traps that are formed by tilted or nontilted horst blocks at the level of the prerift reservoirs; in these traps migration occurs laterally from shales of the Candeias Formation to the adjacent reservoirs of the upthrown block (Figure 30A).

(2) Stratigraphic or combined traps, such as the turbidite reservoirs of the Candeias and Marfim formations, that are in direct contact with source rocks or connected over short distances through open fault planes (Figure 30B).

(3) Structural inversions that are formed by growth faults (rollover structures) at the level of the deltaic Pojuca and Marfim reservoirs in regional lows. This model implies vertical and long-distance migration, through fault planes, of hydrocarbons generated in the pelites of the Candeias Formation (Figure 30C).

Diagenesis

Two main diagenetic sequences can be identified in Recôncavo basin reservoirs, associated with the prerift and synrift sandstones. The distinct diagenetic pathways followed by these two sets of reservoirs make it necessary to develop and describe their diagenetic histories separately. Most of the differences observed between the two paragenetic sequences result from the influence of their depositional environments. The prerift sandstones were deposited in an arid, subaerial environment, whereas those of the synrift section represent deep to shallow lacustrine sediments. The distinct paragenetic sequences of these two zones are discussed in the following sections.

Prerift Zone Diagenetic Sequence

The diagenetic sequences of the prerift section are shown in Figure 31, which was based essentially on data obtained from Sergi Formation reservoirs.

Silcretes, caliches, and dolomite occur as nodules or discontinuous beds associated with paleosoils, and represent an early diagenetic phase resulting from the subaerial exposure of sediments deposited under arid climatic conditions (Figure 32A).

Mechanically infiltrated detrital clays constitute either thick coatings of tangential lamellae surrounding grains or compact masses of lamellae plugging intergranular porosity (Figure 32B).

The original composition of clay coatings (smectite) was transformed into interstratified illite-smectite or illite during progressive burial. At an average burial depth of 300 m, the infiltrated clays are composed predominantly of smectites; at 750 m, illite-smectite

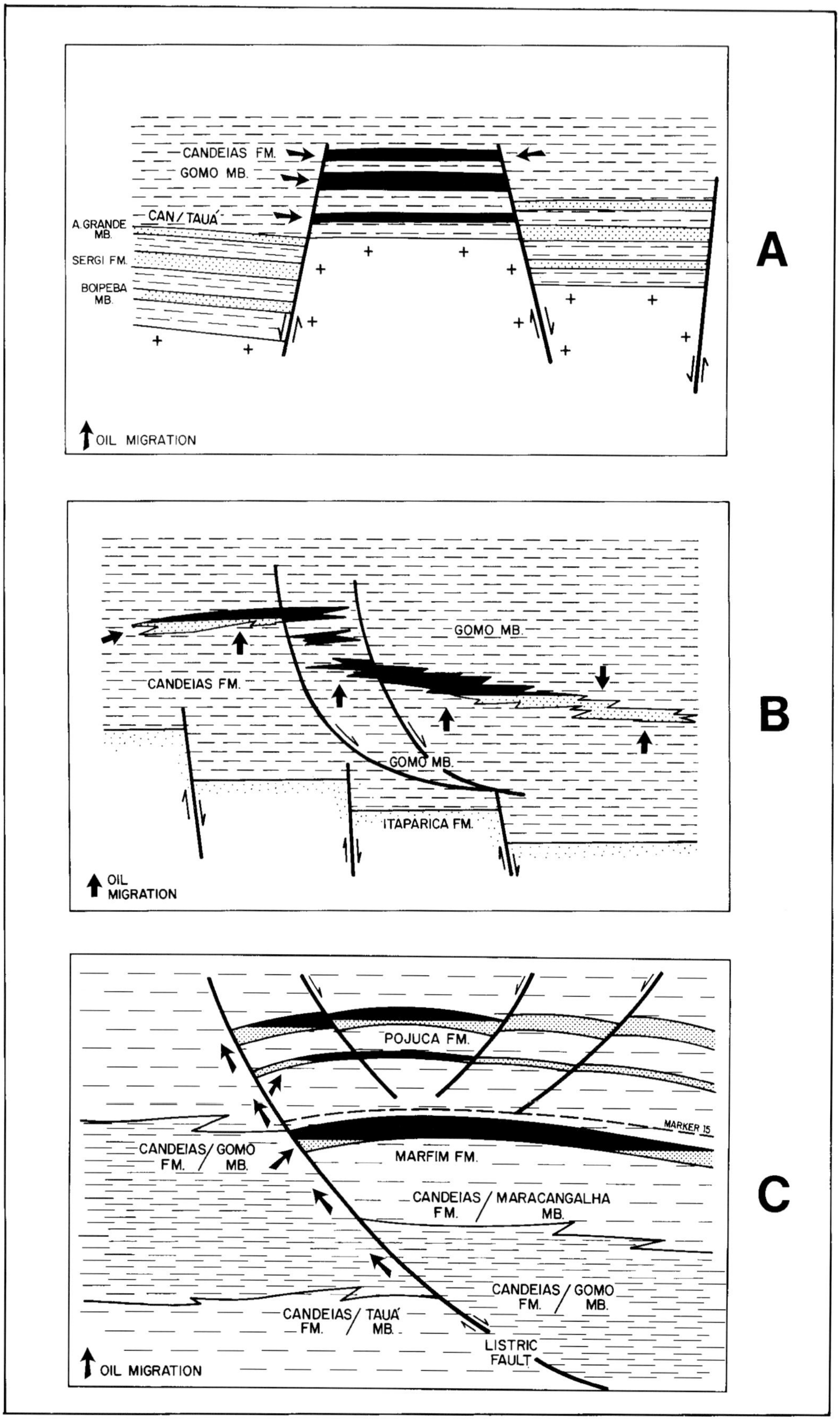

Figure 30. Schematic models of hydrocarbon migration and accumulation, Recôncavo basin. A = Structural traps; B = stratigraphic or combined traps; C = structural inversions.

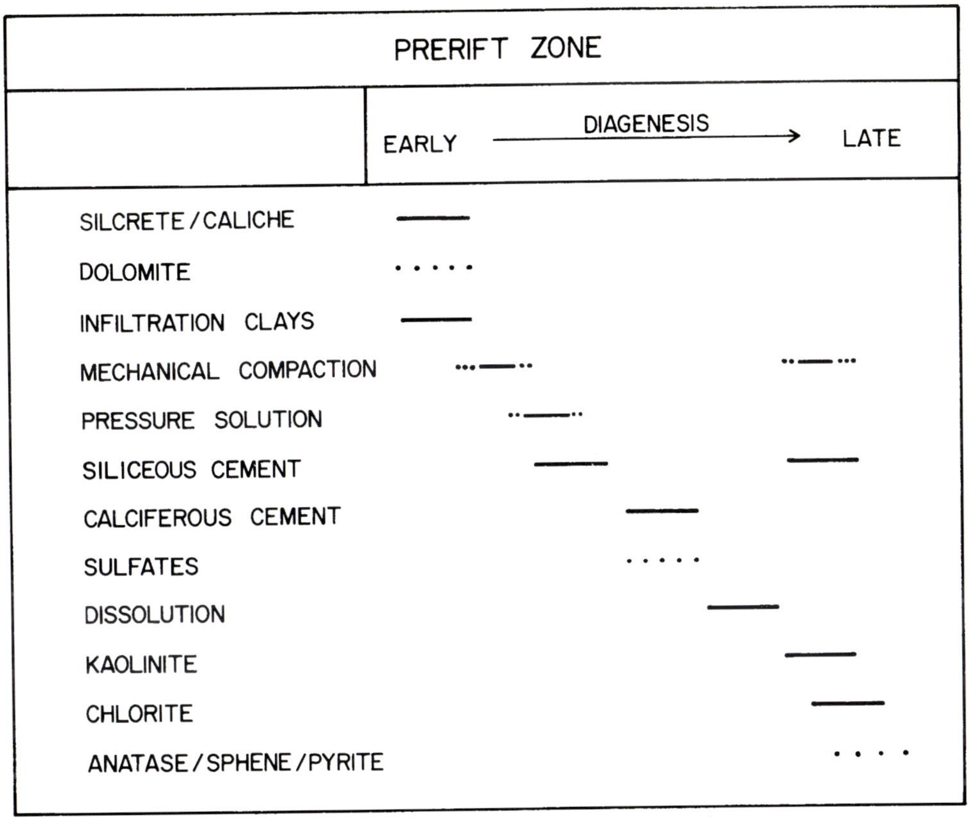

Figure 31. Schematic diagram of diagenetic sequence in prerift reservoirs, Recôncavo basin.

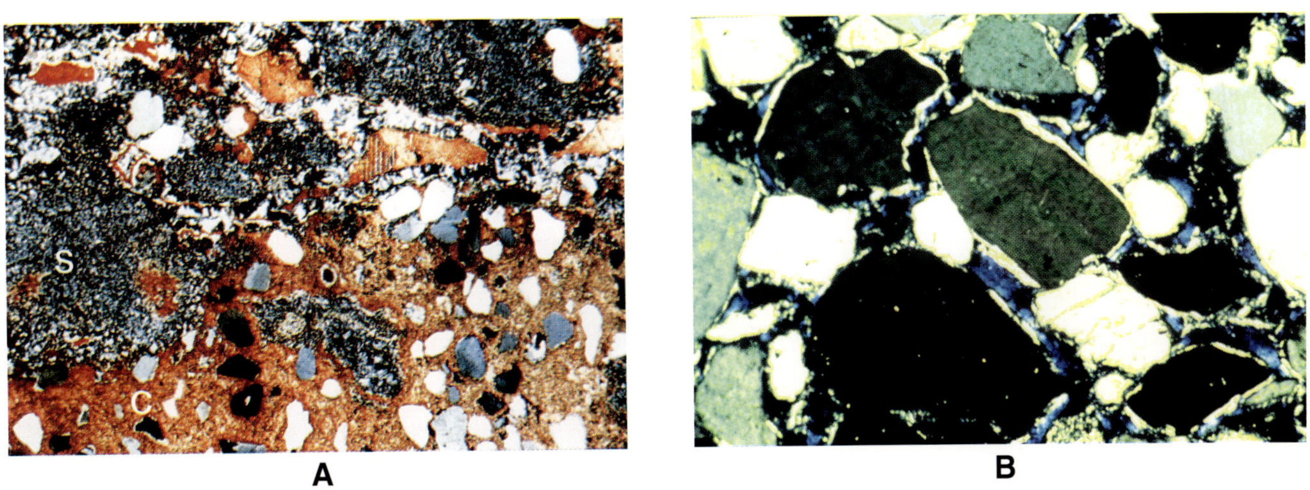

Figure 32. Photomicrographs: A = silcrete (S) and calcrete (C) crusts cementing "floating" grains of quartz and feldspar, Sergi Formation. Crossed nicols (field: 1.5 mm × 2.2 mm); B = thick coating of illite-smectite (pale yellow) surrounding framework grains, Sergi Formation. Crossed nicols (field: 0.7 mm × 1.1 mm).

mixed layers with an I40S60 composition predominate; and at 2700 m, illite-smectite mixed layers also predominate, but with an I80S20 composition (Netto et al., 1982).

These infiltration clays are characteristic of the upper portions of the Sergi Formation. They are associated with coarse-grained facies and represent 7.6% of total rock volume on average.

The origin of these clays is related to the occurrence of episodic torrents, typical in areas with arid climates, to fluctuations of the water table, and to evaporation in the vadose zone, with active infiltration of colloidal solutions and the precipitation of low-crystallinity smectites.

Secondary overgrowths of quartz and feldspar that developed in optical continuity with the host grains

Figure 33. Photomicrographs: A = quartzarenite cemented by quartz overgrowths (Q), Sergi Formation. Crossed nicols (field: 1.5 mm × 2.2 mm); B = calcite cement (C) occupying intergranular pores and replacing grains and quartz (Q) and feldspar overgrowths, Sergi Formation. Crossed nicols (field: 2.4 mm × 3.5 mm).

Figure 34. Photomicrographs: A = secondary porosity (in blue) developed by partial dissolution of a feldspar grain, Sergi Formation. Parallel nicols (field: 0.3 mm × 0.4 mm); B = authigenic chlorite (Cl) filling most of sandstone porosity, Sergi Formation. Crossed nicols (field: 0.4 mm × 0.6 mm).

are common. The distribution of quartz and feldspar cements is very uniform throughout the Sergi Formation, with average content ranging between 1.0 and 2.8%; however, in some thin levels, cement content reaches up to 16% of total rock volume. Even in small amounts, secondary growths effectively reduce reservoir permeability, because their morphology results in preferential obliteration of pore throats (Figure 33A).

Calcite cement occurs as a coarse mosaic or presents a poikilotopic texture, occupying the intergranular pores and replacing grains and quartz/feldspar overgrowths (Figure 33B). The average content of calcite cement is 2.7% in the Sergi Formation, locally reaching 35%. Calcite tends to display higher content in those levels that present a low amount or absence of infiltrated clays.

Sulfates (barite and anhydrite) occur in a few samples, reaching up to 1.0% of total rock volume.

In the development of secondary porosity, acid solutions produced by the maturation of organic matter in rift section shales are believed to have been responsible for the dissolution of calcite cement and feldspar grains, increasing reservoir porosities at greater depths. This dissolution occurred heterogeneously, producing reservoirs with a very complex internal geometry due to the presence of horizons containing remnant calcite (Figure 34A).

Total porosity measured in thin sections of Sergi Formation sandstones averages 16–17%. Note that intragranular porosity represents 20–40% of total porosity. Highest percentages are related to sandstones located at greater depths, indicating that intensity of dissolution increases with depth.

On the average, authigenic clays amount to 1.2% of total rock volume, locally reaching up to 17% (Netto et al., 1982). They occur as discrete, disseminated

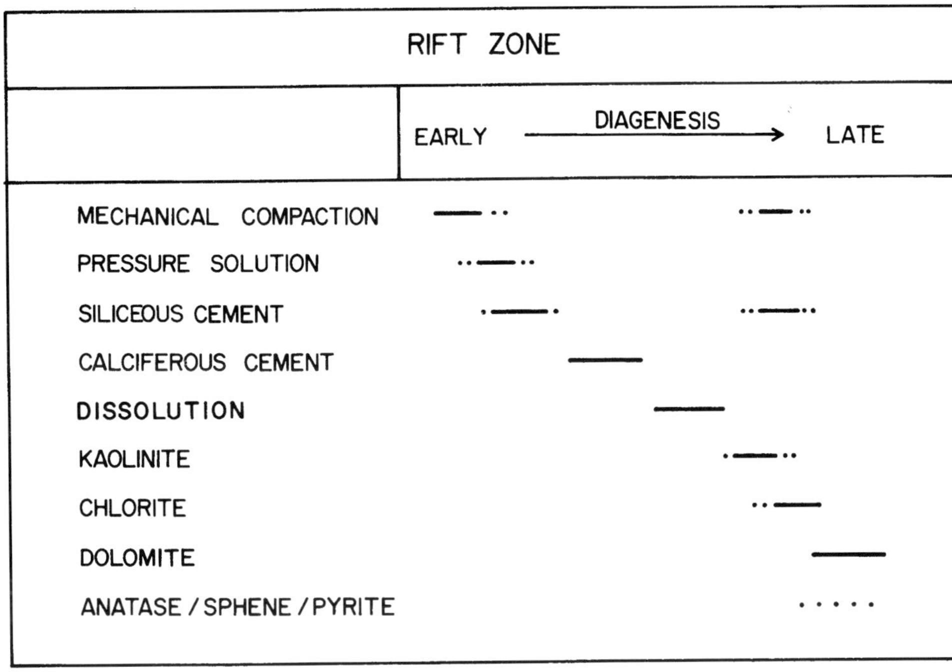

Figure 35. Schematic diagram of diagenetic sequence in rift reservoirs, Recôncavo basin.

aggregates of kaolinite and fringes of chlorite where small crystals are perpendicular to framework grains (Figure 34B). Kaolinite predominates at depths shallower than 1500 m and chlorite, at greater depths (Moraes and De Ros, 1988). Subordinately, other late authigenic phases can be registered as isolated crystals or aggregates of quartz, albite, pyrite, anatase, and dolomite.

Rift Zone Diagenetic Sequence

Figure 35 shows a diagram of the diagenetic sequence in the rift, based on data obtained from the sandstones of the Pojuca, Marfim, and Candeias formations.

The first phase of quartz (Figure 36A) and feldspar cementation in the rift sequence constitutes, on the average, 2% of total rock volume, locally reaching up to 7% (Matto and Souza, 1988). A later phase of quartz and feldspar cementation is represented by small euhedral crystals of quartz and albite and can be distinguished from the previous phase by the fact that these small crystals do not present corrosion features. This second phase is volumetrically less important, representing 0.1% of total rock volume on the average (Figure 36B).

Calcite cement is volumetrically the most important diagenetic phase in the rift reservoirs, constituting 6.5% of total rock volume on the average, locally reaching up to 32% (Matto, 1984). The major varieties, calcite and ferrous calcite, typically occur in the form of a coarse crystalline mosaic, sometimes with a poikilotopic texture, filling the intergranular pore space and replacing framework grains and quartz and feldspar overgrowths (Figure 37).

In the development of secondary porosity, the dissolution of calcite cement and feldspar-rich grains by solutions produced under thermal decarboxylation of the organic matter present in the associated shales, increases the porosity of the synrift reservoirs (Figure 38A). The porosities measured in thin sections range from 23% in the Ilhas Group to 14% in the Candeias Formation, revealing a tendency for porosity to decrease with increasing depth. On the average, intragranular porosity represents 15–18% of total porosity.

Together, kaolinite and ferrous chlorite represent 1% of total rock volume on the average. Kaolinite reaches up to 5% locally, occurring in the form of discrete, disseminated aggregates that occupy intragranular pores. Chlorite usually amounts to less than 1% of total rock volume, occurring as coatings attached to framework grains (Figure 36B). In sandstones of the Pojuca and Marfim formations, chlorite and kaolinite coexist, whereas in the Candeias sandstones only chlorite is present, suggesting that kaolinites are transformed into chlorite with increasing burial, as proposed by Boles and Franks (1979).

Dolomite, ferrous dolomite, ankerite, and zoned crystals with dolomite nuclei and ankerite margins are reported, representing on the average 3% of total rock volume, although contents of up to 35% may be detected (Matto and Souza, 1988). Such cements occur as euhedral crystals occupying intergranular space and replacing calcite cement and mud intraclasts (Figure 38B).

Additional, subordinated cements are represented by isolated aggregates of pyrite, anatase, and sphene crystals.

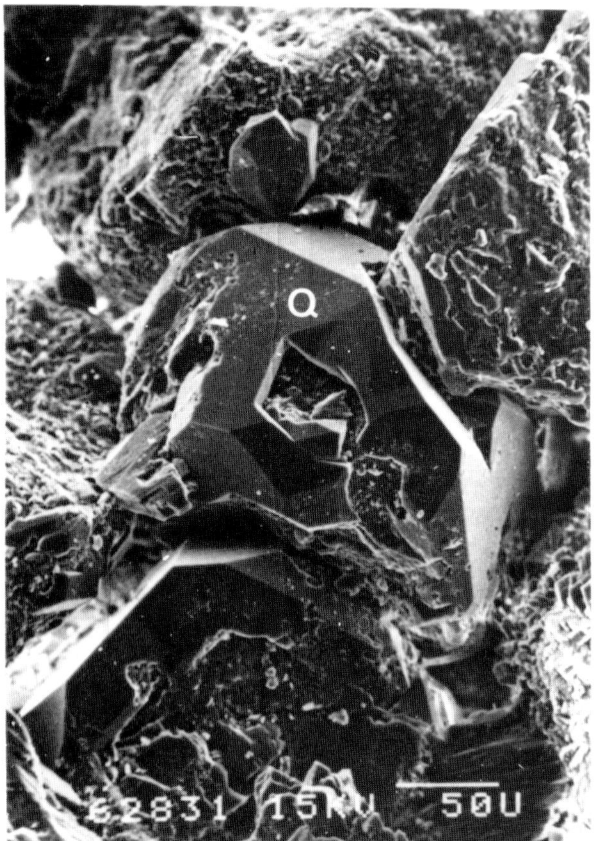

A

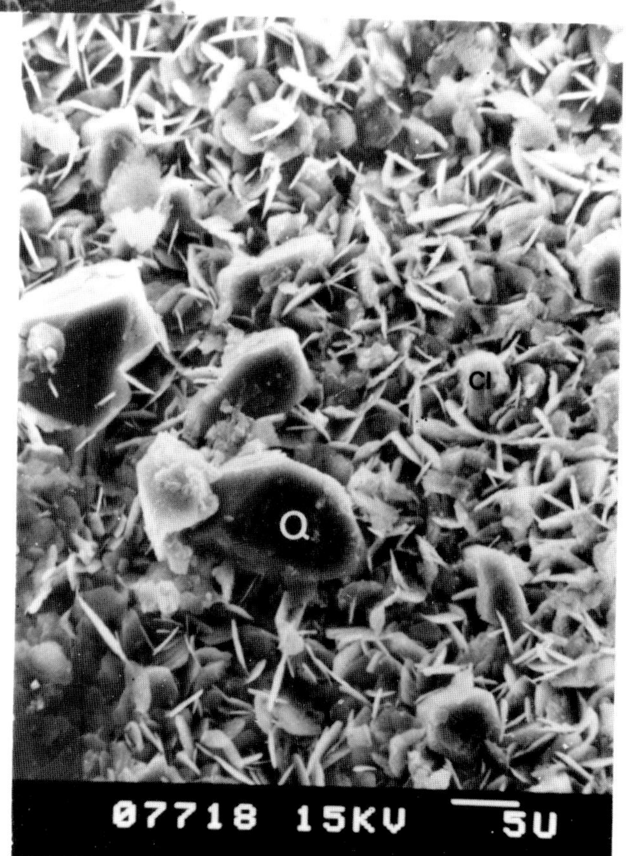

B

Figure 36. Photomicrographs: A: quartz overgrowths (Q), Gomo Member, Candeias Formation (SEM); B: late quartz cementation (Q) and authigenic chlorite (CL), Gomo Member, Candeias Formation (SEM).

Figure 37. Photomicrographs: poikilotopic calcite (C) filling intergranular pores and replacing framework grains and quartz (Q) and feldspar overgrowths, Gomo Member, Candeias Formation. Crossed nicols (field: 0.7 mm × 1.1 mm).

Play Types

Petroleum accumulations in the Recôncavo basin may be grouped into three basic systems: the prerift system and the Candeias and Ilhas systems of the rift (Figure 39), representing more than 80 oil and gas fields discovered in this basin (Figure 40).

Prerift System

The prerift system is associated with the initial extensional tectonic phase, when the basin began its individualization as a rift valley, leading to the development of horsts, tilted blocks, and grabens where oil-prone shales were deposited over, and lateral to, prerift reservoirs (see Figure 30A).

The prerift phase is the main accumulation system in terms of hydrocarbon volume, responsible for 429.7 million m^3 (2702.8 million bbl, 57%) of proven oil volume in the basin, distributed across nearly 25 fields, among which the Dom João, Água Grande, and Buracica stand out. Sandstones of the Sergi Formation are the major reservoirs of the play and contain 301.8 million m^3 (1898.3 million bbl). These are followed by the Água Grande Member, which holds 82.7 million m^3 (520.2 million bbl). Other smaller reservoirs are responsible for the remaining oil volume.

Originating from a braided fluvial system, the Sergi Formation is a very heterogeneous reservoir, presenting a coarsening-upward pattern with sandstone grain size varying from fine to coarse and conglomeratic. Porosity ranges from 10 to 25% and permeability from 20 to 1200 md.

The Água Grande reservoir consists of two bodies; the lower one, trending north–south, displays distinct fluvial characteristics, while the upper, more expressive body is composed of an eolian-reworked deposit trending east–west. Diagenetically, both packages are comparable and have low cement contents, kaolinite being the main diagenetic element. The upper body displays porosities of up to 27% and permeability of up to 2,000 md, whereas in the lower body, porosity reaches 21% and permeability is usually lower than 500 md.

The Água Grande, Buracica, and Dom João oil fields are typical examples of this system. They constitute structural traps formed by horsts associated with major faults. The Água Grande and Buracica oil fields are related to the Mata-Catu transfer fault and are segmented by normal faults into several minor blocks (Figures 41, 42). The Dom João oil field is composed of a narrow horst block controlled by two large synthetic faults close to the western edge of the basin (Figure 43).

Synrift Candeias System

The synrift Candeias system is characterized by both stratigraphic and combined traps that are linked to the genesis of the reservoirs, which were deposited as lacustrine turbidite fans. Accumulations are located preferentially east of the basin axis where, as a function of the high rate of subsidence, a slope-type physiography was established. Unique to this system, accumulations also occur in fractured shales, the most important one being the 4th zone of the Candeias field (Figure 44).

This system contains 117.5 million m^3 (739.1 million bbl) of proven oil, equivalent to 16% of total basin volume. Of this, 60% is concentrated in the Candeias field, while the remainder is spread across 30 other small accumulations. Concerning nonassociated gas accumulation, the Candeias is the main system; the turbidites of the Maracangalha Member contain 5.2 billion m^3 (183.6 bcf) of gas, equivalent to 35% of total basin volume.

The Candeias field, located in the southern compartment and discovered in 1941, is representative of this system (Figure 44). Its main reservoirs consist of cyclic beds of conglomerates and sandstones intercalated with shales, which together may reach thicknesses of 350 m. Reservoir characteristics are considered regular. Porosity ranges from 15 to 18%, and in the best intervals permeability is around 30 md, occasionally reaching 200 md. The good productivity of these reservoirs, in most cases, is due to the presence of natural fractures that provide intercommunication between pore spaces. In this field, production also is reported from fractured shales and limestones, whose main characteristic is high resistivity values on electric logs.

In the northeastern compartment, several oil fields have been discovered in this system during the last decade. Represented by the Riacho da Barra field (Figure 45), they consist of combined traps, where drape-related anticlinal features developed over structural highs that had been produced in the first phase of rifting. The reservoirs consist of laterally discontinuous coarse- to fine-grained sandstones deposited by turbidite currents in lows in front of the clastic wedge of the Salvador Formation.

Figure 38. Photomicrographs: A: partially dissolved feldspar grain with development of intragranular (secondary) porosity, Gomo Member, Candeias Formation (SEM). B: fine crystals of dolomite (D) occupying intergranular pore space, Gomo Member, Candeias Formation (SEM).

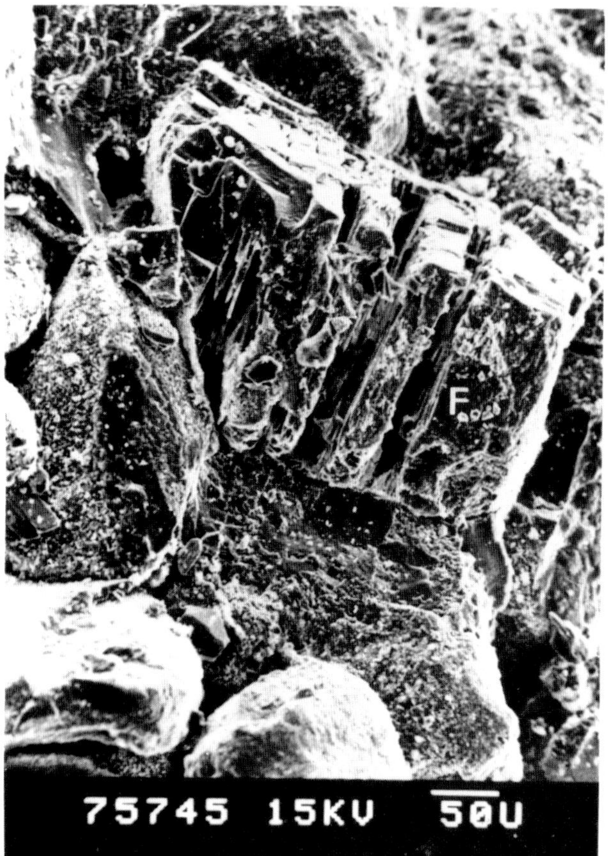

A

B

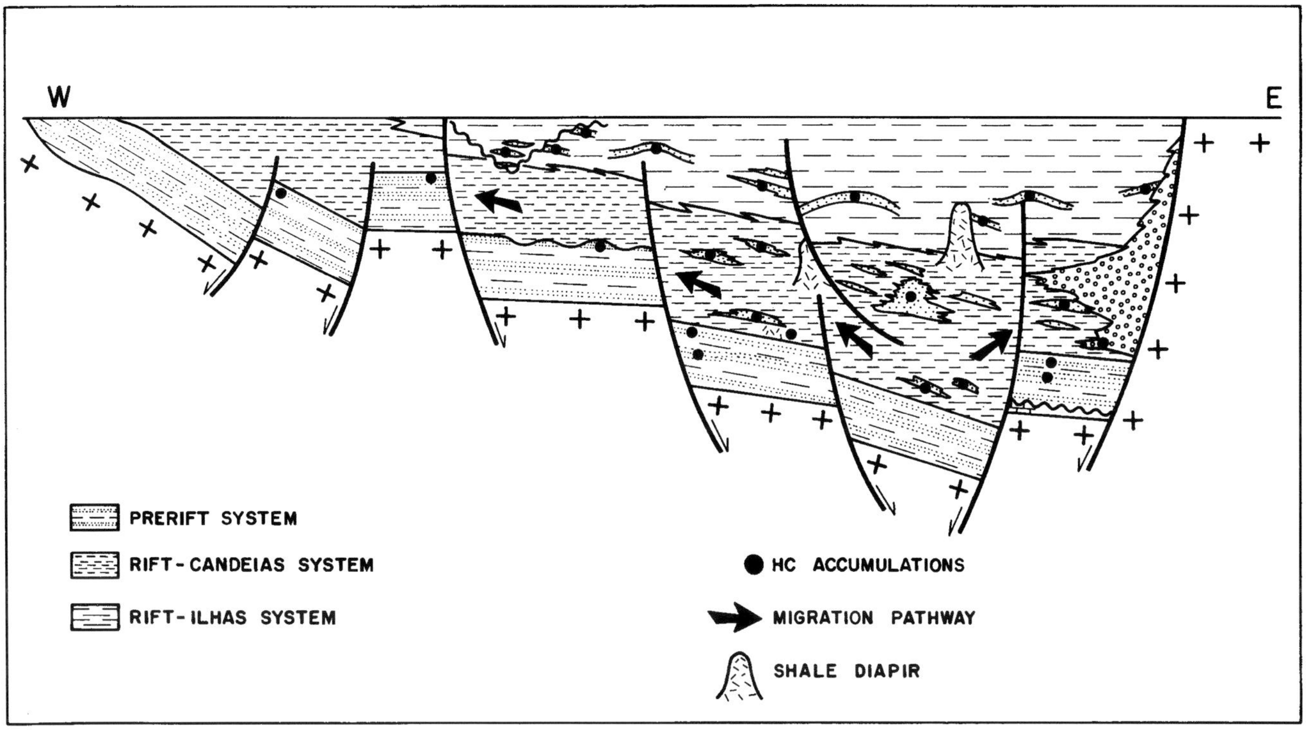

Figure 39. Schematic east-west geologic section showing main types of traps, Recôncavo basin.

Synrift Ilhas System

The synrift Ilhas system, the second most important in the basin, is characterized mainly by anticlinal structures and by stratigraphic and combined traps related to the turbidites of the Taquipe Formation. Despite its complex origin, most of the traps are always associated with the growth-fault system that marks the basin hinge line (Figure 39).

The proven oil volume in this system is 201.2 million m^3 (1,265.5 million bbl), corresponding to 27% of oil in place in the basin and distributed over about 15 accumulations. The Miranga (93.3 million m^3, 586.9 million bbl), Taquipe (37.8 million m^3, 237.8 million bbl), and Araçás (37.3 million m^3, 234.6 million bbl) fields are representative of this system (Figure 40).

The Miranga field, located in the low of the same name, is a multireservoir oil field characteristic of the Ilhas system. It contains 22 hydrocarbon pay zones in the deltaic sandstones of the Marfim and Pojuca formations (Figure 46).

The structural framework of the Miranga field exhibits an anticlinal shape sectioned by normal faults with throws ranging between 10 and 300 m. Its final configuration was strongly influenced by shale diapirism, which practically surrounds the field.

The reservoirs are usually located at intervals of from 900 to 1450 m in depth. These sandstones exhibit uniform petrophysical characteristics, are fine- to very-fine grained with a matrix content lower than 3% of total rock volume, and have porosities from 18 to 24% and permeability between 30 and 300 md.

FINAL CONSIDERATIONS

In the late 1970s, due to the modest results achieved over almost 15 yr of exploration activities, the Recôncavo basin was considered by a significant part of the geologic community to be at a mature exploration stage. Forecasts concerning the incorporation of additional oil volumes were highly conservative. A statistical method known as creaming, applied to the basin data, estimated the discovery of a minimum of 0.4 million m^3 (2.5 million bbl) and a maximum of 5.0 million m^3 (31.5 million bbl) of recoverable oil and gas equivalent during 1982–1985. However, results from the last 10 yr of exploration activities have been much more encouraging; about 60 million m^3 (37.7 million bbl) of oil in-place have been found, adding about 11 million m^3 (69.2 million bbl) of oil to basin reserves.

As a result of the most recent reevaluation of the petroleum potential of the Recôncavo basin, using the play analysis method (Miller, 1982) adapted and improved by PETROBRÁS geologists, an estimated 105 wildcat wells should be drilled over the next 10 yr. Expectations are for the incorporation of around 27 million m^3 (169.8 million bbl) of oil and 10 billion m^3 (353 bcf) of gas, that is, 60% of the remaining petroleum potential of the basin.

The encouraging results obtained in recent years, along with other factors, have served to highlight significant improvement in seismic technology.

The use of high-resolution seismic methods, particularly as an auxiliary tool in surveying combined and

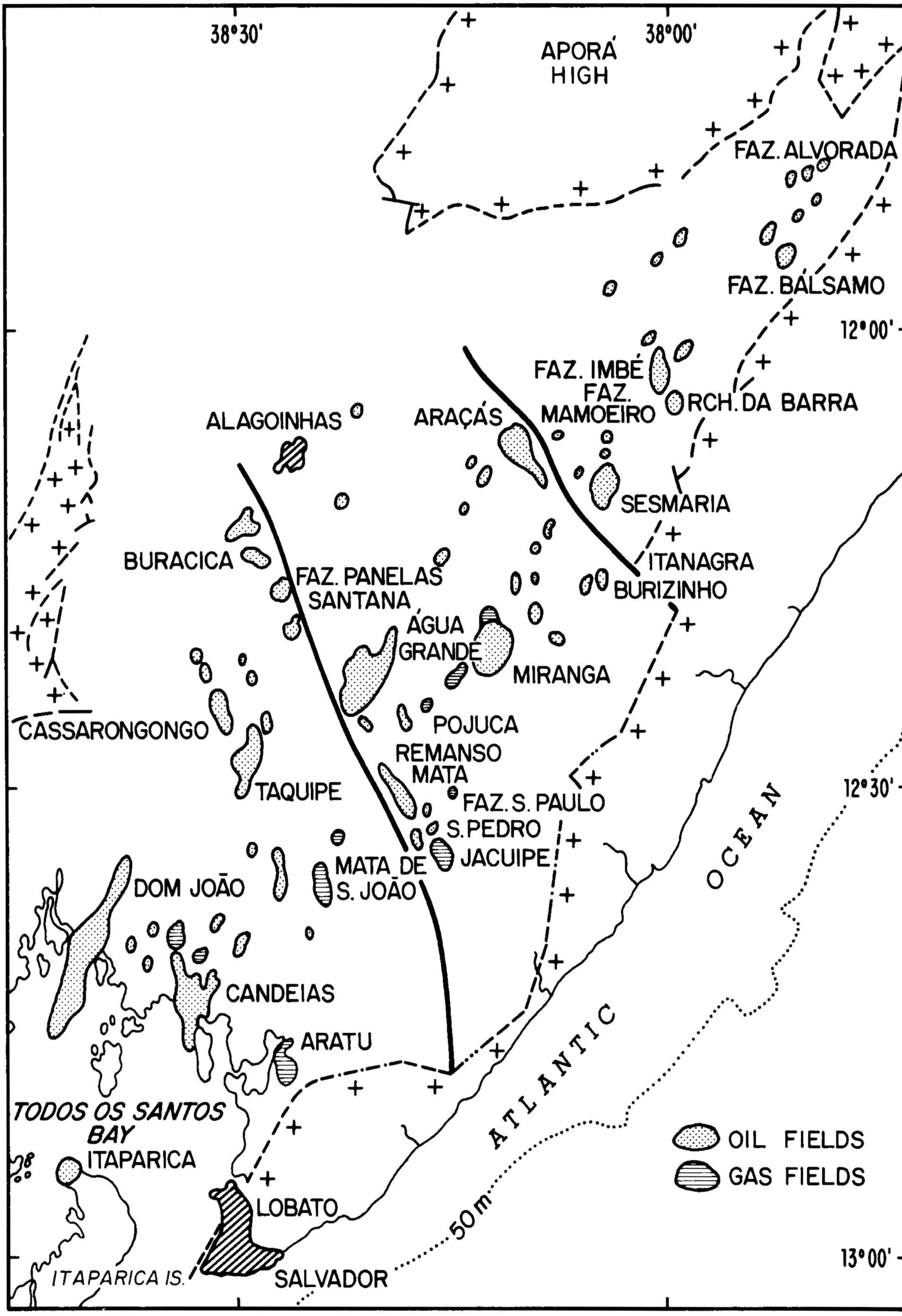

Figure 40. Oil and gas fields, Recôncavo basin.

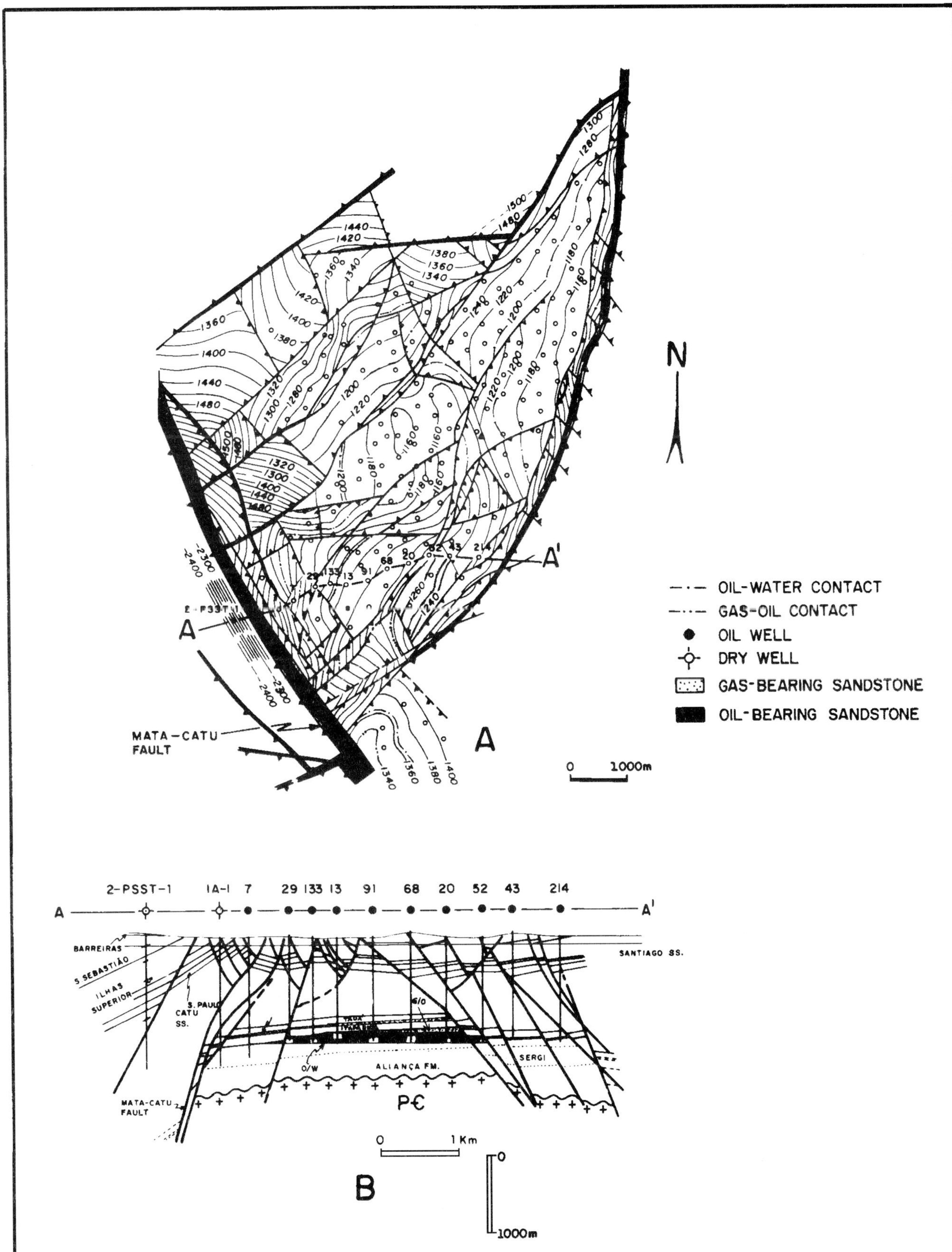

Figure 41. Água Grande field. A = Structural map at top of Sergi Formation; B = geologic cross section A–A´.

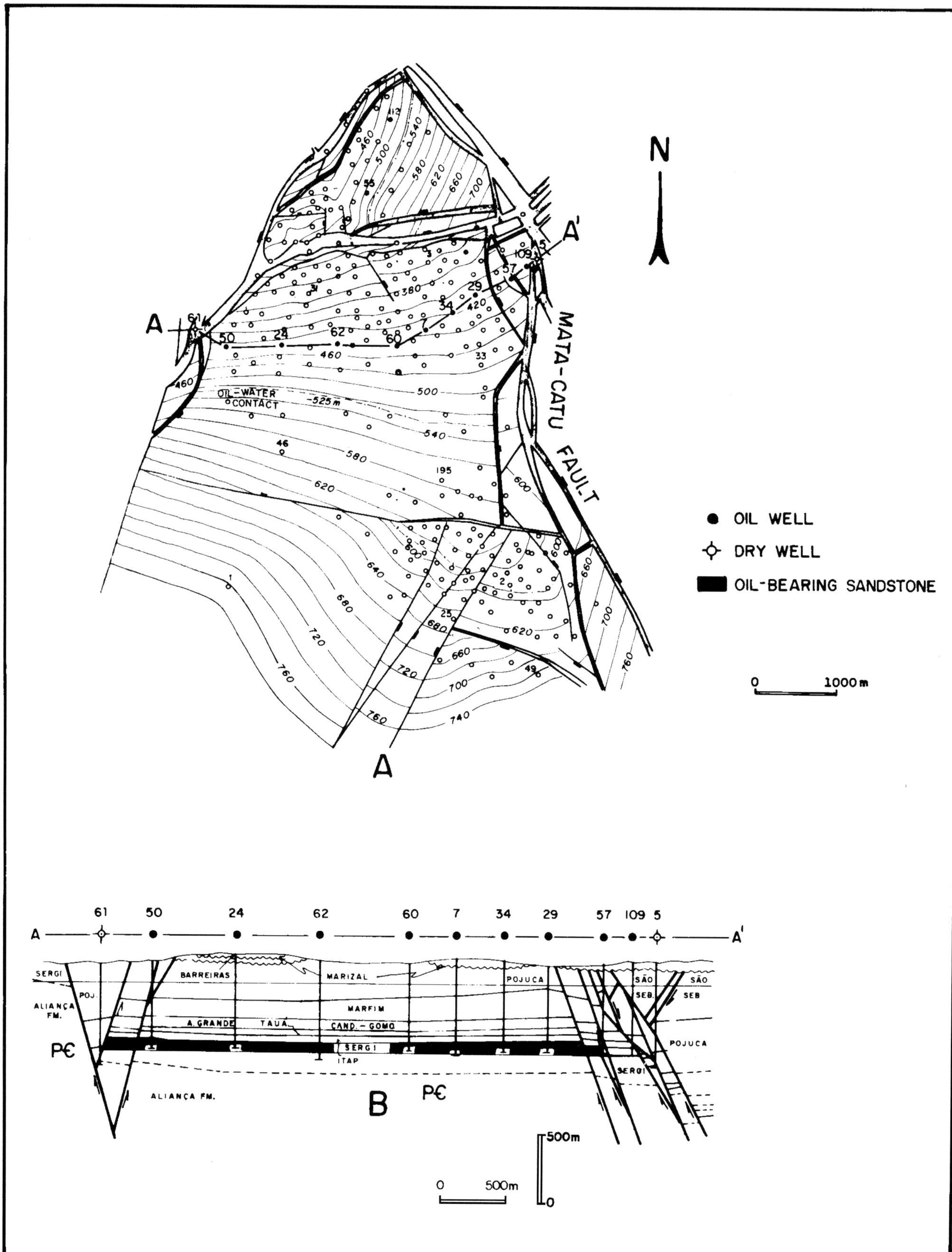

Figure 42. Buracica field. A = Structural map at top of Sergi Formation; B = geologic cross section A–A´.

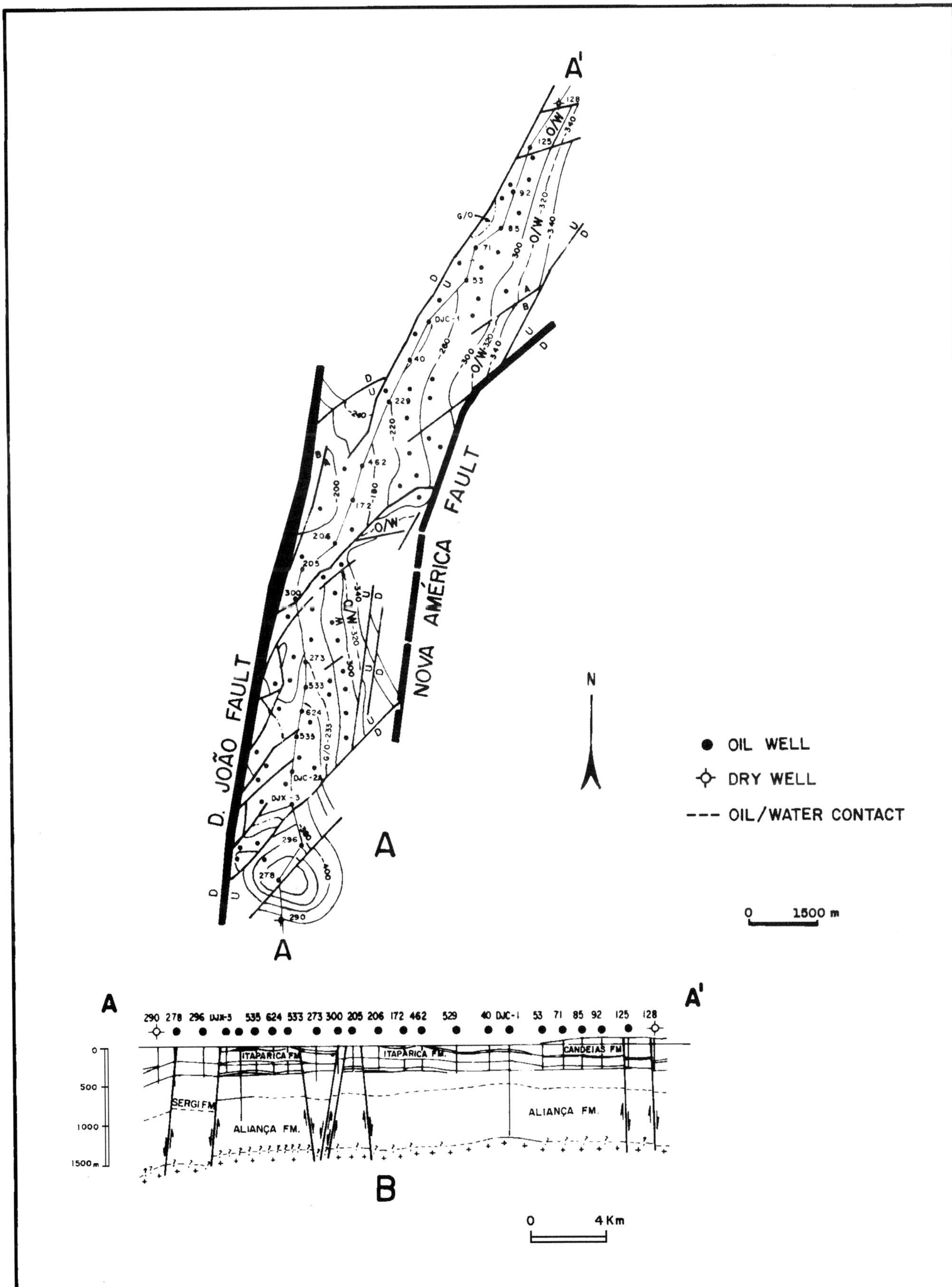

Figure 43. Dom João field. A = Structural map at top of Sergi Formation; B = geologic cross section A–A′.

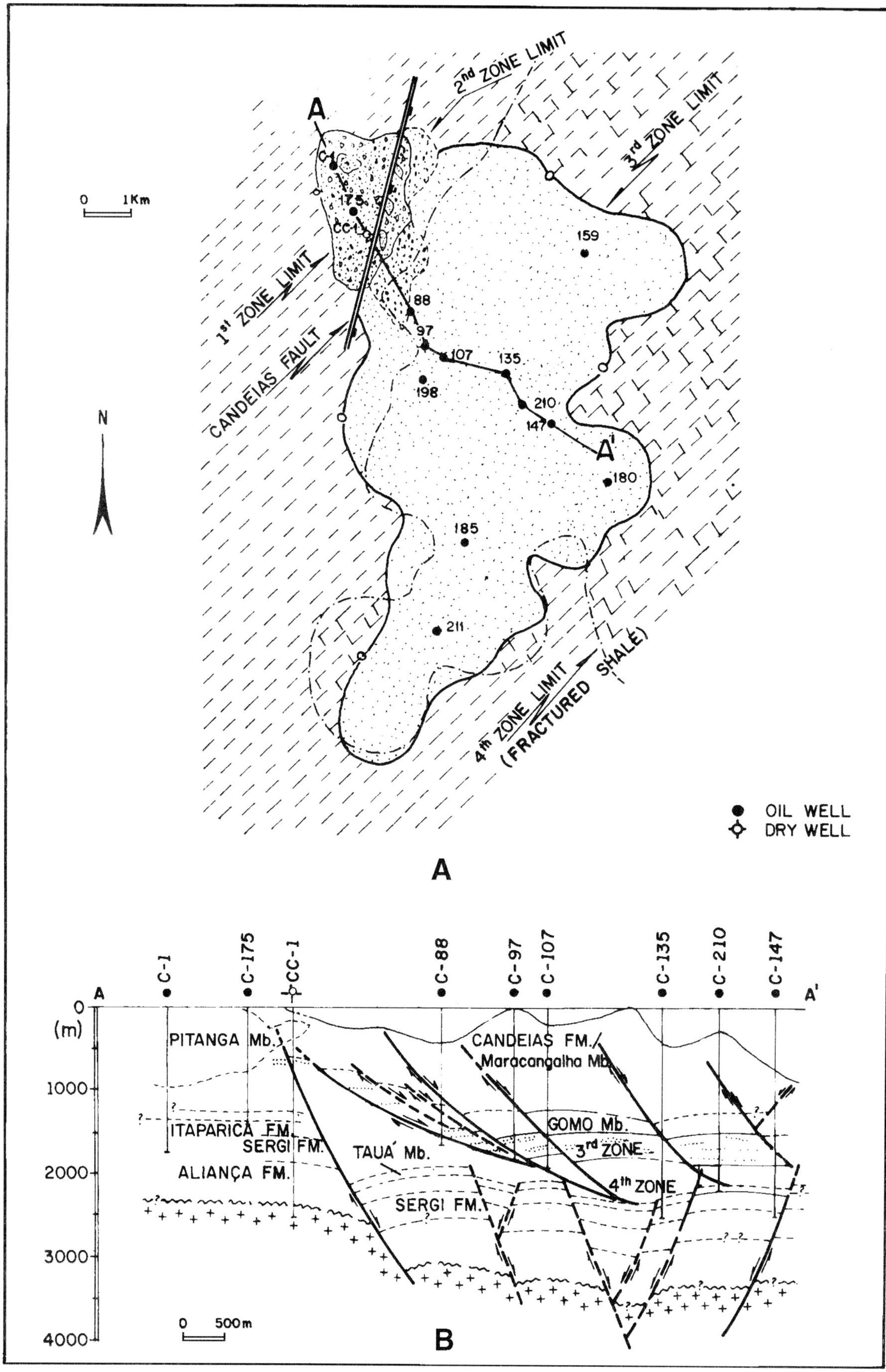

Figure 44. Candeias field. A = General outline of turbidite reservoirs and fractured shale zone; B = geologic cross section A–A´.

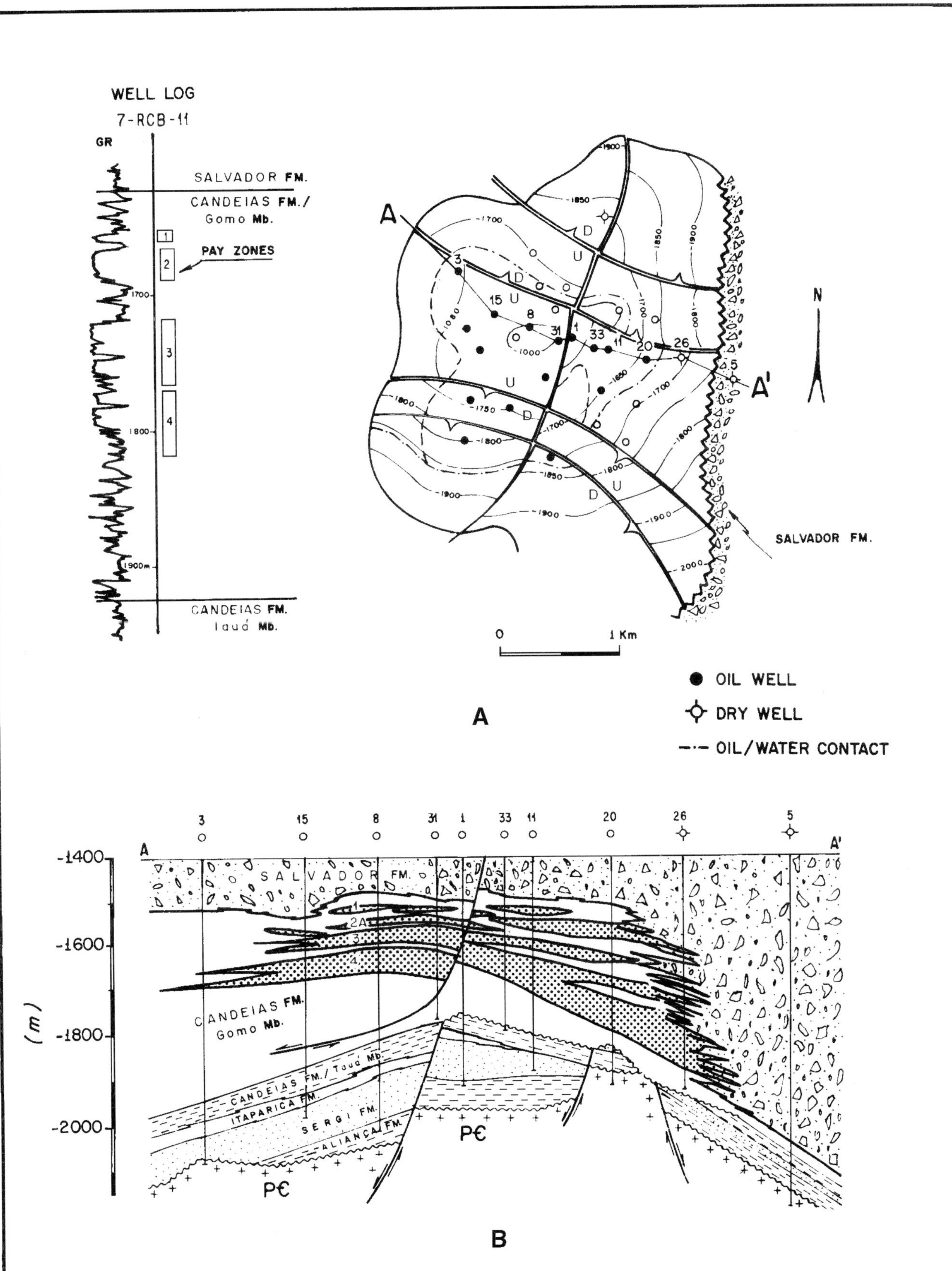

Figure 45. Riacho da Barra field. A = Structural map at top of 4th zone, Gomo Member, Candeias Formation; B = geologic cross section A–A´.

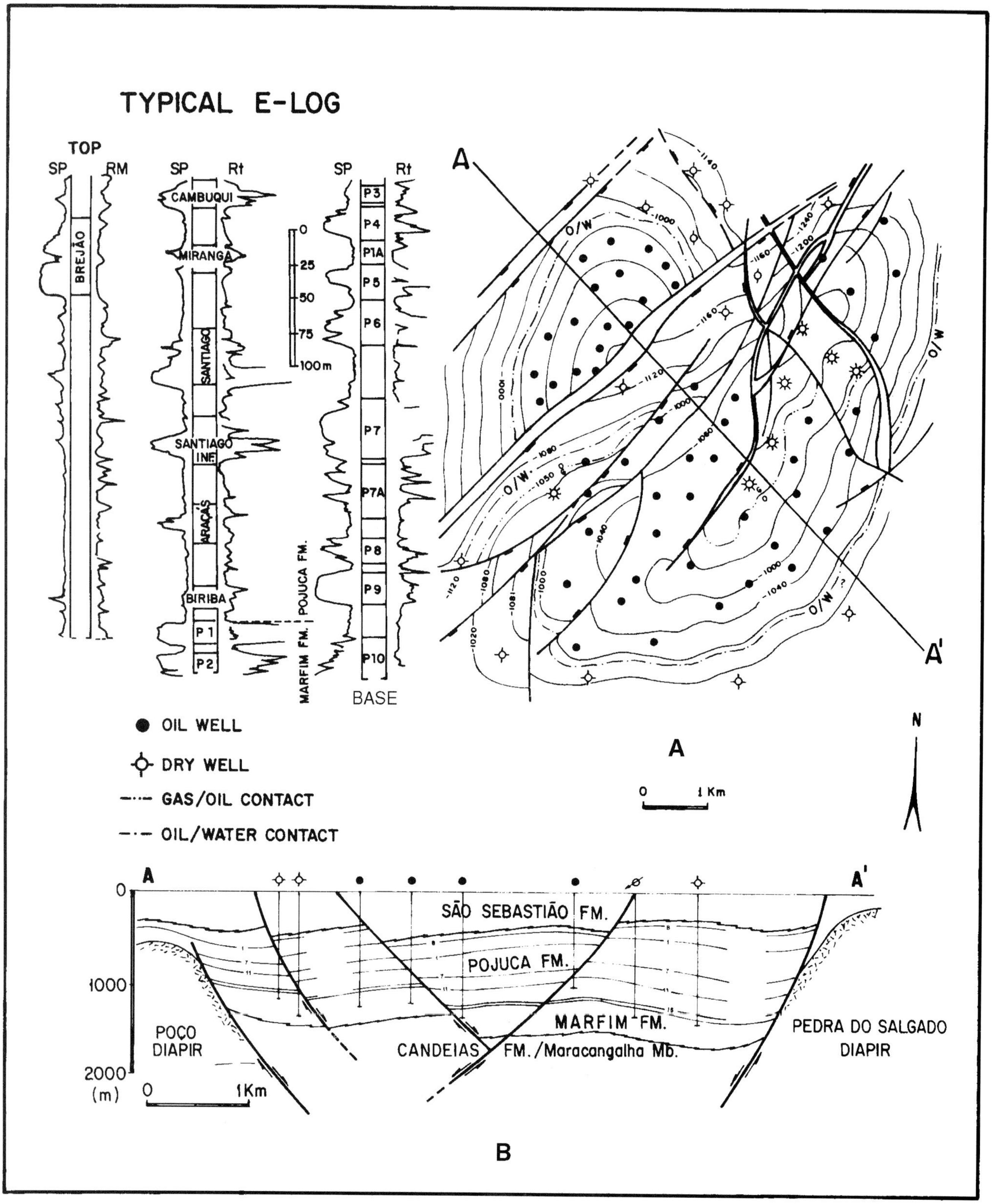

Figure 46. Miranga field, including typical electric log. A = Structural map at top of Santiago Sandstone, Pojuca Formation; B = geologic cross section A–A´.

subtle stratigraphic traps in regional lows, will be intensified. Several three-dimensional seismic surveys are being acquired and the exploration department of PETROBRÁS believes that this strategy will help to maintain exploration efforts in this basin for several years.

REFERENCES CITED

Asmus, H.E., 1975, Structural control for the Mesozoic deposition at the Brazilian continental margin: Revista Brasileira de Geociências, v. 5, p. 160–175.

Baird, W., 1845, Arrangement of the British Entomostraca: Berwickshire Natural Club Proc., no. 2.

Baird, W., 1850, The Natural History of British Entomostraca: Royal Soc. London, 364 p.

Boles, J.R., and S.G. Franks, 1979, Clay diagenesis in Wilcox sandstones of southwest Texas: implications of smectite diagenesis on sandstone cementation: Journal of Sedimentary Petrology, v. 49, p. 55–70.

Estrella, G.O., 1972, O estágio rift nas bacias marginais do leste brasileiro: Anais do XXVI Congresso Brasileiro de Geologia, Belém, Pará, v. 3, p. 29–34.

Ghignone, J.I., 1979, Geologia dos sedimentos fanerozóicos do Estado da Bahia, *in* H.A.V. Inda, (ed.), Geologia e Recursos Minerais do Estado da Bahia textos básicos: Secretaria das Minas e Energia, v. 1.

Krömmelbein, K., 1963, *Ilhasina* n.g. und *Salvadoriella* n.g. zwei neue Ostracoden-Gattungen aus der Bahia Serie Nicht-Mariner Oberjura?—Unterkreide, NE Brasilien: Zoologisher Anzeiger v. 171, no. 9–10, p. 376-390, Leipzig.

Matto, L.F., 1984, Arenitos-reservatório do Membro Santiago superior no Campo de Miranga—Bloco Principal e Bloco Norte, Bacia do Recôncavo: PETROBRÁS/DEPEX/DEXBA internal report, Salvador, 30 p.

Matto, L.F., and E.M. Souza, 1988, Petrografia e diagênese dos arenitos da Formação Candeias/Membro Gomo (Cretáceo Superior) no Campo de Rio do Bu, Bacia do Recôncavo: Boletim de Geociências da PETROBRÁS, v. 2, p. 277–289.

Milani, E.J., 1984, Modelo evolutivo do rift Recôncavo-Tucano-Jatobá: Ciência-Técnica-Petróleo, PETROBRÁS-CENPES, no. 18, 32 p.

Milani, E.J., and I. Davison, 1988, Basement control and transfer tectonics in Recôncavo-Tucano-Jatobá rift, northeast Brazil: Tectonophysics, v. 154, p. 40–70.

Milani, E.J., M.C. Lana, and P. Szatmari, 1987, Mesozoic rift basins around the NE Brazilian Microplate, *in* Manspeizer, W. (ed.), Triassic-Jurassic rifting and the opening of the Atlantic Ocean: Elsevier, ch. 34, p. 1–26.

Miller, B.M., 1982, Application of exploration play-analysis techniques to the assessment of conventional petroleum research by the USGS: Journal of Petroleum Technology, Jan., v. 34, no. 1, p. 55–64.

Moraes, M.A.S., and L.F. De Ros, 1988, Caracterização e influência das argilas de infiltracão mecânica em reservatórios fluviais da Bacia do Recôncavo, nordeste do Brasil: Boletim de Geociências da PETROBRÁS, v. 2, p. 13-26.

Netto, A.S.T., and J.J. Oliveira, 1985, O preenchimento do rift-valley na Bacia do Recôncavo: Revista Brasileira de Geociências, v. 15, p. 97–102.

Netto, A.S.T., A.S. Barroso, C.H.L. Bruhn, J.M. Caixeta, and M.A.S. Moraes, 1982, Projeto Andar Dom João. PETROBRÁS/DEPEX/DEXBA, Salvador, 193 p.

Netto, A.S.T., A.V. Carozzi, M.B. Araujo, P. Cesero, J.R. Fonseca, and U.J.L. Silva, 1984, Projeto prospectos estratigráficos, Bacia do Recôncavo: PETROBÁS/DEPEX/DEXBA internal report, Salvador, Bahia, 145 p.

Ponte, F.C., 1971, Evolução Paleogeográfica do Brasil Orientale da África Ocidental: PETROBRÁS/DEXPRO/DIVEX internal report, Rio de Janeiro.

Regali, M.S.P., and Viana, C.F., 1989, Late Jurassic–Early Cretaceous in Brazilian sedimentary basins: Correlation with the International Standard Scale: Petroles Brasileiro, S.A. Development of Human Resources Special Paper, Rio de Janeiro, 1956.

Sars, G.O., 1866, Oversigt of Norges marine Ostracoder: Norske Vidensk-akad, Forhandl, p. 1–130.

Szatmari, P., E.J. Milani, M.C. Lana, J.C.L. Conceição, and R.P. Lobo, 1985, How South Atlantic rifting affects Brazilian oil reserves distributions: Oil & Gas Journal, v. 83, p. 107–113.

Viana, C.F., A.V. Carozi, M.B. Araujo, P. de Cesero, J.R. Fonseca, and V.J.L. da Silva, 1971, Revisão estratigráfica das bacias do Recôncavo e Tucano: Boletim Técnico da PETROBRÁS, v. 4, p. 157–192.

Chapter 6

Structural Style and Tectonic Evolution of the Albuquerque Basin Segment of the Rio Grande Rift, New Mexico, U.S.A.

L.R. Russell[1]
S. Snelson[2]
Shell Development Company
Bellaire Research Center
Houston, Texas, U.S.A.

ABSTRACT

The Albuquerque basin, located in the central portion of the Rio Grande rift, is filled by 7350 m (24,000 ft) of Tertiary clastic sediments deposited on a "basement" of Mesozoic and Paleozoic sedimentary rocks and Precambrian crystalline rocks. The basin has been a center of Cenozoic volcanic activity over the last 37 Ma.

The geophysical signature of the Rio Grande rift is grossly similar to the Basin and Range province, being characterized by thin crust, anomalously low upper mantle seismic velocity, and velocity inversion in the upper crust. Additionally, there are high levels of shallow seismicity, abnormally high heat flow, regionally low Bouguer gravity values, and anomalously high electrical conductivities at shallow to mid-crustal depths.

Detailed examination of regional seismic reflection data, supplemented by well control and field work, has demonstrated that the Albuquerque basin is asymmetric and structurally complex, consisting of two subbasins downdropped along low-angle to listric, normal faults of opposing structural polarity, some of which flatten at depths of about 10 km (6 mi). The northern subbasin has been downdropped along a major west-dipping, listric normal fault system, whereas the southern subbasin is bordered by a system of major, east-dipping, low-angle normal faults. The two subbasins are separated by a complex, mid-basin transverse structural zone that accommodates the differential extension and polarity change between the basins.

[1]Present address: ARCO Exploration and Production Technology, Plano, Texas.
[2]Present address: Eastsound, Washington.

Palinspastic restorations show that the amount of extension in the Albuquerque basin ranges from 17% in the north basin to at least 28–30% in the south basin.

Seismic and outcrop evidence suggest that preexisting structures in the Precambrian basement rocks, in part, may have controlled the geometry of Tertiary structures in this region.

INTRODUCTION

This study of the Albuquerque basin was conducted between 1978 and 1980, and was undertaken to improve our understanding of the rift stage of passive continental margin development.

Because the Rio Grande rift is a young, active continental rift with abundant subsurface and surface control, we hoped that a study of this segment of the rift would provide answers to some of the fundamental questions of structural style and geometry of rifting during initial stages of continental breakup. Of interest were questions such as:

(1) What is the geometry of rift-controlling normal faults at depth? Are they classic high-angle faults or low-angle and/or listric (downward flattening) in geometry?

(2) If these rift-controlling faults are indeed low-angle or listric, what controls the depth of flattening?

(3) Is continental breakup a symmetrical or asymmetrical phenomenon in terms of the structural configuration of the resultant conjugate margins?

(4) How much crustal stretching or attenuation occurs during rifting and breakup? How variable is it? What are the implications for palinspastic plate tectonic reconstructions and fits?

(5) To what extent is rifting controlled by preexisting basement structures both in terms of the overall geographic location of rift development and the detailed geometry of rift structures?

In the years following the inception of this work in 1978, many of these questions, as well as questions on mechanisms of rifting, have been variously addressed by numerous workers. These include McKenzie (1978) and Wernicke (1981, 1985, 1986), and many others who have discussed the pure shear versus whole-lithosphere simple shear models of rifting; and Bally (1982) and Lister et al. (1986), whose discussions focus on the symmetry versus asymmetry of the rift process and its implications for passive continental margin evolution.

We hope that our interpretations (based upon seismic, outcrop, and well data) from the Albuquerque basin will further add to our understanding of rift phenomena and hopefully stimulate more research into the still highly debated question of rift mechanisms.

MEGATECTONIC FRAMEWORK OF THE RIO GRANDE RIFT AND ITS PLACE WITHIN THE GREATER BASIN AND RANGE PROVINCE

Major Tectonic Elements of the Rio Grande Rift

Extending from just north of Leadville, Colorado, southward at least 1000 km (600 mi) to the El Paso, Texas, area, the Rio Grande rift is the zone of Cenozoic extension resulting from the separation of the Colorado Plateau block from the Great Plains of the North American craton (Figure 1). As such, the rift represents a northerly trending extension of the southern Basin and Range province into the southern Rocky Mountain region.

The northern segment of the Rio Grande rift extends from 20 km (12.5 mi) north of Leadville, Colorado, to the Taos, New Mexico, area and includes the San Luis basin and Upper Arkansas graben (Figures 2, 3). This segment is characterized by north-northwest structural trends, roughly paralleling the late Paleozoic and Laramide structural grain. The southern boundary of this northern rift segment is defined by a northeasterly trending zone of volcanism and contemporary seismicity extending through the area of the Taos and Jemez mountains.

The central segment of the rift runs from Taos, New Mexico, southward to the Socorro, New Mexico, area (Figures 2, 3) and is comprised of both the Espanola and Albuquerque basins (Albuquerque-Belen basin in Figure 2). This segment is distinguished by a north-northeast structural orientation, again parallel to the Laramide tectonic grain, and is bounded on the south by a northeasterly trending zone of volcanism and contemporary seismicity, which trends across the rift through the Socorro area and extends southwestward along the San Agustin basin.

The southern rift segment extends from Socorro, New Mexico, to the El Paso, Texas, area (Figures 2, 3) and is typified by a wide zone of northerly trending basins and ranges, a sharp contrast to the topographically distinct and well defined northern and central segments of the rift. The more diffuse character of the southern rift is likely a consequence of the fact that it has probably undergone a greater amount

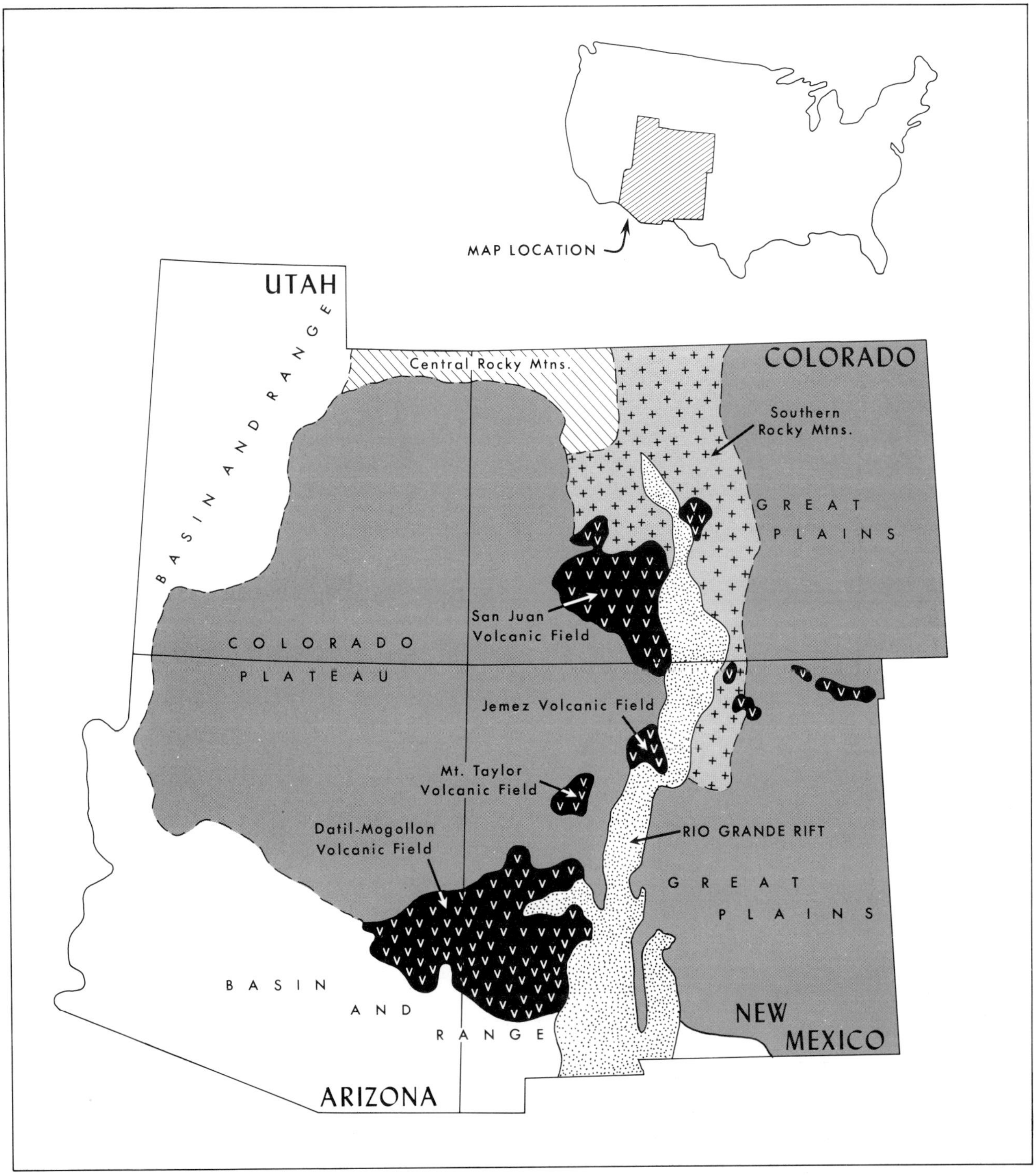

Figure 1. General map of the southwestern United States showing the location of the Rio Grande rift.

of crustal attenuation (crust is approximately 26 km or 16 mi thick; Figures 6, 8), and also that it merges with or overprints northwest-trending Paleocene structures of the southern Basin and Range province (Figures 1 or 5).

The geographic location of the Rio Grande rift very closely follows the distribution of Laramide foreland basement block uplifts in the southern Rocky Mountain region. Although in detail the geometry and orientation of normal faults in the Rio Grande rift system may not correspond precisely with the older Laramide structural grain, the overall spatial coincidence of the rift system with this belt of Laramide deformation is undeniable (Figures 4, 5).

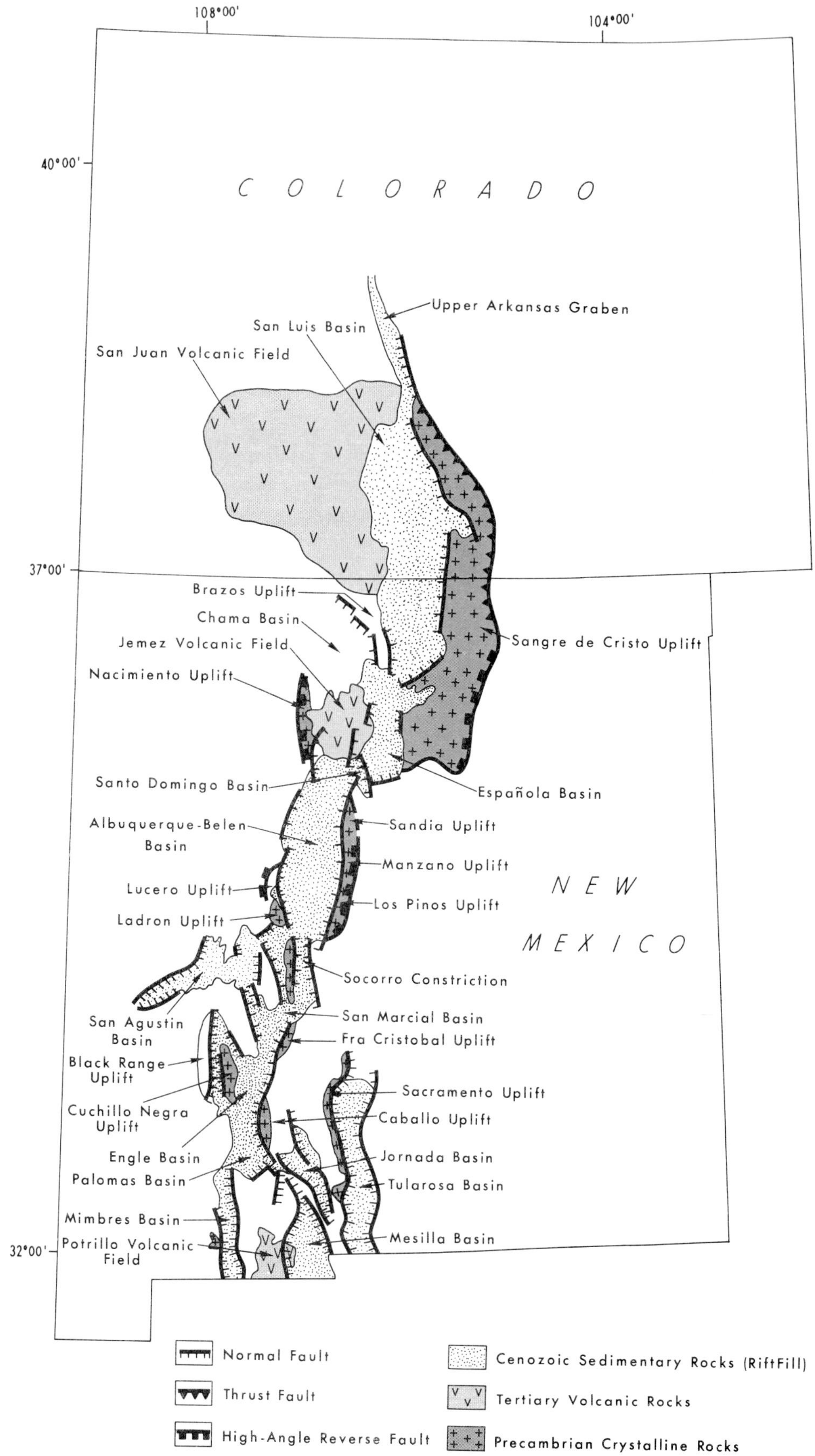

Figure 2. General location map of the basins and uplifts along the Rio Grande rift.

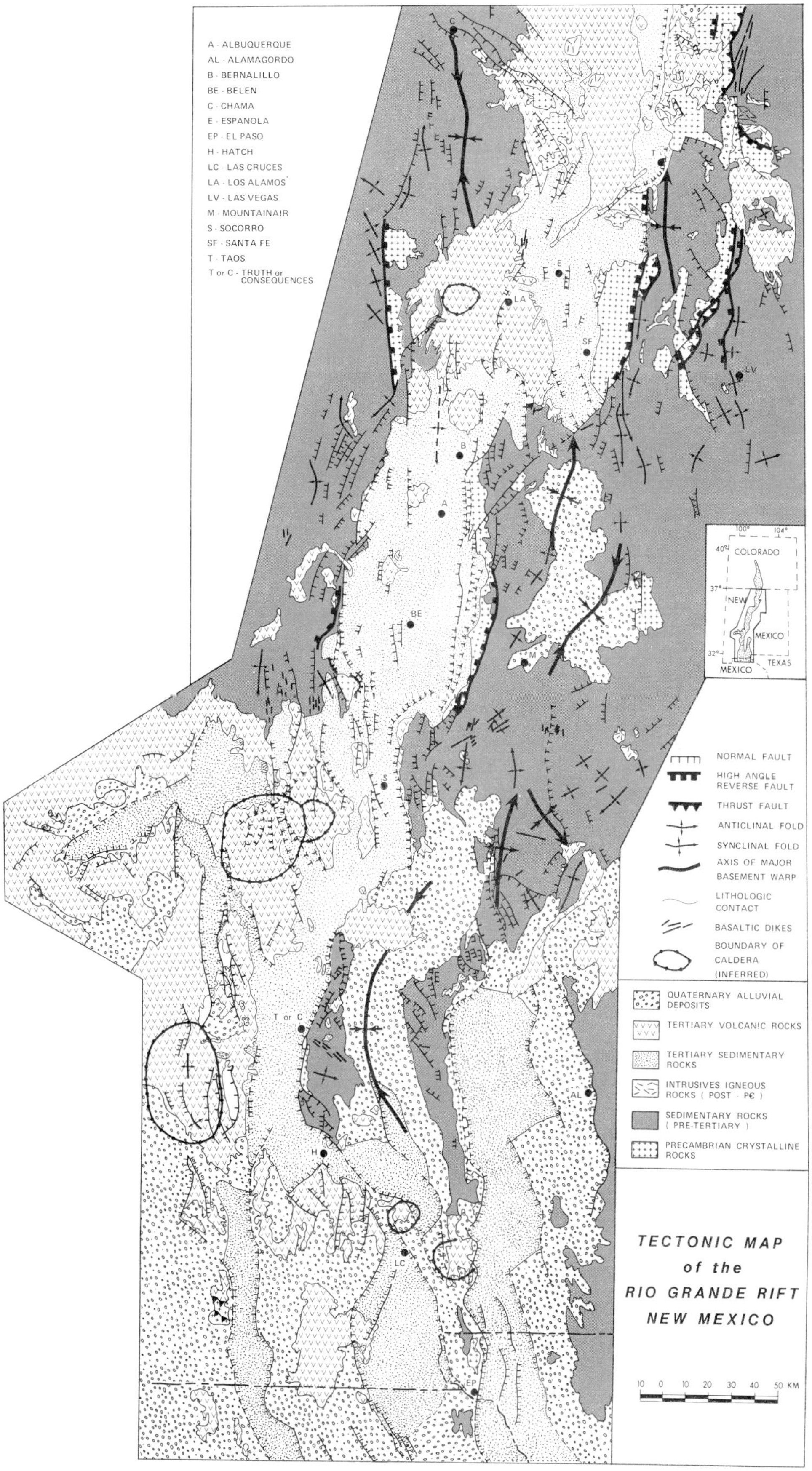

Figure 3. Tectonic map of the Rio Grande rift, New Mexico (modified after Woodward et al., 1978).

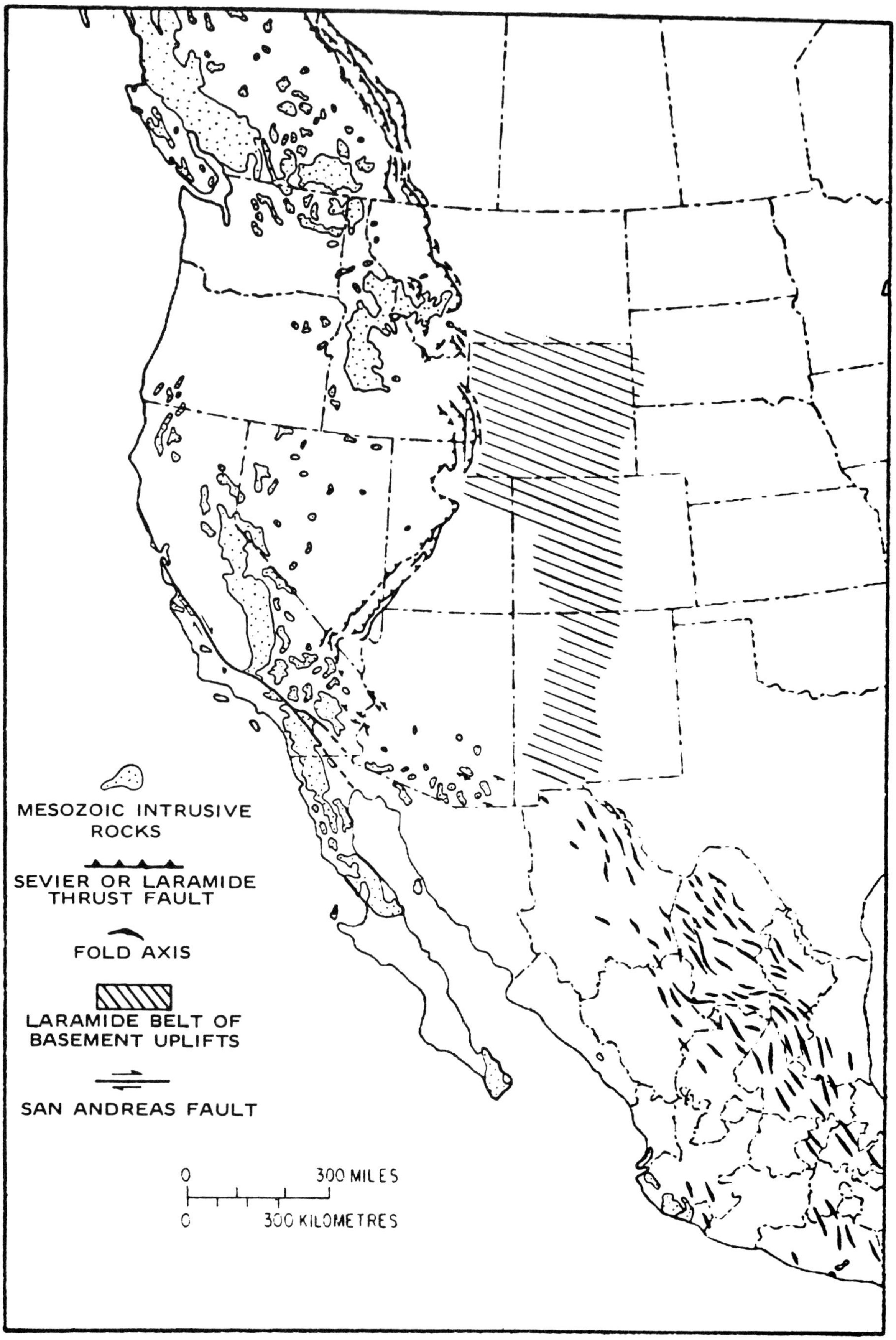

Figure 4. Distribution of belts of Laramide deformation in western North America (after Stewart, 1978).

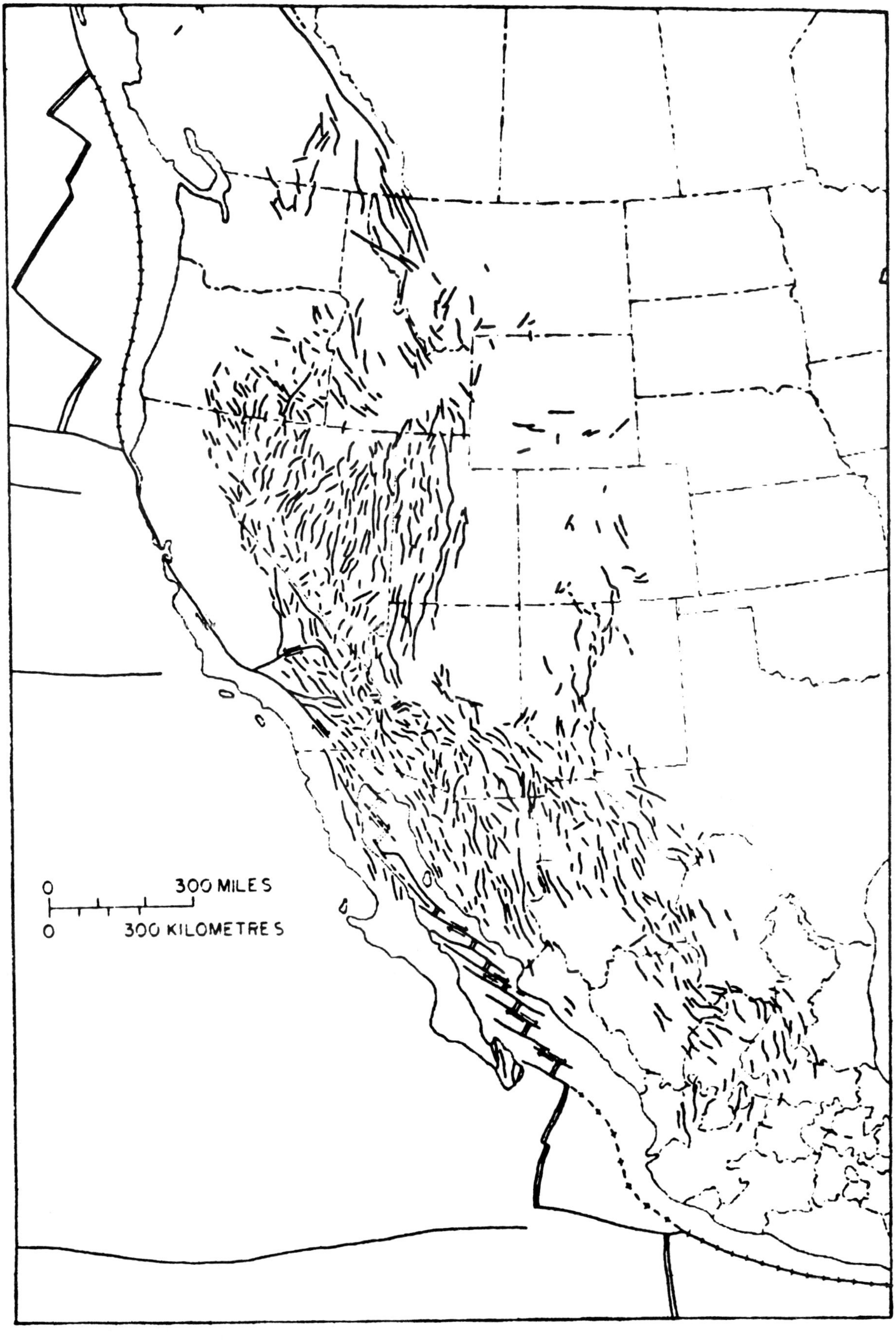

Figure 5. Distribution of Cenozoic extensional faults in western North America (after Stewart, 1978).

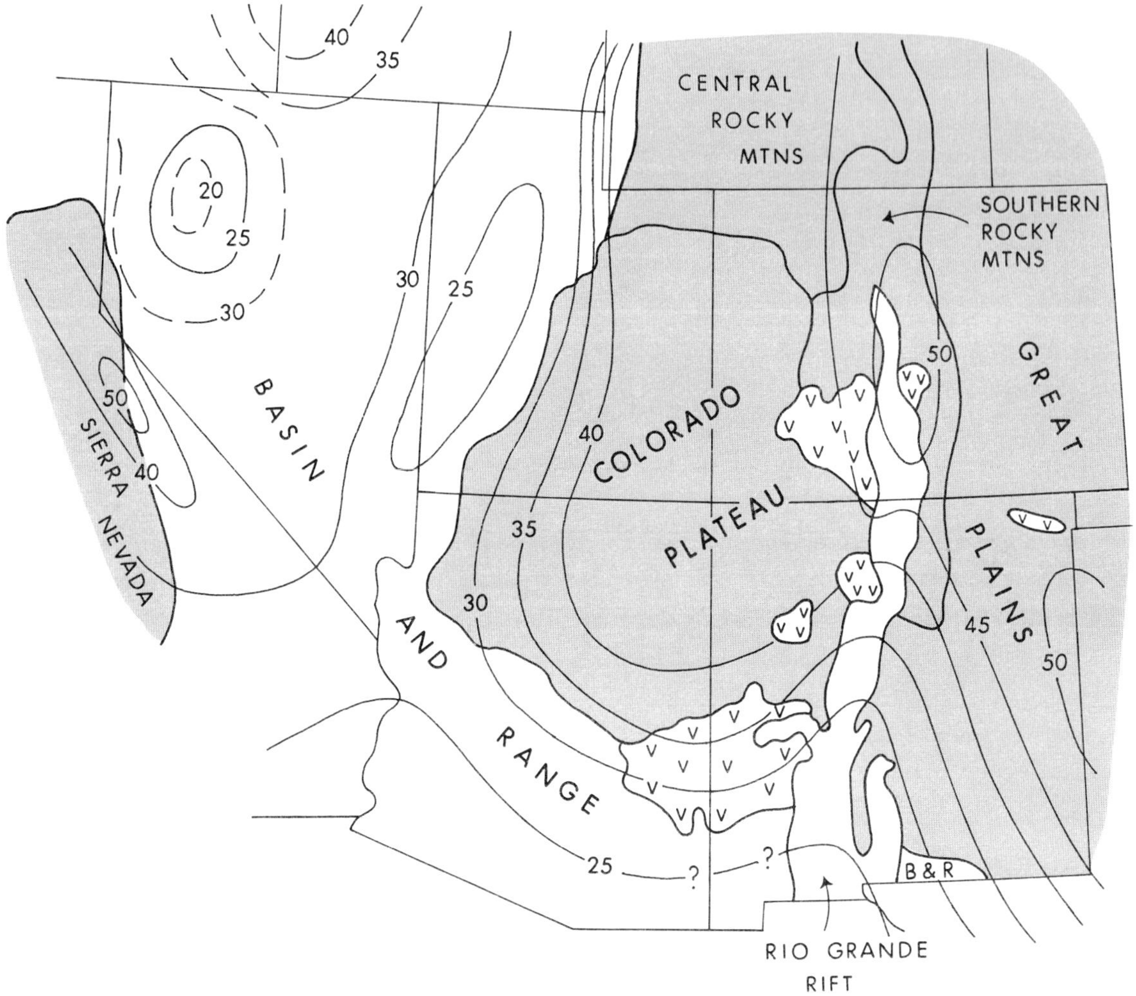

Figure 6. Contour map of crustal thickness in the southwestern United States. Contour interval 5 km (after Smith, 1978; Keller et al., 1979).

Geophysical Signature of the Rio Grande Rift and the Greater Basin and Range Province

Crustal Thickness

Seismic refraction and surface wave dispersion studies have shown that anomalously thin crust is one of the primary characteristics of the Rio Grande rift and Basin and Range provinces (Figure 6). The Rio Grande rift is a zone of abrupt thinning of the crust between the Great Plains to the east, where normal cratonal thicknesses reach 50 km (30 mi), and the Colorado Plateau to the west, where thicknesses are a somewhat transitional 40–45 km (25–28 mi). Within the rift itself the crust ranges in thickness from as low as 25 km (15 mi) in the south to up to 50 km (30 mi) at its northern terminus (Keller et al., 1975, 1979; McCullar and Smithson, 1977; Smith, 1978; Olsen et al., 1979). In the Basin and Range proper, the crust is similarly thin with 25-km (15-mi) thicknesses occurring around the margin of the Colorado Plateau in Utah and Arizona, and a very thin 20 km (12.5 mi) observable in northwestern Nevada.

Upper Mantle Seismic Velocity

The velocity of the compressional wave Pn (the P-wave refracted along the MOHO) derived from regional seismic refraction studies reliably indicates the relative rigidity of the subcrustal lithosphere. Beneath the Great Plains (craton) velocities of 8.0–8.2 km/s are observed, suggesting normal, high-rigidity lithosphere (Figures 7, 8) (Keller et al., 1975, 1979; McCullar and Smithson, 1977; Smith, 1978; Olsen et al., 1979). However, Pn velocities as low as 7.4–7.6 km/s have been recorded within the Rio Grande rift and Basin and Range provinces (Figures 7, 8). These extremely low upper mantle velocities in areas of regional extension suggest anomalously low rigidity and may reflect the upwarp or rise of the asthenosphere-related thermal instability during the rifting process. The Colorado Plateau, on the basis of an

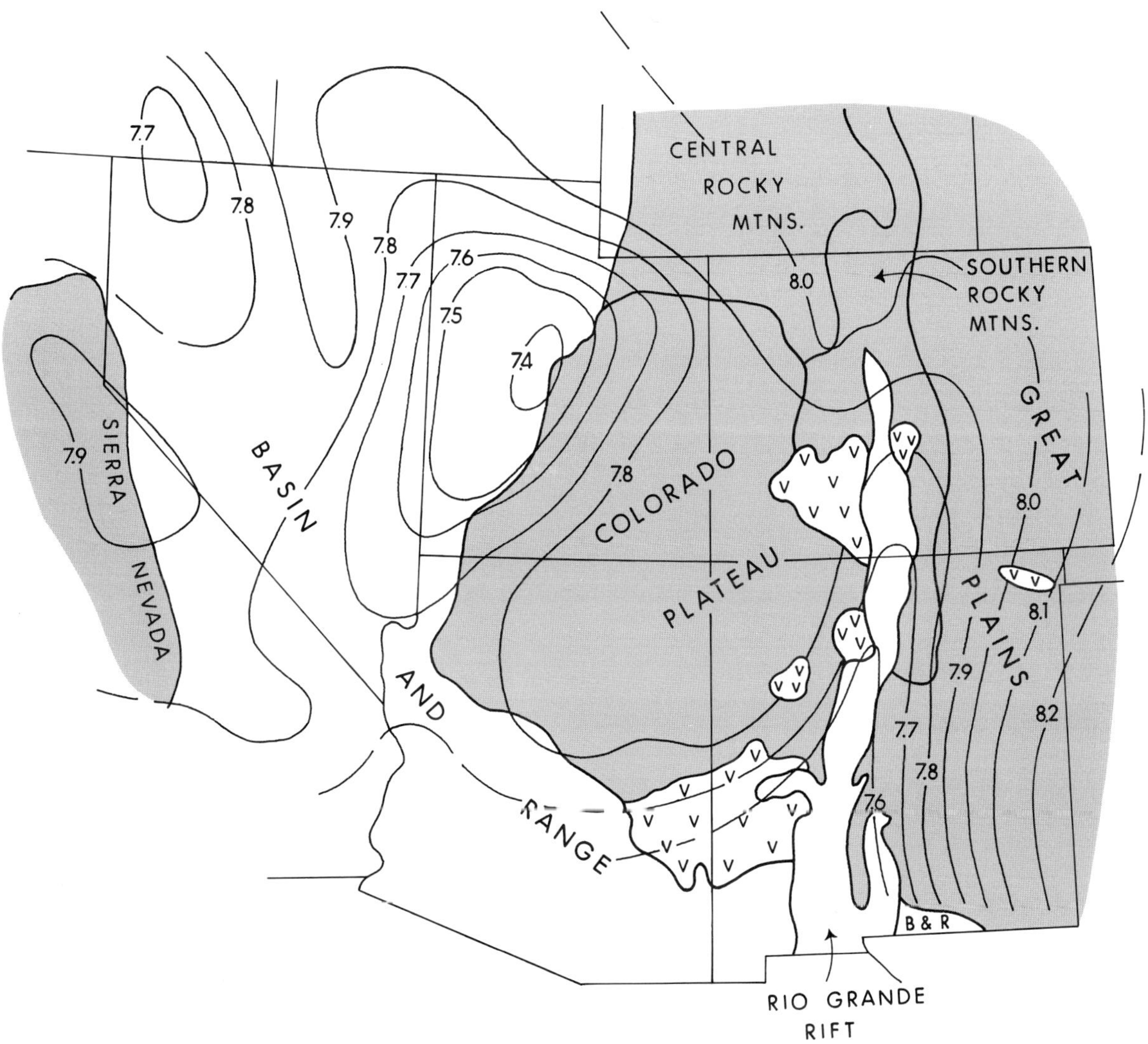

Figure 7. Contour map of compressional wave (Pn) velocities (km/s) in the southwestern United States (after Smith, 1978; Keller et al., 1979).

intermediate Pn velocity of 7.8 km/s, seems to be a somewhat transitional outlier of the craton rifted away along the Rio Grande rift (Figures 7, 8).

Upper Crustal Low-Velocity Channel

Refraction work and surface wave dispersion studies across the boundary between the Colorado Plateau and the eastern Basin and Range province have delineated the presence of an apparent low-velocity channel in the upper crust at a depth of between 10 and 12 km (6 and 7.5 mi), (Shurbet, 1969; Braile et al., 1974; Keller et al., 1975). This apparent low-velocity channel can be observed as a velocity reversal at stations B and C on the regional crustal section in Figure 8. This zone of low velocity has been interpreted to be the seismic manifestation of the brittle/ductile transition in the crust at an abnormally shallow depth due to the thermal effects of active extensional deformation (Smith et al., 1975).

To date, no such low-velocity channel has been detected beneath the Rio Grande rift. However, the detailed regional refraction work necessary to resolve the existence of such a zone has not been done in the area.

Seismicity

The Rio Grande rift and the greater Basin and Range province display high levels of regional seismicity. Activity in the Rio Grande rift is dominantly in the northern and central portions of the rift (Figure 9) (Sanford et al., 1979). Focal mechanisms for these earthquakes clearly document regional extension subnormal to the rift margins (Figure 9) (Sanford et al., 1979). Focal depths for earthquakes in the Rio Grande

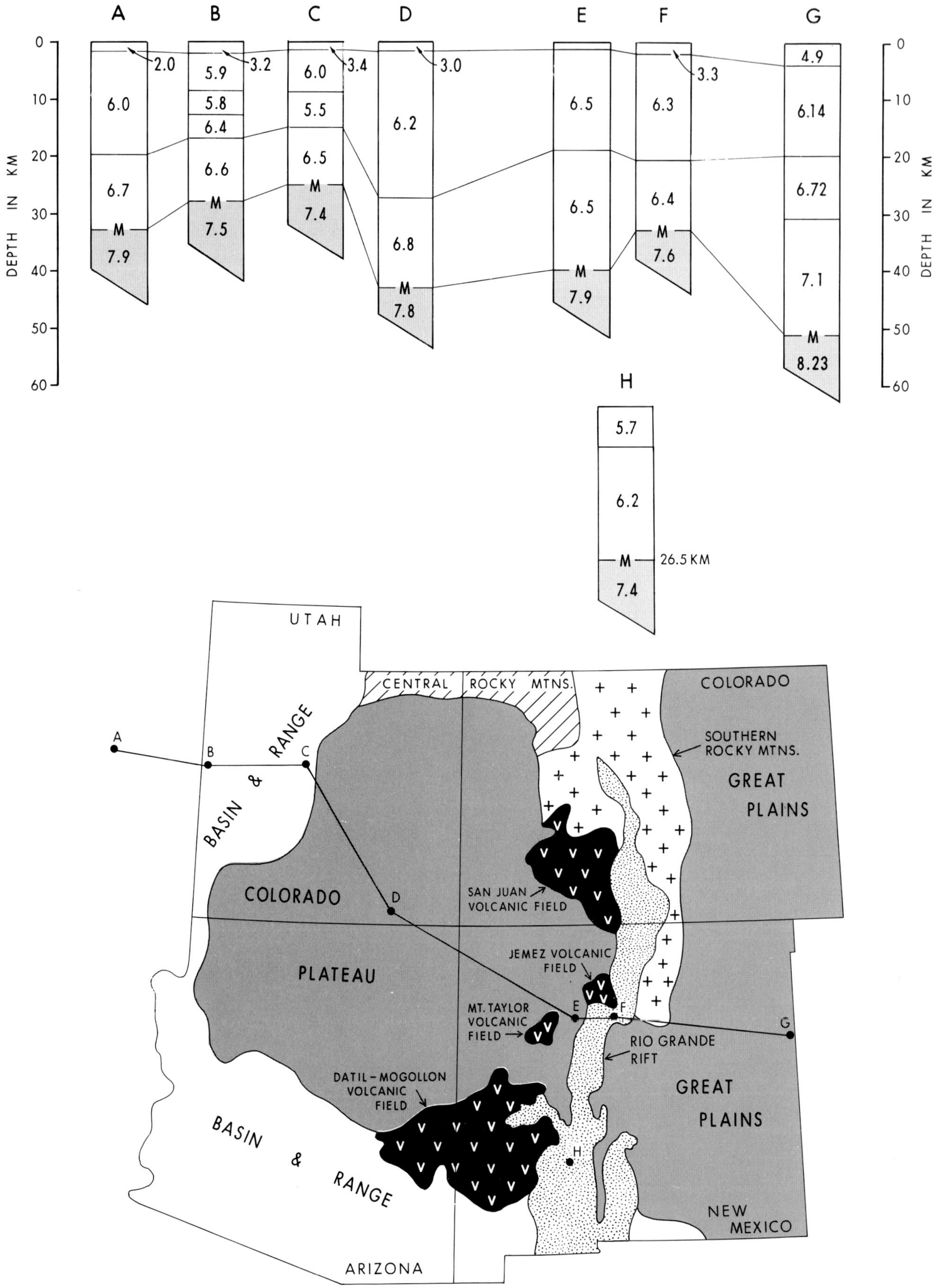

Figure 8. Crustal structure based on refraction data, along a transect across the southwestern United States. Velocities in km/s. Note velocity inversion in upper crust at points B and C. (Data after Keller et al., 1975; McCullar and Smithson, 1977; Olsen et al., 1979.)

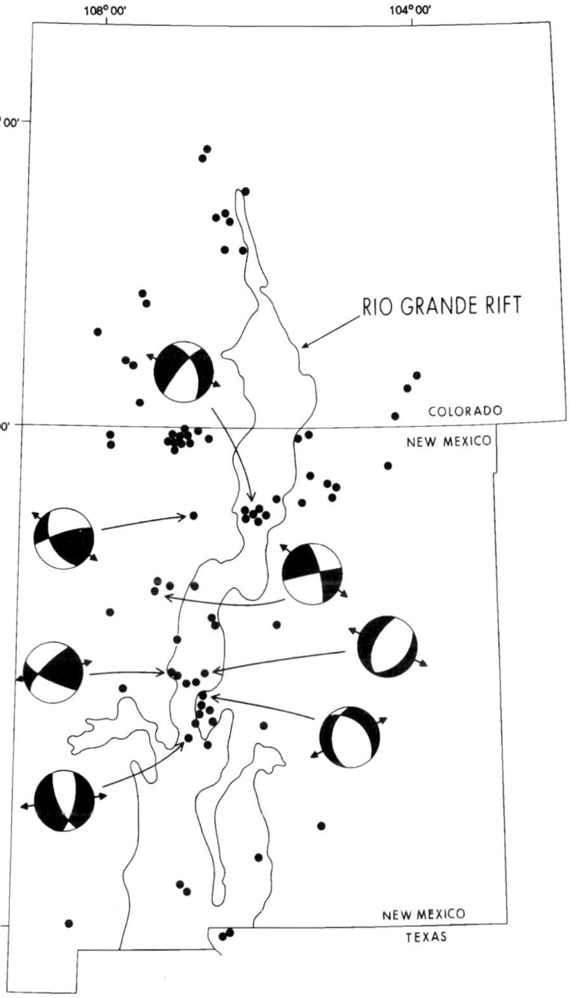

Figure 9. Map of epicenters of seismic events along the Rio Grande rift between 1973 and 1978 showing selected focal mechanisms (after Sanford et al., 1979). Black areas represent compressional quadrants. Short arrows depict tension axes.

rift, and Basin and Range province overall, are dominantly in the upper 10 km (6 mi) of the crust (Figure 10) (Sanford et al., 1979; Smith, 1978). This 10 km (6 mi) depth compares quite favorably with that of the upper crustal low-velocity channel in the eastern Basin and Range (Figure 10) interpreted to represent the brittle/ductile transition in the crust (Smith et al., 1975).

Heat Flow

High heat flow is characteristic of the entire Rio Grande rift/Basin and Range province. The thermal high associated with the Rio Grande rift is especially well defined and parallels the topographic axis of the rift quite closely (Figure 11) (Reiter et al., 1975, 1978, 1979). Locally, heat flow reaches levels as high as 15 heat flow units (normal continental heat flow is about 1.1 heat flow units).

Gravity

Regional Bouguer gravity profiles across the rift (Figure 12) (Cordell et al., 1978) clearly reflect a regional gravity low that extends quite a distance east and west of the topographic axis. Such a gravity low would be expected in a region underlain by a rising low-density asthenosphere that results in a negative density contrast with the flanking normal lower lithosphere. The gravity profiles show local residual positive anomalies directly over the rift axis (Figure 12). These local highs are very likely related to mafic intrusions into the lower crust (Bridwell, 1976).

Magnetotellurics

Regional magnetotelluric studies along the Rio Grande rift by Hermance and Pedersen (1980) have defined conductivity anomalies in the crust beneath the rift (Figure 13). These workers recognized zones of high electrical conductivity in the crust beneath the northern and southern portions of the rift at depths of 13–21 km (8–13 mi) and 23–33 km (14–20 mi), respectively, and interpreted these anomalies to be due to mid- to lower crustal magma bodies. Sanford et al. (1973, 1977) previously postulated such bodies on the basis of earthquake seismograms. In addition, Brown et al. (1980) interpreted deep magma bodies on the basis of reflection seismic data (COCORP [Consortium for Continental Reflection Profiling]). Jiracek et al. (1979) also reported a highly conductive zone in the crust beneath the central portion of the rift at a depth of 10–15 km (6–9.5 mi) (Figure 13), which is shallower than the 18–21 km (11–13 mi) depth of the magma body reported by Sanford et al. (1973, 1977).

Tectonic Evolution of the Rio Grande Rift/Basin and Range Province

The tectonic development of the Rio Grande rift began at approximately 32 Ma when regional crustal extension, stretching, and sagging were superimposed on the southern portion of the Laramide foreland belt of basement uplifts (Eaton, 1979) (Figure 14). Extension occurred along the easternmost edge of an exceptionally wide zone of volcanism that lay to the east of a major low-angle subduction zone along the west coast of North America (Eaton, 1979; Atwater, 1970) (Figure 14). Eruption of abundant rhyolitic and basaltic andesite volcanics and intrusion of local, shallow felsic plutons accompanied this phase of rift development, which lasted from 32–20 Ma (Figure 15).

Between 20 and 10 Ma, extensive crustal rifting, accompanied by voluminous eruptions of alkaline basalts (Figure 16) and active normal faulting, occurred behind (to the east of) the volcanic and plutonic arc (Eaton, 1979) (Figure 14), which by this time had moved significantly westward due to the probable steepening of the subducting slab (Coney and Reynolds, 1977). These authors concluded that regional extension was likely in a rear-arc setting in a region of convective thermal upwelling.

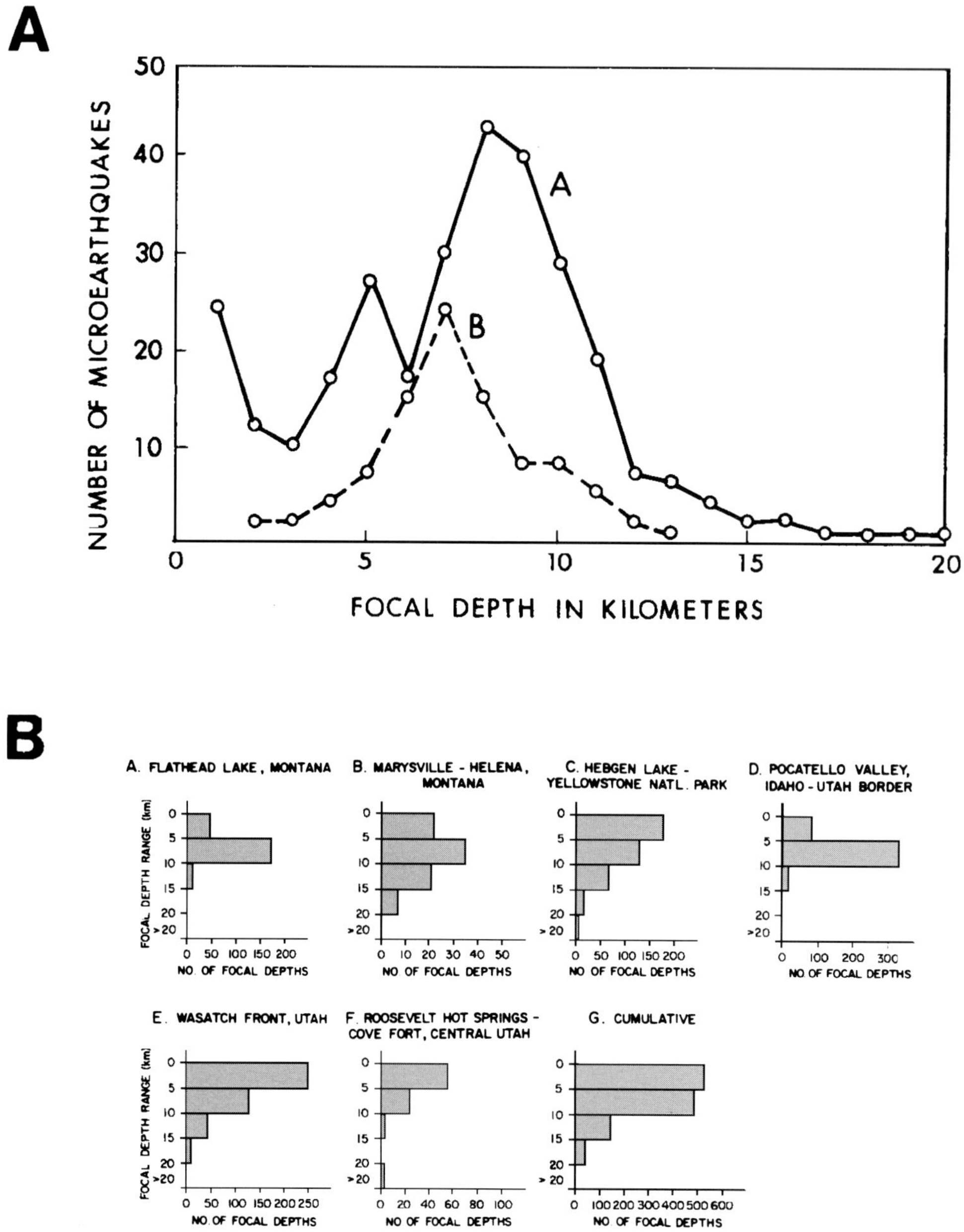

Figure 10. (A) Histogram of earthquake focal depths in the Albuquerque-Socorro portion of the Rio Grande rift. Line A = depths from P-wave determinations; line B – depths from S-wave determinations (after Sanford et al., 1979). (B) Histograms of earthquake focal depths from locations along the eastern Basin and Range and intermountain seismic belt (after Smith, 1978).

From 10 Ma to the present, regional extension has been taking place east of a major transform plate boundary, the San Andreas fault system (Figure 14). During this time, transform faulting progressed from southern California northward to Cape Mendocino, California, in response to the subduction of the East Pacific rise and the attendant cessation of subduction and adjustment of relative plate motion. Transform motion at the plate boundary subjected the western United States to regional dextral shear stresses (Atwater, 1970), resulting in the clockwise rotation of the Colorado Plateau block. Volcanism in the Rio Grande rift during the last 10 Ma (Figure 16) has largely been basaltic in nature and varied in composition from alkali-olivine basalts to more true tholeiitic basalts (Morgan et al., 1986).

Strong uplift took place between 7 and 4 Ma along the Rio Grande rift, perhaps due to the culmination of mantle upwelling in response to accelerated westward motion of the Colorado Plateau (Chapin, 1979).

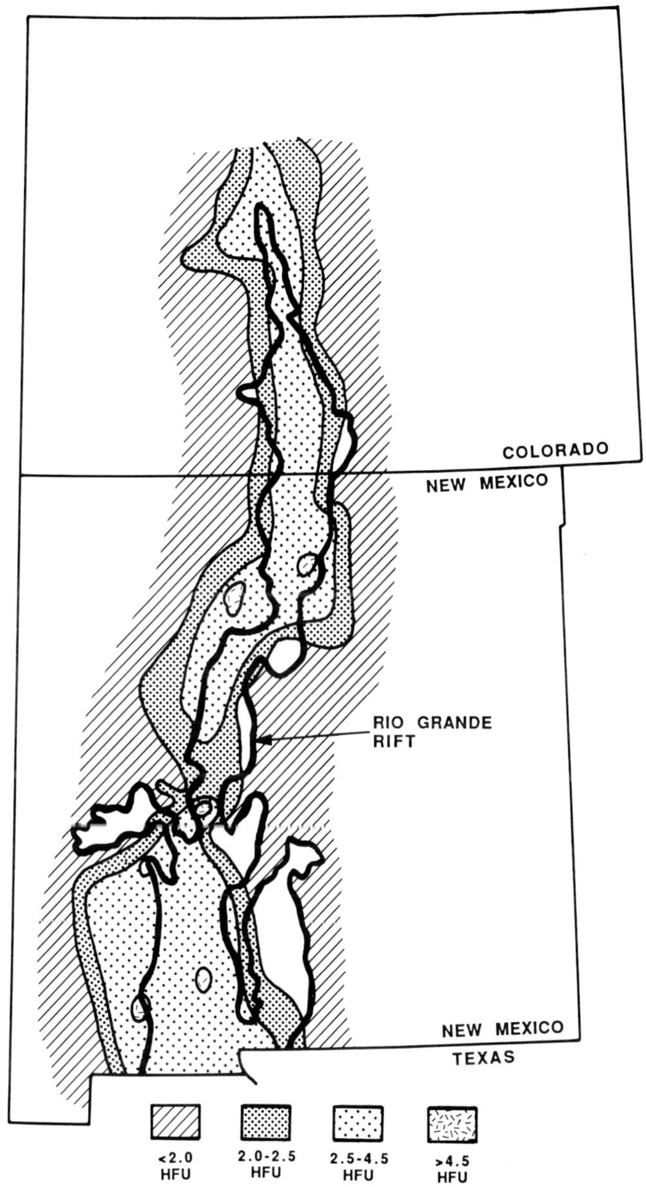

Figure 11. Contour map of heat flow along the Rio Grande rift (after Reiter et al., 1975, 1978, 1979). HFU = heat flow unit.

Increased runoff from uplifted highlands integrated previously internal regional drainage patterns into an axially flowing ancestral Rio Grande river, and widespread erosion surfaces were graded to the local base level of the river (Chapin, 1979; Lozinsky, 1988). Middle Pleistocene capture of the Rio Grande at El Paso (Kottlowski, 1958), diversion of the river to the Gulf of Mexico, and increased runoff due to the Pleistocene pluvial cycle resulted in lower base levels and deep dissection of the basins along the rift (Chapin, 1979). The dissection of basins isolated basalt-capped mesas and elevated pediment surfaces, thus producing the topography seen today (Chapin, 1979).

GEOLOGY AND TECTONICS OF THE ALBUQUERQUE BASIN

Location and Physiography

With approximate dimensions of 160 km (100 mi) north–south and 64 km (40 mi) east–west, the Albuquerque basin is one of the major sedimentary basins of the Rio Grande rift. To the east, the basin is abruptly bounded by the Sandia, Manzanita, Manzano, and Los Pinos mountains (Figures 2, 23), all of which are uplifted blocks of Precambrian crystalline basement rocks. Sandia Crest marks the culmination of the eastern bounding uplifts at 3256 m

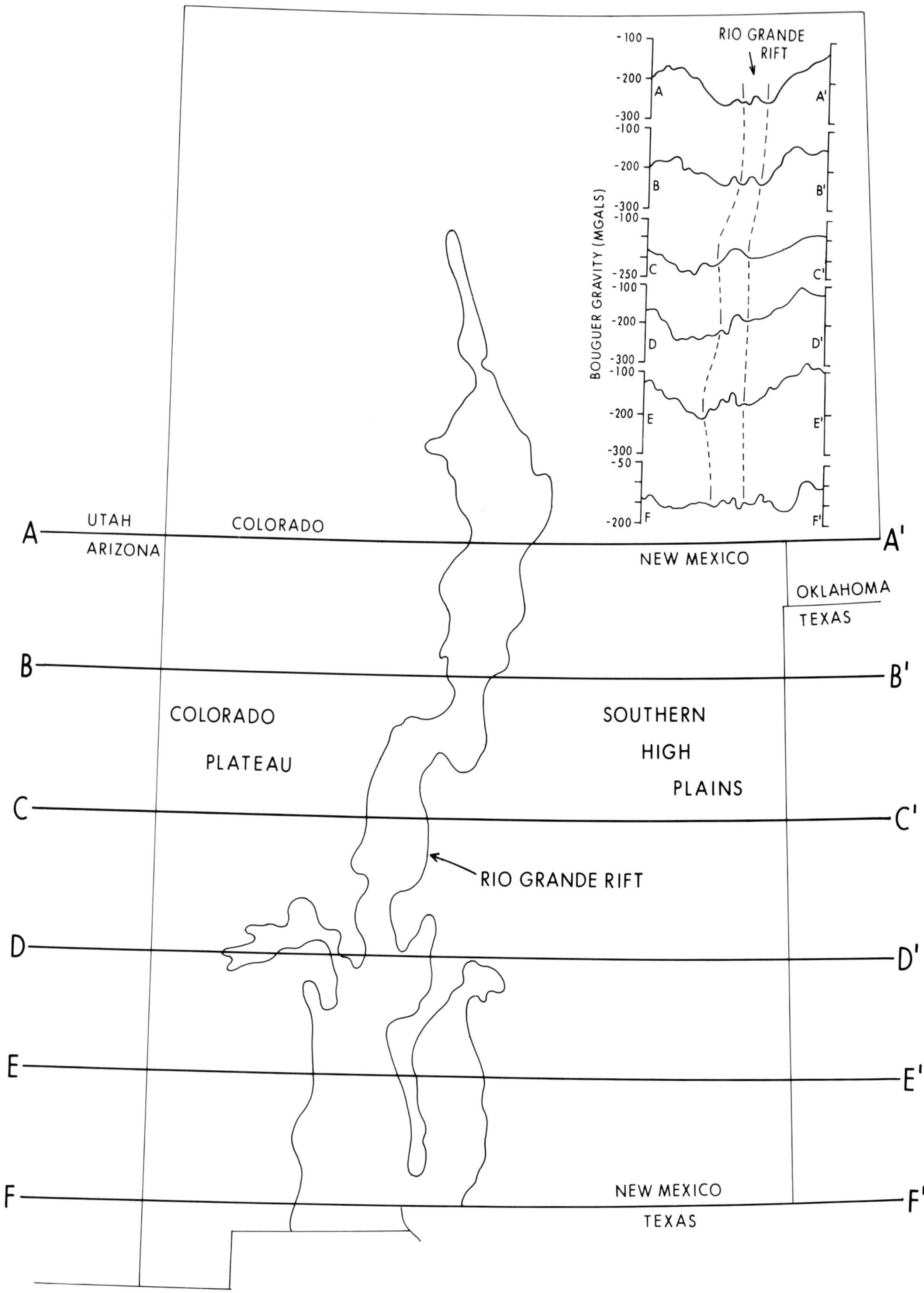

Figure 12. Bouguer gravity profiles across the Rio Grande rift in New Mexico (after Cordell et al., 1978).

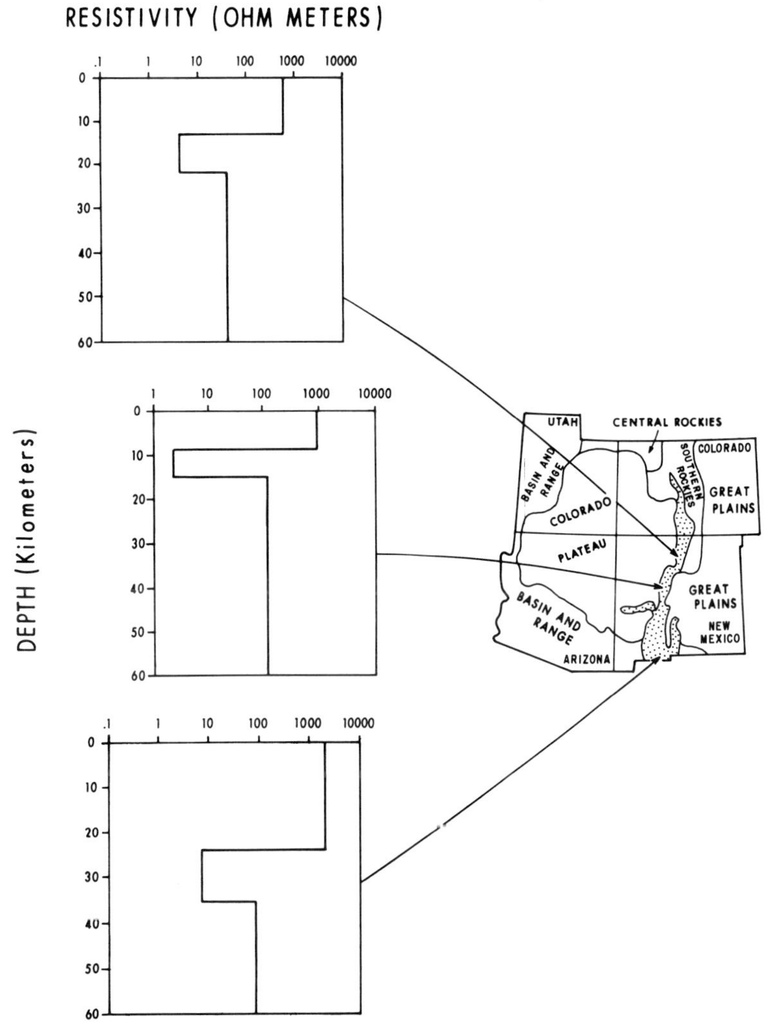

Figure 13. Magnetotelluric profiles from the Rio Grande rift in New Mexico (after Hermance and Pederson, 1980; Jiracek et al., 1979).

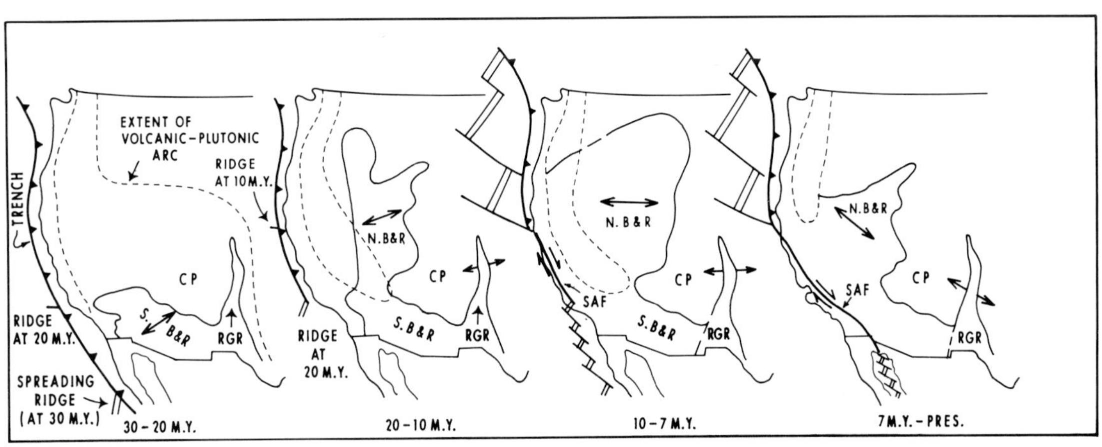

Figure 14. Conceptual model for the Cenozoic tectonic evolution of the western United States (after Eaton, 1979; Atwater, 1970). S. B&R = south Basin and Range, N. B&R = north Basin and Range, CP = Colorado plateau, RGR = Rio Grande rift, SAF = San Andreas fault.

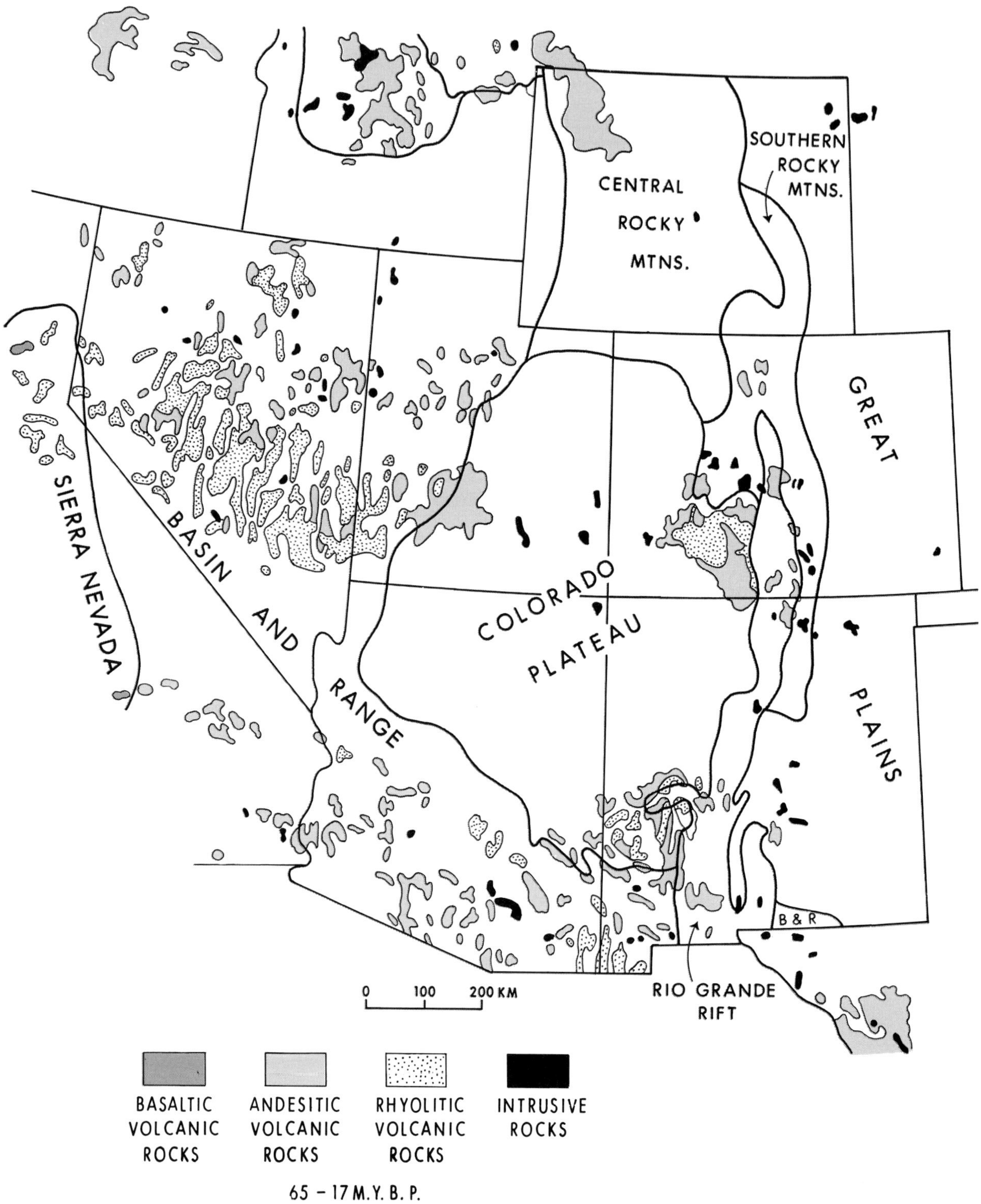

Figure 15. Distribution of volcanic and intrusive igneous rocks in the southwestern United States between 65 and 17 Ma (after Stewart and Carlson, 1978).

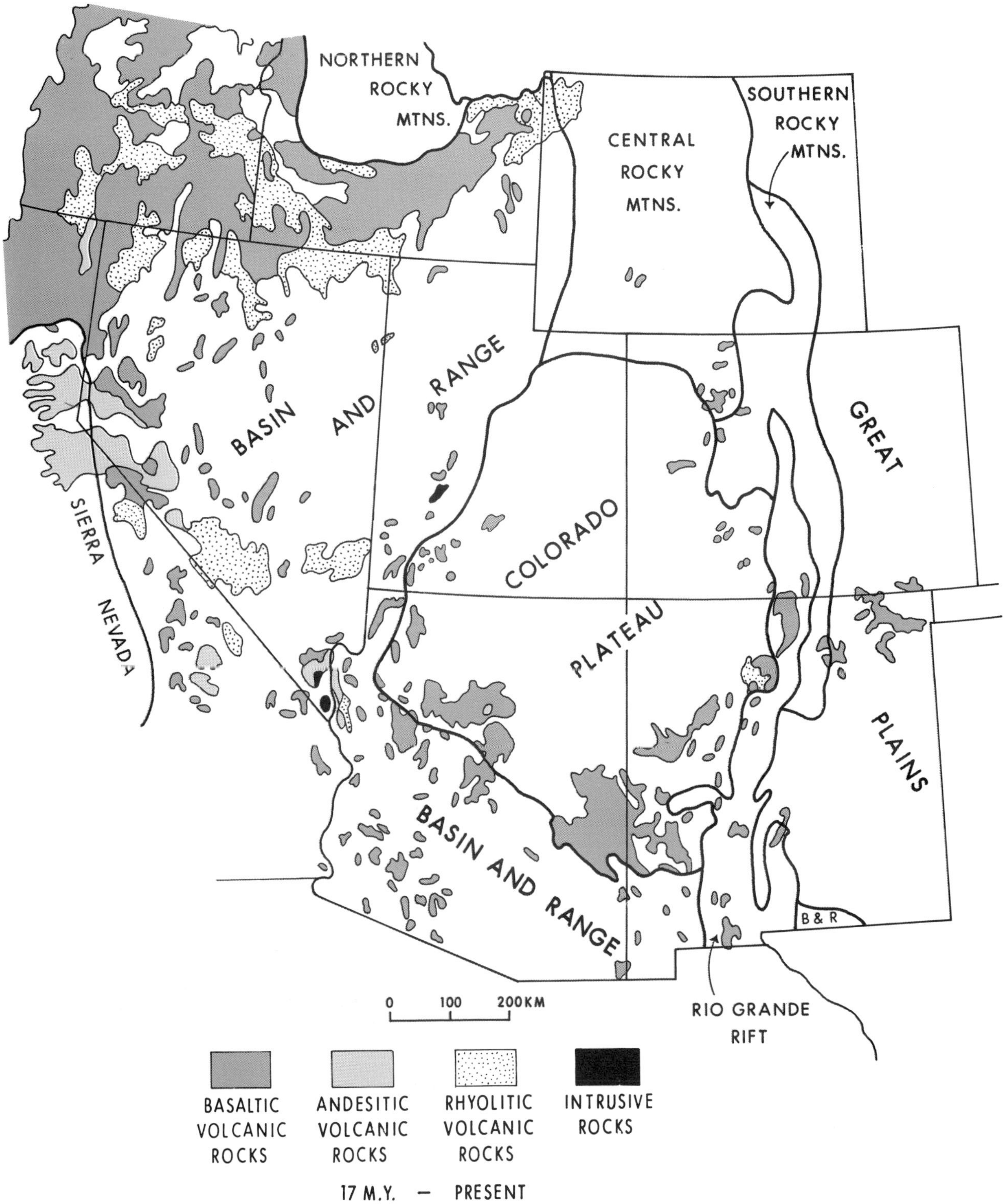

Figure 16. Distribution of volcanic and intrusive igneous rocks in the southwestern United States between 17 Ma and the present (after Stewart and Carlson, 1978).

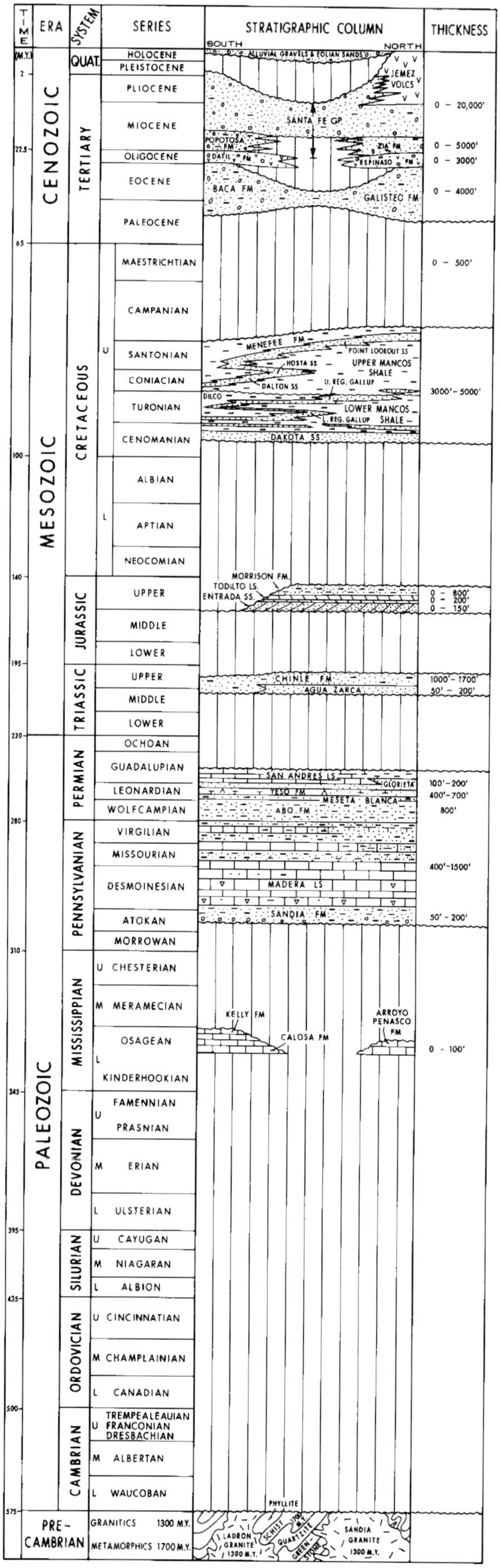

(10,678 ft). To the west lies the Colorado Plateau, capped by the Mt. Taylor Cenozoic volcanic center and a cover of Paleozoic–Mesozoic sedimentary rocks (Figures 2, 23). Rocks of the plateau dip gently eastward into the basin making the western margin of the rift essentially a monoclinal ramp. Thus, the western border of the Albuquerque basin is topographically subdued when compared with the impressive uplifts to the east.

Two exceptions to this rule are the Ladron Mountains and Nacimiento uplift, which are uplifted blocks of Precambrian basement that form the southwest and northwest corners, respectively, of the basin (Figures 2, 23). Both of these ranges attain elevations of greater than 2750 m (9000 ft).

North of the Albuquerque basin lies the Jemez Mountains, a major Cenozoic volcanic center, the southern Sangre de Cristo Range of the Rocky Mountains, and the intervening Espanola basin of the Rio Grande rift (Figure 2). South of the Albuquerque basin, the rift becomes less distinct topographically and is typified by a regional pattern of alternating basins and ranges (Figures 2, 3).

Stratigraphy

As revealed on regional reflection seismic data (Figure 23) and in well penetrations (Figure 24), the Albuquerque basin is filled with up to or greater than 7050 m (23,000 ft) of Tertiary continental clastic sedimentary rocks. Neogene rift fill makes up at least 4900 m (16,000 ft) of this and is composed largely of fanglomerates, alluvial, eolian, and playa deposits of the Santa Fe Formation (Figure 17). Underlying Eocene rocks of the Galisteo and Baca formations were deposited in shallow prerift basins and are dominantly fluvial in nature, although some organic-rich lacustrine facies are observed at places (Figure 17). The Eocene section is typically no greater than 900 m (3000 ft) in thickness, but locally may reach 2150–2450 m (7000–8000 ft) in thickness in the deep basin. Lozinsky (1988), in a study of the basin-fill deposits, determined thicknesses reaching about 4600 m (15,000 ft) for the Santa Fe, 2150 m (7000 ft) of Oligocene volcaniclastics, and 450 m (1500 ft) of Baca/Galisteo.

Up to 3300 m (11,000 ft) of Paleozoic and Mesozoic rocks make up the prerift sedimentary basement on which the Tertiary fill was deposited (Figure 17). Paleozoic rocks range in age from Mississippian to Permian and range from marine limestones to near coastal continent clastics. Exposures of the Paleozoic section occur along the entire uplifted eastern margin of the basin and on the southwestern basin flank (Figure 23). Mesozoic rocks are Triassic–Late Cretaceous in age. Triassic formations are continental in nature and dominated by fluvial sandstones and mudstones. Jurassic rocks also were deposited in continental environments and range from fluvial and

Figure 17. Generalized time-stratigraphic section of the Albuquerque basin region.

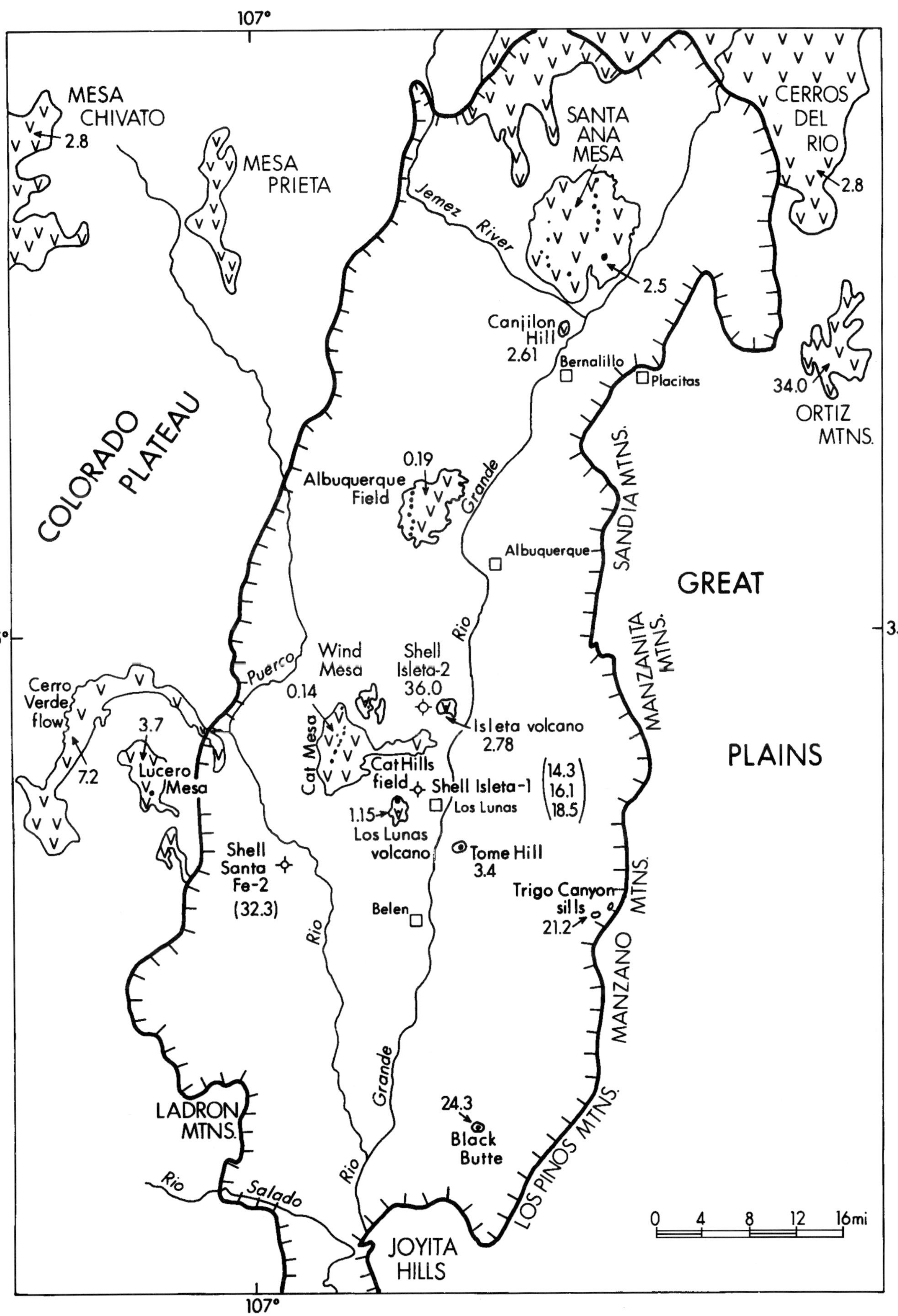

Figure 18. Map of radiometric age dates for Tertiary volcanic and plutonic rocks in the Albuquerque basin and vicinity (after Bachman and Mehnert, 1978; Kudo et al., 1977; White, 1979; and Shell, unpublished data).

Figure 19. Panoramic photograph, looking north, of the southern Manzano Mountains showing the location of Laramide reverse faults.

Figure 20. Panoramic photograph, looking southeast, of the northern Sandia Mountains illustrating the northerly plunge of the range.

MONTOSA
FAULT
IP
P€M

SANDIA MTNS
P€
U
D
RINCON FAULT

Figure 21. Photograph, looking toward the northern end of the Sandia Mountains in the vicinity of Placitas, illustrating the northerly plunge of the range (note the northerly dip on the carbonate-capped hills at the base of the main range).

eolian sandstones to lacustrine limestones and associated evaporites. Upper Cretaceous units of the Albuquerque basin are dominated by regressive coastal-marine sandstones intertonguing with open-marine shales. In addition, Cretaceous deposits also are characterized by coastal-plain fluvial and paludal environments rich in coal.

The Precambrian basement is broadly divisible into 1300-m.y.-old granitic rocks and 1700-m.y.-old metamorphic rocks ranging from pelitic schists and phyllites to quartzites, amphibolites, granitic gneisses, metarhyolites, and greenstones (Condie, 1976; Kelley, 1977; White, 1979) (Figure 17).

Cenozoic Volcanism

Based upon the results of various workers (Kudo et al., 1977; Bachman and Mehnert, 1978; Kelley and Kudo, 1978; Baldridge, 1979; White, 1979; Kudo, 1982) the Cenozoic volcanic history of the Albuquerque basin region can be broken into three broad stages. Prerift volcanism occurred in the middle Oligocene (32–37 Ma) and was dominated by rhyolitic ash flows and shallow, level, felsic plutons of the Ortiz Mountains and nearby Cerrillos Hills, with some activity in the Joyita Hills (Figure 18). Data from two of Shell's exploration wells in the area reveal activity in this time range in the form of a shallow basaltic sill in Shell Santa Fe #2 well and an ash flow tuff in Shell Isleta #2 well (Figure 18). Incipient rifting in the late Oligocene (28–21 Ma) is marked by basaltic-andesite flows and shallow intrusives. Much of this activity took place in the southernmost portion of the basin and surrounding areas (Figure 18). Volcanism associated with the main stage of rifting in the Miocene through Holocene (19 Ma–present) is predominantly basaltic (alkaline) and is concentrated primarily in the northern and central portions of the basin. Eruptive activity has continued almost up to the present day, with the youngest basalt flows being those of the Albuquerque and Cat Mesa fields at 0.19 and 0.14 Ma, respectively (Figure 18).

Prerift Deformational History

Major reverse faulting during the Laramide orogeny was the most significant episode in the prerift deformational history of the Albuquerque basin region. Reverse faults that resulted in the local overturning of Paleozoic and Mesozoic formations are

exposed in the Manzano and Los Pinos mountains (Montosa and Paloma faults; Figures 19, 23), and in the Nacimiento uplift (Nacimiento fault; Figure 23). Upper Cretaceous rocks have been folded along the Tijeras fault zone at the southern end of the Sandia Mountains (Figure 23), but it is unclear whether this deformation was related to Laramide reverse motion along the Tijeras fault, or due to late Tertiary wrenching related to a transcurrent component of movement along the rift in general (J.F. Callender, personal communication, 1979). However, the Sandia uplift broadly has the configuration of a north-plunging antiform (Figure 20) with Eocene rocks of the Galisteo Formation resting in slight angular unconformity on Upper Cretaceous rocks at the north end of the range (Kelley, 1977), suggesting some uplift of probable Laramide age. In the southwest quadrant of the Albuquerque basin is the modest Lucero uplift, whose uplift is related to motion along the Comanche fault, a reverse fault that induced monoclinal upwarp and gravity sliding, probably in the late Laramide (Callender and Zilinski, 1976; Kelley, 1977) (Figures 22, 23).

Another school of thought exists that relates these basin margin uplifts and structures associated with them to dextral wrench faulting of either Laramide or Tertiary (rift related) age (Chapin, 1983; Hayden, 1991; Hayden and Mawer, 1991). Work by Molzer and Erslev (1991) farther north in the Laramide foreland (Owl Creek Range, Wyoming) describes structural patterns very similar to those in the uplifts flanking the Albuquerque basin, and attribute them to Laramide thrusting with a modest component of oblique slip. Lewis and Baldridge (in press) presented arguments that relate the uplift and structures along the Lucero front as largely due to footwall uplift during Tertiary rifting. Considerable room for discussion and debate clearly still exists relative to this topic.

In the region, some late Paleozoic deformation related to the development of the ancestral Rocky Mountains is likely (Chapin and Seager, 1975), although its character is not well defined.

Precambrian basement rocks exposed in the basin flanking uplifts were isoclinally folded and metamorphosed to high grade at approximately 1700 Ma, and later were intruded by major granitic plutons at 1300 Ma (Kelley, 1977; Condie, 1976). In addition, there is strong evidence of significant Precambrian strike-slip motion (postmetamorphic) along the Tijeras fault system (J.F. Callender, personal communication, 1979) that correlates well with similar motion along the Pecos-Picuris fault in the southern Sangre de Cristo Range to the north (Miller et al., 1963).

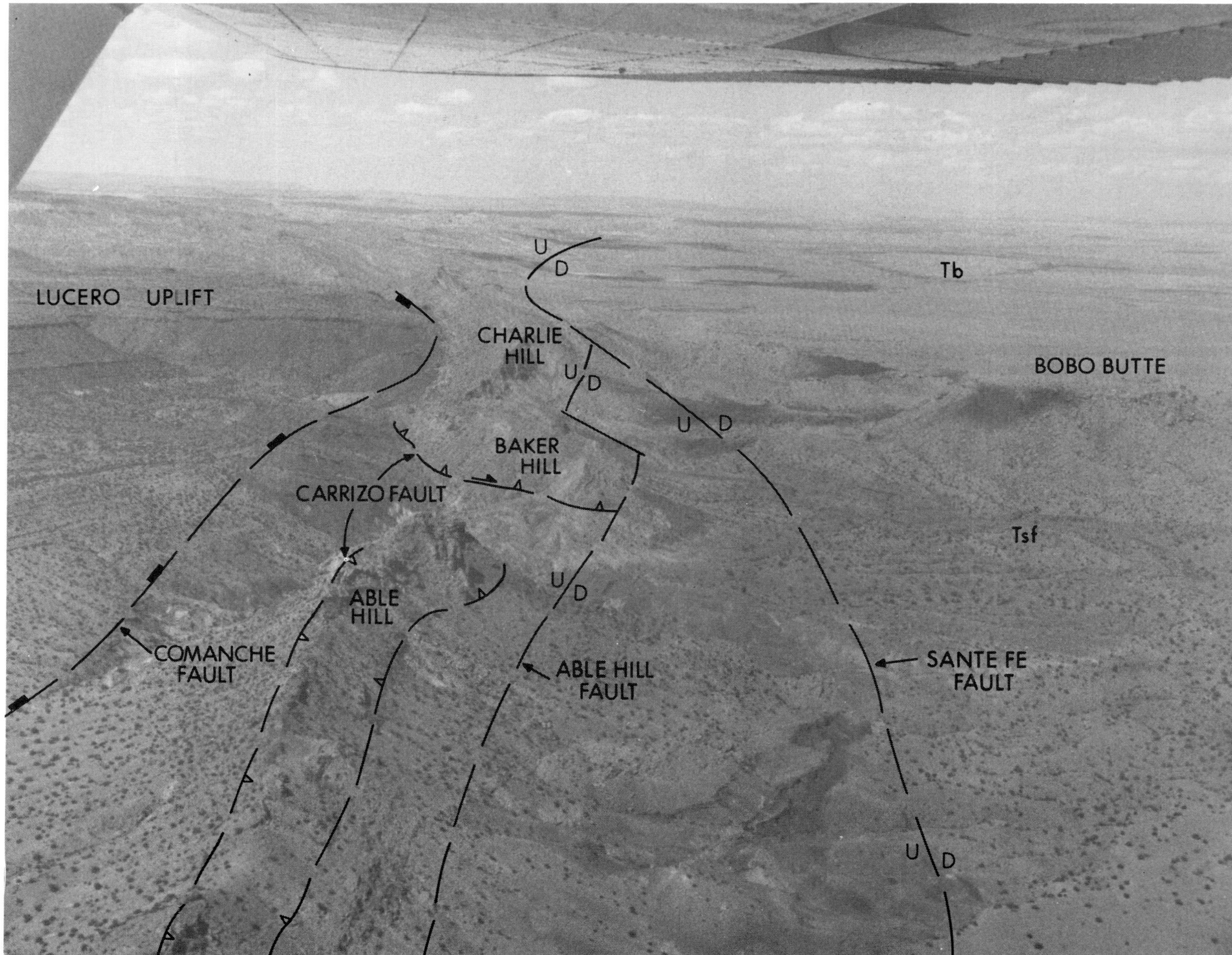

Figure 22. Aerial photograph, looking north, of the western margin of the Albuquerque basin in the vicinity of Carrizo Arroyo, illustrating the major Laramide and Tertiary bounding structures of the basin.

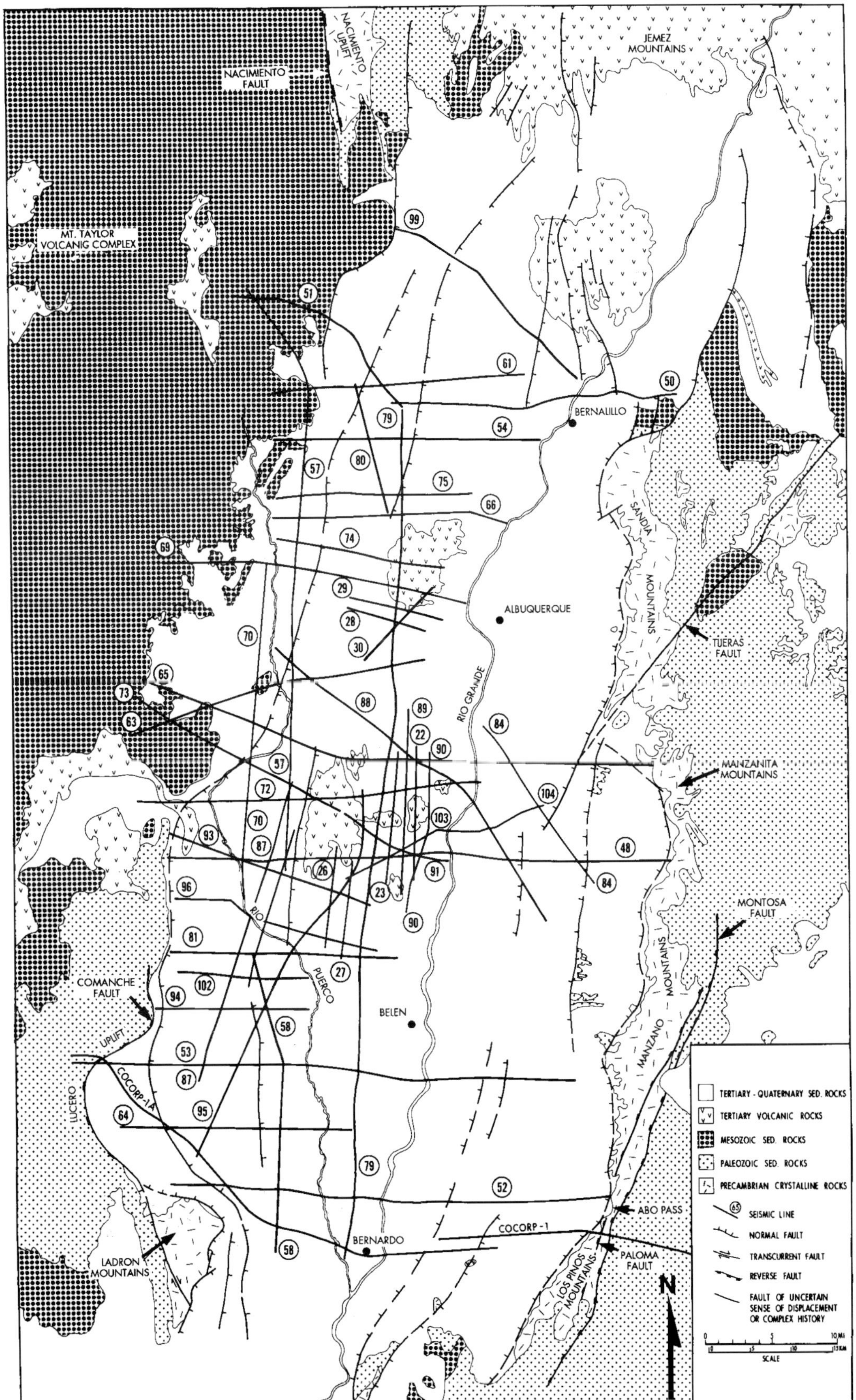

Figure 23. Map showing general outcrop geology of the Albuquerque basin and the reflection seismic control used in this study.

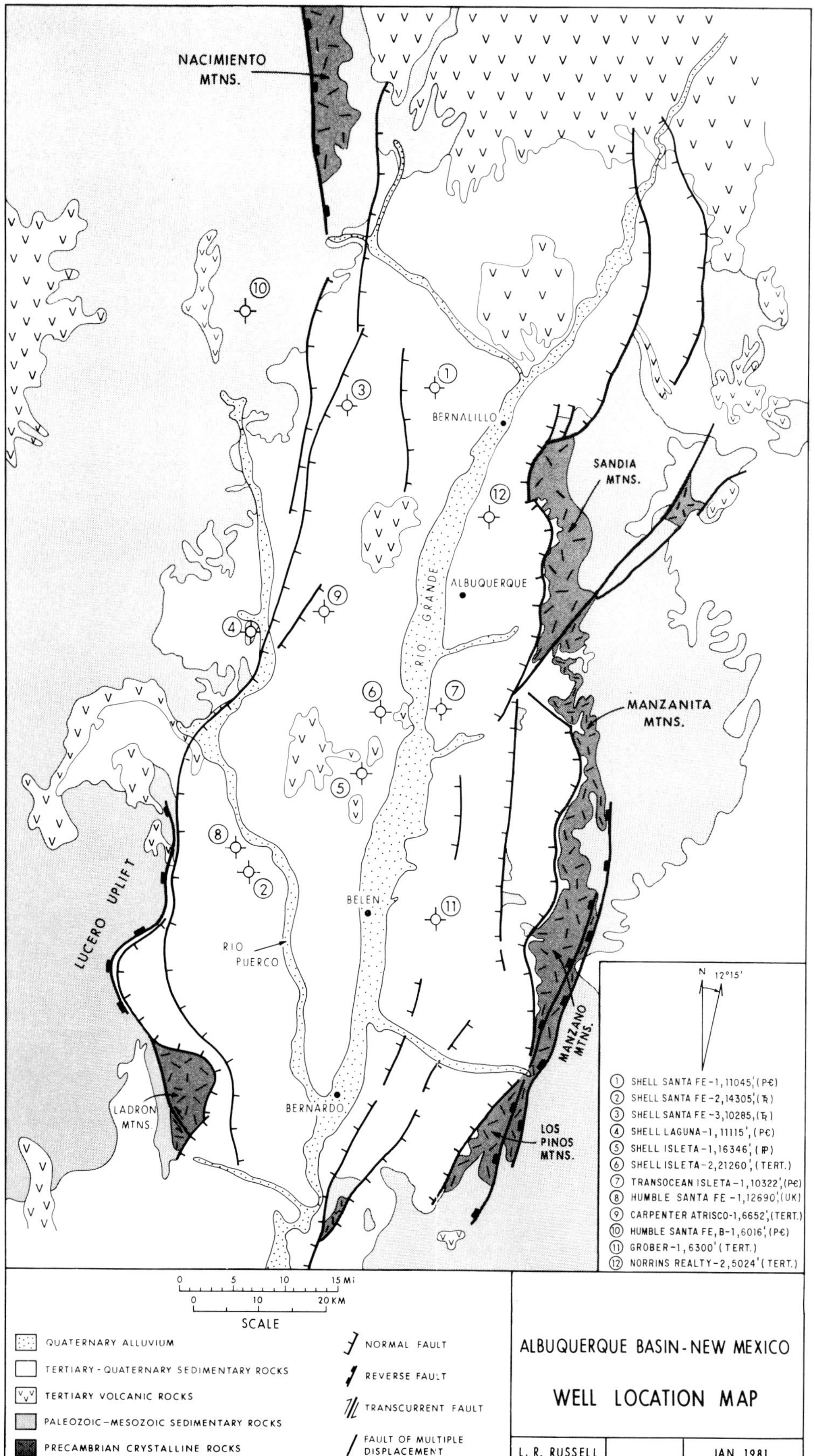

Figure 24. Map showing key exploration wells in the Albuquerque basin.

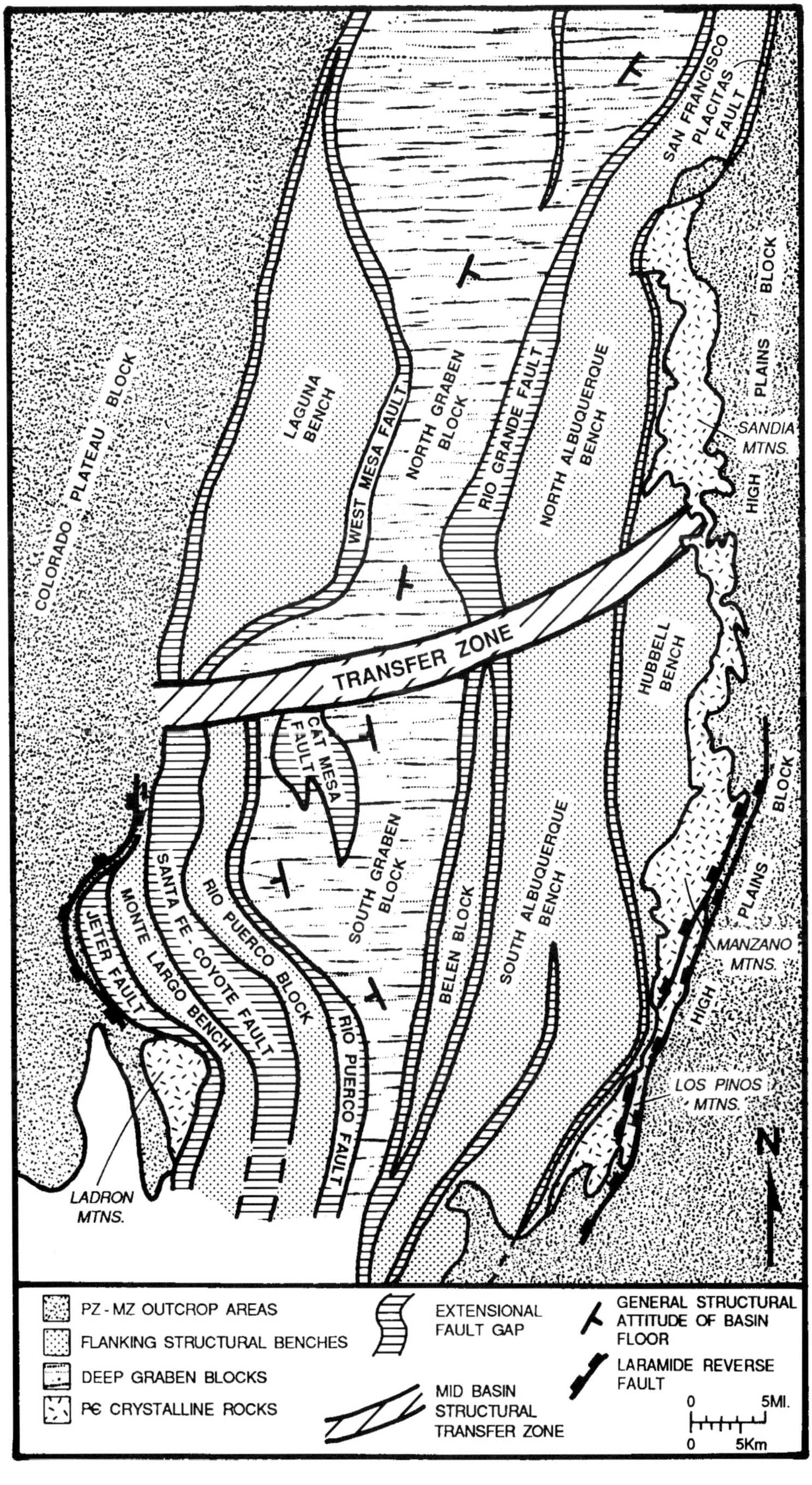

Figure 25. Generalized structural framework map of the Albuquerque basin depicting the structural configuration of the pre-Tertiary "basement" (mostly top of Upper Cretaceous).

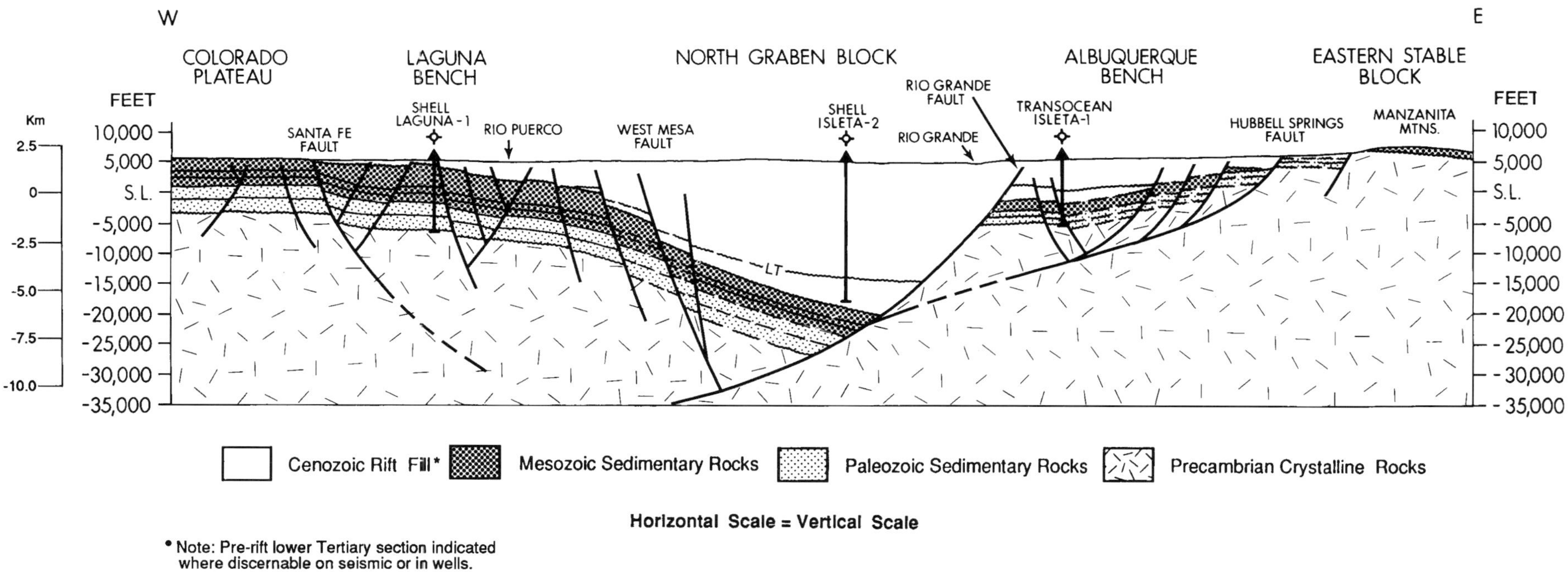

Figure 26. Geologic and seismic sections illustrating the structural configuration of the southern portion of the North Albuquerque basin.

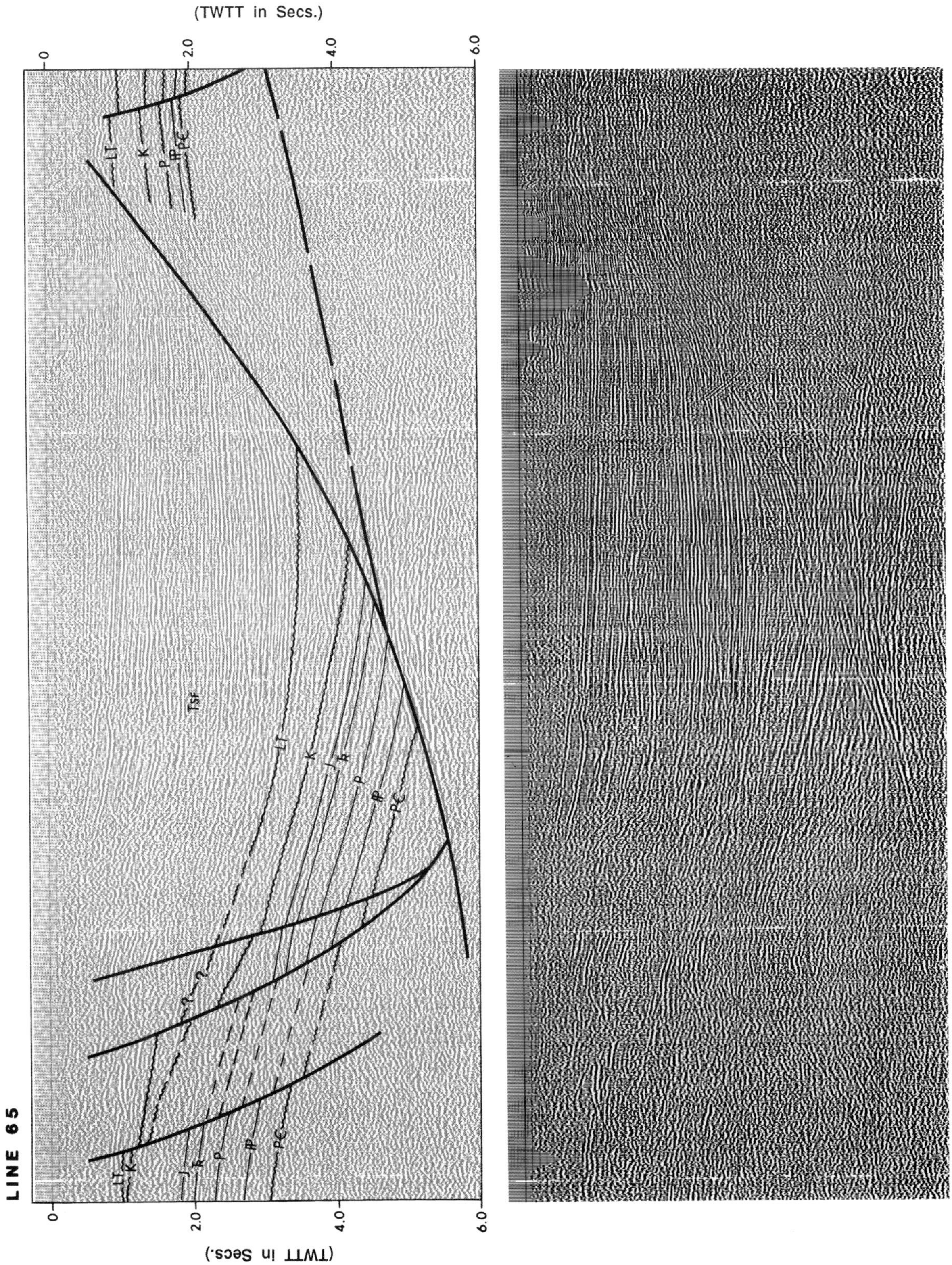

Figure 26 (continued). TWTT = two-way traveltime.

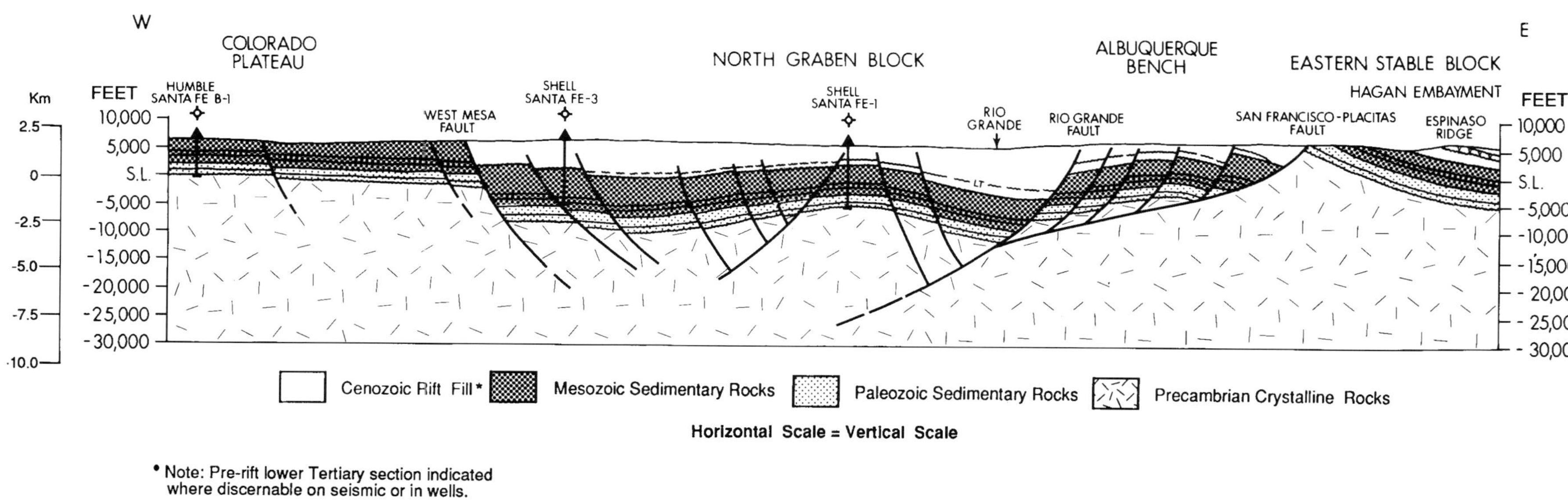

Figure 27. Geologic and seismic sections, as well as outcrop relationships, at the northern end of the Sandia Mountains illustrating the structural configuration of the northern portion of the North Albuquerque basin.

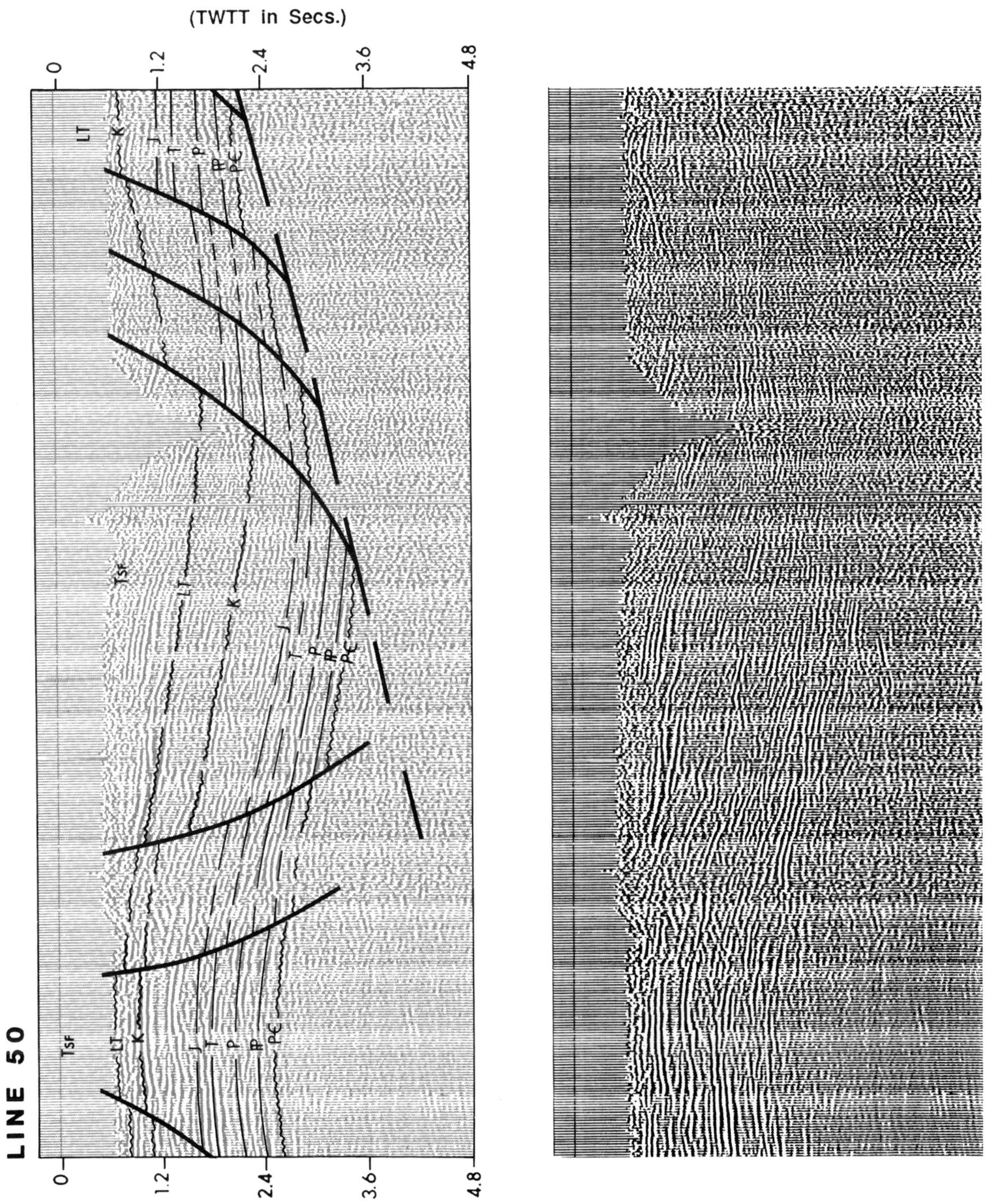

Figure 27 (continued).

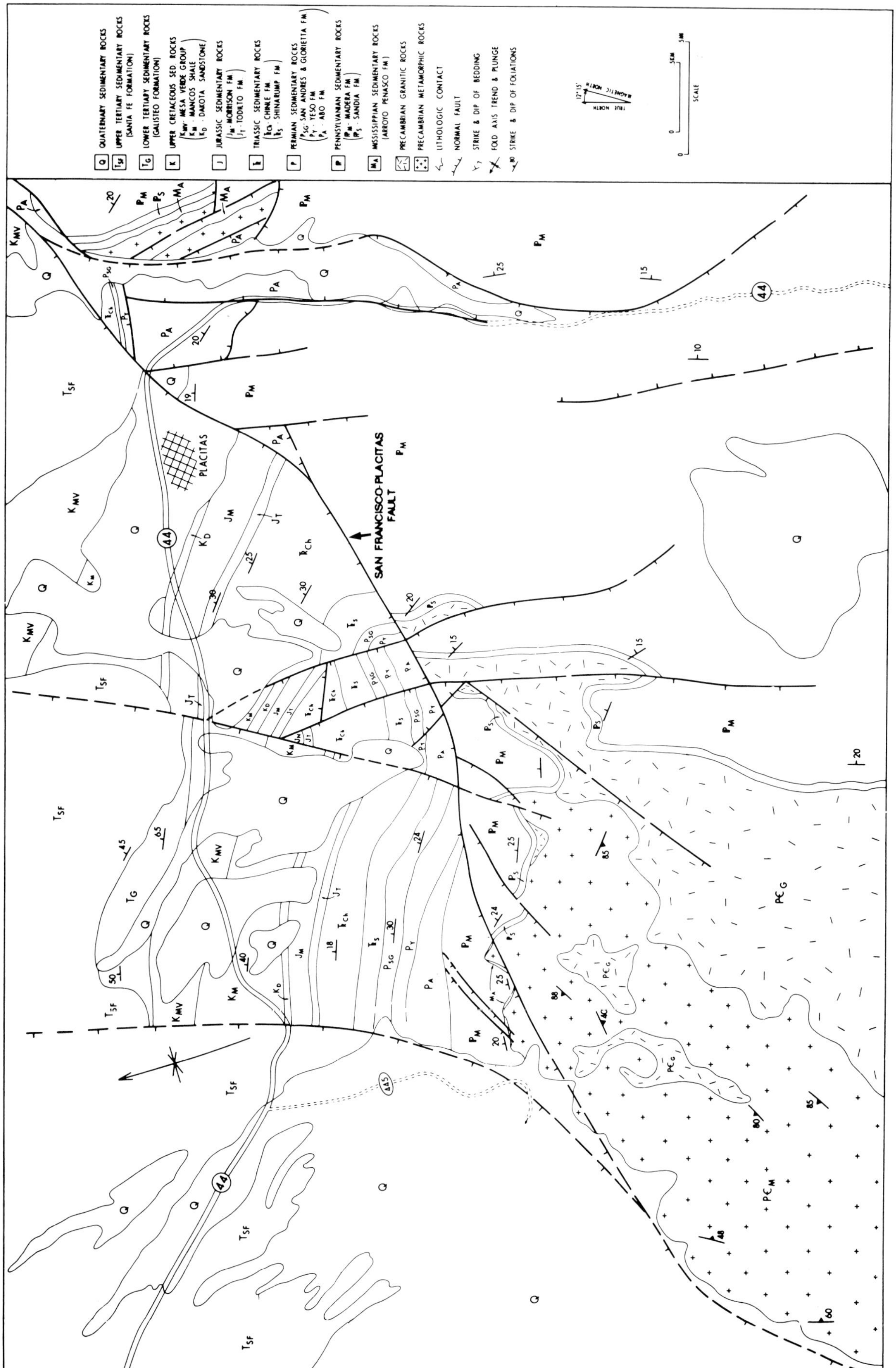

Figure 27 (continued).

Furthermore, evidence is growing for probable reverse movement along these faults at some point during the Precambrian (J.F. Callender, personal communication, 1979).

Structural Framework of the Basin

The structural framework map shown in Figure 25 is based upon regional reflection seismic data (Figure 23) supplemented by well control (Figure 24). This map depicts the structural complexity of the prerift "basement" (mainly top of Upper Cretaceous) and illustrates that the Albuquerque basin is made up of two subbasins (North and South graben blocks; Figure 25) downdropped along major low-angle normal faults of opposing polarity. The North graben is controlled by the west-dipping Rio Grande fault, whereas the South graben is principally controlled by the east-dipping Santa Fe–Coyote fault (Figure 25).

The basin deeps are quite narrow and reach maximum depths (to Mesozoic and older "basement") of approximately 7350 m (24,000 ft) in the North graben and 7050 m (23,000 ft) in the South graben. Flanking the narrow basin deeps to the east and west are a variety of structural benches and detached blocks. To the east, both the North and South grabens are bounded by a broad structural bench extending the entire length of the basin (North and South Albuquerque bench; Figure 25). The North graben is bounded to the west by a major east-dipping antithetic normal fault (West Mesa fault; Figure 25), which separates the deep basin from a wide, shallow structural bench (Laguna bench; Figure 25). In contrast, the South graben is flanked to the west by two smaller blocks or benches (Monte Largo bench and Rio Puerco block) produced by displacements along the Jeter and Santa Fe–Coyote faults (Figure 25).

One conspicuous aspect of basin geometry is that the physiographic margins of the basin do not correspond with the location of the major basin-controlling normal faults. The main structural depressions occur much farther into the basin, with structural relief at the topographic margins generally being comparatively subdued. Rigid and relatively undeformed basement terraces flank the basin to the east and west (High Plains block and Colorado Plateau block; Figure 25).

The northern and southern terminations of the basin are quite different in structural character. The north basin floor rises gradually northward, essentially in ramplike fashion, out of the structural deep. In the south, the pattern is more complex and obscure, with the major characteristic being a distinct narrowing and shallowing of the basin. The southern terminus has been termed the "Socorro Constriction" (Kelley, 1952).

Major Basin-Controlling Normal Faults, Northern Albuquerque Basin

The Rio Grande fault, the major basin-controlling structural element of the northern Albuquerque basin, is a west-dipping listric normal fault having a throw of between 4600–6100 m (15,000 and 20,000 ft; Figures 25, 26, 27). The fault is expressed seismically by strong fault-plane reflections and prerift (Cretaceous and older) basement reflections that distinctly tilt into the fault plane (Figures 26, 27). Dips range from greater than 60° on the upper portion of the fault to as low as 17° along its lower reaches. This master fault appears to be flattening at approximately 10 km (33,000 ft) when the seismic data are depth corrected using an average velocity-depth function for the basin extrapolated from exploration-well velocity surveys.

A related low-angle normal fault along the northeastern flank of the Albuquerque basin is the San Francisco–Placitas fault (Figure 25). To the west, this low-angle fault merges with, or is truncated at depth by, the higher angle Rio Grande fault (Figure 27). The low-angle character of the San Francisco–Placitas fault is suggested by faint fault-plane reflections on seismic data (Figure 27) and by viewing geologic map relationships of the northern Sandia Mountains (Figure 27) using the down-structure method of Mackin (1950). As discussed previously, the Sandia uplift is a northerly plunging antiformal structure (Figures 20, 21, 28), as is illustrated on the geologic map (Figure 27) by dips in the Pennsylvanian strata south of the arcuate surface trace of the fault. Dips change from due east, to northeast, and then northerly to slightly northwesterly at the northern end of the range. If one views this geologic map (Figure 27) employing the techniques of Macklin (1950), the result is an approximate cross sectional view of the area. This graphically supports the interpretation that the San Francisco–Placitas fault is low angle and listric in geometry, as suggested on seismic, and that the fault was uplifted and exposed as a result of the late Tertiary uplift of the Sandia range. In addition, geologic map relationships suggest rollover or reverse drag of the Mesozoic section into the San Francisco–Placitas fault, geometries common to listric fault systems.

This joint use of both seismic data and surface geologic map relationships allows for the more confident completion of the geologic section (Figure 27). The fact that the San Francisco–Placitas fault merges with or is truncated by the Rio Grande fault suggests that extension initially began on the San Francisco–Placitas fault to the east, and that with time, the main focus of extension shifted westward to the Rio Grande fault.

The overall listric, downward-flattening geometry of the Rio Grande fault system is of a style compatible with the pure shear model of extension (McKenzie, 1978). The depth of flattening of the fault apparently is in the range of 10–12 km (6–7.5 mi), which correlates well with the base of earthquake activity in the basin (Sanford et al., 1979), suggesting the fault system is decoupling at the top of the brittle-ductile transition in the crust.

Major Basin-Controlling Normal Faults, Southern Albuquerque Basin

In the southern Albuquerque basin, the east-dipping Santa Fe–Coyote, Cat Mesa, and Jeter faults form

Figure 28. Photograph looking at the northern end of the Sandia Mountains illustrating the northerly plunge of the range and the location of the San Francisco-Placitas fault.

the key structural elements of basin extension and have virtually the same geometric relationship to one another as the Rio Grande and San Francisco–Placitas faults in the northern basin (Figure 29). The Santa Fe–Coyote fault, with over 3050 m (10,000 ft) of throw, appears to be older than the Cat Mesa fault, which has 4600–5200 m (15,000–17,000 ft) of throw. Both fault systems are delineated seismically by strong fault-plane reflections, synthetic faults rooting into the master fault, and prominent tilt of prerift reflections into the fault planes (Figure 29). In outcrop, surface dips on the Santa Fe fault are about 65°, whereas dips as low as 15° are suggested on seismic at depths of approximately 10 km (33,000 ft).

In the southern part of the basin, the Coyote fault is the younger, dominant structure and truncates the earlier, lower angle Jeter fault (Figure 31). Seismic reflections from the Coyote fault suggest 15–17° dips at the depth of maximum seismic resolution.

The Jeter fault is well exposed along the northeastern and eastern fronts of the Ladron Mountains (Figure 6), where it dips basinward at angles as low as 26°. Features such as strongly slickensided fault planes, sheared Precambrian basement rocks in the footwall (Figures 32, 33), and slivers or horses of lower Tertiary, Permian, and Triassic strata (Black, 1964) collectively support the seismic interpretation that the major basin-controlling faults, and the Jeter fault specifically, involve basement rocks.

In contrast to the apparently pure shear geometry of the north subbasin, the low-angle, ramplike basin-controlling faults in the south subbasin are strongly analogous to the simple shear model of full lithospheric extension proposed by Wernicke (1985, 1986). The basin-bounding faults of the south Albuquerque basin show no signs of flattening or soling at mid-crustal depths. Instead, they appear to be continuing to perhaps full-crustal or lithospheric depths when seismic resolution is lost.

Palinspastic Analysis of the Basin

To test our structural interpretations of the seismic data, we attempted to volumetrically balance and restore our cross sections palinspastically. The reconstructions were accomplished using the computer program PREP, discussed in detail in Worrall and Snelson (1989). All reconstructions were made on the top of the Precambrian basement surface.

Palinspastic restorations across the northern and southern halves of the Albuquerque basin (Figure 11) have revealed a distinct variation in the amount of Tertiary extension. The northern Albuquerque basin has undergone a minimum of 17% extension, whereas the southern part of the basin has been stretched by at least 28% (Figure 34). These minimum extension values apply to reconstruction of a preextension basement assumed to be horizontal. We suspect, however, that after Laramide thrusting, the basement underlying the southern Albuquerque basin was structurally higher than the basement in the Manzano and Los Pinos mountains to the east as a result of reverse motion along the Montosa and Paloma faults. In this likely scenario, a greater, yet unconstrained, amount

of extension would have been necessary for the basement to reach present-day depths beneath the rift.

Mid-Basin Zone of Structural Compensation/Accommodation Zone

The differential amount and polarity of extension between the two subbasins necessitates the presence of a transfer zone or zone of structural accommodation in the basin center (Figures 25, 35). This accommodation zone between the northern and southern subbasins is a relatively narrow and enigmatic zone. Unfortunately, the seismic quality at this critical point in the basin, in general, is quite poor because of the effects of the extensive surface volcanic cover localized in this portion of the basin. However, a recently reprocessed key dip line across this mid-basin zone, interpreted by L.G. Garmezy (Figure 35), reveals geometries similar to those in negative flower structures associated with zones of transcurrent tectonics (Harding, 1985). This interpretation is structurally compatible with the zone being the locus of differential lateral motion accommodating the opposing extensional regimes of the northern and southern subbasins.

The location of this accommodation zone seems to correspond roughly with the basinward extension of the Tijeras fault zone at the southern end of the Sandia Mountains (Figure 23). The Tijeras system continues northeast of the basin into the High Plains, and perhaps southwest onto the Colorado Plateau. This continuation allows for the geometrically required transferring of the differential extension within the basin to structures that extend out of the basin boundaries, with the displacements dying out to the northeast and southwest.

Control of Tertiary Extensional Geometries by Older Structures

The question as to whether the geometry of Tertiary extension was influenced by preexisting zones of structural weakness in the crust can be addressed somewhat by the data presented here. As discussed earlier in the chapter, the Rio Grande rift is located along the zone of Laramide basement uplifts of the southern Rocky Mountains, arguing for some element of spatial control on the location of the rift. More detailed relationships can be seen when examining the Montosa and Paloma faults, which are west-dipping Laramide reverse faults of the southern Manzano and Los Pinos mountains (Kelley, 1977) (Figures 2, 19, 23). Compressional deformation associated with these two faults is best expressed in the Abo Pass area, where overturned bedding in Pennsylvanian rocks is observed.

Shell seismic data shot in the vicinity of Abo Pass display two west-dipping zones of faint basement reflections, which, when projected a short distance updip, correlate well with the outcrop location of these two reverse faults (Figure 31). West-dipping Tertiary normal faults in the basin apparently are rooting into the older reverse faults. Thus, the pattern of Tertiary extension seems to have been controlled by the preexisting structural grain in the basement. In this case, the roots of Laramide reverse faults have provided a zone of weakness or structural anisotropy into which the Tertiary normal faults sole. This interpretation also has

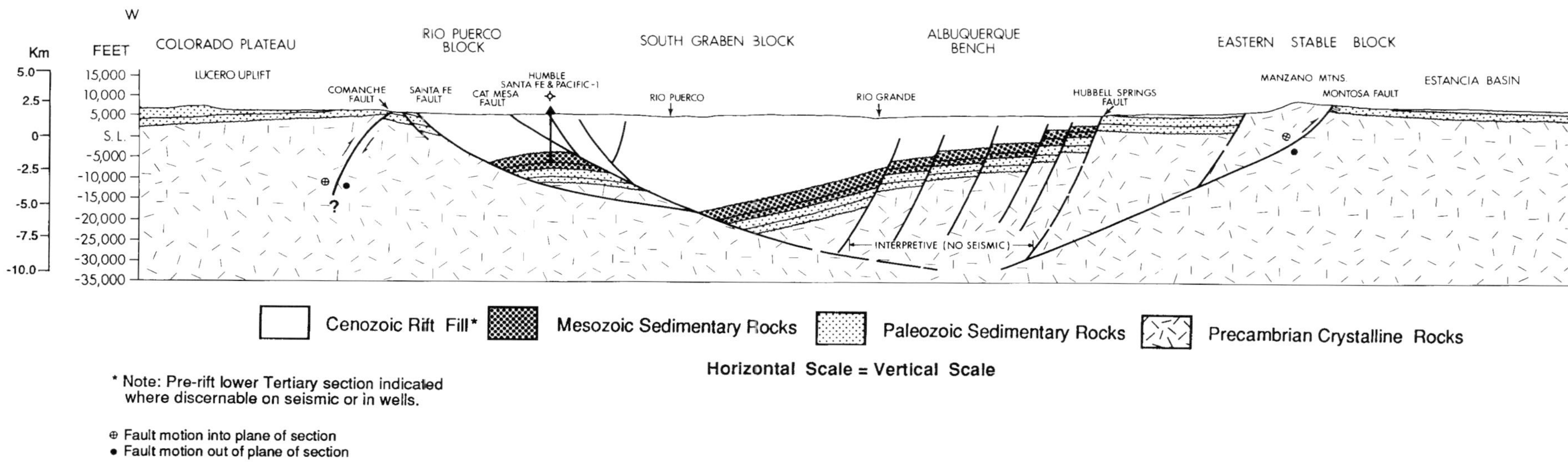

Figure 29. Geologic and seismic sections illustrating the structural configuration of the northern portion of the South Albuquerque basin.

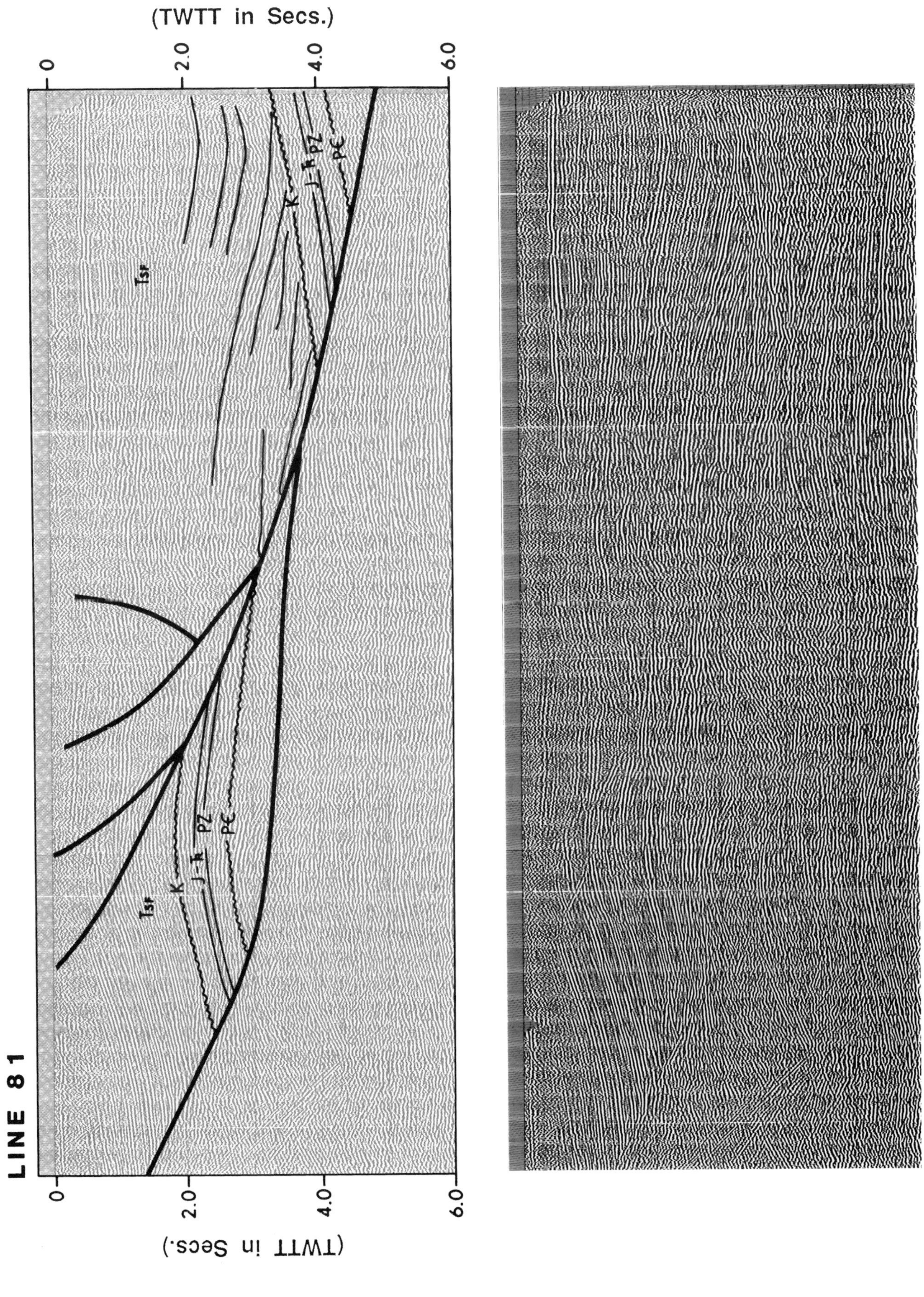

Figure 29 (continued).

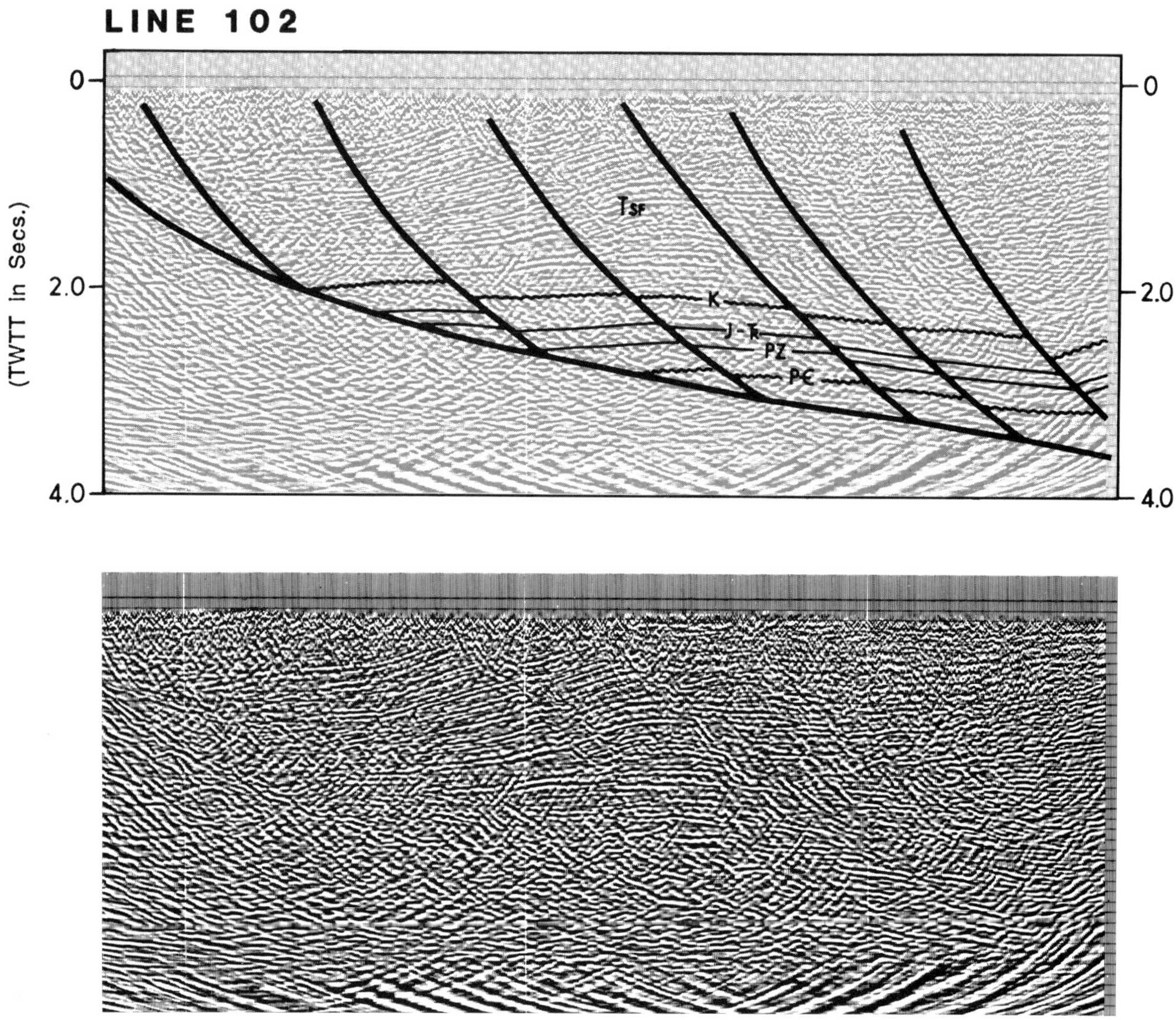

Figure 29 (continued).

been proposed by the COCORP group in reexamining the data from the Rio Grande rift (de Voogd et al., 1988).

Another probable example of basement fabric controlling younger extensional faulting occurs in the Ladron Mountains on the southwestern flank of the basin. The Jeter fault is well exposed along the eastern and northeastern fronts of the Ladron Mountains and dips basinward at angles of between 26 and 30°. Precambrian metamorphic units exposed in the range in close proximity to the Jeter fault are mainly well foliated and locally mylonitic quartzites, schists, gneisses, and granitic gneisses with easterly dips as low as 20–30° (Figures 36, 37). Seismic data shot immediately basinward of the exposures of the Jeter fault show it to be distinctly low angle in nature with dips between 20 and 30° (Figure 30). These relationships suggest the possibility that the geometry of Tertiary extension was controlled, at least in part, by the preexisting structural fabric in the basement.

In addition, the position of the mid-basin transfer or accommodation zone relative to the long-lived Tijeras fault zone (Figure 25) also argues for tectonic inheritance of older structural trends. At the very least, the presence of zones such as the Tijeras fault zone provided a convenient crustal weakness or anisotropy that was used during rifting.

Implications of This Study for Previous Interpretations of COCORP Data

Over the past decade, a number of interpretations of the COCORP deep-seismic profiles in this area have been published; many of these interpretations differ to varying degrees with the interpretations proposed in this chapter (Figure 38). Early interpretations of the COCORP lines in the southern Albuquerque basin suggested that a highly faulted and fractured, uplifted basement horst existed beneath the central portion of the basin. This horst block was attributed to be the

Figure 30. Photographs of the structure along the western margin of the Albuquerque basin at Carrizo Arroyo. Top: Southward view of the low-angle Laramide(?) age Carrizo fault underlying Able and Baker hills. Bottom: Northward view of Tertiary, rift-bounding Santa Fe fault.

downplunge, basinward expression of the Ladron uplift (Brown et al., 1980) (Figure 38). In later interpretations of reprocessed versions of the same data, better seismic resolution allowed for discussion of the role of listric normal faults, including the postulate that the Ladron horst block is detached above a low-angle normal fault (Cape et al., 1983; de Voogd et al., 1988) (Figure 38).

Several factors need to be addressed in reconciling the differing interpretations of this area. The first of these factors is the orientation of the seismic lines themselves. Shell line 52 is a nearly true dip section across the basin; however, COCORP line 1-A is largely a strike line relative to the major basin controlling structures and becomes a dip section only in its eastern one-third (Figure 38). Previous interpretations of these data apparently did not account for this fact. Consequently, whereas the apparent mid-basin horst block is in the middle of the line, it actually lies very near the western margin of the basin.

Another key relationship important to interpreting these data is the arcuate trace of the Jeter and Santa

Fe–Coyote faults, the two major western boundary faults in this portion of the basin. The trace of these faults, broadly arcuate regionally, becomes much more tightly arcuate in the vicinity of the Ladron Mountains and conforms closely to the northern and eastern flanks of the range (Figures 23, 25). This relationship suggests that the faults were probably folded late in their history during the uplift of the Ladron structural block. This uplift is constrained at 10–15 Ma by apatite fission-track cooling ages (Kelley et al., 1992). The folded nature of these faults is not nearly as pronounced on seismic dip transects across the basin as on strike lines. Thus, Shell line 52 reveals a mildly folded geometry. COCORP line 1-A, however, closely follows the strike of these faults. Our interpretation of these data shows a distinctly folded nature of the structures. One inherent distortion is created by this strike section: the footwall cutoff in the Paleozoic–Mesozoic section of the Santa Fe–Coyote fault on the eastern edge of the Monte Largo Bench (Figure 38) is much longer in length than in the hanging wall. This difference is the result of the footwall being imaged on that portion of the COCORP line that is strike to the fault and the hanging wall being imaged on the dip-oriented section of the line.

Despite these differences, the interpretations of all these authors share some fundamental and important similarities. All show the basin to be quite asymmetric and also interpret the western boundary faults of the rift to be low-angle to listric in character. Finally, all interpretations of basin-margin faults show basement involvement.

Implications for Plate Tectonic Reconstruction of Passive Continental Margins

In the past, the gaps and overlaps produced by plate reconstructions have been major reasons for modifications and rearrangements in the fits between continental plates. Palinspastic analysis of Albuquerque basin cross sections suggests that variations in extension along a rifted continental margin could be quite significant. Accounting for these variations might reduce problems with gaps and overlaps.

This study also suggests that portions of a continental margin that experience differential amounts of crustal attenuation and exhibit opposing polarity during rifting are likely to be separated by transcurrent structural zones or accommodation zones. Thus, when analyzing a margin for stretching variations of this sort, one must pay particular attention to evidence of transcurrent features possibly correlative with oceanic fracture zones. These transform zones could then be used as tentative flow lines to guide the restoration of the margin to its original position and structural configuration.

In addition, the geometric lessons acquired when studying a continental rift basin, such as the Albuquerque basin, lead one to the conclusion that conjugate passive margins that evolve after rifting progresses to sea floor spreading will not be mirror images of one another. On the contrary, they will be asymmetric, just as the initial rift was, and display very different structural geometries, subsidence histories, thermal structures, and perhaps hydrocarbon potential.

CONCLUSIONS

The ideas presented in this chapter are not unlike those of other workers, who in recent years have discussed the asymmetric nature of rifting, continental breakup, and the role of low-angle detachment tectonics along passive continental margins (Bally, 1982; Lister et al., 1986). Furthermore, the concepts related to accommodation zones have been abundantly illustrated by Rosendahl in his studies of variations along the East African rift lakes (Rosendahl, 1987). The uniqueness and value of our study of the Albuquerque basin is in the completeness of the data set and the clarity with which the geometry of the rift zone is expressed both on seismic and the outcrop. In this sense, our study provides one of the most comprehensive views of a rift basin yet published.

The lessons to be drawn from this study, and their implications for oil and gas exploration, are as follows.

(1) Rifting, as a process, appears to be inherently asymmetric and controlled by listric to low-angle normal faults.

(2) Rift zones should be expected to be segmented into tectonic domains of reversing structural polarity and differential extension separated by transcurrent fault zones and accommodation zones. Along passive continental margins these transcurrent zones may be linked to oceanic fracture zones.

(3) Because of the asymmetric nature of rifting, one should reasonably expect the rift-stage portion of passive margins to be dominated by half-grabens.

(4) Further application of these concepts suggests that the exploration potential of conjugate continental margins may well be dissimilar because of the fundamental differences in structural style and subsidence histories of the opposing "hanging wall" and "footwall" margin segments. These dissimilarities result in varied trap potential, as well as differences in thermal maturation of potential source rocks. Thus, a strategy of exploring the conjugate margin to one with major oil and gas accumulations may be flawed.

(5) Any exploration effort in rift basins must recognize the controlling role of low-angle and listric normal faults in basin formation. In particular, trapping geometries and thermal maturity within prerift objective sections are profoundly influenced by the rifting process.

EPILOG: OIL AND GAS IN THE ALBUQUERQUE BASIN

Although the details of Shell's exploration efforts in the Albuquerque basin remain confidential, the

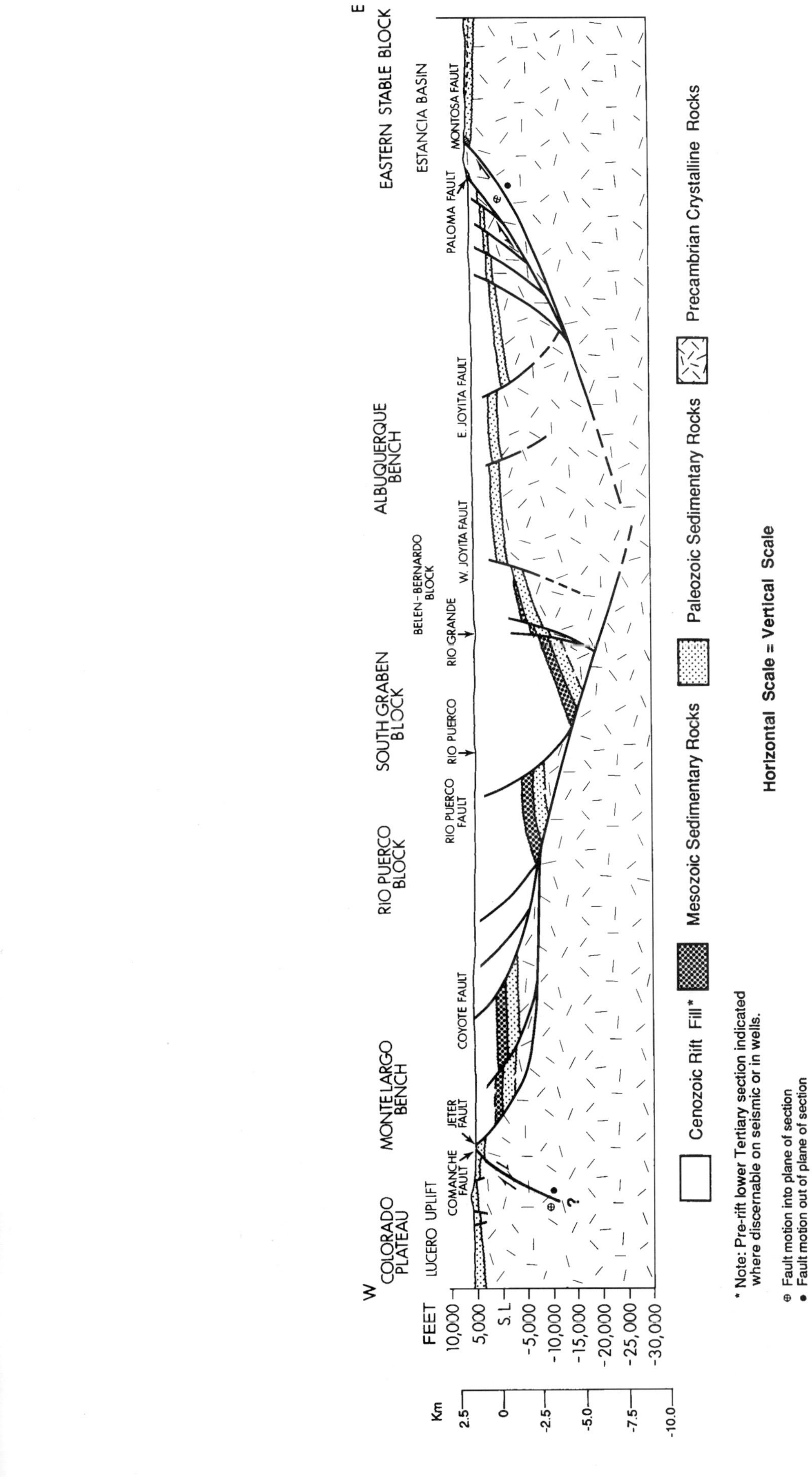

Figure 31. Geologic and seismic sections illustrating the structural configuration of the southern portion of the South Albuquerque basin.

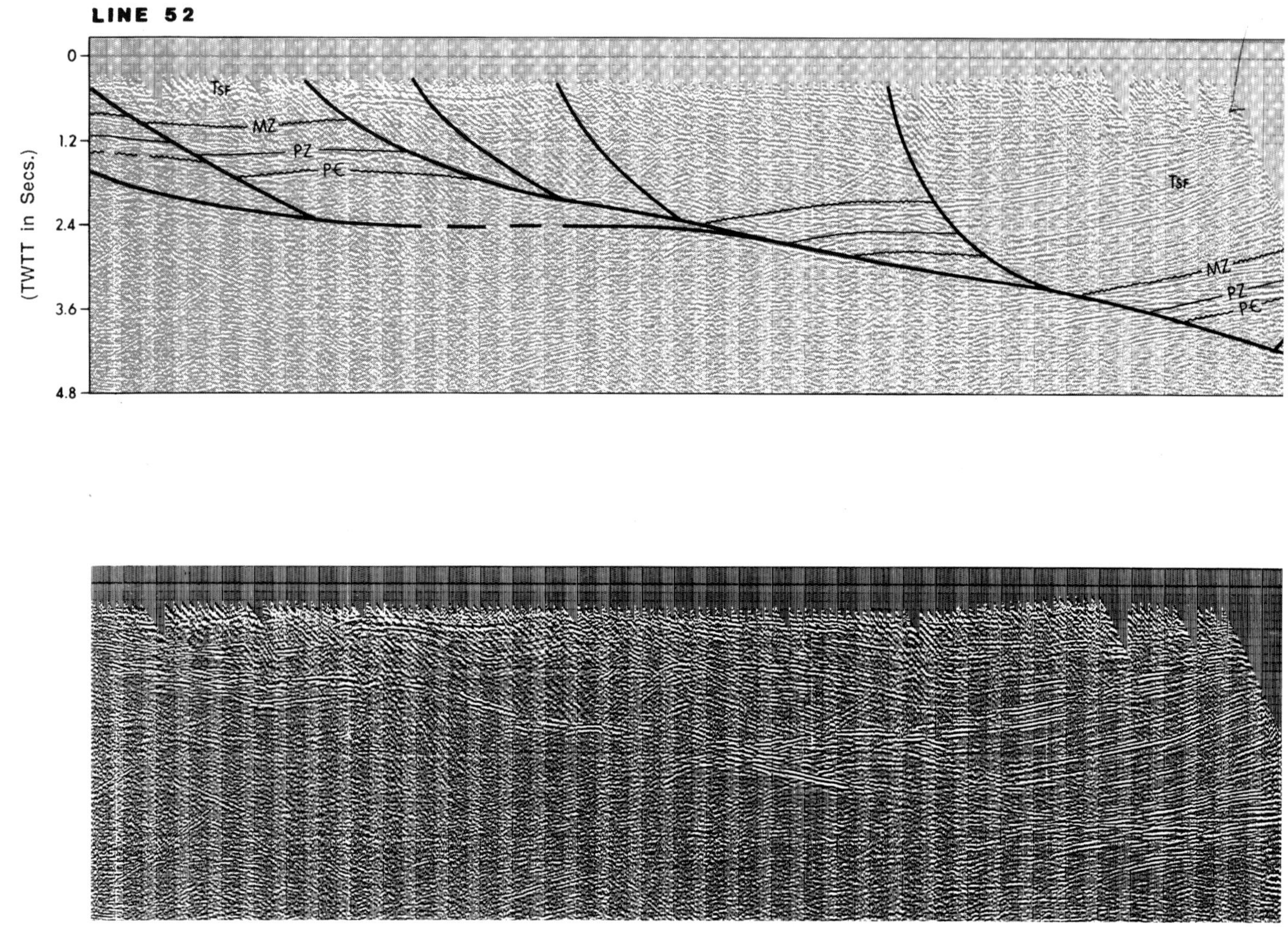

Figure 31 (continued).

hydrocarbon potential of the basin has been discussed by Black (1982) and Molenaar (1988). Both of these authors point out that the primary potential in the basin lies within Upper Cretaceous transgressive and regressive sandstone reservoirs (Figure 17). These potential reservoirs conceivably could be sourced by oil- and gas-prone Upper Cretaceous marine shales and marginal marine shales and coals, as well as by organic-rich Jurassic lacustrine limestones (Figure 17), all presumably matured to various levels by the Tertiary rift-related sedimentary loading and elevated thermal gradients (Black, 1982; Molenaar, 1988). Trap potential is varied with structural, stratigraphic, and combination traps possible. Although numerous shows have been reported and hydrocarbons have been recovered on production tests, no commercial accumulations have been confirmed to date, with the last significant deep Cretaceous test drilled in 1984 (Black, 1982; Molenaar, 1988). The potential and allure of the Albuquerque basin, and other basins within the northern segment of the Rio Grande rift, lies in the presence of the equivalent section that is so productive in the San Juan basin immediately to the west.

ACKNOWLEDGMENTS

The authors wish to thank Shell Oil Company for permission to publish this paper, and to acknowledge the contributions and insights provided by our many colleagues. A.W. Bally, in particular, was instrumental in the conception of the study.

REFERENCES CITED

Atwater, T., 1970, Implications of plate tectonics for Cenozoic tectonic evolution of western North America: Geological Society of America Bulletin, v. 81, p. 3513–3536.

Bachman, G.O., and H.H. Mehnert, 1978, New K-Ar dates and the late Pliocene to Holocene geomorphic history of the central Rio Grande region, New Mexico: Geological Society of America Bulletin, v. 89, p. 283–292.

Baldridge, W.S., 1979, Petrology and petrogenesis of Plio-Pleistocene basaltic rocks from the central Rio Grande rift, New Mexico, and their relation to rift structure, *in* R.E. Riecker, ed., Rio Grande rift, tec-

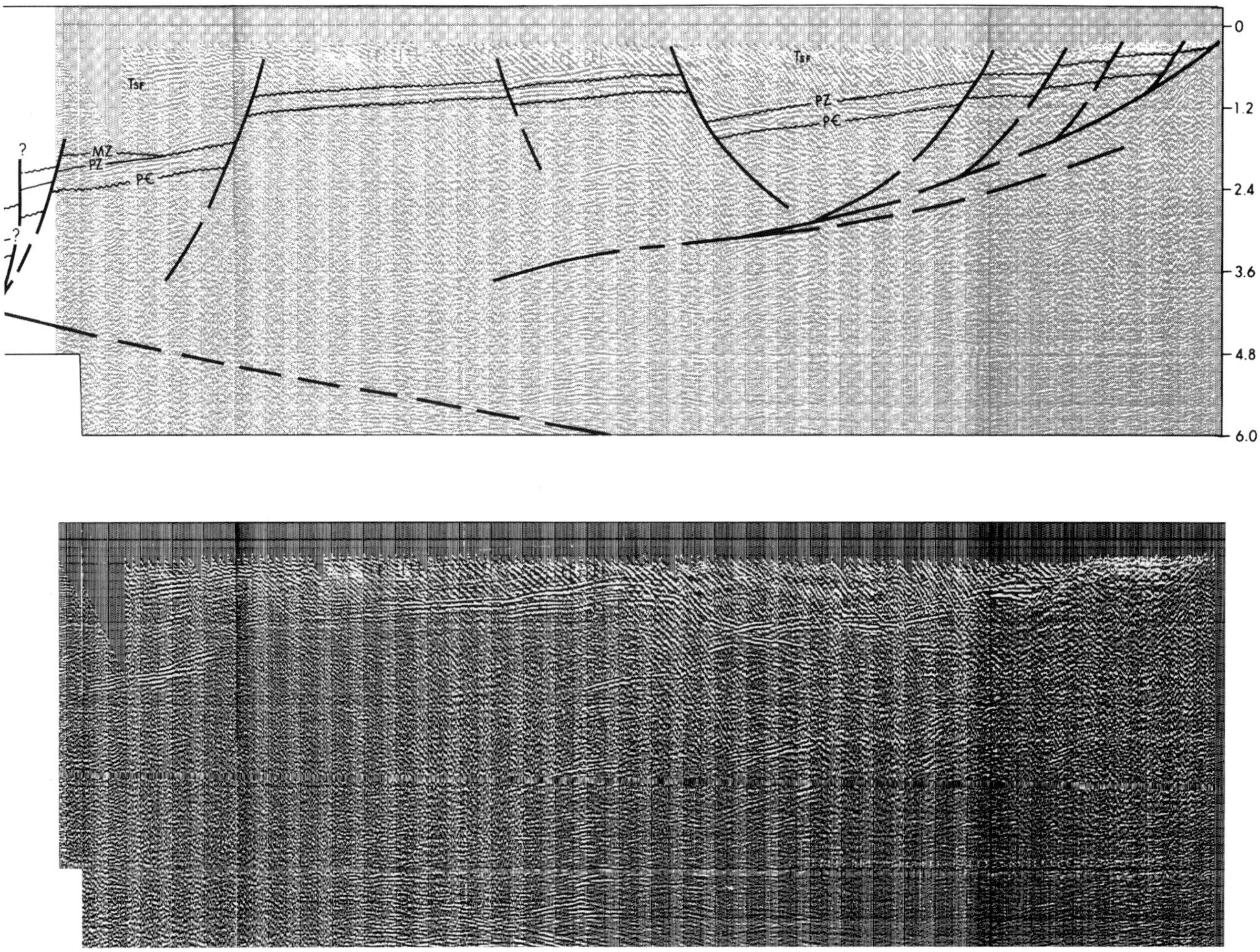

Figure 31 (continued).

tonics and magmatism: American Geophysical Union Monograph Series, p. 323–353.

Bally, A.W., 1982, Musings over sedimentary basin evolution: Royal Society of London Philosophical Transactions, A305, p. 325–328.

Black, B.A., 1964, The geology of the northern and eastern parts of the Ladron Mountains, Socorro County, New Mexico: M.S. thesis, University of New Mexico, Albuquerque, New Mexico, 117 p.

Black, B.A., 1982, Oil and gas exploration in the Albuquerque basin: New Mexico Geological Society Guidebook, Albuquerque County, p. 313–324.

Braile, L.W., R.B. Smith, G.R. Keller, and R.M. Welch, 1974, Crustal structure across the Wasatch Front from detailed seismic refraction studies: Journal of Geophysical Research, v. 79, p. 2669–2677.

Bridwell, R. J., 1976, Lithospheric thinning and the late Cenozoic thermal and tectonic regime of the northern Rio Grande rift: New Mexico Geological Society, 27th Field Conference Guidebook, p. 283–292.

Brown, L.D., C.E. Chapin, A.R. Sanford, S. Kaufman, and J. Oliver, 1980, Deep structure of the Rio Grande rift from seismic reflection profiling: Journal of Geophysical Research, v. 85, p. 4773–4800.

Callender, J.F., and R.E. Zilinski, 1976, Kinematics of Tertiary and Quaternary deformation along the eastern edge of the Lucero uplift, central New Mexico, *in* L.A. Woodward and S.A. Northrop, eds., Tectonics and mineral resources of southwestern North America: New Mexico Geological Society Special Publication no. 6, p. 53–61.

Cape, C.D., S. McGeary, and G.A. Thompson, 1983, Cenozoic normal faulting and the shallow structure of the Rio Grande rift near Socorro, New Mexico: Geological Society of America Bulletin, v. 94, p. 3–14.

Chapin, C.E., 1979, Evolution of the Rio Grande rift, a summary, *in* R.E. Riecker, ed., Rio Grande rift, tectonics and magmatism: American Geophysical Union Monograph Series, p. 1–5.

Chapin, C.E., 1983, An overview of Laramide wrench faulting in the southern Rocky Mountains with emphasis on petroleum exploration: Rocky Mountain Association of Geologists, p. 169–179.

Chapin, C.E., and W.R. Seager, 1975, Evolution of the Rio Grande rift in the Socorro and Las Cruces areas: 26th Field Conference Guidebook, New Mexico, Geological Society, p. 297–321.

Figure 32. Southward views of the Jeter fault at its type locality at the Jeter mine. Note the sheared nature of the Precambrian basement in the footwall of the fault and the reverse drag into the fault of rift fill bedding in the hanging wall.

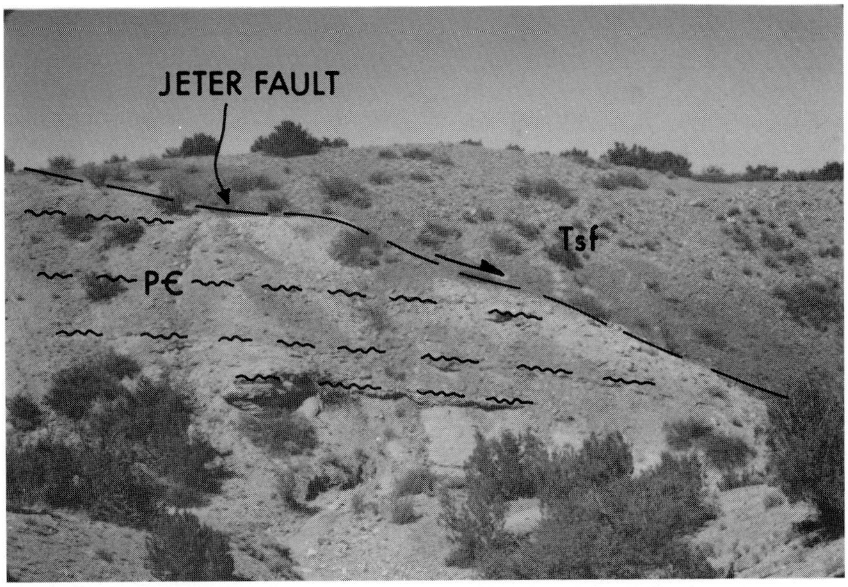

Figure 33. Northwestward views of the Jeter fault on the northeastern flank of the Ladron Mountains. Top photo illustrates an exhumed portion of the polished and slickensided Jeter fault surface itself; bottom photo illustrates the low-angle dip on the fault and the highly sheared nature of the Precambrian in the footwall.

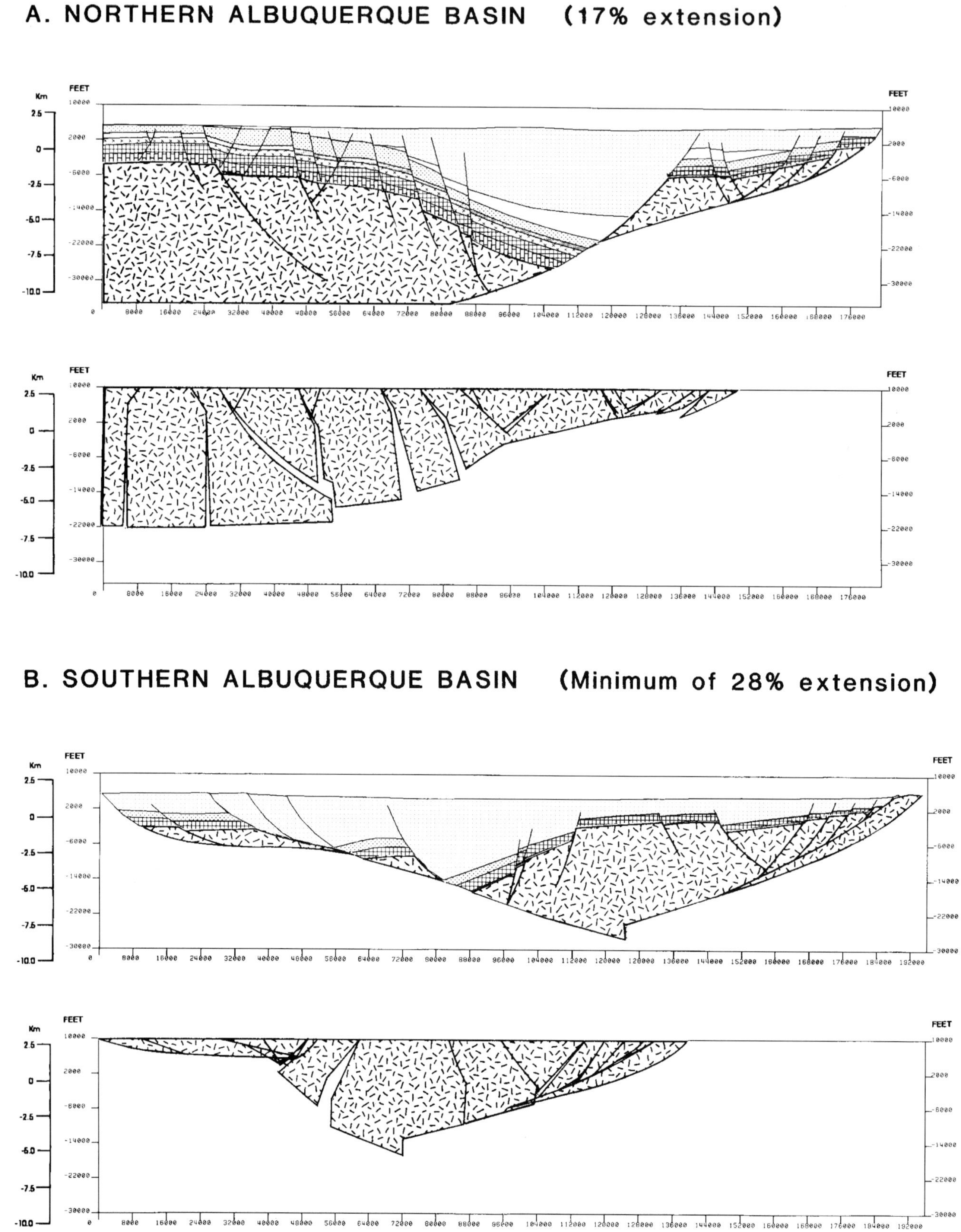

Figure 34. Palinspastic reconstructions of cross sections illustrating the variation in amount of extension between the (A) Northern and (B) Southern Albuquerque basin (horizontal scale = vertical scale). Sections reconstructed on the top of Precambrian basement.

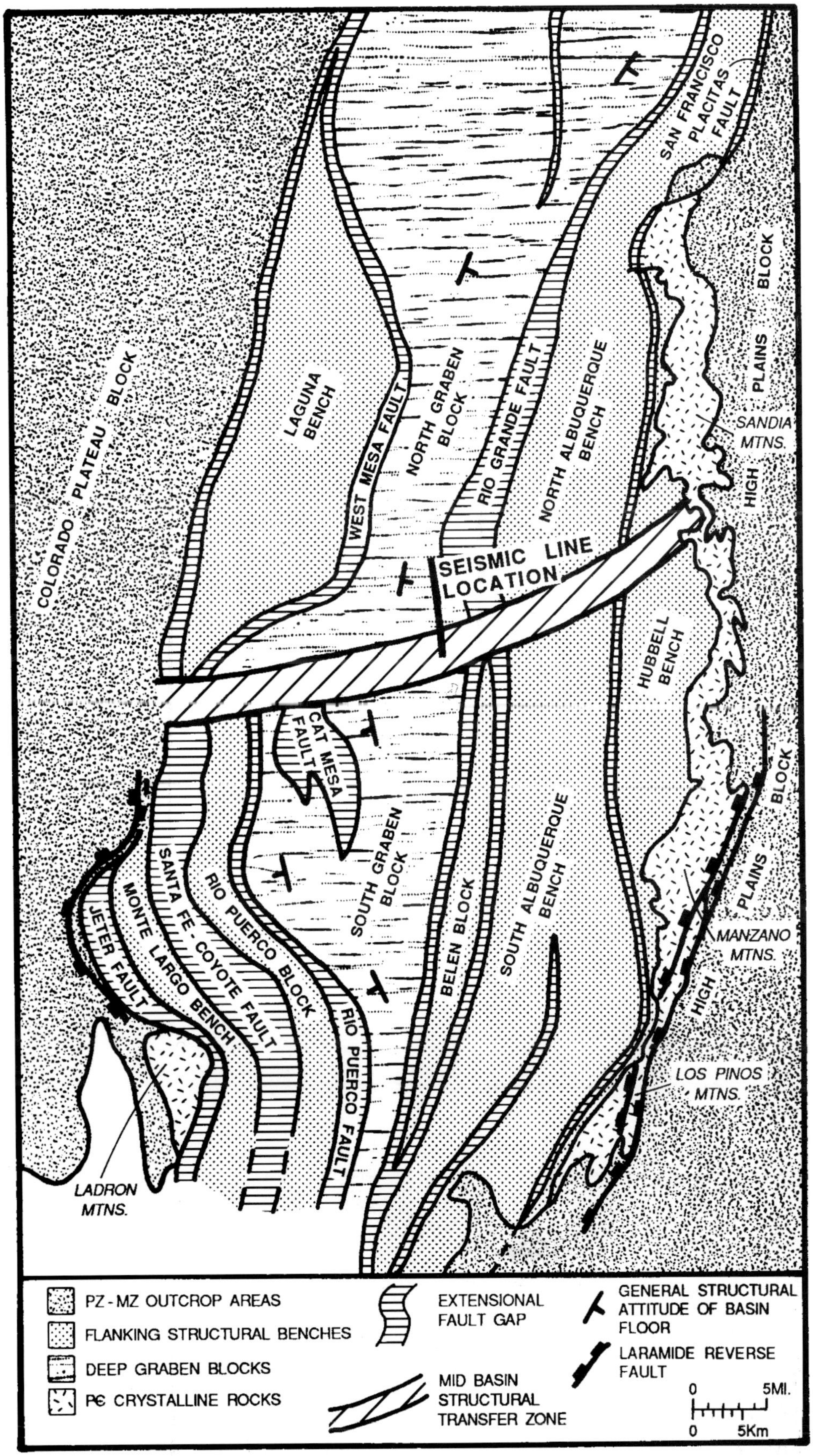

Figure 35. Seismic section across the mid-basin structural transfer zone illustrating the possible negative flower structure interpretation (unpublished interpretation after L. Garmezy). TWTT = two-way traveltime, Td = total depth.

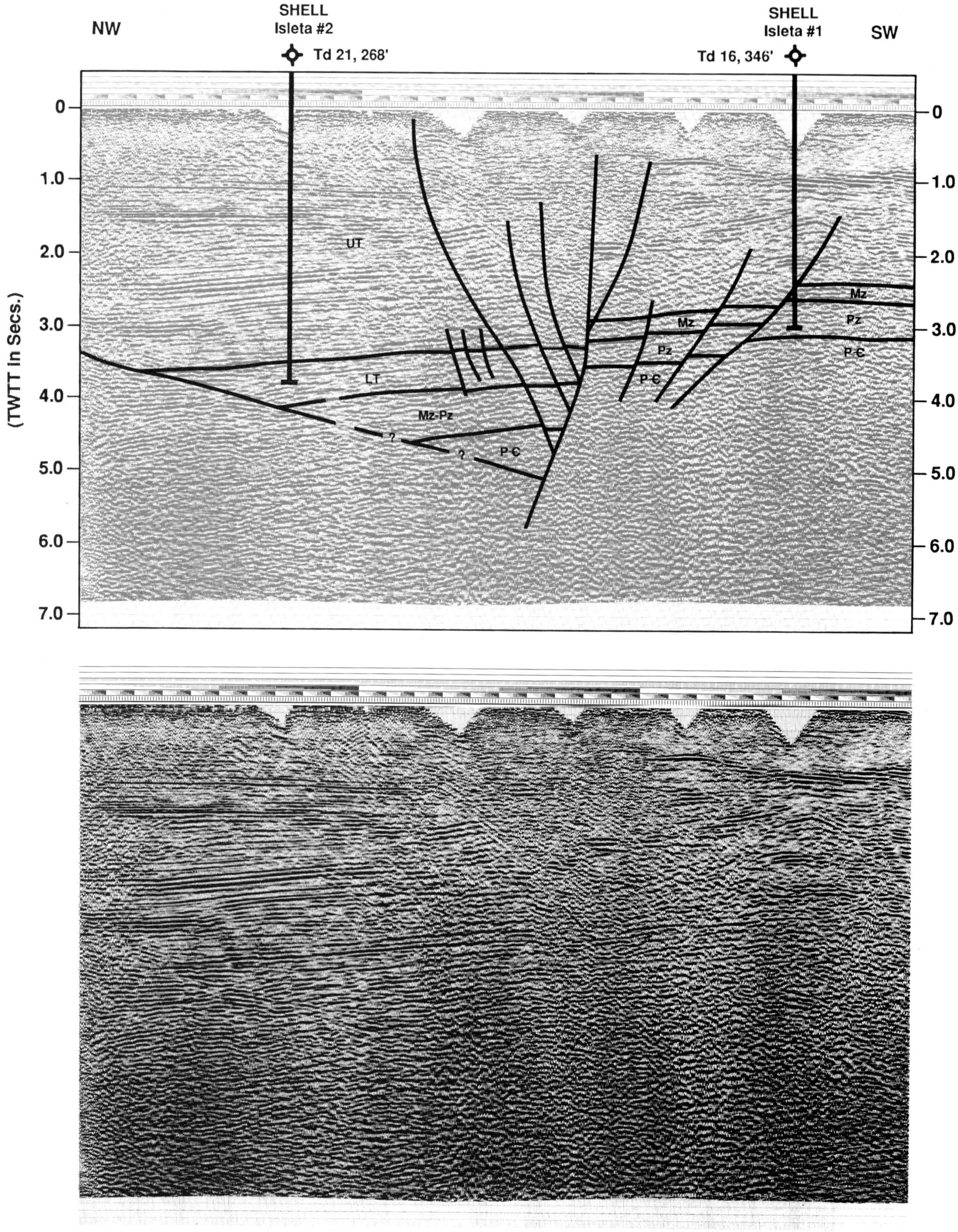

Figure 35 (continued).

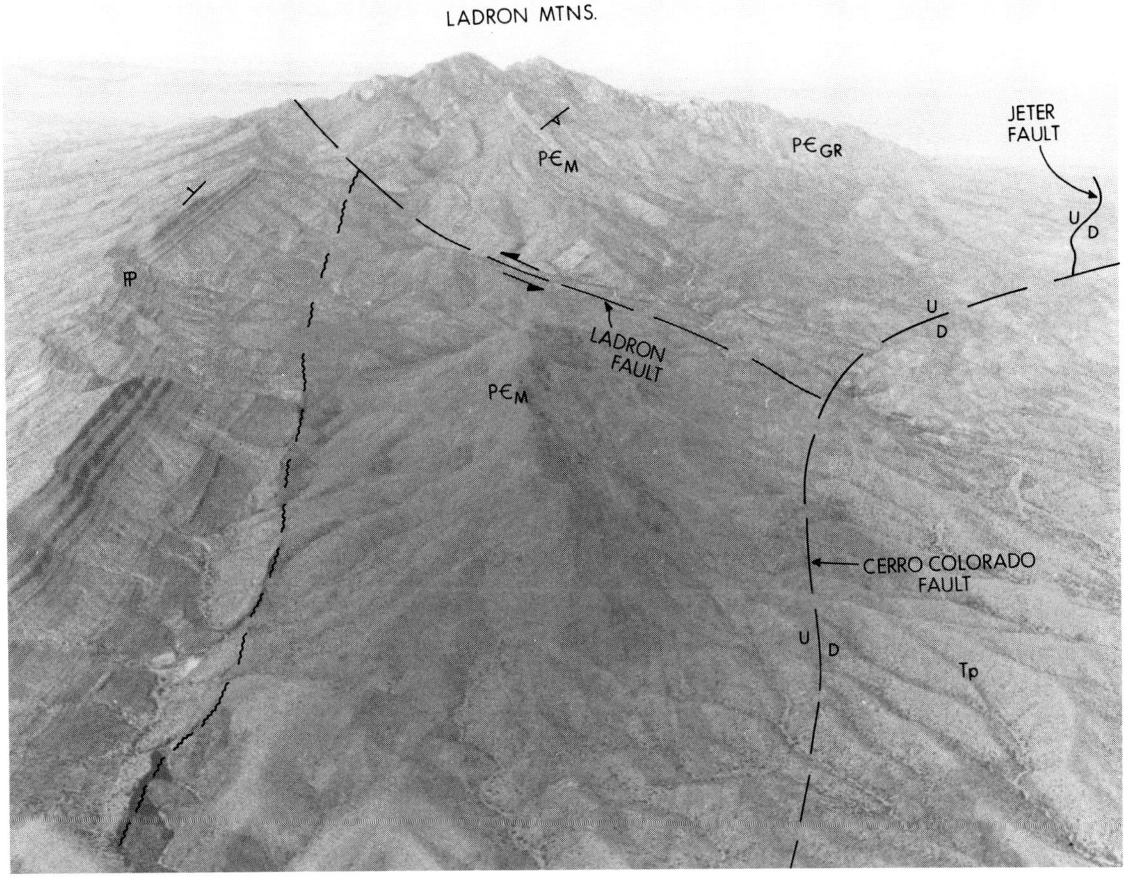

Figure 36. Aerial photograph, looking north, of the Ladron Mountains illustrating major structural features and the distribution of Precambrian metamorphic and igneous rocks, as well as Pennsylvanian sedimentary rocks.

Condie, K.C., 1976, Precambrian rocks of the Ladron Mountains, Socorro County, New Mexico: New Mexico Bureau of Mines and Mineral Resources, Geological Map 38, 1:24,000.

Coney, P.J., and S.J. Reynolds, 1977, Cordilleran Benioff zones: Nature, v. 270, p. 403–406.

Cordell, L., 1978, Regional geophysical setting of the Rio Grande rift: Geological Society of America Bulletin, v. 89, no. 7, p. 1073–1090.

Cordell, L., G.R. Keller, and T.G. Hildenbrand, 1978, Complete Bouguer gravity anomaly map of the Rio Grande rift: U.S. Geological Survey Open-File Map, 1:1,000,000.

de Voogd, B., L.D. Brown, and C. Merey, 1986, Nature of the eastern boundary of the Rio Grande rift from COCORP surveys in the Albuquerque basin, New Mexico: Journal of Geophysical Research, v. 91, p. 6305–6320.

de Voogd, B., L. Serpa, and L.D. Brown, 1988, Crustal extension and magmatic processes, COCORP profiles from Death Valley and the Rio Grande rift: Geological Society of America Bulletin, v. 100, p. 1550–1567.

Eaton, G.P., 1979, A plate tectonic model for Late Cenozoic crustal spreading in the western United States, *in* R.E. Riecker, ed., Rio Grande rift, tectonics and magmatism: American Geophysical Union Monograph Series, p. 7–32.

Everson, J.E., and L.T. Silver, 1978, Lead systematics of Late Cenozoic basalts from the Rio Grande rift (abs.): International Symposium on the Rio Grande Rift, Santa Fe, New Mexico, p. 36–37.

Harding, T.P., 1985, Seismic characteristics and identification of negative flower structures, positive flower structures, and positive structural inversion: AAPG Bulletin, v. 69, p. 582–600.

Hayden, S., 1991, Dextral oblique slip deformation along the Montosa fault zone at Abo Pass, Valencia and Socorro counties, New Mexico: New Mexico Geology, v. 13, p. 64.

Hayden, S., and C.K. Mawer, 1991, Dextral transtensional deformation along the western margin of the Rio Grande rift, central New Mexico: Geological Society of America, Abstracts with Programs, v. 23, no. 4, p. 31.

Hermance, J.F., and J. Pedersen, 1980, Deep structure of the Rio Grande rift—a magnetotelluric interpretation: Journal of Geophysical Research, v. 85, p. 3899–3912.

Jiracek, G.R., M.E. Ander, and H.T. Holcombe, 1979, Magnetotelluric soundings of crustal conductive zones in major continental rifts, *in* R.E. Riecker, ed.,

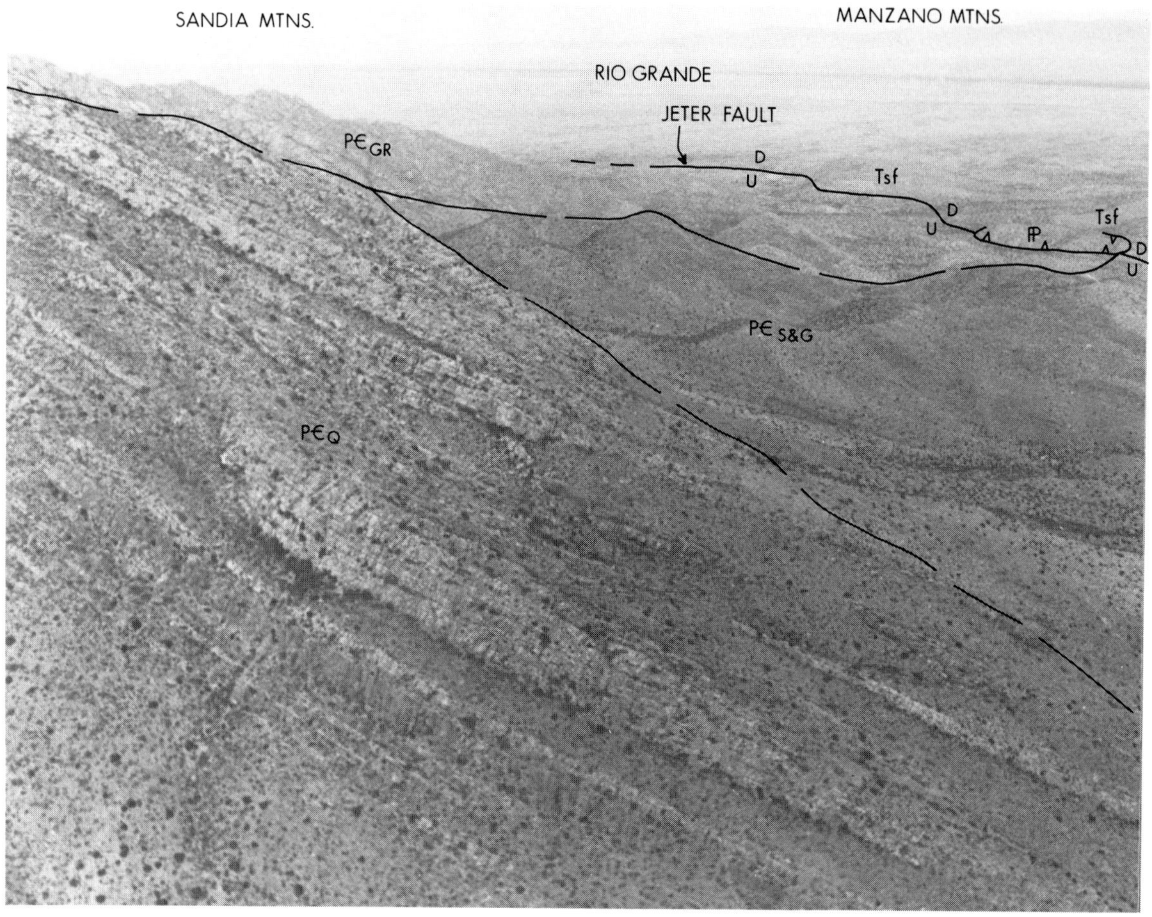

Figure 37. Closeup aerial photograph of the southeastern front of the Ladron Mountains showing the low-angle structural fabric of the Precambrian basement complex and the outcrop location of the Jeter fault.

Rio Grande rift, tectonics and magmatism: American Geophysical Union Monograph Series, p. 209–222.

Keller, G.R., R.B. Smith, and L.W. Braile, 1975, Crustal structure along the Great Basin–Colorado Plateau transition from seismic refraction studies: Journal of Geophysical Research, v. 80, no. 8, p. 1093–1098.

Keller, G.R., L.W. Braile, and P. Morgan, 1979, Crustal structure, geophysical models and contemporary tectonism of the Colorado Plateau: Tectonophysics, v. 61, p. 131–147.

Kelley, S.A., C.E. Chapin, and J.D. Corrigan, 1992, Late Mesozoic to Cenozoic cooling histories of the flanks of the northern and central Rio Grande rift: New Mexico Bureau of Mines and Mineral Resources Bulletin 145, 40 p.

Kelley, V.C., 1952, Tectonics of the Rio Grande depression of central New Mexico: New Mexico Geological Society Guidebook no. 3, p. 92–105.

Kelley, V.C., 1977, Geology of Albuquerque basin: New Mexico Bureau of Mines and Mineral Resources Memoir 33, 60 p.

Kelley, V.C., and A.M. Kudo, 1978, Volcanoes and related basalts of the Albuquerque basin, New Mexico: New Mexico Bureau of Mines and Mineral Resources Circular 156, 30 p.

Kottlowski, F.E., 1958, Geologic history of the Rio Grande near El Paso: West Texas Geological Society Guidebook to the Franklin and Hueco Mountains, p. 46–54.

Kudo, A.M., 1982, Rift volcanics of the Albuquerque basin: overview with some new data: New Mexico Geological Society, 33rd Field Conference Guidebook, p. 285–289.

Kudo, A.M., V.C. Kelley, P.E. Damon, and M. Shafiquillah, 1977, K-Ar ages of basalt flows at Canjilon Hill, Isleta volcano, and Cat Hills volcanic field, Albuquerque basin, central New Mexico: Isochron/West, no. 18, p. 15–16.

Lewis, C.J., and W.S. Baldridge, in press, Crustal extension in the Rio Grande rift, New Mexico: Half graben, accommodation zones, and shoulder uplift in the Ladron Peak-Sierra Lucero area, *in* G.R. Keller and S. Cather, eds., Structure and Stratigraphy of Basins along the Rio Grande Rift: Geological Society of America Special Paper 291.

Lister, G.S., M.A. Etheridge, and P.A. Symonds, 1986, Detachment faulting and the evolution of passive continental margins: Geology, v. 14, p. 246–250.

Lozinsky, P.R., 1988, Stratigraphy, sedimentation, and sand petrology of the Santa Fe Group and pre-

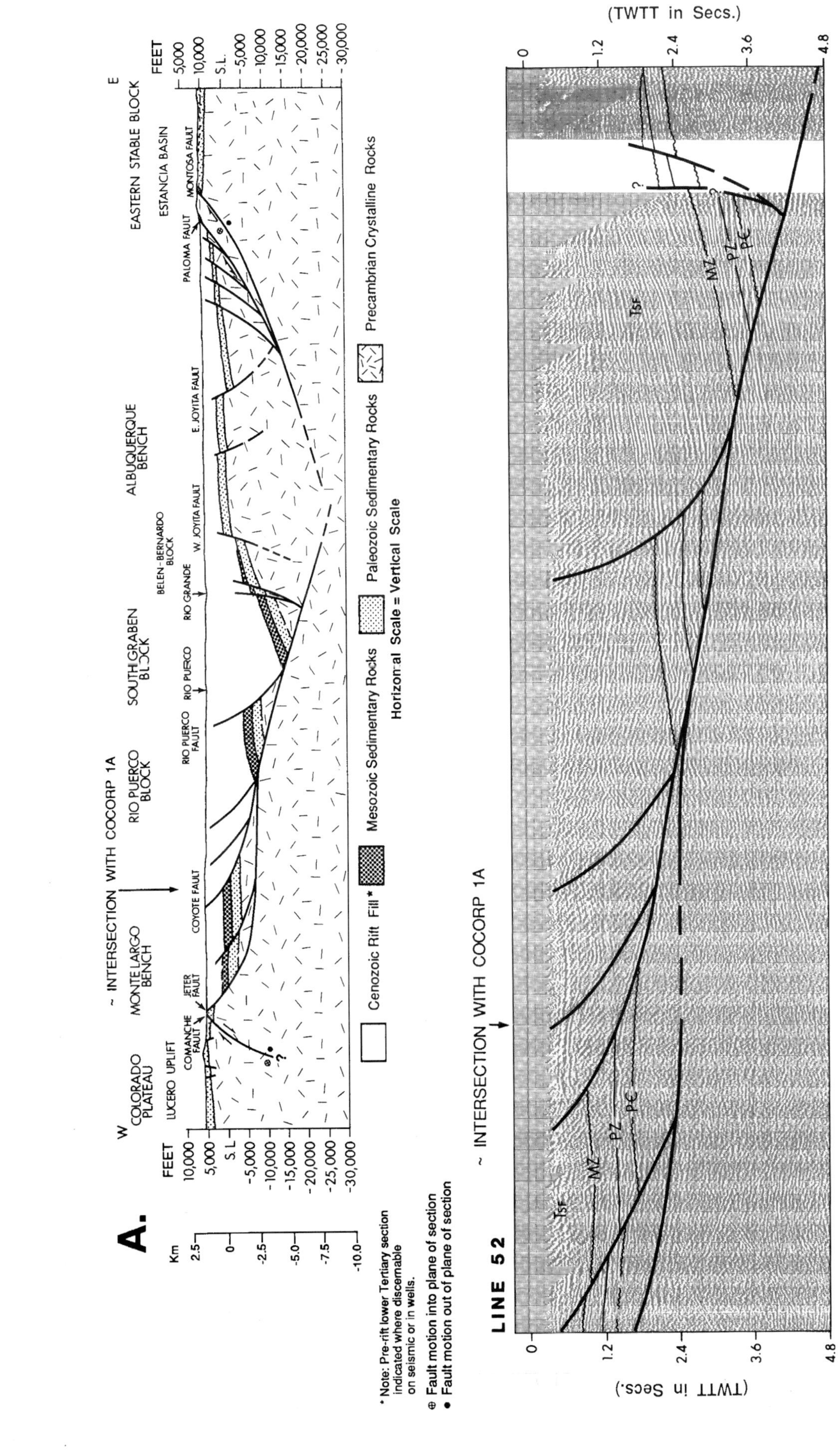

Figure 38. (A) Geologic and seismic sections illustrating the structural interpretation from this study in the vicinity of COCORP line 1A. Location of geologic section shown by dashed line on index map; position of seismic lines indicated by solid lines.

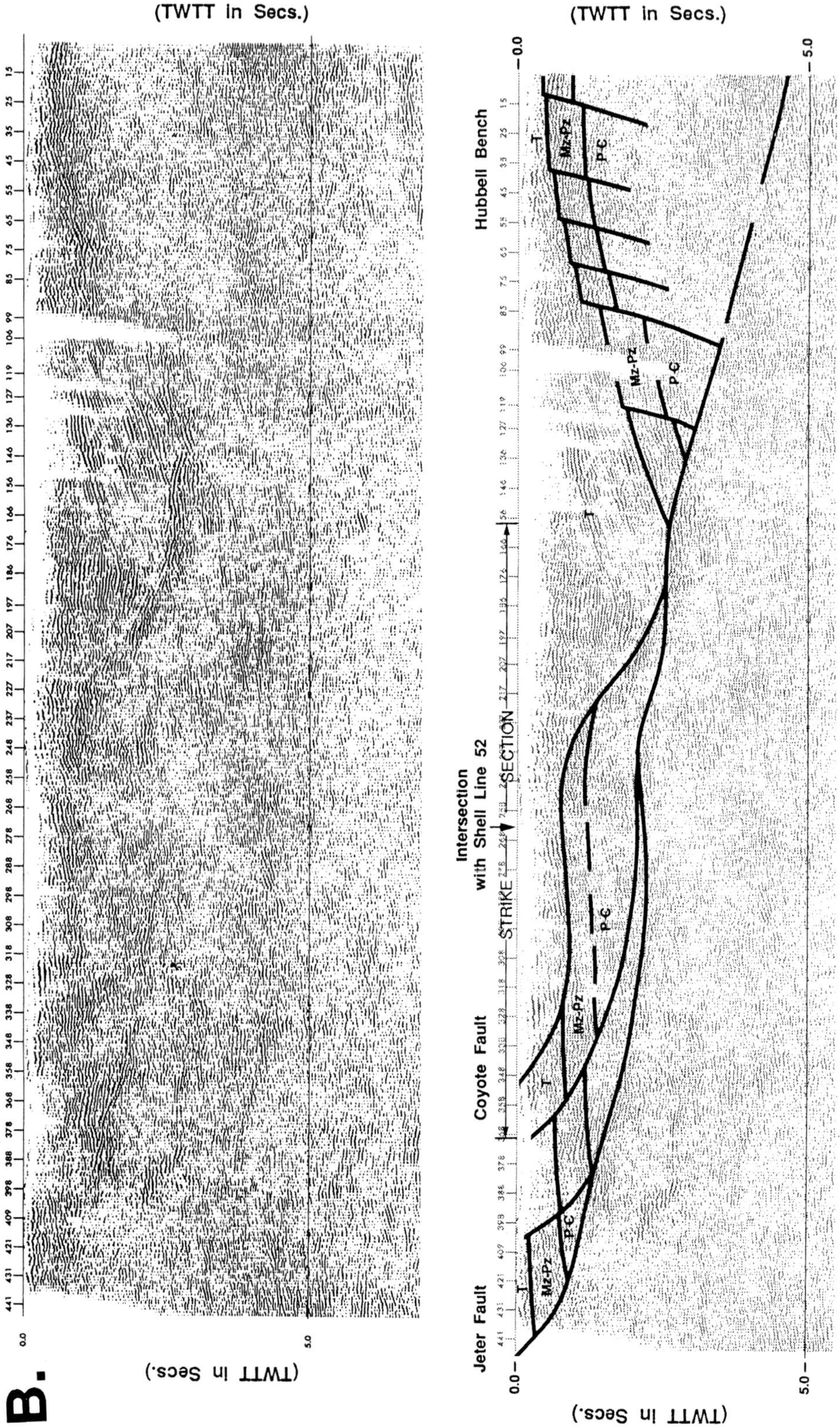

Figure 38. (B) COCORP line 1A uninterpreted and interpreted based on this study. Note from the index map that the western portion of the COCORP line is largely a strike line relative to basin-controlling structures.

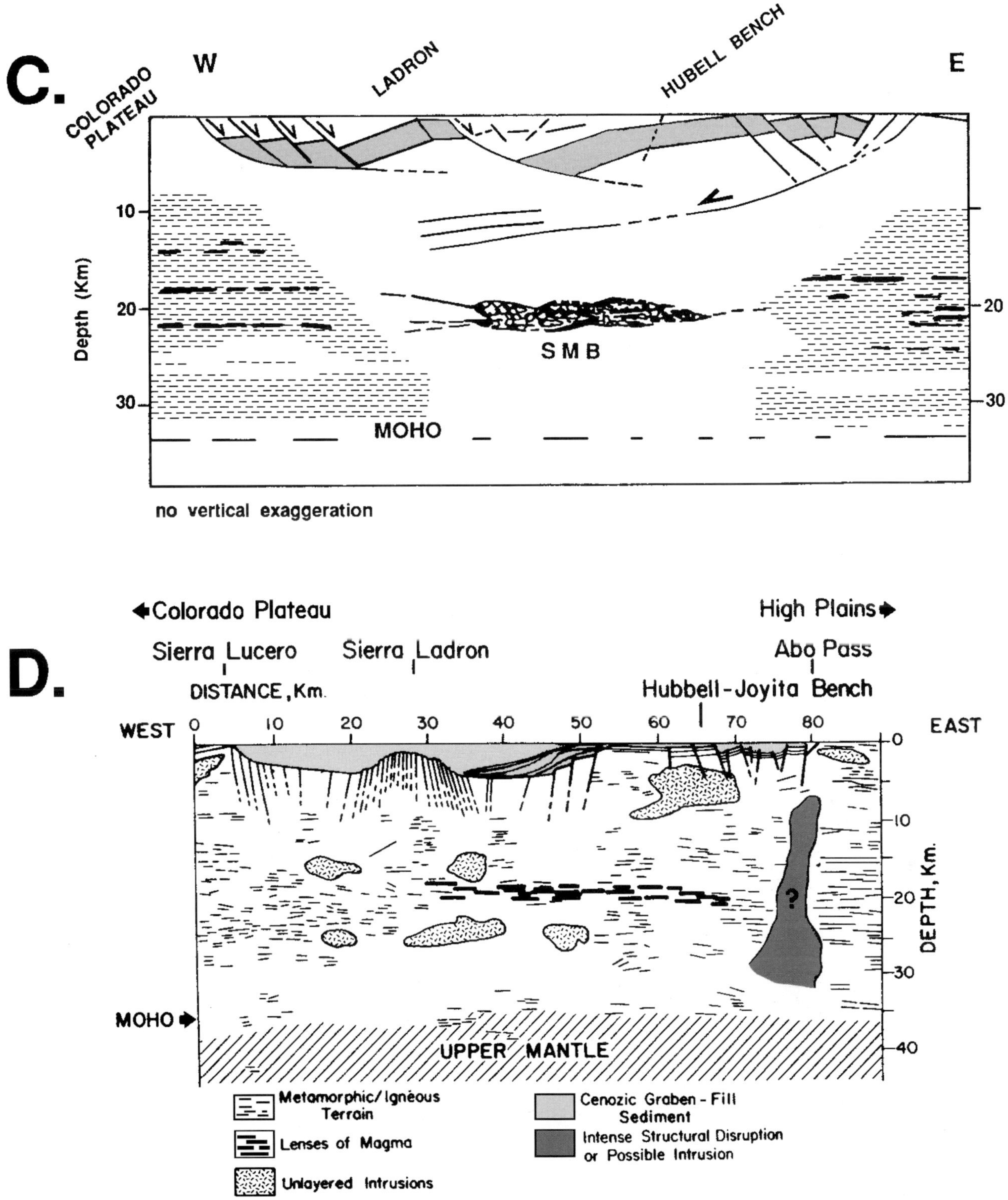

Figure 38. (C) A contrasting interpretation of COCORP line 1a by COCORP workers (after de Voogd et al., 1988). (D) An early interpretation of COCORP workers (after Brown et al., 1980).

Santa Fe Tertiary deposits in the Albuquerque basin, central New Mexico: Ph.D. dissertation, New Mexico Institute of Mining and Technology, Socorro, New Mexico, 298 p.

Mackin, J.H., 1950, The down-structure method of viewing geologic maps: Journal of Geology, v. 58, p. 55–72.

McCullar, D.B., and S.B. Smithson, 1977, Unreversed seismic crustal refraction profile across the southern Rio Grande rift (abs.): EOS, v. 58, p. 1184.

McKenzie, D.P., 1978, Some remarks on the development of sedimentary basins: Earth and Planetary Sciences Letters, v. 40, p. 25–32.

Miller, J.P., A. Montgomery, and P.K. Sutherland, 1963, Geology of part of the southern Sangre de Cristo Mountains, New Mexico: New Mexico Bureau of Mines and Mineral Resources, Memoir 11, 106 p.

Molenaar, C.M., 1988, Petroleum geology and hydrocarbon plays of the Albuquerque-San Luis rift basin, New Mexico and Colorado: USGS Open-File Report 87-450-S, 26 p.

Molzer, P.C., and E.A. Erslev, 1991, Oblique thrusting in east-west Laramide foreland uplifts: Geological Society of America, Abstracts with Programs, v. 23, no. 4, p. 50.

Morgan, P., W.P. Seager, and M.P. Golombek, 1986, Cenozoic thermal, mechanical, and tectonic evolution of the Rio Grande rift: Journal of Geophysical Research, v. 91, no. B6, p. 6263–6276.

Olsen, K.H., G.R. Keller, and J.N. Stewart, 1979, Crustal structure along the Rio Grande rift from seismic refraction profiles, *in* R.E. Riecker, ed., Rio Grande rift, tectonics and magmatism: American Geophysical Union Monograph Series, p. 127–143.

Reiter, M., C.L. Edwards, H. Hartman, and C. Weidman, 1975, Terrestrial heat flow along the Rio Grande rift, New Mexico and Colorado: Geological Society America Bulletin, v. 86, p. 811–818.

Reiter, M., C. Shearer, and C.L. Edwards, 1978, Geothermal anomalies along the Rio Grande rift: Geology, v. 6, p. 85–88.

Reiter, M., A.J. Mansure, and C. Shearer, 1979, Geothermal characteristics of the Rio Grande rift with the southern Rocky Mountains complex, *in* R.E. Riecker, ed., Rio Grande rift, tectonics and magmatism: American Geophysical Union Monograph Series, p. 253–267.

Rosendahl, B.R., 1987, Architecture of continental rifts with special reference to East Africa: Annual Review of Earth and Planetary Science, v. 15, p. 445–503.

Sanford, A.R., O.S. Alptekin, and T.R. Toppozada, 1973, Use of reflection phases on microearthquake seismograms to map an unusual discontinuity beneath the Rio Grande rift: Seismic Society of America, v. 63, p. 2021–2034.

Sanford, A.R., et al., 1977, Geophysical evidence for a magma body in the crust in the vicinity of Socorro, New Mexico, *in* J.G. Heacock, ed., The Earth's Crust and Upper Mantle: American Geophysics Union Monograph no. 20, p. 385–403.

Sanford, A.R., K.H. Olsen, L.H. Jaksha, 1979, Seismicity of the Rio Grande rift, *in* R.E. Riecker, ed., Rio Grande rift, tectonics and magmatism: American Geophysics Union Monograph Series, p. 145–168.

Shurbet, D.H., 1969, A low-velocity layer in the Earth's crust: Geological Society of America Bulletin, v. 80, p. 895–898.

Smith, R.B., 1978, Seismicity, crustal structure, and intraplate tectonics of the interior of the Western Cordillera, *in* R.B. Smith and G.P. Eaton, eds., Cenozoic tectonics and regional geophysics of the Western Cordillera: Geological Society of America Memoir 152, p. 111–144.

Smith, R.B., L.W. Braile, and G.R. Keller, 1975, Upper crustal low-velocity layers: possible effects of high temperatures over a mantle upwarp at the Basin and Range–Colorado Plateau transition: Earth and Planetary Sciences Letters, v. 28, p. 197–204.

Stewart, J.H., 1978, Basin and Range structure in western North America: a review, *in* R.B. Smith and G.P. Eaton, eds., Cenozoic tectonics and regional geophysics of the Western Cordillera: Geological Society of America Memoir 152, p. 1–31.

Stewart, J.H., and J.E. Carlson, 1978, Generalized map showing distribution, lithology and age of Cenozoic igneous rocks in the western United States, *in* R.B. Smith and G.P. Eaton, eds., Cenozoic tectonics and regional geophysics of the Western Cordillera: Geological Society of America Memoir 152, p. 263–264.

Wernicke, B., 1981, Low-angle normal faults in the Basin and Range Province: nappe tectonics in an extending orogen: Nature, v. 291, p. 645–647.

Wernicke, B., 1985, Uniform-sense normal simple shear of the continental lithosphere: Canadian Journal of Earth Sciences, v. 22, p. 108–125.

Wernicke, B., 1986, Whole-lithosphere normal simple shear: an interpretation of deep reflection seismic profiles in Great Britain, *in* M. Barazangi and L. Brown, eds., Reflection seismology—the continental crust: American Geophysical Union Geodynamics Series, v. 14, p. 331–339.

White, D.L., 1979, Rb-Sr age and isotopic data of some Precambrian plutons in central New Mexico: Mountain Geologist, v. 16, p. 131–137.

Woodward, L.A., et al., 1978, Tectonic map of the Rio Grande rift region in New Mexico, Chihuahua, and Texas, *in* J.W. Hawley, ed., Guidebook to the Rio Grande rift in New Mexico and Colorado: New Mexico Bureau of Mines and Mineral Resources, Circular 163, 241 p.

Worrall, D.M., and S. Snelson, 1989, Evolution of the northern Gulf of Mexico, with emphasis on growth faulting and the role of salt, *in* A.W. Bally and A.R. Palmer, eds., The geology of North America—an overview: Geological Society of America, The Geology of North America, v. A, p. 97–138.

Summary

Susan M. Landon
Thomasson Partner Associates
Denver, Colorado, U.S.A.

Rift basins are not single, elongate grabens, but a series of linear or relay-pattern (Harding and Lowell, 1979), usually asymmetric, grabens. These grabens form as a result of pockets of extension that propagate along the rift and, in some areas such as the Red Sea, ultimately create oceanic crust and an ocean basin (Figure 1). Rift basins represent extension in the brittle crust in response to ductile deformation in the upper mantle. The processes that cause ductile deformation in the upper mantle, and the resulting crustal extension, have been debated since rift basins were first described in the last century. The process or combination of processes may be interpreted from the physical characteristics of an active or inactive rift. Each process results in different basinal elements and generates various basinal elements. Examination of these elements, the pattern of volcanism, the presence or absence of uplift, etc., provide a clue to basin origin.

Rosendahl (1987) pointed out that brittle fracturing and faulting are symptoms of processes that involve ductile deformation of the lithosphere. Two general categories of rift processes have been proposed: active and passive (Sengor and Burke, 1978; Morgan and Baker, 1983) (Table 1). Active rifting is a direct response to tension generated by uplift or doming. Passive rifting results from tensional stresses in the crust causing thinning and passive upwelling of hot asthenosphere. Allen and Allen (1990) summarized the various models of rift processes, which are outlined in Table 1. Stresses resulting in crustal extension may occur as a result of uplift (active) or lithospheric thickness changes and resulting isostatic compensation. Therefore, thermal history, lithospheric thinning, and topographic uplift are closely related. Each scenario assumes, in general, that the crust will behave brittlely and the lithospheric mantle ductilely. Figure 2 diagrammatically illustrates the major processes that create rift basins. Actual rift systems may exhibit elements of both active and passive mechanisms, which are end members of idealized models (Allen and Allen, 1990).

During formation, rift basins are characterized by negative Bouguer gravity anomalies, high heat flow, and volcanic activity. These characteristics indicate a deep thermal anomaly. Refraction seismic has shown that rifts overlie thinned crust (perhaps 25 km instead of an average of about 40 km, but actual values vary depending on the particular rift) (Figure 1). Active crustal extension is reflected by fault-bounded half-grabens and full grabens. Often these rift grabens are flanked by topographically high shoulders indicating uplift. Presence and timing of uplift (before or contemporaneous with graben formation) is a major factor in determining the process responsible for extension.

Processes that create extension are important to the petroleum geologist because of their impact on sediment distribution and heat flow history. The presence and timing of uplift control the presence or absence of prerift sediments in a graben, facies distribution of synrift sediments, and the distribution of unconformities in the section. The processes responsible for continental rifting will dictate distribution of faults that, in turn, define basin geometry and hydrocarbon traps. Source and distribution of heat flow control volcanism and source rock maturation and also are determined by the rift process; therefore, processes responsible for crustal extension are important in defining the exploration strategy that may be most effective in a rift basin.

Rift basins develop over a relatively very short period of time, geologically, with very rapid subsidence and rapid rates of sedimentation. Active extension and associated sedimentation in basins discussed in this volume range in duration from about 10 to 30 m.y. (Table 2). In some areas, such as in the North Sea, active extension does occur over a longer period of time. Volcanism can occur throughout the rift process as a result of thermal activity and thinning crust. Thick piles of basalt and other extrusive and intrusive igneous rocks are common. Postrift sags result from cooling of the asthenosphere and accompanying thermal contraction after extension has ceased.

AULACOGENS

Rift basins, occurring at a high angle to plate margins, are termed aulacogens and, generally, are the failed third arm of a triple junction associated with the early stages of ocean formation. The Benue rift in west Africa (Petters and Ekweozor, 1982) and the Anadarko basin, Oklahoma, are examples of this type of rift basin. Aulacogens occur at doglegs or reentrants in the continental margin because each arm originally was at an angle to the other two. Aulacogens originally were described in the Russian literature, and Salop and Scheinmann (1969) used the Donets basin as a typical example. Ulmishek et al. (this volume) provide evidence to suggest that this

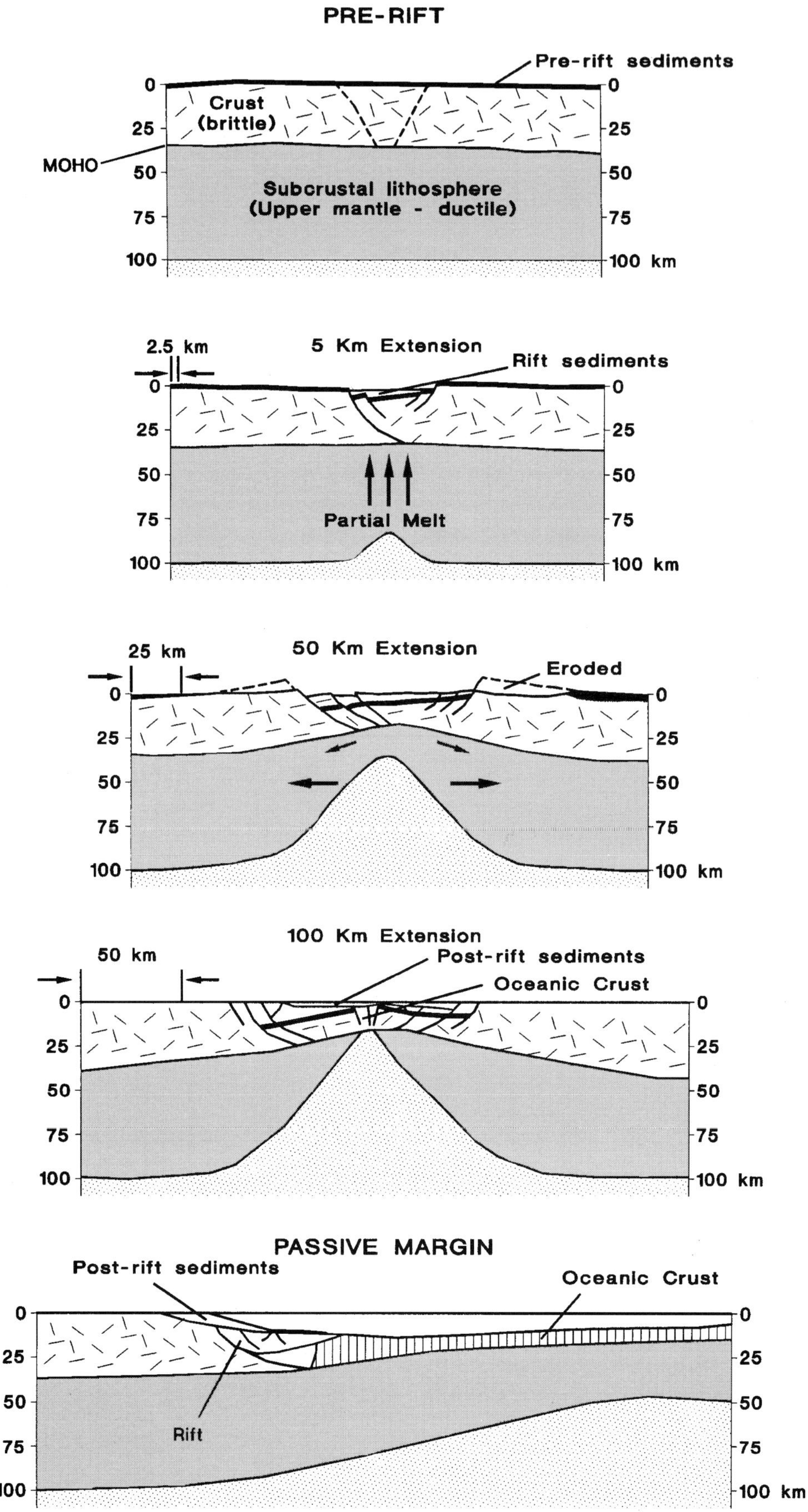

Figure 1. Diagrammatic model of rift basin evolution (from Salveson as modified by Allen and Allen, 1990).

Table 1. Tabulation of rift mechanisms with selected references (after Allen and Allen, 1990; refer to Figure 2)

ACTIVE: UPLIFT PRECEDES RIFTING

Tensile response to doming or uplift (Gregory, 1896): impingement of a thermal plume on the base of the lithosphere causing convective thinning, domal uplift, and crustal extension; conductive heating from the mantle plume, heat transfer from magma generation, or convective heating may result in lithospheric thinning; enough heat transfer will result in isostatic uplift, providing tensional stresses that would result in rifting.

- **Gravity spreading:** Uplift (lithospheric swell) is analogous to ridge push forces that help drive motion of oceanic plates; rifting results from lateral pressure gradients (Housemann and England, 1986; Neugebauer, 1978; Bott, 1981).
- **Diapirism:** Cold, dense mantle lithosphere overlying less dense, hot asthenosphere may result in diapirism in the asthenosphere (Woidt and Neugebauer, 1981).

PASSIVE: SUBSIDENCE IS FIRST EXPRESSION OF RIFTING

Tensional stresses in the continental lithosphere cause thinning and the passive upwelling of hot asthenosphere; rifting is a response to a regional stress field and doming and volcanic activity are secondary.

- **Pure shear:** Symmetrical stretching of crust and subcrust lithosphere (Figure 1).
  - **Uniform extension:** Crustal and lithospheric extension are equal (McKenzie, 1978).
  - **Uniform extension with induced mantle convection:** Uniform extension does not cause rift flank uplift observed in many rifts; evidence suggests that uplift occurs contemporaneously with extension rather than before rifting; additional heating has been proposed as the mechanism resulting in uplift, and additional heat may result from convective flow induced by the large temperature gradients created by rifting (Steckler, 1985; Buck, 1986).
  - **Nonuniform extension:** Changes in rheology (elasticity, viscosity, and plasticity) within the lithosphere provide for variation in extension.
    - **Discontinuous:** Evidence of decoupling at a mid-crustal layer suggest two rheological layers than can extend by different amounts (Hellinger and Sclater, 1983).
    - **Continuous:** Mantle responds to extension as a function of depth (Rowley and Sahagian, 1986).
- **Simple shear** (Coward, 1986; Wernicke, 1981, 1985)
  - **Asymmetrical with relay:** Lithospheric extension results from displacement on a large-scale, gently dipping shear zone that traverses the lithosphere, resulting in a physical offset of the zone of fault-controlled extension and the zone of upwelling asthenosphere; does not explain geographically superimposed postrift sag resulting from thermal contraction.

basin is not an aulacogen, but is similar to the Red Sea: created by rotation of the Ukrainian shield relative to the Russian craton.

STRUCTURAL STYLE

As shown by the rift basins documented in this volume, there is some diversity in structural style, but the overall framework is the same, that is, each basin is defined by major boundary faults commonly shown by reflection seismic to be listric normal faults (Figure 3). These fault systems define usually asymmetric grabens linked by various geometries along the trend of the rift, creating subbasins. A variety of authors illustrated structural styles in rift basins. Detailed interpretations have been made in the currently active

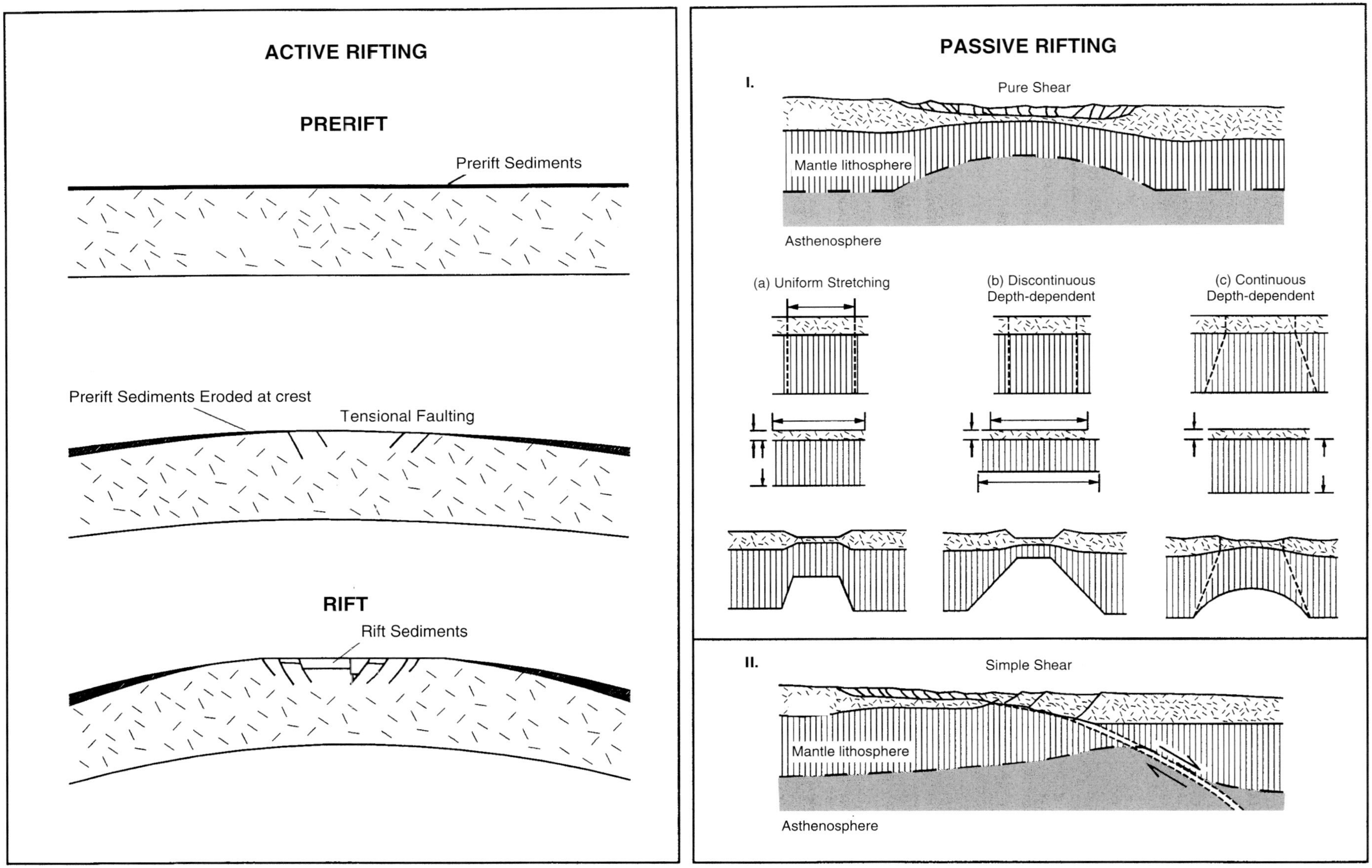

Figure 2. Models of active and passive rifting. This figure illustrates the models tabulated in Table 1. Active rifting is initiated by uplift. Passive rifting may be accomplished by pure shear (I) or simple shear (II). For pure shear, (a) illustrates lithospheric geometry of uniform stretching (no uplift) and nonuniform stretching is illustrated in (b) and (c). (b) Discontinuous depth-dependent stretching (crust and lithospheric mantle stretch at different amounts) and (c) continuous depth-dependent stretching are both characterized by uplift. (After Allen and Allen, 1990.)

Table 2. Summary of the characteristics of rift basins discussed in this volume

Basin and Location	Age	Size	Prerift Stratigraphy	Synrift Stratigraphy	Postrift Sag Stratigraphy
GULF OF SUEZ EGYPT	OLIGOCENE–MIOCENE 27–16 Ma	25,000 km^2 310 × 80 km Portion of longer trend (Red Sea)	Pre-Carboniferous–Cenomanian: predominantly continental sandstones, few thin-marine incursions Cenomanian–Eocene: marine limestones with minor sands and shales Oligocene–Miocene: minor volcanics	Miocene: fluvial and shallow-marine overlain by deeper marine shales with turbidites	M. Miocene–Holocene: Evaporites with interbedded shales, sands, and limestones
PRIPYAT AND DNIEPER-DONETS BYELARUS, UKRAINE	DEVONIAN (Frasnian/Famennian) 370–360 Ma	130,000 km^2 1300 × 70–130 km	Middle Devonian–Lower Frasnian platorm clastics overlain by massive reefal carbonates (Proterozoic rift present in Western Pripyat and possibly southeastern Dnieper-Donets) 460–1200 ft (140–360 m), locally absent in the Dnieper-Donets	Pripyat: Frasnian-Famennian marine clastics, carbonates (shallow, reefal to deep water with turbidites) and volcanics, two evaporite units 13,000 ft (4000 m) Dnieper-Donets: deep, poorly known 16,000–20,000 ft (5000–6000 m)	Pripyat: 5,000 ft (1.5 km) primarily continental sandstones, shales, and marls Dnieper-Donets: > 26,000–33,000 ft (8–10 km) complex continental and marine
RECÔNCAVO BRAZIL	LATE JURASSIC–EARLY CRETACEOUS 150–120 Ma	12,000 km^2 160 × 75 km Aborted portion of longer trend (opening of the South Atlantic)	Upper Jurassic arid alluvial fan deposits	Lower Cretaceous (Neocomian) continental and lacustrine sediments	Thin Cretaceous (Aptian) and Tertiary coarse clastics
ALBUQUERQUE SEGMENT RIO GRANDE RIFT USA	HOLOCENE–PRESENT 32 Ma	6400 km^2 160 × 40 km Portion of longer trend (Rio Grande Rift)	3300 m of Paleozoic and Mesozoic sediments, 900 m (up to as much as 2450 m) Eocene fluvial-lacustrine sediments Mid-Oligocene rhyolite ash flows and intrusives	16,000 ft (4900 m) of alluvial, eolian, playa sediments Upper Oligocene basalts and andesite flows and shallow intrusives Miocene–Holocene basalt	None

Table 2. (Continued)

Basin	Reservoir	Source	Seal	Trap Type	Reserves and Discovery Date
GULF OF SUEZ Hot, arid	Prerift: Carboniferous and older Nubia "C" continental sandstones; Cretaceous Nubia "A" sandstones; Nezzazat Group sandstones and limestones; Eocene Thebes limestones Synrift: Nukhul sandstones and conglomerates; Rudeis sandstones; Kareem sandstones; Belayim sandstones	Prerift Cretaceous (Campanian) Sudr Formation (Brown Limestone) Oil source Generation: Peak generation and onset of migration about 8 Ma (Late Miocene) in Ramadan area	Primary seal: evaporites (conformable and uncomformable over eroded tilted fault blocks) Secondary seals: shales	Primarily tilted fault blocks Alluvial fan and fan delta sandstones on downthrown blocks	> 5.25 billion bbl proven recoverable; > 7 billion bbl estimated ultimate recoverable (Vargo, et al., 1993) 1886
PRIPYAT AND DNIEPER-DONETS	Pripyat: prerift and synrift carbonates Dnieper-Donets: primarily postrift sag clastics, new discoveries mainly in deeper Lower Carboniferous	Pripyat: not identified Dnieper-Donets: not identified	Pripyat: evaporites Dnieper-Donets: primarily Lower Permian salt	Pripyat: tilted fault blocks Dnieper-Donets: primarily salt-assisted structural traps, fault traps, and minor stratigraphic traps	Pripyat: reserves not available 1964 (oil) Dnieper-Donets: reserves not available 1950 (gas)
RECÔNCAVO Hot, humid	Prerift: 57% reserves in braided fluvial Sergi sandstones and fluvial overlain by eolian sandstones of the Agua Grande Synrift: 43% reserves in turbidites and fractured shales of the Candeias Fm and deltaic sandstones of the Ilhas Fm	Lacustrine shales of the Gomo and Tauá members of the Candeias Fm: average TOC ≈1%, very mature Oil source Generation in early Aptian	Shales	Tilted & nontilted fault blocks Turbidites Rollover structures formed by growth faults in deltas	4.79 billion bbl (761 million m^3) 2.9 tcf 82 billion m^3 1939
ALBUQUERQUE SEGMENT RIO GRANDE RIFT Arid	Potentially prerift: Upper Cretaceous sandstones	Prerift: Upper Cretaceous marine shales, marginal marine shales, and coals Generation contemporaneous with rift sedimentation	Shales	Fault blocks and stratigraphic traps	None to date Limited exploration

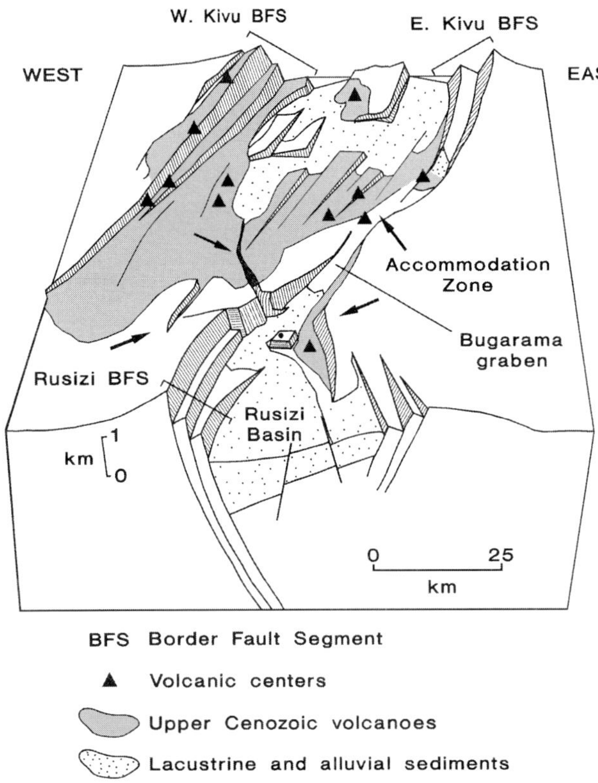

Figure 3. Model of the geometry of rift basins and an accommodation (or transfer) zone is illustrated in this block diagram of a portion of the East African rift system (from Ebinger, 1989). The characteristic half-graben geometry and reversal in asymmetry are shown.

and well-exposed East African rift system (Morley et al., 1990; Ebinger, 1989; Rosendahl, 1987).

Figure 3 is a three-dimensional model of asymmetric rift grabens in the East African rift. The individual basins are separated by structurally high areas termed accommodation zones (Ebinger, 1989; Rosendahl, 1987) or transfer zones (Patton et al., this volume). This basic geometry of curvilinear listric faults that define inconsistently overlapped elements aligned parallel or subparallel with each other and with the zone of deformation (Harding and Lowell, 1979), generally asymmetric half-grabens and full grabens, is represented by each of the rifts discussed in this volume.

Preexisting zones of weakness may have a major impact on the geometry resulting from extension as shown by Patton et al. (this volume). Late Proterozoic and Cambrian structural grain in the Gulf of Suez was important in controlling the final geometry of the Miocene rift, especially by influencing the position and orientation of cross-trending features. Preexisting structural grain may influence the location of the major boundary faults or cross faults. However, there are other areas where preexisting structural grain has had little or no impact on younger structures.

INVERSION

Rift basins are the product of extensional tectonics, but after extension, cooling, thermal contraction, and associated sedimentation have ended, these basins are often subjected to subsequent tectonic events. Structural inversion is reversal of the original sense of motion in structural deformation. The term is most commonly applied to normal faults and resulting basins that are uplifted as a result of compression. Compression and convergent wrenching (transpression) invert rift structures, reactivating normal faults as high-angle reverse faults. Just as extension can be influenced by preexisting crustal weakness, structures resulting from compression or transpression will be strongly influenced by the preexisting faults in a rift. Ulmishek et al. (this volume) describe the inversion of the southeastern portion of the Pripyat and Dnieper-Donets trend to form the Donbas fold belt at the end of Early Permian. The structure of northwestern Europe and mainly the southern North Sea also has been modified as a result of reactivation during the Alpine orogeny.

Inversion has a major impact on hydrocarbon occurrences in a rift basin. New traps may form in response to compression and uplift. Accumulations may be subjected to remigration. Significant uplift may alter the thermal history of the source rock and influence subsequent maturation and expulsion history. Inversion also influences porosity and permeability in reservoirs and seismic velocities.

DEPOSITIONAL SYSTEMS

Three stratigraphic packages usually occur in a rift basin: prerift, synrift, and postrift sag sediments. Prerift rocks are those that were in place prior to rifting and may be crystalline basement or sediments deposited prior to extension. Synrift rocks, both sedimentary and igneous, are deposited (or intruded) contemporaneously with crustal extension. Postrift sediments are deposited in the sag formed as a result of cooling and lithospheric contraction. Overlying sediments unrelated to rifting also may be present and be a part of the hydrocarbon system, providing seals or reservoirs for hydrocarbons generated from synrift source rocks.

Prerift

The prerift section, when preserved, is composed of rocks older than rift-related extension. If a rift forms as a result of uplift, prerift sediments may be eroded and not preserved in the grabens. Prerift rocks may be of any age predating the initiation of rifting. In the Gulf of Suez, uplift occurred after the initiation of rifting and a significant section of prerift sediments are preserved in the rift basin. The presence of thick prerift sediments is a strong line of evidence to suggest the mechanism by which extension was initiated. As discussed, the presence of prerift sediments within grabens and erosion of prerift sediments on the rift flanks indicate lack of initial doming and, therefore, passive rifting. Presence of prerift sediments is critical to hydrocarbon potential in the gulf because the primary source rock is the prerift Cretaceous Brown Limestone Member of the Sudr Formation and several of the significant reservoir units occur in these prerift rocks. The prerift sediments are affected by normal faulting associated with rifting that may create traps very similar to traps in the synrift section. The rift-related tilted fault blocks at Ramadan field in the Gulf of Suez provide the trap for the Carboniferous and Cretaceous prerift reservoirs (Abdine et al., this volume).

Synrift

The boundary between prerift and synrift rocks usually is marked by an unconformity. Most synrift sediments are dominated by siliciclastics that reflect the active tectonic framework of the rift. Coarse-grained alluvial sediments commonly reach significant thicknesses adjacent to bounding faults and grade laterally into finer grained sediments of fluvial and lacustrine or marine systems. These sediments usually are characterized by facies with lateral continuity in a strike direction, but with significant facies changes in a dip direction. In lacustrine or marine phases, very fine grained sediments are deposited. Commonly, these fine-grained sediments are deposited in anoxic conditions providing for excellent preservation of organic matter and excellent potential for hydrocarbon source rock. In lower latitudes, carbonates and evaporites may be deposited. Evaporites are not formed exclusively in proximity to marine conditions. Lake Natron, on the Kenyan–Tanzanian border, is an example of a basin having major evaporite deposits in a lacustrine environment controlled by the internal drainage.

Most rifts are characterized by siliciclastic deposition, including those rifts with a significant marine history (Gulf of Suez), because of the active tectonism and rift shoulder uplift. There are exceptions, such as the carbonate-dominated Jurassic High Atlas rift. This rift is associated with extension in the Tethys and provides an excellent example of the interrelationship of tectonics and sea level change in marine rifts. Initial rift-related rocks are basalts and subaerial alluvial-fluvial sediments grading rapidly upward, as well as laterally, into shallow-marine carbonates and deep-water sediments, including carbonate turbidites (Warme, 1989). The distribution of rift-related facies is well exposed as a result of inversion of the High Atlas rift. Potential for marine conditions in a rift depends on proximity to an ocean, initiation as a sag versus uplift, and rate of extension and amount of subsidence.

Postrift

Postrift sag sediments are deposited in basins formed as a result of cooling and thermal contraction after extension ceases. This type of basin is areally larger than the original grabens and has been called a steer-horn basin. Sediments deposited in a sag basin may be continental or marine, depending on the regional setting at the end of rifting. Although sediments deposited in the sag are usually not affected by rift-related faulting, source and reservoir rocks may occur and younger structures and stratigraphic traps may be present. In the Dnieper-Donets basin (Ulmishek et al., this volume), postrift sag clastics are the primary reservoir, and synrift salt-controlled structures provide the major traps with postrift sag salt providing the seal.

Older rift basins, as zones of major crustal weakness, may influence much younger basin formation as documented in cratonic sag basins, such as the Michigan basin (Leighton et al., 1990).

HYDROCARBON POTENTIAL

Traps

Most of the hydrocarbon reserves discovered in rift basins occur on the highest part of a tilted fault block. As a result of syn-depositional faulting, these locations also tend to be the site of facies changes, reef development, and unconformities that also may have an impact on reservoir quality and trap. A much smaller percentage of hydrocarbons have been discovered on the downthrown sides of faults. An example occurs in the Gulf of Suez at October field, which is a large field producing from the crest of a tilted fault block (Schütz, this volume). The discovery of a hydrocarbon trap in a downthrown position at October field was made as a result of a well that tested 4900 BOPD from a 100-ft-thick basal Miocene sandstone and conglomerate (Abdine, 1981). In the Recôncavo basin, production has been established from coarse fanglomerates on the downthrown side of faults (de Figueiredo et al., this volume). Brea field in the North Sea is a similar trap type (Harms et al., 1981) and these examples indicate the potential for significant reserves in coarse siliciclastics on downthrown blocks in rift basins.

Stratigraphic traps are more difficult to find with current exploration methods and represent the greatest potential for undiscovered reserves in rift basins that are currently productive. Of the field studies from rift basins published in the AAPG *Atlas of Oil and Gas Fields* (Table 4), 24 fields are structural traps and one is stratigraphic. This discrepancy, which reflects the division of known reserves, is primarily the result of exploration technology. There are a variety of potential stratigraphic traps, but turbidite reservoirs may be one of the most significant trap types. Normal faults with large amounts of offset, which are common in rift settings, provide potential for significant topographic relief and a source of coarse clastic turbidites that may be encased in fine-grained lacustrine or marine shales. Hence, these traps should occur on the downthrown side of faults providing the topographic high for a source of the

Table 3. Kerogen types, their origin, and hydrocarbon potential (from Merrill, 1991)

<table>
<tr><th>Environment</th><th>Kerogen Type</th><th>Kerogen Form/Maceral</th><th>Origin</th><th>Hydrocarbon Potential</th></tr>
<tr><td rowspan="3">Aquatic</td><td rowspan="2">I</td><td>Alginite</td><td>Algal bodies</td><td rowspan="4">Oil</td></tr>
<tr><td rowspan="2">Amorphous Kerogen</td><td>Structureless debris of algal origin</td></tr>
<tr><td rowspan="2">II</td><td>Structureless, planktonic material, primarily of marine origin</td></tr>
<tr><td rowspan="3">Terrestrial</td><td>Exinite</td><td>Skins of spores and pollen, Cuticle of leaves and herbaceous plants</td></tr>
<tr><td>III</td><td>Vitrinite</td><td>Fibrous and woody plant fragments and structureless, colloidal humic matter</td><td>Gas and some Oil
Mainly Gas</td></tr>
<tr><td>IV</td><td>Inertinite</td><td>Oxidized, recycled woody debris</td><td>None</td></tr>
</table>

Table 4. Rift basin field studies included in AAPG *Atlas of Oil and Gas Fields, Treatise of Petroleum Geology*

Volume	Field	Location	Basin
Structural Traps I	Leman	U.K.	North Sea
Structural Traps II	Brent	U.K.	North Sea
	Magnus	U.K.	North Sea
	Ninian	U.K.	North Sea
	Sarir	Libya	Sirte
Structural Traps III	Hibernia	Canada	Jeanne d'Arc
	Unity	Sudan	Muglad
Structural Traps IV	Ekofisk	Norway	North Sea
	Eldfisk	Norway	North Sea
	West Ekofisk	Norway	North Sea
	Tommeliten Gamma	Norway	North Sea
	Fulmar	Norway	North Sea
	Midgard	Norway	North Sea
	Sleipner Ost	Norway	North Sea
	Sleipner Vest	Norway	North Sea
Structural Traps V	Snapper	Australia	Gippsland
	Tirrawarra	Australia	Cooper
Structural Traps VI	Cormorant	U.K.	North Sea
	Draugen	Norway	North Sea
	Gullfaks	Norway	North Sea
	Piper	Norway	North Sea
	Cod	Norway	North Sea
	Maureen	U.K.	North Sea
	Dan	Denmark	North Sea
Structural Traps VII	Raguba	Libya	Sirte
Structural Traps VIII	NONE		
Stratigraphic Traps I	Frigg	U. K. & Norway	North Sea
Stratigraphic Traps II and III	NONE		

coarse sediment. Turbidites pose significant possibilities for stratigraphic traps in many rifts as shown by production in the Recôncavo basin (de Figueiredo et al., this volume). Production also has been established from turbidites in the Gulf of Suez. Turbidites have been identified in the Dnieper-Donets and Pripyat basins, but so far are nonproductive.

Reservoirs

As a result of the stacked depositional sequences represented in rifts, there are usually multiple opportunities for reservoirs in prerift, synrift, sag, and younger sediments. Fields commonly have multiple reservoirs and a field in the Dnieper-Donets basin has 27 pay zones.

Source Rock

The stacked architecture of rift basins is conducive to multiple source rocks that may be pre-, syn-, or postrift. Synrift lacustrine and marine sediments deposited in rift basins commonly are deposited under restricted conditions because of the series of elongate, deep depocenters created by extension. Anoxic conditions commonly enhance the preservation of organic matter in the sediments in both marine and lacustrine environments. In the Gulf of Suez, the primary source rock is the marine, prerift Campanian Brown Limestone Member of the Sudr Formation. The source rocks of the Recôncavo basin are an example of the potential of lacustrine source rocks. The organic matter in synrift source rocks is predominately type I (lacustrine) and type II (marine and lacustrine), an oil source (Table 3). Klemme (1980) indicated that rift basins provide 12% of the world's oil reserves, but only 4% of the gas reserves, further documenting the oil-prone nature of rift-related source rocks. As indicated by Ulmishek et al. (this volume), the significant gas reserves present in the Pripyat and Dnieper-Donets basins were the result of advanced thermal maturity rather than original generative potential.

Maturity

Because of the elevated heat flow associated with rifting, pre- and synrift source rocks normally mature fairly early in the basin history. Hydrocarbons may be generating contemporaneously with extension, but because of the variation in heat flow and sediment thickness along the rift trend, generation may occur at different times in different locations along a rift.

Abdine et al. (this volume) indicate that peak generation in the Gulf of Suez began about 8 Ma (late Miocene), postextension. In the northern gulf, Vargo et al. (1993) reconstructed the burial history and determined that generation began in the late Pliocene, approximately 3 Ma, at depths between 2898 and 3050 m (9500 and 10,000 ft). In deeper parts of depocenters and areas of highest heat flow, generation began significantly earlier.

In general, trap formation is syndepositional as tilted fault blocks form in response to extension. Generation also may occur very early in rift development, concurrent with trap formation. Generation may not be confined to the early history of rift formation because generation will occur at different times based on burial depth and heat flow history. Traps created by subsequent structural inversion may be charged by remigration and/or newly generated hydrocarbons.

Field Studies

The *Atlas of Oil and Gas Fields* in the *Treatise of Petroleum Geology* series published by the AAPG contains a number of field studies from rift basins worldwide. (One of these field studies, Ramadan field, has been reprinted in this volume [Abdine et al.].) These papers are an excellent companion to this volume and are listed in Table 4. In addition, a comprehensive description of October field, Gulf of Suez, was recently published in AAPG *Memoir 54* (Lelek et al., 1992).

FUTURE EXPLORATION POTENTIAL

Several areas have been identified for future exploration potential by the authors in the basins discussed in this volume. Deeper drilling and more effective exploration for stratigraphic traps are most frequently mentioned. As seismic technology continues to expand, there is significant potential for new reserves in rift basins that have been considered as being in a mature exploration phase. The recent discovery of Warda field in the northern part of the Gulf of Suez, with the assistance of three-dimensional seismic, is evidence of this potential (Vargo et al., 1993).

Although most of the known reserves in rift basins are in structural traps, seismic data are not totally effective in defining deep basin structure. One of the keys to a successful exploration program in rift basins is the ability to map structures, but in the case of the Gulf of Suez, thick evaporites often are barriers to acquiring usable data below them (Schütz, this volume; Patton et al., this volume). Thick basalts also may be a detriment to obtaining deeper seismic data. Continuously evolving seismic technology will aid in the discovery of new reserves in rift basins. The ability to map stratigraphic traps will also improve.

Prediction of facies, especially source rock and reservoir, also is an area where advancements will provide for new discoveries.

Another area for future potential is the discovery of unidentified rift basins in the subsurface. Rift basins in the Sudan and Yemen have been identified only within the last few years. Gravity, magnetics, and seismic, with an understanding of the tectonic framework of rifts, will provide the exploration tools to successfully locate potential new basins.

The Mid-continent rift system (MRS) in North America is an example of an unexplored rift basin that might yield new reserves (Dickas, 1986). The MRS is a Proterozoic rift with documented source rock, potential reservoir, and a variety of documented trap types. Exploration has been limited to regional seismic and only a few significant wells.

The termination of industry interest in the MRS coincided with the decline of the industry in the mid-1980s; attractive targets could await renewed exploration.

REFERENCES CITED

Abdine, A.S., 1981, Egypt's petroleum geology: good grounds for optimism: World Oil, Dec., v. 202, p. 99–106.

Allen, P.A., and J.R. Allen, 1990, Basin analysis: principles and applications: Oxford, England, Blackwell Scientific Publications, 451 p.

Bott, M.H.P., 1981, Crustal doming and the mechanism of continental rifting: Tectonophysics, v. 73, p. 1–8.

Buck, W.R., 1986, Small-scale convection induced by passive rifting: the cause for uplift of rift shoulders: Earth and Planetary Science Letters, v. 77, p. 362–372.

Coward, M.P., 1986, Heterogeneous stretching, simple shear and basin development: Earth Planetary Science Letters, v. 80, p. 325–336.

Dickas, A.B., 1986, Comparative Precambrian stratigraphy and structure along the Mid-continent rift: AAPG Bulletin, v. 70, p. 225–238.

Ebinger, C.J., 1989, Geometric and kinematic development of border faults and accommodation zones, Kivu-Rusizi rift, Africa: Tectonics, v. 8, p. 117–133.

Gregory, J.W., 1896, The Great Rift Valley: John Murray, London, 422 p.

Harding, T.P., and J.D. Lowell, 1979, Structural styles, their plate tectonic habitats, and hydrocarbon traps in petroleum provinces: AAPG Bulletin, v. 63, p. 1016–1058.

Harms, J.C., P. Tackenberg, E. Pickles, and R. Pollock, 1981, The Brae oilfield area, *in* L.V. Illing and G.D. Hobson, eds., Petroleum geology of the continental shelf of north-west Europe: London, England, Institute of Petroleum, p. 352–357.

Hellinger, S.J., and J.G. Sclater, 1983, Some comments on two-layer extensional models for the evolution of sedimentary basins: Journal of Geophysical Research, v. 88, p. 8251–8270.

Housemann, G., and P.C. England, 1986, A dynamical model of lithosphere extension and sedimentary basin formation: Journal of Geophysical Research, v. 91, p. 719–729.

Klemme, H.D., 1980, Petroleum basins—classifications and characteristics: Journal of Petroleum Geology, v. 3, p. 187–207.

Leighton, M.W., D.R. Kolata, D.F. Oltz, and J.J. Eidel, eds., 1990, Interior cratonic basins: AAPG Memoir 51, 819 p.

Lelek, J.J., D.B. Shepherd, D.M. Stone, and A.S. Abdine, 1992, October field: the latest giant under development in Egypt's Gulf of Suez, *in* M.T. Halbouty, ed., Giant Oil and Gas Fields of the Decade: AAPG Memoir 54, p. 231–249.

Lowell, J.D., 1985, Structural styles in petroleum exploration: Tulsa, Oklahoma, Oil and Gas Consultants International, Inc., 477 p.

McKenzie, D.P., 1978, Some remarks on the development of sedimentary basins: Earth and Planetary Science Letters, v. 40, p. 25–32.

Merrill, R.K. 1991, Preface, *in* R.K. Merrill, ed., Source and Migration Processes and Evaluation Techniques: AAPG Treatise of Petroleum Geology, p. xiii–xviii.

Morgan, P., and B.H. Baker, 1983, Introduction—processes of continental rifting: Tectonophysics, v. 94, p. 1–10.

Morley, C.K., R.A. Nelson, T.L. Patton, and G.G. Munn, 1990, Transfer zones in the East African rift system and their relevance to hydrocarbon exploration in rifts: AAPG Bulletin, v. 74, p. 1234–1253.

Neugebauer, H.J., 1978, Crustal doming and the mechanism of rifting. Part 1, rift formation: Tectonophysics, v. 45, p. 159–186.

Petters, S.W., and C.M. Ekweozor, 1982, Petroleum geology of Benue Trough and southeastern Chad Basin, Nigeria: AAPG Bulletin, v. 66, p. 1141–1149.

Rosendahl, B.R., 1987, Architecture of continental rifts with special reference to East Africa: Annual Review Earth Planetary Science, v. 15, p. 445–503.

Rowley, D.B., and D. Sahagian, 1986, Depth-dependent stretching: a different approach: Geology, v. 14, p. 32–35.

Salop, L.I., and Y.M. Scheinmann, 1969, Tectonic history and structures of platforms and shields: Tectonophysics, v. 7, no. 5–6, p. 565–597.

Sengor, A.M.C., and K. Burke, 1978, Relative timing of rifting and volcanism on Earth and its tectonic implications: Geophysical Research Letters, v. 5, p. 419–421.

Steckler, M.S., 1985, Uplift and extension at the Gulf of Suez: indications of induced mantle convection: Nature, v. 317, p. 135–139.

Vargo, J.M., M.J. Greer, B.P. Collins, C.B. Davis, and M.H. Miller, Jr., 1993, Gulf of Suez plays center on tilted fault blocks: World Oil, v. 214, p. 69–74.

Warme, J.E., ed., 1989, Rift tectonics and carbonate facies response: exploration models from the Jurassic High Atlas rift, Morocco: AAPG Field Seminar Guidebook, Exploration Geosciences Institute, Colorado School of Mines, Publication 9, 332 p.

Wernicke, B., 1981, Low angle normal faults in the Basin and Range province: nappe tectonics in an extending orogen: Nature, v. 291, p. 645–648.

Wernicke, B., 1985, Uniform-sense normal simple shear of the continental lithosphere: Canadian Journal of Earth Science, v. 22, p. 108–125.

Woidt, W.D., and H.J. Neugebauer, 1981, Lithospheric thinning and the dynamics of density instabilities: EOS, Transactions American Geophysical Union, v. 62, p. 814.

Index

Abu Darag, 78
Abu Durba, 38, 40, 73, 78
Abu Durba block, 33, 34
Abu Durba Formation, 18, 78, 79
  isopach map, 19
Abu Gharadig basin, 21, 62, 66
Abu Gharadig Marine Basin, 80
Abu Roash Formation, 81
Abu Roash structure, 21
Abu Rudeis, 40
  location map, 58
Abu Zenima, 26, 40, 69, 72, 77, 83
Abu Zenima Basin, 62, 69, 71–72, 90, 92
  type log correlation, 70
Abu Zenima fault, 38, 39, 72
Abu Zenima Formation, 9, 24, 45
Active rift mechanisms, 261
  model of, 262
Afligidos Member, 165
African craton, 61
African shield, 61
African trend, 34
Afro-Brazilian Depression, 183
Água Grande field, 157, 164, 197
Água Grande Member, 168, 169, 170, 185
Ain Sukna fault, 58, 66, 69
Alagoinhas low, 164, 180, 181, 186
Alamein fault, 58, 62
Alam el Bueib fault, 58, 62
Alam el Bueib Member, 79
Albuquerque basin, 205, 206, 263–264
  basin-controlling faults, 237–238
  COCORP seismic data, 215, 242–244, 256–257
  cross sections, palinspastic reconstructions of, 250
  general outcrop geology of, 229
  key exploration wells, 230
  location and physiography of, 2–3, 217, 222
  oil and gas reserves in, 244, 246
  palinspastic analysis of, 238–239, 244, 250
  prerift deformational history, 226–227, 237
  seismic sections, 232–236, 240–242, 245–247, 251–252, 255
  stratigraphy of, 222, 226
  structural framework, 237
    map of, 231
  time-stratigraphic section of, 222
Albuquerque-Belen basin, 206
Albuquerque field, 226
Aliança Formation, 165, 167, 181
Amal field, 89
Amer-Bakr-Gharib high, 71–73, 86, 88
Amer field, 84
Anadarko basin, 259
Ancestral Rocky Mountains, 227
Anorogenic phase of development, Arabian-Nubian shield, 14–15
Aporá high, 157, 164
Aqaba/Dead Sea rift, 61
Aqaba–Dead Sea wrench system, 31, 32
Aqaba rift system, 84
Aqaba trend, 34, 63
Araba block, 33
Araba Formation, 15, 16, 78
Arabian Gulf salt basin, 15
Arabian-Nubian shield, 12, 14–15
Arabian Peninsula, 14, 23
Arabian plate, 22, 23, 31, 61
Arabian Platform, 19
Arabian shield, 61
Arabo-African craton, 57
Araçás field, 157, 164, 195
Archaediscidae, 78
Archeocyathidae, 78
Areif El Naqa structure, 21
Asl field, 7, 84
Asl Formation, 30
Asl Member, 89
Ataqa Formation, 18
Aualitic trend, 34
Aulacogens, 3, 128, 139, 259, 261
Ayun Musa structural block, 32

Baba Member, 69, 90
Baba Plain, 40
Baca Formation, 222
Bahariya Formation, 80
Bahia Supergroup, 181
Bakr field, 72, 84
Baltic basin, 134
Barreiras Formation, 165, 180, 181
Basaltic volcanism, in Gulf of Suez, 9
Basement and onshore plateaus, Gulf of Suez, 62–67
Basin and Range province, 205, 207, 212, 213, 215, 219
  tectonic evolution of, 215–217, 219
Basin classification schemes, 1
Basin-controlling faults, Albuquerque basin, 237–238
Basin development, Donbas-Pripyat rift, 139–143
Bathyal benthonic fauna, 87
Belayim, location map, 58
Belayim area, Gulf of Suez basin, 7
Belayim basin, 75
Belayim fault, 62
Belayim field, 7, 66, 78, 88, 93
Belayim Formation, 10, 32, 59, 66, 69, 72, 75, 85, 90–92, 99, 106–109
  isopach map, 91
Belayim horizon, 39
Belayim Marine Basin, 76
Belayim Marine field, 7, 66, 71, 88, 90, 92
  location map, 58
Belorussian high, 127, 140
Benue rift, 259
Berezin field, 134
  cross section and contour map, 146
Biostratigraphy, Recôncavo basin, 181
Boa Esperança field, 164
Boipeba Member, 165, 167
Bouguer gravity map, Recôncavo basin, 162
Bouguer gravity profiles, Rio Grande rift, 215, 218
Bragin-Loev high, 127, 135
Bragin-Loev structural saddle, 127
Bragin-Loev uplift, 130, 132, 141
Brea field, 266
Brent field, 267
Brest basin, 126
Brotas Group, 181, 183
Brown Limestone Member, 7, 80, 81, 82, 265, 266
Buracica field, 157, 198
Burg el Arab Formation, 79

Cairo-Suez trough, 44, 45
Camaçari low, 164, 181, 185, 186
Candeias field, 157, 193, 200
Candeias Formation, 165–168, 170–175, 177, 181, 183, 185–187, 191–194, 201
Candeias sandstones, 191
Candeias shales, 187
Capianga Member, 165, 181
Cardita-bearing limestones, 82
Carrizo Arroyo, 228, 243
Carrizo fault, 243
Caruaçu beds, 173, 174, 177
  isopach map, 175
Caruaçu sandstones, 175
Cat Mesa, 237
Cat Mesa fault, 238
Cat Mesa field, 226
Catu Member, 174–176, 185
  isopach map, 176
Cerrillos Hills, 226
Chronostratigraphy, Recôncavo basin, 181

Clysmic faults, 37–38, 40, 99
normal, 100–101
"Clysmic Gulf," 57
Clysmic orientation, 18
Clysmic trends, 33, 34, 111
COCORP, seismic data, Albuquerque basin, 215, 242–244, 256–257
Cod field, 267
Colorado Plateau, 212–213, 219, 239
Colorado Plateau block, 206, 216
Comanche fault, 227
Conchostraca, 181
Conjugate transfer zones, 38–39
Cormorant field, 267
Coyote fault, 238
Crocodylidae, 181
Cross fault, Gulf of Suez, defined, 37
Crustal structure, Rio Grande rift, 212, 214
Cruziana sp., 78
Cyprididae, 181
Cytheridae, 181

Dan field, 267
Darag Basin, 32, 37–38, 40, 44, 62, 68–69, 76, 79, 88–89, 92–93
geological cross section, 69
type log correlation, 70
Dara Gebel, 14
Dara ridge, 21, 36, 37, 39
Darag fault, 38, 39
Darat Formation, 82
Dead Sea wrench system, 10, 31, 32, 47
Diagenesis, Recôncavo basin, 187–192
Diapirism, 261
Dibba transform, 23
Displacement transfer, Gulf of Suez, 38–41
Dnieper-Donets basin, 263–264, 266, 268
contour map, 137, 140
cross section, 138, 142
future potential of, 153–154
location of, 2, 127
maturation modeling in, 149–153
petroleum exploration in, 143
postrift platform sequence, 133–134
postrift sag sequence, 132–133
prerift sequence, 128–130
source rocks in, 148–149
stratigraphy of, 128–134
structure of, 135–139
synrift sequence, 130
Dokhan Volcanics, 14
Domanik Formation, 149
Dom João field, 157, 199
Dom João high, 164
Donbas fold belt, 125, 126, 128, 134, 142, 152, 265
cross section, 142
Donbas-Pripyat rift, 126
basin development, 139–143
Donets basin, 259. *See also* Dnieper-Donets basin
Draugen field, 267
Duwi trend, 36

East African rift lakes, 244
East African rift system, 141, 265
East Brazilian microplate, 159, 160, 185
Eastern Desert, 12, 14, 34
Eastern Desert basement, 62, 75
Eastern Desert mountains, 57, 79
EE 82-FF/GG 83 high, 72–73
Egma Limestone, 82–83
Ekofisk field, 267
Eldfisk field, 267
El Morgan field, 7
Elphidium sp., 87
El Qaa basin, 62, 69, 75, 76, 86, 88, 89, 92
El Qaa Marine Basin, 73
type log correlation, 74
El Qaa plain, 90
El Tih Plateau, 40
El Tor, 40, 66, 73
El Tor fault, 74–75
El Tor ridge, 89
Endothyridae, 78
En echelon normal faults, 38
Erythrean trend, 34
Esh el Melaha, 34, 39, 63
Esh el Melaha Basin, 75
Esh el Melaha block, 33, 37
Esh el Melaha Range, 62, 65–66, 74, 75, 89
Esna Formation, 62, 82, 111, 112, 114, 117
Esna shales, 104
Espanola basin, 206, 222
Exploration analogs, 1
Extensional model, for development of Gulf of Suez, 15
Extension estimates, Gulf of Suez, 41–44

Fault controlled seeps, in Suez rift, 7
Fault map, simplified, Gulf of Suez, 34
Fault pattern, Gulf of Suez, 61–62
postulated, 60
Faults
clysmic, 37–38, 40, 99, 100–101, 111
en echelon normal, 38
transfer, 38
transform, 38
Fauna
bathyal benthonic, 87
Indo-Pacific, in Gulf of Suez, 47
Feiran gneiss, 13
Feiran Member, 69, 75, 90, 109
Field studies, rift basins, 267–268
Fractal dimensions, in Gulf of Suez faults, 41
Frigg field, 267
Fulmar field, 267

Galala fault segment, 34
Galala Plateau, 19, 38, 66
Galisteo Formation, 222
Gas fields, Recôncavo basin, 196
Gas reserves
Albuquerque basin, 244, 246
Pripyat basin, 144–146
Gastropods, 181

Gebel, defined, 11
Gebel Abu Alaqa, 30
Gebel Abu Durba, 38, 63, 66–67, 78, 79
Gebel Abu Durba basement block, 33
Gebel Abu Durba fault, 73
Gebel Abu Gerfan, 75
Gebel Abu Shaar el Gibli, 89, 91
Gebel Abu Suweira, 91
Gebel Araba, 34, 40, 62, 66–67, 73, 74, 78, 89
location map, 58
Gebel Araba block, 33, 44, 63
Gebel Ataqa, 82, 83
Gebel Durba, 88
Gebel Duwi, 34
Gebel Ekma, 66, 91
Gebel Esh el Melaha, 36, 37, 75
Gebel Galala, 66, 82
Gebel Gharamul, 72, 73, 74, 84
Gebel Hadahid, 26, 38
Gebel Hammam Faraun, 22, 24, 26, 27, 33, 38
Gebel Katharina assemblages, 14
Gebel Maghara, 81
Gebel Maghara fault, 58, 62, 79
Gebel Magmar, 81
Gebel Nazzazat, 63, 66, 73, 80, 81
Gebel Nukhul, 35
Gebel Qatrani, 58, 84
Gebel Tanka, 82
Gebel Zeit, 7, 26, 34, 37–39, 58, 62–65, 72, 74, 75, 79, 80, 88–93
geological cross section, 65
geological map, 65
location map, 58
Gebel Zeit basin, 76, 88
type log correlation, 74
Gebel Zeit fault, 73
Gebel Zeit Marine Basin, 62, 73, 74, 75, 89, 90
Geisum field, 76
Gemsa field, 7, 75, 91
Gemsa Peninsula, 7, 57, 75
Geology of Egypt, 60
Gerofit Formation, 81
Ghara Member, 28–29
Gharamul Member, 28–29
Gharandal Group, 85–86
Gharib Gebel, 14
Gharib Granite, 15
Glinsk-Rozbyshev field, 144
Globigerina, 87
Globigerina limestones, 58, 64, 89
Globigerina marls, 7, 29–30
Globigerina-rich marls, 90
Globigerina trilobus trilobus, 88
Globigerinoides, 83
Globigerinoides mitrus, 90
Globigerinoides sicanus, 88
Globorotalia cf. *scitula*, 90
Globorotalia siakensis, 90
Gnedinstev field, 143, 144
Gomo Member, 170, 171, 185, 186, 192–194, 201
isopach map, 170
Gondwana, 159, 181
Gravity map, Bouguer, Recôncavo basin, 162

Gravity profiles, Bouguer, Rio Grande rift, 215, 218
Gravity spreading, 261
Gulf of Aqaba, 12, 34, 57, 61, 63, 84, 93
Gulf of Aqaba faults, 72
Gulf of Suez, 261, 266, 268
 "Clysmic Gulf," 57
 displacement transfer, 38–41
 exploration activity in, 10
 extension estimates, 41–44
  published, 42
 fault orientation data, summary, 36
 fractal dimensions in, 41
 geological definition of, 57
 geothermal gradient map, 77
 Hercynian growth, 16–18
 igneous rocks, 20
 integrated model, 2
 location map, 11
 mature source rocks in, 77
 postulated fault pattern, 60
 prerift tectonic elements, 16
 rift abandonment, 32–33
 rift development, 29–32
  summary of, 45–47
 rift structure, 33–44
 rift tectonics, 24–33
 sequential development of, 46
 serial cross sections, 35
 simplified fault map, 34
 stratigraphic column, 60
 stratigraphic heaves, fractal nature of, 41
  size distribution of, 43
 stratigraphy
  postrift, 89–93
  prerift, 77–85
  synrift, 85–89
 structural features, 62–76
  map of, 59
 structure of, 61–77
  northern, 63
  southern, 64
 subbasins, 62, 68–76
 tectonostratigraphic events, charts, 13, 25
 as type rift basin, 1
Gulf of Suez basin, 263–264, 265
 introduction to exploration in, 5–8
 location of, 2
 structural setting of, 33–47
 tectonic evolution of, 12–33
Gulf of Suez event, 9
Gullfaks field, 267

Hadahid block, 33, 34
Hammamat Formation, 14
Hammam Faraun, 21, 23, 40, 69, 82
Hammam Faraun block, 24
Hammam Faraun fault segment, 34
Hammam Faraun Member, 69, 90, 109
Heat flow
 in rift basins, 268
 Rio Grande rift, 215, 217
Helez geanticline, 16, 18
Hercynian collision, 126, 140
Hercynian event, 9
Hercynian growth, Gulf of Suez, 16–18
Hercynian orogeny, 44
Hercynian tectonism, 17
Herodotus Basin, 61
Heterohelix globulosa, 80
Hibernia field, 267
High Atlas rift, 266
Hurghada, 57
Hurghada field, 7
Hurghada/Islands area, Gulf of Suez, 62, 75–76, 86–87
Hydrocarbon potential
 of kerogen types, 267
 of rift basins, 266–268
Hydrocarbon production, in rift basins in this study, 2–3

Igneous rocks
 distribution of, southwestern United States, 220–221
 Gulf of Suez, 20
Ilhasina, 181
Indo-Pacific fauna, in Gulf of Suez, 47
Integrated model, Gulf of Suez, 2
Interior rift basins, 1
Intracontinental rift basin, example of, 157–203
Inversion, in rift basins, 261, 265
Issaran Basin, 62, 69, 71–72, 76, 92
Itanagra-Araças fault, 161, 162, 164
Itaparica Formation, 165, 168, 169, 170, 181, 183, 185

Jatobá basin, 157, 159
Jemez Mountains, 222
Jeter fault, 237–238, 242–244, 248–249, 254
Joyita Hills, 226
July field, 7, 29, 73, 78, 79, 83, 97–99, 108
 location map, 58

Kairakozov field, 143, 144
Kareem Formation, 31, 32, 47, 59, 61, 66, 67, 69, 72, 75, 85, 88, 90, 91, 93, 99
Karpinsky ridge, 140
 cross section, 142
Kattanyia fault, 62
Kattanyia horst, 58, 66
Kerogen types, hydrocarbon potential of, 267
Khaboba Formation, 82
Kharga arch, 20
Khatatba Formation, 79, 80
Khoynin uplift, 134, 135
Koshelev uplift, 136

Ladron Mountains, 222, 238, 242, 244, 249, 253, 254
Laguna bench, 237
Lake Natron, 266
Lan horizon, 130
Laurasia, 140
Lelyakov field, 143, 144
Leman field, 267
Levantine Basin, 61
Levantine Sea, 61
Los Pinos Mountains, 217, 227, 238, 239
Lucero uplift, 227
Lyuten depression, 152

Maghara Basin, 80
Magnetotellurics, Rio Grande rift, 215, 219
Magnus field, 267
Malha Formation, 20–21, 79
Mansour Member, 81
Manzanita Mountains, 217
Manzano Mountains, 217, 227, 238, 239
Maracangalha Member, 170, 172, 185, 193
 isopach map, 173
Maracangalha shales, 175
Marfim Formation, 165, 167, 173, 175, 176, 181, 185, 187, 191
 electric log, 174
Marfim Member, 174–177
Marizal Formation, 165, 179, 180
Markha Member, 69, 90
Masajid Formation, 79
Mashev field, 144
 cross section and contour map, 146
Mata-Araças platform, 164
Mata-Catu fault, 161, 164
Matarma field, 84
Matulla Formation, 21, 80, 81, 93, 113, 114, 117
Maturation modeling, Dnieper-Donets and Pripyat basins, 149–153
Maureen field, 267
Mediterranean Ridge, 61
Mediterranean Sea, 61
Mheiherrat Formation, 34
Miaranga low, 181
Michigan basin, 266
Mid-continent rift system (MRS), 268
Midgard field, 267
Midland basin, 153
Migration and accumulation models, Recôncavo basin, 187, 188
Mikashevich uplift, 127
Minia Formation, 83, 83–84
Miogypsina sp., 87, 91
Miranga field, 157, 195, 202
Miranga low, 164, 185
Moiseev bed, 129, 149
Mokattam Formation, 83
Monte Largo bench, 237, 244
Montosa fault, 227, 238, 239
Morgan Basin, 62, 66, 73–75, 88–90
Morgan fault, 62
Morgan field, 62, 72–74, 89, 90, 93, 97
 location map, 58
Musandam Peninsula, 23

Nacimiento fault, 227
Nacimiento uplift, 222, 227
Najd fault system, 15
Naqus Formation, 15, 16
 isopach map, 19
Natash basin, 20

Nazzazat Group, 80, 98, 104, 109, 112, 114
Negev Desert, 21
Neo-Tethyan closure, 21, 22
Neo-Tethyan extensional event, 37
Neo-Tethyan rift event, 9, 21
Neo-Tethys Ocean, 12, 18–19, 21
Nile Cone, 61
Ninian field, 267
North American craton, 206
North Caspian basin, 134
Northern Galala Plateau, 33–34, 58, 66, 78, 79, 82
  location map, 58
North graben, Albuquerque basin, 237
North Kattania Basin, 80
North Qattra ridge, 66
Nubia "A," 79–80
Nubia "B," 79
Nubia Formation, 66, 117
Nubia sandstones, 7, 15, 27–28, 62, 66, 67, 72–78, 88, 98, 99, 106, 111–115
Nubia trend, 34
Nukhul Basin, 29
Nukhul Formation, 9, 24–26, 29, 34, 45, 58, 69, 72, 86–87, 93, 99, 106, 108
  facies, 76
  isopach map, 28, 87, 111
"Nullipore" carbonates, 75
Nullipore limestones, 67, 68, 91
Nummulites, 66, 82, 83, 84
Nummulitic limestones, 84

Oblique-slip motion, in clysmic fault, 33
Ocean basins, 3
October Basin, 62, 69, 71–72, 73, 76, 77, 90, 92
  geological cross section, 73
October field, 26, 72, 78, 266, 268
  location map, 58
October high, 71, 72, 88
  type log correlation, 70
October Member, 28–29
Oil fields
  formation of, Dnieper-Donets and Pripyat basins, 149–153
  Recôncavo basin, 196
Oil reserves
  Albuquerque basin, 244, 246
  Pripyat basin, 144–146
  Recôncavo basin, 159
Oktyabr field, 134
Oman Mountains, 23
Oman salt basin, 15
Oppel's zones, 181
Orbulina suturalis, 88, 89, 90
Oritz Mountains, 226
Ostashkovich field, 134, 143, 149
  cross section and contour map, 146
Ostracods, 181, 182
Owl Creek Range, 227

Pachelma aulacogen, 139
Pakistan salt basin, 15
Paleo-Tethys Ocean Basin, 18
Palinspastic analysis, Albuquerque basin, 238–239, 244
Palinspastic reconstructions, of Albuquerque basin cross sections, 250
Palmyra-Sinjar trough, 23
Palmyra trough, 19
Paloma fault, 227, 238, 239
Palynomorphs, Recôncavo basin, 181, 182
Pan-African event, 9, 12, 44
Pangea, 18
Passive margin development, 3
Passive rift mechanisms, 261
  model of, 262
Patioba fault, 162
Patioba Platform, 162, 164, 185
Pecos-Picuris fault, 227
Pedra do Salgado platform, 164
Pedras fault, 164
Pelagic foraminfera, 87
Pelusium fault, 58
Pelusium line, 62, 84
Petrikov horizon, 130, 131
Petrikov uplift, 134, 135
Petrobrasia, 181
Petroleum geology, of Dnieper-Donets basin, 143–149
Pink Granites, 14
Piper field, 267
Pitanga Member, 170, 171, 173, 185
  isopach map, 172
Plate tectonics, 1
  in eastern Mediterranean, 61
    reconstruction of, 23
  in passive continental margins, reconstruction of, 244
Play types, Recôncavo basin, 193–195
Plutonic rocks, Albuquerque basin, map of, 223
Podlasie-Brest basin, 126
Podocopina, 181
Pogarshchin field, 143, 144, 152
Pojuca Formation, 165, 177–181, 185–187, 191, 202
  electric log, 174
Poles horizon, 132
Poles saddle, 127
Poltava depression, 152, 153
Postrift sag stage, 142, 266
Postrift stratigraphy, Gulf of Suez, 89–93
Prerift deformational history, Albuquerque basin, 226–227, 237
Prerift section, in rift basins, 265
Prerift sequence
  diagenetic, Recôncavo basin, 187, 189–191
  Dnieper-Donets basin, 128–130
  Pripyat basin, 128–130
Prerift stratigraphy, Gulf of Suez, 77–85
Prerift system, Recôncavo basin, 193
Prerift tectonic evolution, Gulf of Suez, 12–24
Pripyat basin, 263–264, 266, 268
  contour map, 136, 137
  cross section, 141
  future potential of, 153–154
  location of, 2, 127
  main structural units and oil fields, 134
  maturation modeling in, 149–153
  oil and gas reserves in, 144–146
  postrift platform sequence, 133–134
  postrift sag sequence, 132–133
  prerift sequence, 128–130
  source rocks in, 148–149
  stratigraphic section, 129
  stratigraphy of, 128–134
  structure of, 134–135
  synrift sequence, 130
Pripyat and Dnieper-Donets trend, 265
"Productus" fossils, 78
Pure dip-slip motion, in clysmic fault, 33
Pure shear, 261
  model, 262

Qaa basin, 66
Quiambina low, 164, 185
Quiricó Platform, 185
Quseib Formation, 18

Radchenkov field, 143, 144
Raguba field, 267
Raha Formation, 21, 80
Rajim Formation, 82
Ramadan field, 7, 73, 78, 79, 83, 86, 88, 265
  characteristics of, 114–116
  contour map, 105, 106
  description of, 119–121
  development concepts, 116
  discovery, 98
    method, 99
  exploration concepts, 116
  facies, 111–112
  geologic evolution, 99–107
    Northern Zone, 107, 109
    Southern Zone, 107, 109
  geo-seismic cross section, 108
  geothermal radient map, 118
  history, 97–99
  location, 58, 97
  maturation and migration of hydrocarbons in, 113–114
  post-discovery, 98–99
  production data, 121
  representative gas chromatogram, 118
  reservoir characeristics, 111–112
  schematic cross section, 112, 113
  seismic structure map, 98
  source rocks, 112–113
  stratigraphic column of area, 104
  stratigraphy, 109–111
  structural setting, present-day, 106–107
  subcrop map, 114
  tectonic history, 99
  trap, 111–116
Ranim fault, 62
Ras Abu Darag, 69
Ras Budran field, 71, 77, 78, 79

location map, 58
Ras Fanar field, 91
Ras Gharib field, 7 , 72, 86, 88, 91
Ras Malaab Group, 85
Ras Mohamed, 57
Rechitsa field, 134, 143, 149
Rechitsa horizon, 128, 130
Recôncavo basin, 263–264, 266
biostratigraphy of, 181
Bouguer gravity map, 162
chronostratigraphy of, 181
evolution of, 181–186
location map, 2, 158
oil and gas fields, 196
oil and gas systems, 186–195
petroleum exploration in, 157
play types, 193–195
prerift system, 193
proven oil reserves and exploratory wells, 159
regional setting of, 157–158
schematic geologic evolution of, 184
seismic section of, 163
stratigraphic column of, 165
stratigraphy of, 165–168
structural configuration of, 160–164
synrift Candeias system, 193
synrift Ilhas system, 194
tectonics, plate setting, 159
typical geochemical log of, 186
Reconcavona, 181
Recôncavo-Tucano basin (NRT), 181
Recôncavo-Tucano-Jatobá rift, 159
Red Sea, 12, 61, 62, 91–92, 261
rift type, 140
Red Sea Basin, 24
Red Sea rift, 45
Reservoirs, in rift basins, 266
Rhine graben, 1
Rhynchouella sp., 79
Riacho da Barra field, 162, 193, 201
Rift abandonment, Gulf of Suez, 32–33
Rift basins
as attractive areas for hydrocarbon accumulations, 1
evolution, 259
diagrammatic model of, 260
field studies, 268
listing of, 267
hydrocarbon potential of, 266–268
intracontinental, example of, 157–203
inversion in, 261, 265
model of geometry of, 265
postrift sag sediments, 266
prerift section, 265
reservoirs in, 266
source rocks in, 266–267
structural style of, 261
summary of characteristics of examples, 263–264
synrift section, 265–266
type, Gulf of Suez as, 1, 5–55
Rift development, Gulf of Suez, 29–32
summary of, 45–47
Rift faults, Gulf of Suez, and clysmic orientation, 37
Rifting
geometry of, 206
and hydrocarbon potential, 3
as part of spectrum of extension, 3
Rift mechanisms
active, 261
passive, 261
tabulation of, 261
understanding, 1
Rifts, 1, 3
and plate tectonics, 7
transition to postrift sag stage, 142
Rift sequence, diagenetic, Recôncavo basin, 191
Rift structure, Gulf of Suez, 33–44
Rift tectonics, Gulf of Suez, 24–33
Rio Grande fault, 237, 238
Rio Grande rift, 205, 239, 263–264
basins and uplifts, location of, 208
Bouguer gravity profiles, 215, 218
crustal structure, 212, 214
geophysical signature of, 212–215
heat flow, 215, 217
location of, 207
magnetotellurics, 215, 219
major tectonic elements of, 206–207
seismicity of, 213, 215, 216
tectonic evolution of, 215–217, 219
tectonic map, 209
Rio Puerco block, 237
Rocky Mountain region, 206, 207
Rocky Mountains, 222, 239
ancestral, 227
Romen field, 144
Romen salt dome, 143
Rudeis Formation, 10, 29–30, 45, 58, 61, 64–69, 72, 75, 85–88, 91, 93, 99
isopach map, 87
large-scale foreset strata, 31
lower, 88–89
Rudenkov field, 144, 147
cross section and contour map, 147
Russian craton, 126, 128, 139, 140, 149, 261
structure of, 128

Sabiá Formation, 165, 181
St. Anthony Monastery, 66, 67
location map, 58
St. Paul Monastery, 37, 66, 67, 80, 82, 84
location map, 58
Salt domes, major, in Dnieper-Donets basin, 143
Salvador fault, 157, 161, 162, 164, 177, 185
Salvador Formation, 165, 174, 177, 180, 185, 193
isopach map, 180
Salvadoriella, 181
San Agustin basin, 206
San Andreas fault, 216, 219
Sandia Crest, 217
Sandia Mountains, 217, 226, 227, 237–239
Sandia uplift, 237
San Francisco–Placitas fault, 237, 238
Sangre de Cristo Range, 222, 227
San Luis basin, 206
Santa Fe–Coyote fault, 237, 238, 243–244
Santa Fe Formation, 222
Santiago Sandstone, 202
São Domingos platform, 164
São Sebastião Formation, 165, 177–181, 186
electric log, 174
Sargay horizon, 130
Sarir field, 267
Sarmatian shield, 139
Seismicity, of Rio Grande rift, 213, 215, 216
Semiluki horizon, 130, 144, 148
Sergi Formation, 165–169, 181, 189, 190, 197–199
Sesmaria Member, 177, 180
Shadwan Island, 57
Shagar Member, 32, 69, 90
Shebelinka field, 143, 144, 152
cross section and contour map, 146
Shebelinka structure, 147
Shoab Ali Member, 28, 45
Shukheir Bay, 72
Sidri Member, 69, 90
Simple shear, 261
model, 262
Sinai/Levant plate, 61
Sinai massif, 14
Sinai Peninsula, 14, 35, 36, 61
Sinjar trough, 19
Sleipner Ost field, 267
Sleipner Vest field, 267
Snapper field, 267
Solokhov field, 144, 152
Source rocks
in Dnieper-Donets and Pripyat basins, 148–149
Recôncavo basin, 186–187
in rift basins, 266–267
spent, in Ramadan field, 114
South Belayim Basin, 62, 66, 69, 72, 73
Southern Galala Plateau, 58, 66, 82, 84
location map, 58
South Gharib Formation, 10, 32, 47, 69, 72, 85, 91–93, 106
isopach map, 89
South graben, Albuquerque basin, 237
Spent source rock, in Ramadan field, 114
Spraberry Sandstone play, 153
Srebnin depression, 139, 148, 153
Stratigraphic heaves, Gulf of Suez
fractal nature of, 41
size distribution of, 43
Stratigraphy
Albuquerque basin, 222, 226
Dnieper-Donets basin, postrift, 132–134
prerift, 128–130
synrift, 130
Gulf of Suez, postrift, 89–93
prerift, 77–85
synrift, 85–89
Pripyat basin, postrift, 132–134
prerift, 128–130

synrift, 130
Recôncavo basin, 165–181
Subbasins, Gulf of Suez, 62, 68–76
Sudovits field, 134, 147
Sudr Chalk, 21, 66, 72, 80, 81
Sudr field, 7, 84
Sudr Formation, 7, 62, 81, 82, 104, 111, 112, 117, 265, 266
Suez rift. *See* Gulf of Suez
Synrift Candeias system, Recôncavo basin, 193
Synrift Ilhas system, Recôncavo basin, 194
Synrift section, rift basin, 265–266
Synrift sequence
Dnieper-Donets basin, 130
Pripyat basin, 130
Synrift stratigraphy, Gulf of Suez, 85–89
Synthetic transfer zones
defined, 38
examples of, 39
Syrian Arc, 9, 22, 36, 37, 38, 42, 45
Syrian Arc folds, 21, 23

Taha Formation, 98
Tanka Formation, 82
Taquipe Canyon, 164
tectonic and depositional model of, 185
Taquipe field, 157, 195
Taquipe Formation, 165, 177–179, 185, 195
isopach map, 179
Tauá Member, 165, 170, 183, 186
"Tayiba salt," 92
Tectonic elements, of northern Egypt and Sinai, 58
Tectonic evolution, Rio Grande rift/Basin and Range province, 215–217, 219
Tectonic framework, Gulf of Suez region, 61–62
Tectonics
Hercynian, 17
plate, 1
of Recôncavo basin, 159
prerift, Gulf of Suez, 16
Tectonostratigraphic events, Gulf of Suez, charts, 13, 25
Tectono-thermal event, Pan-African, 44
Tetrataxidae, 78
Thebes Formation, 62, 82–84, 104, 111–114, 117
Thelemet, 66
Tijeras fault system, 227, 239, 242
Tirrawarra field, 267
Tommeliten Gamma field, 267
Transfer faults, 38
Transfer zones
conjugate, 38–39
synthetic, 38–39
Transform faults, 38
Transverse fault, Gulf of Suez, defined, 37
Traps
in rift basins, 266
types of
in Pripyat basin, 146–148
in Recôncavo basin, 2–3, 195
Tucano basin, 157
Tucano Sul basin, 164
Type basin, rift, Gulf of Suez as, 1, 5–55

Ukrainian shield, 126–131, 134, 139, 140, 147, 153, 261
Umm Bogma Formation, 16, 18
isopach map, 19
Unity field, 267
Upper Arkansas graben, 206
Uvigerina, 87, 88
Uweinat-Aswan uplift, 17–18

Velascoensis event, 21
Vetella ushbasica, 78
Viking Graben, North Sea, 41
Vishan field, 134
Volcanic boulder beds, Nukhul Formation, 27
Volcanic rocks
Albuquerque basin, map of, 223
distribution of, southwestern United States, 220–221
Volcanism, in Albuquerque basin, 226
basaltic, in Gulf of Suez, 9
Volyn-Orsha aulacogen, 128, 139
Voronezh high, 128, 134, 139
Voronezh horizon, 130

Wadi, defined, 11
Wadi Aheimer, 78
Wadi Araba, 37–44, 58, 66–67, 77–79, 82, 84, 89, 93
geological cross section, 67
Wadi Araba fault, 62
Wadi Araba high, 83
Wadi Araba structure, 21, 32, 37
Wadi Baba, 23, 34, 36, 89
Wadi Dakhal, 79
Wadi Feiran, 12, 14, 34, 62, 69, 78, 79
Wadi Gharandal, 62
Wadi Khaboba–Gebel Nukhul High, 16
Wadi Kharaza, 91
Wadi Malha, 79
Wadi Matulla, 81
Wadi Natrun, 84
Wadi Natrun fault, 62
Wadi Nukhul, 24, 82
Wadi Rod el Hamal, 79
Wadi Shellal, 78
Wadi Sudr, 30, 31, 81
Wadi Tanka, 82
Wadi Tayiba, 24, 25, 26, 84
Wadi Thal, 21, 24
slope breccias, 22
Warda field, 268
"Wardan clastics," 92
Wata Formation, 21, 80, 81, 98, 114, 117
West Bakr basin, 62, 71, 72, 76, 88
West Bakr Marine Basin, 69
West Ekofisk field, 267
Western Desert, 15, 21, 58, 61, 62, 79, 80, 81, 84, 93
Western Desert basement, 72
Western Desert structural trends, 66
West Krestishchen field, 143, 144
West Mesa fault, 237
West Qarun block, 21
Wrench faults, 62

Yablunov field, 144, 152
Yefremov field, 143, 144
Yelets horizon, 130, 146
Yevlanov horizon, 130
Younger Granites, 14

Zadon horizon, 130, 131, 144
"Zafarana anhydrite," 92
Zafarana Platform, 66, 67, 69, 71, 88
Zeit Bay basin, 62, 72, 76, 86, 88
type log correlation, 74
Zeit Bay field, 91
Zeit Bay Marine Basin, 74, 75
Zeit Formation, 10, 32, 47, 71–75, 85, 92–93, 111
Zenima Basin, 76
Zihor Formation, 81